DENSE
Z-PINCHES

Related Titles from AIP Conference Proceedings

650   BEAMS 2002:   14th International Conference on High-Power Particle Beams
Edited by Thomas A. Mehlhorn and Mary Ann Sweeney, December 2002, 0-7354-0107-1

625   High Energy Density and High Power RF: 5th Workshop on High Energy Density and High Power RF
Edited by B. E. Carlsten, August 2002, 0-7354-0078-4

409   Dense Z-Pinches: Fourth International Conference
Edited by Nino R. Pereira, Jack Davis, and Peter E. Pulsifer, December 1997, 1-56396-610-7

To learn more about these titles, or the AIP Conference Proceedings Series, please visit the webpage **http://proceedings.aip.org/proceedings**

DENSE Z-PINCHES

5th International Conference on
Dense Z-Pinches

Albuquerque, New Mexico   23-28 June 2002

EDITORS
Jack Davis
Naval Research Laboratory, Washington, DC

Christopher Deeney
*Sandia National Laboratories
Albuquerque, New Mexico*

Nino R. Pereira
Ecopulse, Inc., Springfield, Virginia

SPONSORING ORGANIZATIONS
Department of Energy
Sandia National Laboratories
Defense Threat Reduction Agency
Air Force Research Laboratory

Melville, New York, 2002
AIP CONFERENCE PROCEEDINGS ■ VOLUME 651

Editors:

Jack Davis
Plasma Physics Division, Code 6720
Naval Research Laboratory
Washington, DC 20375
USA

E-mail: davisj@ppdmail.nrl.navy.mil

Christopher Deeney
Sandia National Laboratories
P.O. Box 5800, MS 1168
Albuquerque, NM 87185-1168
USA

E-mail: cdeene@sandia.gov

Nino R. Pereira
Ecopulse, Inc.
P.O. Box 528
Springfield, VA 22150
USA

E-mail: pereira@speakeasy.org

The articles on pp. 388–391, 392–395, 400–403, and 404–407 were authored by U. S. Government employees and are not covered by the below mentioned copyright.

The United Kingdom Government retains copyright in the article on pp. 79–82, © Crown Copyright (2002).

Authorization to photocopy items for internal or personal use, beyond the free copying permitted under the 1978 U.S. Copyright Law (see statement below), is granted by the American Institute of Physics for users registered with the Copyright Clearance Center (CCC) Transactional Reporting Service, provided that the base fee of $19.00 per copy is paid directly to CCC, 222 Rosewood Drive, Danvers, MA 01923. For those organizations that have been granted a photocopy license by CCC, a separate system of payment has been arranged. The fee code for users of the Transactional Reporting Service is: 0-7354-0108-X/02/$19.00.

© 2002 American Institute of Physics

Individual readers of this volume and nonprofit libraries, acting for them, are permitted to make fair use of the material in it, such as copying an article for use in teaching or research. Permission is granted to quote from this volume in scientific work with the customary acknowledgment of the source. To reprint a figure, table, or other excerpt requires the consent of one of the original authors and notification to AIP. Republication or systematic or multiple reproduction of any material in this volume is permitted only under license from AIP. Address inquiries to Office of Rights and Permissions, Suite 1NO1, 2 Huntington Quadrangle, Melville, N.Y. 11747-4502; phone: 516-576-2268; fax: 516-576-2450; e-mail: rights@aip.org.

L.C. Catalog Card No. 2002115280
ISBN 0-7354-0108-X
ISSN 0094-243X

CD-ROM available separately: ISBN 0-7354-0109-8

Printed in the United States of America

CONTENTS

Preface xv
Letters from the Chairmen of Beams/DZP2002 xvii
Sponsors xix
Organizing Committees xx
Brief History of the Conferences xxii
Statistics xxiii

NATIONAL PERSPECTIVES

Fast Z-Pinch Study in Russia and Related Problems 3
    E. V. Grabovskii
History of Z-Pinch Research in the U.S. 9
    M. A. Sweeney
Southern Dense Pinch Discharges 15
    H. Chuaqui

MULTI-MEGA-AMPERE Z-PINCH PULSED POWER DRIVES

The ZR Refurbishment Project 23
    D. H. McDaniel, M. G. Mazarakis, D. E. Bliss, J. M. Elizondo,
    H. C. Harjes, H. C. Ives, III, D. L. Kitterman, J. E. Maenchen,
    T. D. Pointon, S. E. Rosenthal, D. L. Smith, K. W. Struve, W. A. Stygar,
    E. A. Weinbrecht, D. L. Johnson, and J. P. Corley
Pulse Power System Development for Megajoule X-Ray Facility
BAIKAL 29
    E. A. Azizov, V. V. Alexandrov, S. G. Alikhanov, V. H. Bachtin,
    V. I. Chetvertkov, V. A. Glukhikh, E. V. Grabovskii, A. N. Gribov,
    Y. A. Hallimullin, V. A. Levashov, A. P. Lotocky, A. M. Zhitlukhin,
    E. P. Velikhov, G. I. Dolgachev, J. G. Kalinin, A. S. Kingsep,
    A. I. Kormilitcin, V. G. Kouchinsky, S. L. Nedoseev, O. P. Pechersky,
    V. D. Pismenniy, G. P. Rikovanov, and V. P. Smirnov
Experiments Aimed at the "Baikal" Program 33
    A. Kingsep, Y. Bakshaev, A. Bartov, P. Blinov, A. Chernenko, S. Danko,
    G. Dolgachev, Y. Kalinin, I. Kovalenko, A. Lobanov, D. Maslennikov,
    V. Mizhiritsky, A. Shashkov, and V. Smirnov
Study of the Plasma Focus as a Driver for the Magnetic Compression
of Liners 37
    V. E. Fortov, M. A. Karakin, E. Y. Khautiev, V. I. Krauz,
    S. F. Medovschikov, A. N. Mokeev, V. V. Myalton, S. L. Nedoseev,
    V. P. Smirnov, and V. P. Vinogradov
Conceptual Design of Decade Half, a 15-MA, 300-ns PRS Driver 43
    P. Spence, P. Corcoran, J. Douglas, T. Tucker, B. Altes, K. Childers,
    P. Sincerny, L. Whitehead, V. Kenyon, M. Babineau, T. Cotter, P. Kurucz,
    and R. Davis

New IES Scheme for Power Conditioning at Ultra-high Currents:
From Concept to MHD Modeling and First Experiments 47
      A. S. Chuvatin, L. I. Rudakov, V. A. Kokshenev, L. E. Aranchuk, D. Huet,
      V. A. Gasilov, A. Y. Krukovskii, N. E. Kurmaev, and F. I. Fursov

ECF2: A Pulsed Power Generator Based on Magnetic Flux
Compression for K-Shell Radiation Production 51
      P. L'Eplattenier, F. Lassalle, C. Mangeant, F. Hamann, M. Bavay, F. Bayol,
      D. Huet, A. Morell, P. Monjaux, G. Avrillaud, and B. Lalle

Performance of Llampüdkeñ with Short Circuit and Plasma Loads 55
      H. Chuaqui, I. H. Mitchell, R. Aliaga-Rossel, M. Favre,
      and E. S. Wyndham

Load Current Sharpening and Liner Implosions on a Capacitor Bank 59
      S. A. Sorokin and S. A. Chaikovsky

WIRE ARRAY EXPERIMENTS

Effect of Discrete Wires on the Implosion Dynamics of
Wire Array Z-Pinches 65
      S. V. Lebedev, J. P. Chittenden, S. N. Bland, D. J. Ampleford,
      C. Jennings, and M. G. Haines

Ablation Rate of Wire Cores in Wire Array Z-Pinch Experiments............ 71
      S. V. Lebedev, D. J. Ampleford, S. N. Bland, J. P. Chittenden,
      and M. G. Haines

Implosion Dynamics and X-Ray Characteristics of Nested
Wire Array Z-Pinches 75
      S. N. Bland, S. V. Lebedev, J. P. Chittenden, and M. G. Haines

The Effect of Precursor Plasma Flow on Foam Targets in
Wire Array Z-Pinch Experiments 79
      J. B. A. Palmer, S. V. Lebedev, S. N. Bland, J. P. Chittenden,
      and D. J. Ampleford

The Effect of Array Configuration on Current Distribution in a
Wire Array Z-Pinch 83
      S. N. Bland, S. V. Lebedev, F. Beg, H. Kwek, J. P. Chittenden, and
      M. G. Haines

Prolonged Plasma Production and Dynamics of Implosion of
Multiwire Arrays... 87
      V. V. Alexandrov, I. N. Frolov, M. V. Fedulov, E. V. Grabovskii,
      K. N. Mitrofanov, S. L. Nedoseev, G. M. Oleinik, I. Y. Porofeev,
      A. A. Samokhin, P. V. Sasorov, V. P. Smirnov, G. S. Volkov, M. V. Zurin,
      G. G. Zukakischvili, and K. Struve

Experimental Study of Wire Array Implosion in Presence of
Prolonged Plasma Production on Angara-5-1 Facility...................... 91
      V. V. Alexandrov, M. V. Fedulov, I. N. Frolov, E. V. Grabovskii,
      K. N. Mitrofanov, S. L. Nedoseev, G. M. Oleinik, I. Y. Porofeev,
      A. A. Samokhin, P. V. Sasorov, V. P. Smirnov, G. S. Volkov, M. V. Zurin,
      and G. G. Zukakischvili

Wire Array Z-Pinch Experiment on Inductive Energy Storage
Generator ASO-X 95
    Y. Teramoto, H. Urakami, S. Kohno, N. Shimomura, S. Katsuki,
    and H. Akiyama

GAS PUFFS EXPERIMENTS

Progress in Double-Shell Gas Puff Z-Pinches.............................. 101
    H. Sze, J. Banister, B. H. Failor, A. Fisher, J. S. Levine, Y. Song,
    J. P. Apruzese, R. W. Clark, J. Davis, D. Mosher, J. W. Thornhill,
    A. L. Velikovich, B. V. Weber, C. A. Coverdale, C. Deeney,
    T. L. Gilliland, J. McGurn, R. B. Spielman, K. W. Struve, W. A. Stygar,
    D. Bell, P. Coleman, M. Babineau, C. Enis, V. Kenyon, and T. Worley

K-Shell Yield Performance Assessment of Argon Gas Puff Loads
Imploded on Double-Eagle, Z, and Decade Quad 105
    J. W. Thornhill, J. P. Apruzese, J. Davis, A. L. Velikovich, K. G. Whitney,
    H. Sze, J. S. Levine, B. H. Failor, C. Deeney, C. A. Coverdale, and D. Bell

Z-Pinch Experiments on Decade Quad Using a Double-Shell Gas Puff........ 109
    J. S. Levine, M. A. Babineau, J. Banister, D. Bell, C. Enis, B. H. Failor,
    V. Kenyon, H. M. Sze, and T. Worley

Gas Puff Radiation Performance as a Function of Radial
Mass Distribution 113
    P. L. Coleman, M. Krishnan, R. Prasad, N. Qi, E. Waisman, B. H. Failor,
    J. S. Levine, and H. Sze

Double Gas Puff Z-Pinch with Axial Magnetic Field for K-Shell
Radiation Production 117
    A. V. Shishlov, R. B. Baksht, S. A. Chaikovsky, A. Y. Labetsky,
    V. I. Oreshkin, A. G. Rousskikh, and A. V. Fedunin

Effect of an Axial Magnetic Field on the K-Shell Radiation of a Neon
Double Gas Puff 123
    S. A. Chaikovsky, A. Y. Labetsky, A. V. Shishlov, A. V. Fedunin,
    V. I. Oreshkin, R. B. Baksht, and A. G. Rousskikh

Soft X-Ray Production and Spectrum Measurements in Imploded
Krypton Liners 127
    S. A. Sorokin and S. A. Chaikovsky

Spatial Structure of X-Ray Emission of a Gas-Puff Z-Pinch Plasma 131
    K. Takasugi, S. Narisawa, and H. Akiyama

Puff-Gas Z-Pinch Experiment on "Yang" Accelerator 135
    J. Deng, L. Yang, Y. Gu, Z. Li, X. Huang, Z. Yang, N. Ding, C. Ning,
    B. Ding, and X. Peng

Z-PINCH, X-PINCH, FOCUS AND CAPILLARY DISCHARGE PLASMA DYNAMICS AND DEVICES

The X Pinch as a Point Source for Point-Projection
X-Ray Radiography 141
    T. A. Shelkovenko, S. A. Pikuz, D. B. Sinars, K. M. Chandler,
    M. D. Mitchell, and D. A. Hammer

X Ray Emission From X Pinch Experiments on the
Llampüdkeñ Generator 145
    I. H. Mitchell, R. Aliaga-Rossel, J. A. Gomez, H. Chuaqui, M. Favre,
    and E. S. Wyndham

Regimes of Energy Input in the Pseudospark Discharge in the
Sources of EUV Radiation.. 149
    Y. D. Korolev, O. B. Frants, V. G. Geyman, R. V. Ivashov, N. V. Landl,
    and I. A. Shemyakin

Optical Observation of a Pseudospark Discharge Development in a
Source of EUV Radiation 153
    Y. D. Korolev, I. M. Datsko, O. B. Frants, V. G. Geyman, R. V. Ivashov,
    N. V. Landl, N. A. Ratachin, and I. A. Shemyakin

Experimental Setup for Investigation of the Pseudospark Discharge as
Applied to Generation of EUV Radiation............................... 157
    Y. D. Korolev, I. M. Datsko, O. B. Frants, V. G. Geyman, R. V. Ivashov,
    N. V. Landl, and I. A. Shemyakin

Fast Capillary Discharge Experiments in a Small Device; Spectra in
the VUV Region 161
    L. Soto, A. Nazarenko, C. Pavez, P. Silva, J. Moreno, K. Aubel,
    and T. Vucina

Soft X-Ray Radiation of Fast-Capillary-Disharge CAPEX 2.............. 165
    J. Schmidt, K. Kolacek, V. Bohacek, M. Ripa, O. Frolov, P. Vrba,
    A. Jancarek, and M. Vrbova

Plasma Dynamics and Lasing Condition of Fast Capillary Discharges 169
    N. Sakamoto, G. Niimi, Y. Hayashi, M. Masnavi, M. Nakajima, E. Hotta,
    and K. Horioka

X-Ray Spectroscopic Studies of X-Pinch Plasma Micropinches with
~10 ps Resolution... 173
    S. A. Pikuz, T. A. Shelkovenko, D. B. Sinars, I. Y. Skobelev,
    K. M. Chandler, M. D. Mitchell, and D. A. Hammer

Spectroscopic Analysis of 1MA X-Pinch Implosions at the Nevada
Terawatt Facility... 177
    A. S. Shlyaptseva, S. B. Hansen, V. L. Kantsyrev, D. A. Fedin,
    N. D. Ouart, K. B. Fournier, and U. I. Safronova

X-Ray Wide-Band and Time-Resolved Imaging and Spectroscopic
Characterization of Hot Spots and Jets in 0.9-1.0 MA X-Pinches.............. 181
    V. L. Kantsyrev, D. A. Fedin, A. S. Shlyaptseva, S. B. Hansen,
    and N. D. Ouart

**Analysis of Anisotropy of Spectra and Spatial Distribution of Hard
X-Ray Emission from 0.9-1.0 MA High-Z X-Pinches** 185
    V. L. Kantsyrev, D. A. Fedin, A. S. Shlyaptseva, S. B. Hansen,
    D. Chamberlain, and N. D. Ouart

**Analysis of Time Evolution of Z-Pinch Lines Using an Advanced
5-Channel Spectrometer Polychromator** 189
    D. A. Fedin, V. L. Kantsyrev, A. S. Shlyaptseva, S. B. Hansen,
    and S. Fuelling

**Emission Produced at Compression of Deuterium Current-Sheath
with Wire in Plasma Focus Discharge** 193
    P. Kubeš, J. Kravárik, D. Klír, M. Scholz, M. Paduch, K. Tomaszewski,
    I. Ivanova-Stanik, B. Bienkowska, L. Karpinski, L. Ryć, L. Juha, J. Krása,
    M. J. Sadowski, L. Jakubowski, A. Szydlowski, A. Banaszak, H. Schmidt,
    and V. M. Romanova

**Energy Transformation in Z-Pinch and Plasma Focus Discharges with
Wire and Wire-in-Liner Loads** 197
    P. Kubeš, J. Kravárik, D. Klír, M. Scholz, M. Paduch, K. Tomaszewski,
    L. Karpinski, Y. L. Bakshaev, P. I. Blinov, A. S. Chernenko, S. A. Dan'ko,
    V. D. Korolev, A. Y. Shashkov, and V. I. Tumanov

Multi-Wire Z Pinch Experiments 201
    M. Hu and B. R. Kusse

Plasma Formation Around Single Wires 205
    P. U. Duselis and B. R. Kusse

**Investigation of the Initial Stage of Electrical Explosion of
Fine Metal Wires** 209
    G. S. Sarkisov, S. E. Rosenthal, K. W. Struve, D. H. McDaniel,
    E. M. Waisman, and P. V. Sasorov

Joule Energy Deposition in Exploding Wire Experiments 213
    G. S. Sarkisov, S. E. Rosenthal, K. W. Struve, D. H. McDaniel,
    E. M. Waisman, and P. V. Sasorov

**Results of Experiment on Explosion of W and Al Wires in
Water and Vacuum** 217
    A. G. Rousskikh, R. B. Baksht, V. I. Oreshkin, and A. V. Shishlov

Wire Explosion with 0-1kA Per Wire 221
    B. R. Kusse, M. Hu, K. M. Chandler, and D. A. Hammer

**Layering of an Annular Z-Pinch Sheath in the Presence of an Axial
Magnetic Field** 225
    S. A. Chaikovsky and A. Y. Labetsky

**Dense Transient Plasmas Driven by a Mega-Ampere Device in the
Chilean Nuclear Energy Commission** 229
    L. Soto, W. Kies, G. Sylvester, G. Ziethen, M. Zambra, J. Moreno, P. Silva,
    L. Birstein, R. M. Muñoz, and R. Saavedra

Laser Initiated Hollow Gas-Embedded Z-Pinch 233
    C. Pavez, H. Chuaqui, R. Aliaga-Rossel, M. Favre, I. Mitchell,
    and E. Wyndham

Discharge Formation in Fast Pulsed Capillary Discharges 237
    M. Favre, P. Choi, A. M. Leñero, F. Castillo, F. Susuki, H. Chuaqui,
    I. Mitchell, and E. Wyndham

Experimental Observation in a High Current Capillary Discharge 241
    E. Wyndham, H. Chuaqui, M. Favre, I. Mitchell, R. Aliaga-Rossel,
    and P. Choi

Properties of Hot-Spots in Plasma Focus Discharges Operating in
Hydrogen-Gas Mixtures.. 245
    P. Silva and M. Favre

Energy Dissipation in the Run-Down Phase of Plasma
Focus Discharge 249
    M. A. M. Kashani and T. Miyamoto

FUSION, NEUTRON PRODUCTION AND OTHER APPLICATIONS

X-Ray and Neutron Emission from PF-1000 Facility 255
    M. Scholz, B. Bienkowska, I. Ivanova-Stanik, L. Karpinski,
    R. Miklaszewski, M. Paduch, K. Tomaszewski, E. Zielinska,
    M. J. Sadowski, L. Jakubowski, A. Szydlowski, A. Banaszak,
    H. Schmidt, P. Kubes, J. Kravarik, V. Romanova, and S. Vitulli

Neutron Emission Characteristics of a High-Current Plasma Focus:
Initial Studies 261
    B. L. Freeman, J. C. Boydston, J. M. Ferguson, B. Lindeburg,
    A. D. Luginbill, J. C. Rock, T. E. Tutt, E. C. Hagen, and L. Ziegler

A Very Small Plasma Focus Operating at Tens of Joules.................. 265
    L. Soto, P. Silva, J. Moreno, A. Clausse, and W. Kies

Characteristics and Dynamics of a 215-eV Dynamic-Hohlraum X-Ray
Source on Z 269
    T. W. L. Sandford, D. L. Peterson, R. W. Lemke, R. C. Mock,
    G. A. Chandler, J. P. Chittenden, R. E. Chrien, G. C. Idzorek, R. J. Leeper,
    C. L. Ruiz, and R. G. Watt

Direct Drive Inertial Confinement Fusion in a Z-Pinch Plasma 275
    R. W. Clark, J. Davis, A. Velikovich, L. Rudakov, and J. L. Giuliani, Jr.

The Inverse Z-Pinch as a Physics Test Bed, and a Possible Target
Plasma for Magnetized Target Fusion (MTF) 279
    I. Lindemuth, B. Bauer, S. Fuelling, R. Kirkpatrick, V. Makhin, R. Presura,
    P. Sheehey, and R. Siemon

Ignition of Fusion Burn Wave by Nonthermal Plasma of Z-Pinch 283
    V. V. Vikhrev

On Perspectives of Creation of High-Power Neutron Source at
Deuterium Plasma Compression in Z-Θ Pinch Geometry 287
    V. D. Selemir, A. V. Ivanovsky, A. P. Orlov, V. F. Yermolovich,
    G. V. Dolgoleva, V. A. Demidov, V. I. Karelin, and P. B. Repin

The Production of Hypersonic, Radiatively Cooled Plasma Projectiles
of Extremely High Energy Density in Imploding Z-Pinches 291
    J. P. Chittenden, A. M. Dunne, M. Zepf, S. V. Lebedev, A. Ciardi,
    and S. N. Bland

Fusion Criteria for Dense Cylindrical and Sheet Z-Pinches................. 295
    T. Miyamoto

Considerations for Generating up to 10 Mbar Magnetic Drive
Pressures with the Refurbished Z-Machine (ZR) 299
    R. W. Lemke, M. D. Knudson, A. C. Robinson, T. A. Haill, K. W. Struve,
    T. A. Mehlhorn, and J. R. Asay

3-D Modeling of Modifications to the Z Accelerator for Generating
Shaped Pulses 305
    T. D. Pointon, M. E. Savage, and H. C. Harjes, III

The RT Instability in Cylindrical Implosion of a Jelly Ring 309
    Y. Libing, L. Haidong, S. Chengwei, O. Kai, L. Jun, and H. Xianbin

Phase Transitions in Metal under Fast Selfheating by High-Power
Current Pulse 313
    K. V. Khishchenko, S. I. Tkachenko, V. E. Fortov, P. R. Levashov,
    I. V. Lomonosov, and V. S. Vorob'ev

Experiments with Radiatively Cooled Supersonic Plasma Jets
Generated in Conical Wire Array Z-Pinches 317
    S. V. Lebedev, D. J. Ampleford, S. N. Bland, J. P. Chittenden, A. Ciardi,
    N. Naz, M. G. Haines, A. Frank, E. Blackman, and T. Gardiner

Deflection of Supersonic Plasma Jets by Ionised Hydrocarbon Targets 321
    D. J. Ampleford, S. V. Lebedev, S. N. Bland, A. Ciardi, M. Sherlock,
    J. P. Chittenden, and M. G. Haines

Plastic Deformation and Perforation of Metal Using Metallic Jet............. 325
    P. Sarkar, S. Chaturvedi, A. Shyam, R. Kumar, D. Lathi, V. Chaudhari,
    R. Verma, J. Sonara, K. Shah, and B. Adhikary

Anomalous Resistivity Change in $NiFe_2O_4$ Nanosized Powders
Synthesized by Pulsed Wire Discharge 329
    H. Suematsu, K. Ishizaka, Y. Kinemuchi, T. Suzuki, W. Jiang,
    and K. Yatsui

Reduction of Micrometer Size Al Particles in Nanosize AlN Powder
Synthesized by Pulsed Wire Discharge 333
    C. Cho, Y. Kinemuchi, H. Suematsu, W. Jiang, and K. Yatsui

THEORY AND MODELING

Pitfalls in Radiation Modeling of Z-Pinch Plasmas 339
    J. Davis, J. L. Giuliani, J. P. Apruzese, R. W. Clark, J. W. Thornhill,
    K. G. Whitney, A. Velikovich, Y. K. Chong, C. A. Coverdale, C. Deeney,
    and P. D. LePell

Why do Wire-Array Z-Pinches Give Such a Sharp and Efficient
X-Ray Pulse? 345
    M. G. Haines, S. V. Lebedev, J. P. Chittenden, F. N. Beg, S. N. Bland,
    and M. Sherlock

Calculated Evolution of Side-On and End-On X-Ray Images of Wire
and Gas Puff Implosions on Z 350
    J. P. Apruzese, J. W. Thornhill, C. Deeney, C. A. Coverdale, J. Davis,
    A. L. Velikovich, H. Sze, P. L. Coleman, B. H. Failor, J. S. Levine,
    and K. G. Whitney

How 3D Effects Limit X-Ray Power in Wire Array Z-Pinches 354
    J. P. Chittenden, S. V. Lebedev, M. E. Cuneo, C. A. Jennings,
    and A. Ciardi

**Modeling Enhanced Energy Coupling of Z-Pinches to
Pulsed-Power Generators.** 358
    K. G. Whitney, J. W. Thornhill, C. Deeney, C. A. Coverdale,
    J. P. Apruzese, J. Davis, A. L. Velikovich, and L. I. Rudakov

Numerical Studies of Neon Gas-Puff Z-Pinch Dynamic Process 364
    C. Ning, Z. Yang, and N. Ding

**Computational Assessment of the Effect of Nozzle Geometry on the
Performance of Gas-Puff Plasma Radiation Sources.** 368
    J. J. Watrous and M. H. Frese

Computational MHD on Lagrangian Grids 372
    C. L. Rousculp and D. C. Barnes

**Three Dimensional Resistive Wire Array Implosion Simulations
Continued from Two Dimensional R-Θ Initiation Simulations** 376
    M. H. Frese and S. D. Frese

**Recent Improvements to MACH2 and MACH3 for Fast
Z-Pinch Modeling** 380
    S. D. Frese and M. H. Frese

Simulation of Electric Explosion of Metal Wires 384
    V. I. Oreshkin, R. B. Baksht, A. G. Rousskikh, A. V. Shishlov,
    P. R. Levashov, I. V. Lomonosov, K. V. Khishchenko, and I. V. Glazyrin

**Improved Neon L-Shell Physics in MHD Modeling of Hawk Gas-Puff
Z-Pinch Implosions.** 388
    J. Schumer, D. Mosher, A. Starobinets, V. Fisher, and Y. Maron

**Cooperative Relaxation Methods for Multigroup Radiation Diffusion
in Radiation Hydrodynamics.** 392
    R. E. Terry, J. L. Giuliani, and J. P. Apruzese

**Hybrid Simulations of Current-Carrying Instabilities in Z-Pinch
Plasmas with Sheared Axial Flow.** 396
    V. I. Sotnikov, V. Makhin, B. S. Bauer, P. Hellinger, P. Travnicek, V. Fiala,
    and J.-N. Leboeuf

A Comparison of Radiation Transport Models for a Ti Z Pinch. 400
    J. L. Giuliani, R. W. Clark, J. W. Thornhill, and J. Davis

Z-Scaled K-Shell Dielectronic Recombination Rate Coefficients 404
    A. Dasgupta, P. Kepple, and J. Davis

**A Kinetic Description of Ions in Aluminium Wire-Array
Precursor Plasma** 408
    M. Sherlock, J. P. Chittenden, S. V. Lebedev, and M. G. Haines

**Theoretical Development of M-Shell Spectroscopy for Z-Pinch
Plasma Diagnostics** 412
    A. S. Shlyaptseva, S. M. Hamasha, S. B. Hansen, N. D. Ouart,
    and U. I. Safronova

Modeling of Capillary Discharge Plasma for X-Ray Lasers, XUV Lithography and Other Applications 416
    V. N. Shlyaptsev, J. Dunn, S. J. Moon, K. B. Fournier, A. L. Osterheld,
    J. J. Rocca, J. Filevich, M. Marconi, E. Jankowska, E. C. Hammarsten,
    S. Sakadzic, A. Rahman, M. Frati, F. G. Tomasel, N. Fornaciari,
    D. Buchenauer, H. A. Bender, S. Karim, M. Kanouff, J. Dimkoff,
    G. Kubiak, G. Shimkaveg, and W. T. Silfvast

The Z-Pinch Structure Generation by the Evolution of the Nonquasineutral Electron Vortex 420
    A. V. Gordeev and T. V. Losseva

On Stabilization of the Rayleigh-Taylor Instability for the Imploding Liner on Account of Ion-Ion Collisions 424
    A. V. Gordeev

The Tentative Opinion of Modeling Plasma Formation in Metallic Wire Z-Pinch 428
    N. Ding

APPENDICES

Group Photo 435
Conference Photographs 437
List of Participants 445
Author Index 455

Preface

The 5th Dense Z-pinch (DZP) meeting was held from 23 June to 28 June, 2002, in Albuquerque, NM. In contrast to previous Conferences in this series, a combined Conference was organized in close cooperation with the Beams 2002 and was consequently much larger than earlier DZP Meetings, with 450 attendees. A foreword is included in the proceedings from the BEAMS chairperson, Dr. Jeff Quintenz of Sandia National Laboratories.

This DZP Proceeding summarizes the highlights of technical and scientific advances made since the last DZP Conference: the program was arranged by the DZP's technical editors. A companion volume contains the papers from the Beams part of the technical program: its technical chairperson was Dr. Tom Mehlhorn (SNL). The editors would like to acknowledge the great support of Sara Cordova, Melissa Murray, and Lisa Mattox in preparing these proceedings.

It is the opinion of the Editors that our understanding of experimental z-pinches has advanced steadily over the last few years. Diagnostics and spectral analyses are better, and so are computations. However, the z-pinch is too complicated for any one computation to include all of the many effects that are known to be important. As an example, it is not yet possible to accurately calculate time-dependent radiation production and transport in more than just the radial dimension over a broad spectrum of frequencies. And even in one dimension it is not yet possible to include some of the important traditional plasma physics.

Active Z-pinch facilities are also changing. Almost everywhere the diagnostics continually improve, producing new results. As yet there is no thermonuclear fusion with z-pinches, but it is becoming increasingly clear that what is needed to make this happen is about two to three times the 20 MA peak current of the present 'Z' machine. Nevertheless, there is a growing interest in Z-pinch based Inertial Fusion Energy, and Sandia is in the process of designing a refurbishment of Z that will take it to 28 MA.

One of the interesting new topics at this DZP was the use of pulsed-power generators to compress without shocks (quasi-isentropic) or to produce multi-Mbar shocks in materials through the magnetic pressure pulse generated during the pulsed-power discharge. A Z-pinch implosion and the resulting x-rays are no longer wanted for this application. The future will tell whether dynamic material property measurements with z-pinch drivers will expand in the next DZP.

The 6th DZP will be organized by an Executive Committee that will be elected by a representative Advisory Committee. This committee's members were elected during the Conference, and they are listed in these proceedings. Please feel free to contact them if you want to share your opinion about the upcoming DZP and how this meeting should be organized. At the moment the options are to collocate the 6th DZP with Beams 2004, in St. Petersburg, Russia, or to hold a stand-alone DZP in 2005 somewhere in Europe.

Jack Davis, Naval Research Laboratory
Chris Deeney, Sandia National Laboratory
Nino R. Pereira, Ecopulse, Inc.

Letter from the Chairman of Beams/DZP 2002

The 14th International Conference on High-Power Particle Beams and the 5th International Conference on Dense Z-Pinches (BEAMS/DZP 2002) took place at the Hilton Albuquerque in Albuquerque, New Mexico, on June 23-28, 2002. This year we tried the experiment of combining the programs of these two historically separate conferences because of the significant overlap in subject matter and interests. By most measures the experiment was deemed a success, and we will likely co-locate some future meetings of the two conferences.

As in previous separate Beams and DZP conferences, papers were given as oral (invited and contributed) and poster presentations. The Beams part of the meeting included electron and ion beam physics, free electron lasers, x-ray lasers, high-power microwaves, imploding liners, and radiography, and the DZP topics included magnetohydrodynamics, radiation hydrodynamics, z-pinch physics, and z-pinch diagnostics. The combined topics were pulsed power technology, electrical diagnostics, x-ray spectroscopy, electrodynamic modeling, plasma instabilities, high energy density physics, astrophysical applications, industrial applications, inertial or magnetized target fusion, plasma focus, and x-ray lithography. The 394 participants from 19 countries presented 358 papers, of which 85 were orals and 273 were posters. Most of these papers are included in the conference proceedings.

The 15th International Conference on High-Power Particle Beams will be held in Russia in 2004. This will be the third time the conference has been there. Discussion is underway to determine if the 6th International Conference on Dense Z-Pinches will also meet at this time.

In addition to the outstanding technical aspects of the meeting, the participants took advantage of the unique New Mexico environment, which offers spectacular scenery and a melting pot of Native American, Spanish and Anglo cultures. Social events included a welcome reception on Sunday, a western barbecue on the Sandia Indian Reservation on Tuesday, the conference banquet on Wednesday, and the DZP reception on Thursday.

We would like to acknowledge the technical contributions of all the participants. Furthermore, we thank all the members of the international steering committees, the local steering committee, and the technical program committees for their contributions to the success of the meeting. We would also especially like to thank the many sponsors that financially supported the Beams/DZP 2002 conference.

Jeff Quintenz
Conference Chairman

Dear Colleagues,

It is our great pleasure to welcome you to the 14th International Conference on High-Power Particle Beams and the 5th International Conference on Dense Z-Pinches in Albuquerque, New Mexico.

When Gerry Yonas chaired the first Beams conference in Albuquerque in 1975, the technologies were exciting and young. Now that the Conference returns to Albuquerque 27 years later, with Gerry as its honorary chairman, pulsed power technologies are more mature and applied, but exciting still. The 2002 Conference will be a worthy successor to the 2000 Beams Conference held in Nagaoka, Japan.

With the increased prominence of research on z pinches, the Beams Conference has been co-located in 2002 with the 5th International Conference on Dense Z-Pinches. Since 1988 that Conference has gathered the world's z-pinch researchers approximately every four years. The present Conference skipped one year to be hosted together with Beams 2002 in Albuquerque, home of Sandia National Laboratories' Z accelerator and its z-pinch, which generates the world's most powerful x-ray burst in the laboratory.

The two Conferences received more than 360 abstracts and expect participation from scientists in more than 15 countries. The Technical Chairmen, Tom Mehlhorn and Chris Deeney, combined the two programs in such a manner that it will be possible to attend both Beams and DZP talks and posters during the 4-1/2 day technical program.

Albuquerque is located in the wide valley between the Sandia Mountains and the Rio Grande. It is the largest city in New Mexico and near the center of the Southwest. Albuquerque has become a major research and development center that offers spectacular scenery, casual elegance, and a melting pot of Native American, Spanish, and Anglo cultures. The conference hotel, Hilton Albuquerque, reflects the flavor and history of the Land of Enchantment, with Indian rugs, pueblo art, clay pottery, Kachina dolls, and arched doorways.

The social events include a Western barbecue on the Sandia Indian Reservation on Tuesday evening and the conference banquet at the Hilton on Wednesday evening. Ravi Sudan and Valentin Smirnov will receive Beams Awards during the banquet. In addition, a full schedule of activities for companions is planned in Albuquerque, Santa Fe, and Taos.

We wish all of you a rewarding scientific meeting and an enjoyable time in the Southwest tradition.

Sincerely,

Jeff Quintenz
Conference Chairman

Mary Ann Sweeney
Conference Co-Chairman

Nino Pereira
Jack Davis
Dense Z-Pinches Co-Chairmen

The Organizing Committees thank the following sponsors for their support of Beams/DZP 2002.

Sponsors

Department of Energy

Sandia National Laboratories

Defense Threat Reduction Agency

Air Force Research Lab

12th International Conference on High-Power Particle Beams in Israel

13th International Conference on High-Power Particle Beams in Japan

4th International Conference on Dense Z-Pinches in Vancouver

Ecopulse, Inc.

Ktech

Pulsed Power Conference Inc.

Science Applications International Corporation

Team Specialty Products

Titan Pulse Sciences Division

University of New Mexico

Technical Co-Sponsor

IEEE Electron Devices Society

Organizing Committees

Beams/DZP Local Steering Committee

Mary Ann Sweeney, *Chairman*	Sandia National Laboratories, USA
Lisa Mattox	Sandia National Laboratories, USA
Fran DiMarco	Fran DiMarco Meeting and Event Planning, Albuquerque, NM, USA
Rab Freeman	Albuquerque, NM, USA
Chuck Gilman	Science Applications International Corporation, USA
Judy Gilman	Albuquerque, NM, USA
Pat Helles	Air Force Research Laboratory, USA
Dan Jobe	Ktech, USA
Melissa Murray	Sandia National Laboratories, USA
Steve Nickerson	Sandia National Laboratories, USA
Edl Schamiloglu	University of New Mexico, USA
Bill Shoup	ITT Industries, USA
Dick Siemon	Los Alamos National Laboratory, USA
Marshall Sluyter	Rockville, MD, USA
Al Toepfer	Science Applications International Corporation, USA
Jo Toepfer	Albuquerque, NM, USA

Beams International Advisory Committee

Edek Blaugrund	Weizmann Institute of Science, Israel
Hans Bluhm	Forschungszentrum Karlsruhe, Germany
Donald Cook	Sandia National Laboratories, USA
Gerald Cooperstein	Naval Research Laboratory, USA
Vladimir Fortov	Academy of Sciences, Russia
Timothy Goldsack	Atomic Weapons Establishment, UK
Karel Jungwirth	Czech Academy of Sciences, Czech Republic
Vasili Koidan	Budker Institute of Nuclear Physics, Russia
Alan Kolb	ACK Associates, USA
Gennady Mesyats	Academy of Sciences, Russia
Jeffrey Quintenz	Sandia National Laboratories, USA
Donald Rej	Los Alamos National Laboratory, USA
Joseph Shiloh	Rafael, Israel
Valentin Smirnov	Kurchatov Institute, Russia
Charles Stallings	Stallings and Associates, USA
Ravi Sudan	Cornell University, USA
Roger White	Titan Pulse Sciences Division, USA
Kiyoshi Yatsui	Nagaoka University of Technology, Japan
Gerold Yonas	Sandia National Laboratories, USA

Dense Z-Pinches International Advisory Committee

Hernan Chuaqui	Pontificia Universidad Catolica de Chile, Chile
Jack Davis	Naval Research Laboratory, USA
Malcolm Haines	Imperial College, UK
Thomas Hopkins	Defense Threat Reduction Agency, USA
Hans-Joachim Kunze	Ruhr-Universitat, Germany
Keith Matzen	Sandia National Laboratories, USA
Tetsuo Miyamoto	Nihon University, Japan
Valentin Smirnov	Kurchatov Institute, Russia

Beams Technical Program Committee

Thomas Mehlhorn, *Chairman*	Sandia National Laboratories, USA
Gerold Yonas, *Honorary Chairman*	Sandia National Laboratories, USA
Hans Bluhm	Forschungszentrum Karlsruhe, Germany
Ronald Davidson	Princeton Plasma Physics Lab, USA
Ronald Gilgenbach	University of Michigan, USA
Timothy Goldsack	Atomic Weapons Establishment, UK
Koichi Kasuya	Tokyo Institute of Technology, Japan
Meir Markovits	Rafael, Israel
Craig Olson	Sandia National Laboratories, USA
Paul Ottinger	Naval Research Laboratory, USA
Markus Roth	GSI, Germany
John Sethian	Naval Research Laboratory, USA

Dense Z-Pinches Technical Program Committee

Christopher Deeney, *Chairman*	Sandia National Laboratories, USA
John Apruzese	Naval Research Laboratory, USA
David Bell	Defense Threat Reduction Agency, USA
Melissa Douglas	Los Alamos National Laboratory, USA
Julio Herrera	Inst de Ciencias Nucl, Mexico
Konstantin Koshelev	Institute for Spectroscopy, Russia
Sergei Lebedev	Imperial College, London, UK
Yitzak Maron	Weizmann Institute of Science, Israel
Alexander Shishlov	High Current Electronics Institute, Russia
Marshall Sluyter	Rockville, MD, USA
Alexander Velikovich	Naval Research Laboratory, USA

Brief Histories of the Conferences

The first International Conference on Dense Z-Pinches (DZP), in 1984 in Alexandria, Virginia, was organized by the Naval Research Laboratory's John Sethian in large part to discuss the z pinch prospects for thermonuclear fusion. The meeting turned out to be very lively, with the 30 participants from various countries even discussing optical radiation from z pinches. Shortly thereafter it became possible to discuss production of x rays from z pinches. Hence, Professor Norman Rostoker of the University of California Irvine and Nino Pereira and Jack Davis of NRL decided to add x rays to the conference topics for the second DZP, in Laguna Beach, California, in 1988. The second conference was even more successful, with over 120 participants from all over the world including, for the first time, many from the Soviet Union. Subsequently, the DZP Conference kept this format, in 1993, in London, United Kingdom, and in 1997 in Vancouver, Canada. The fifth DZP Conference maintains the tradition, but is co-located with the Beams Conference in the location of the world's largest z pinch.

Nino Pereira

The first International Conference on High-Power Particle Beams (Beams) was held in Albuquerque, New Mexico, in 1975. The conference, organized by Gerold Yonas of Sandia National Laboratories, attracted scientists from around the world. The success of this first conference prompted the formation of the Beams International Advisory Committee and the organization of subsequent conferences on an approximately biannual basis. In addition to the United States, the Beams conference has been held in Russia, Germany, France, the Czech Republic, Israel, and Japan. As the field has matured, the conference has expanded the list of topics covered. The number of papers on the contributions of z pinches to high energy density physics has grown significantly in recent Beams Conferences, and we are pleased to be co-locating Beams with the Dense Z-Pinches Conference in 2002. Recognizing our beginnings, we are returning to Albuquerque this year and are happy to honor Gerry Yonas' contributions to the Conference by having him serve as Honorary Chairman of Beams 2002.

Jeff Quintenz

Beams Conference Statistics (Location, Dates, Attendance)

Location	Dates	USA	Russia (+ other fSU)	Japan	Germany	France	Israel	Others	Total	Host Country
Albuquerque, NM	Nov. 3-5, 1975	178	4	1	3	4	4	7	201	178
Ithaca, NY	Oct. 3-5, 1977	192	6	2	0	5	3	3	211	192
Novosibirsk, Russia	July 3-6, 1979	38	183	8	5	6	0	3	243	177
Palaiseau, France	June 29-July 3, 1981	81	15	9	14	81	6	16	222	81
San Francisco	Sept. 12-14, 1983	293	0	15	8	14	5	7	342	293
Kobe, Japan	June 9-12, 1986	77	17	155	5	8	5	14	281	155
Karlsruhe, Germany	July 4-8, 1988	99	16	19	75	21	8	21	259	75
Novosibirsk, Russia	July 2-5, 1990	52	149	14	10	6	0	13	244	149
Washington, DC	May 25-29, 1992	286	65	15	10	13	12	22	423	286
San Diego, CA	June 20-24, 1994	173	58	19	13	10	12	10	295	173
Prague, Czech Rep.	June 10-14, 1996	60	113	17	25	19	12	63	309	38
Haifa, Israel	June 7-12, 1998	58	80	17	8	9	62	37	209	62
Nagaoka, Japan	June 25-30, 2000	20	42	161	4	2	9	29	267	161

Dense Z Pinches Conference Statistics (Location, Dates, Attendance)

Location	Dates	USA[2]	Russia (+ other FSU)	Total
Alexandria, VA	March 1984[1]			32
Laguna Beach, CA	April 26-28, 1989	69		120
London	April 19-23, 1993	25		117
Vancouver	May 28-31, 1997	34		129

[1] Precise dates of the 1st conference are unknown.
[2] Number of U.S. attendees is unknown for the 1st conference; all but a few attendees are assumed to be from the USA.

Combined Beams and Dense Z Pinches Conference Statistics (Location, Dates, Attendance)

Location	Dates	USA (Host Country)	Russia (+ other FSU)	Japan	Germany	France	Israel	Others	Total
Albuquerque, NM	June 23-28, 2002	231	72	17	9	6	7	51	394

NATIONAL PERSPECTIVES

Fast Z - Pinch Study in Russia and Related Problems

E. V. Grabovskii

SSC Troitsk Institute for Innovation and Fusion Research,(TRINITI),
142092 Troitsk, Moscow Region, Russia

Abstract

The fast Z pinches are considered as a perspective source of powerful soft x-ray emission for the ICF pellet ignition. The physical phenomena which take place in process fast of Z-pinch implosion are under investigation in the TRINITI (Troisk), in the RSC Kurchatov Institute (KI, Moscow) and the HCEI (Tomsk).

In the KI the possibility of terawatt electrical power transfer in small volume hohlraum during nanosecond time duration is studied. In the TRINITI the physics of multi wire arrays implosion, the rate of plasma production in current-driven wire arrays, the conversion of pulsed power energy into x-ray emission are studied. In the HCEI (Tomsk) the stability of double gas puff implosion and the influence of gas puff regime and current pulse duration on the implosion and emitted x-ray spectrum are under investigation.

The HCEI develops the new components of pulse power multi spark switches and the generators of impulse currents (LTD) with duration of an energy supply less than 100 ns.

As available way to get the pulsed power generator with multi tens megaampere current the joint team of scientists from the laboratories of the TRINITI, the Efremov Institute, RFNC VNIITF (Snezinsk) have developed the concept of the Baikal facility. The KI designs, creates and tests the plasma erosion switches for the module of the Baikal facility. The inductive storage, the systems of magnetic field compression and the explosive open switches are developed in the TRINITI and the Efremov Institute. The development of new design of the pulse power generators and physics of fast Z-pinch implosion aims to create next advanced generation of powerful driver for ICF.

The investigation of dense Z-pinches in Russia are take place at several scientific centres and are directed both on clearing up of fundamental properties of

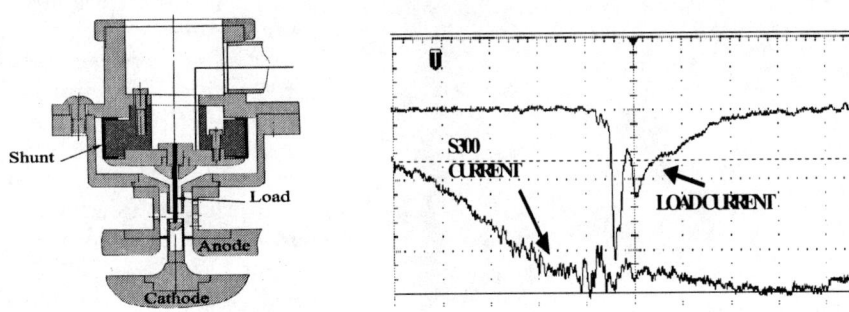

FIGURE 1. Left- scheme of nanosecond plasma flow switch for S-300. Right - S-300 total current (Amplitude 2,5 MA) and current through wire array load (Amplitude 2 MA). 10ns/div

substance, and on application of powerful Z-pinches as a powerful source of a X-rays for the appendices. The considerable effort is given to physics of Z-pinches himself. In this report I shall concern only results made per last several years and including, so-called, fast Z-pinches - pinches, which are in a nonequilibrium dynamic state during majority of the time, being imploded by a magnetic field of a current. For the most of facilities such Z-pinches have time of implosion of the order 100 ns at currents I more than of megaamper. The main scientific centers where the these researches are developed are KI, (Moscow), TRINITI (Troitsk), VNIIEF (Sarov), VNIITF (Snejinsk), HCEI (TOMSK), Lebedev Institute (Moscow).

The investigations in KI [1,2,3] take place on S-300 facility. Its parameters are: $U = 1,3$ MV, $I = 3,5$ MA, current rise time is 100 ns. The heating of substance by electronic thermal stream from the end of a dense Z-pinch is under investigation on this facility. These scheme is known as "liner-converter scheme".

The second problem under investigation of S-300 is fast electrical power transfer to a small size hohlraum. The scheme of a current transfer in volume on installation S-300 is presented in Fig. 1. The shunt ensuring a time resolution 1-2 ns was used to measure current. In a fig. 1 the profiles of a total current of S-300 load and the shunt current are shown. The given scheme produce known earlier plasma-flow switch scheme in a new scale of times, when the switching occurs during some nanoseconds. It allows to pass in new unexplored earlier area of parameters. The carried out experiments show that the magnetic isolation ensuring introduction of power in a gap with the size of 1 mm during about to ten nanosecond, works for described experimental requirements.

In TRINITI on installation "Angara-5-1" the investigation of double liner scheme offered in the beginning of the 90-th years is prolonged. At these scheme the irradiation of a target occurs at a collision by the liner imploded by a current, to fixed one [4-7]. Using the suitable substances of the liners it is possible to capture of radiation inside volume of the inner liner and to increase the temperature on a surface of a pellet. This scheme is named also " Dynamic hohlraum ".

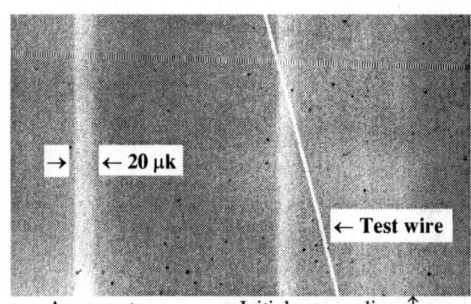

FIGURE 2. X-ray shadow photo of area (1*1 mm) of wire array.

Simultaneously with "Dynamic hohlraum" the physics of dense Z-pinches formed by wire arrays implosion is studied. The phenomenon of prolong plasma production in wire arrays is revealed. This phenomenon shows that despite of the value of the energy deposition in a pinch many times exceeding necessary one for sublimation the tungsten wire in array up to almost final part of an implosion are dense, not moving from an initial standing. The dense kernels of wires produce plasma with frozen magnetic field. This plasma flows to the center of a pinch.

In a fig. 2 the x-ray shadow photo of wire array with the resolution less than 2 microns and with a exposure time 1 ns at the moment of 80 ns from a beginning of a current through array is presented. The wires are visible at the initial standing. Their diameter was increased from 6 microns up to 20 microns and the dense kernels are well visible on the x-ray images. The measuring with the help of microprobes of a magnetic field inside the multiwire liner shows that in 30-40 ns after the very beginning of a current through the array there is a plasma precursor on an axis of a pinch. This precursor carries few percents of a array current. The measuring of a magnetic field in some point inside a pinch allows to estimate the thickness of a skin - shell of the imploded array and to show, that in case of prolong plasma production it considerably exceeds skin length calculated from classical assumptions. It apparently gives in the greater stability of the array and to deriving a narrow and powerful X-ray impulse. The electrotechnical measuring of a current and a voltage near the array allow to calculate an electrical energy inlet into a Z-pinch and inductance of array.

The power balance which is based on these measure and measure of energy of a X-rays, gives us a deduction, that energy of a X-rays exceeds a kinetic energy of imploded plasma. The considerable contribution of energy of a magnetic field in a pinch take place. Prolong plasma production specifies a importance of one of earlier

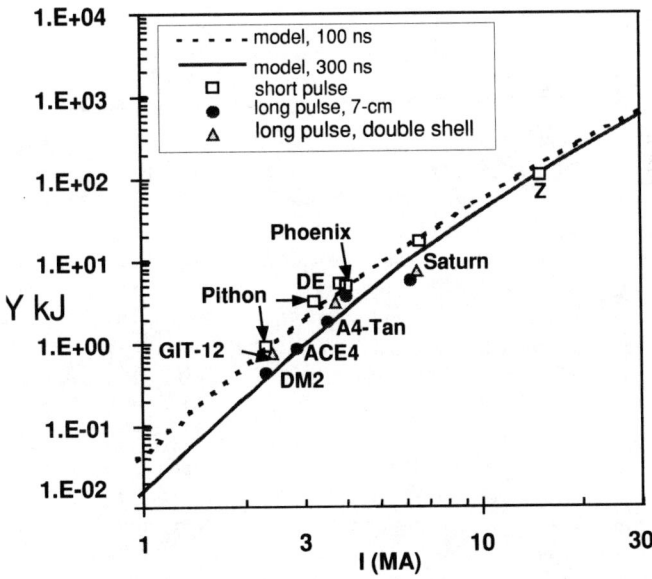

FIGURE 3. Comparison data of X-ray yield (kJ) for Ar gas puff from current for different models (calculations) and installations. GIT 12 result for 300 ns current rise time is close to calculation and results for 100 ns current rise time.

not taken into account parameters of a Z-pinch - a intensity of plasma production from dense kernels. The management of this parameter, probably, will allow to effect on the shape of a X-ray impulse.

Plasma of gas pinches at currents up to 6 MA with current rise time 100 ns is under investigation on installation "Pirit" located in VNIIEF. The results and scheme of experiments on plasma flow switch in a current scale time about a microsecond are shown in [8]. The obtained velocity of a motion of a current shell was equal to 20 + -5 cm/mks.

In HCEI investigations of Z-pinches formed by double gas liners are conducted [9-12]. They are aimed to produce the power peak of radiation in a high-temperature part of a spectrum and to research dynamics of cascade imploded gas puffs and wire arrays. It was shown on installation GIT-16 that after a collisions in appropriate way fitted liners RT perturbations taking place in an outside shell diminish the amplitude. Despite of loss of energy in collision it gives a magnification of power of a X-ray impulse including in a high-temperature part of a spectrum.

In the course of experiments on the implosion of large-diameter gas puffs it has been established that during the implosion the wavelength of RT oscillation increases, reaching at the final stage a value of 0.8–1 cm. As this takes place, bubbles appear in the plasma shell bulk. An increase in the atomic number of the imploding gas is accompanied by the appearance of a two-mode implosion: the lighter layer formed by bubbles arrives at the gas puff axis and implodes 40–50 ns earlier than the major portion of the gas puff material. The unstable implosion of a gas puff with a high atomic weight is accompanied by an increase in the resistance of the gas puff in the run-in phase, to a value exceeding the Spitzer resistance by two or three orders of magnitude. An increase in the resistance of the load plasma enhances the diffusion of the magnetic field, which in turn leads to a more intense acceleration of the low-density regions of the plasma and a less compact implosion.

The important contribution to diagnostics methods of wire array with the high time and spatial resolution was developed by S. Pikuz group from Lebedev Institute. The X-pinch, studied already long time was effectively applied by this group as a sours of emission for wire array backlighting. The high-quality images of blowing up wires from different substances has cleared physical pattern of multiwire liners explosion in an initial stage of the discharge (fig. 10). This diagnostics applied already on several installations, - Imperial College (Great Britain), TRINITI, (Russia), Rino University (USA). [9]

The new elements for powerful installations is develop at HCEI which has developed a set of elements of a new generation - multispark switches, LTD systems, multicascade megavolt scale switches. Multispark switch with parameters 100 kV and 500 kA provide of twenty separate sparks with high reliability and stability. LTD system consisting from set of low inductance capacities, switches and inductances. It possible to assemble an megaamper facilities as set of identical LTD systems. In HCEI LTD with energy output time from 0,5 mks up to 0,1 mks are developed. The installations based on LTD works in HCEI and in France. Multicascade megavolt scale switch has high reliability because of constructive simplicity and very perspective for next generation of large facilities. [10-12]

In KI the plasma open switch (POS) is under developing. The features the POS is that the current density is small to increase the POS resistance and maximum working voltage. Additional external magnetic field is used with a configuration hindering electron current from cathode to the anode in POS area. The method of a measurement of voltage on POS by nuclear responses independently confirms voltage value achieved on POS. This development is made for the task of creation of installation "Baikal" that aimed to generate by multiwire liners implosion the power pulse of a x-ray radiation with energy more than 10 MJ. This installation is based on use already being available in TRINITI of inductive engines with stored energy 3.5 GJ and additional complicated system of a peaking of electrical pulse. Such installation is developed by cooperation from KI, TRINITI, Efremov Institute and VNIITF. In TRINITI such kind installation with an inductive accumulator on 12 MJ is under developing.

Thus, in Russia, despite of problems with financing of a science, the investigation are continued in the field of Z pinch physics, the new results are obtained and are developed new elements for installations of the following generation.

REFERENCES.

1. Chernenko, A.S., Gorbulin, Yu.M., Kalinin, Yu.G., et al. Proceedings of the 11th International Conference on High Power Beams, Prague, 1996, v. 1, p. 154.
2. Bakshaev, Yu.L., Bartov, A.V., Blinov, P.I., et al, Proceedings of the 12th International Conference on High Power Beams, Haifa, 1998, p. 244.
3. Kingsep, A., Blinov, P., Chernenko, A., et al, Proceedings of the 1-st International Conference on Inertial Fusion Science and Applications, Bordeaux, 1999, MO1c5_141.
4. Grabovski E. V. , Smirnov V. P., Zakharov S. V. At al. , JETF. 82(3), 445 (1996).
5. A.V.Branitskii E.V.Grabovskii M.V.Frolov et. al. . Proceedings of 12th Int. Conference on High-Power Particle Beams. BEAMS'98. Haifa, Israel, June 7-12 1998, p. 599-602.
6. V. V. Alexandrov, A.V. Branitsky, G. S. Volkov et.al., Inertial Fusion Sciences and Applications 99. Edited by C.Labaune, W.J. Hogan, K. A. Tanaka, Superfast multi-wire liner implosion physics study on Angara-5-1. p. 591, 2000.
7. E. V. Grabovskij, S. L. Nedoseev, G. M. Oleinik et.al.,Proceedings of the sixteenth International Conference on fusion energy organized by the International Atomic Energy Agency (FUSION ENERGY 1996) and held in Montreal, 7 - 11 October 1996, in three volumes, Intenational Atomic Energy Agency, Vienna, 1997, IAEA-CN-64/BP-4, p.129.
8. Proceedengs of 7 Int. Conf. On MegaGauss Magnetik Feield, Megagauss 7, Sarov, 1996 r. N. F. Popkov, E. A. Riaslov, V. I. Kargin at al. 10 TW pulse facility Pirit...p.399-401.
9. S. A. Pikuz, T. A. Shelkovenko, V. M. Pomanova, D. A. Hammer, A. Ya. Faenov, and T. A. Pikuz, Rev. Sci. Instrum. 68, 740 ~1997!.
10. Multigap, multichannel spark switches / Kim A.A., Kovalchuk B.M., Kremnev V.V., Kumpjak E.V., Novikov A.A., Etlicher B., Frescaline L., Leon J.F., Roques B.,

Lassalle F., Lample R., Avrillaud G., Kovacs F. // Proc. XI IEEE Intern. Pulsed Power Conf., Baltimore, 1997, p. 862-867.
11. Multi gap switch for Marx generators / B. Kovalchuk, A. Kim, E. Kumpjak, N. Zoi, J. Corley, D. Johnson, K. Struve // Abstr. XIII IEEE Intern. Pulsed Power Conf., Las Vegas, 2001, P4-H02.
12. Volkov S.I. Kim A. A. Kovalchuk B. M. at al. Izvestija vuzov, Fizika, 1999, т. 42, №12, с. 91-99.
13. Development of X-Ray Facility Baikal based on 900 MJ Inductive Store and Related Problems E. A. Azizov, V. V. Alexandrov, S. G. Alikhanov, e. et. al. Digest of technical papers of PPPS-2001p. 773. (Pulse power Science Conference, Las-Vegas 2001)

History of Z-Pinch Research in the U.S.

Mary Ann Sweeney

Sandia National Laboratories, P. O. Box 5800, Albuquerque, NM 87185-1191

Abstract. Over the years, the scientific community has been fascinated with z pinches. Z-pinch references include papers on the quest for fusion, on applications for radiation effects testing, lithography, x-ray microscopy, and pumping x-ray lasers, and on the production of intense magnetic fields. Because much of the research has been pursued elsewhere–in the USSR, Russia, England, Germany, and Chile, among other countries–we must place the U.S. work in an international context. We assert here that the z pinch is a valuable asset for its applications, chiefly those related to the production of x rays, but it is a tool that has sometimes deceived us with its seeming simplicity.

In the 2001 high-tech crime comedy film *Ocean's 11*, an apparatus called 'the pinch' plays a key role in the success of Danny Ocean and his ten accomplices in pulling off a $160M heist of three Las Vegas casino vaults. Fit into a van and set off in mid Las Vegas, 'the pinch' is a take-off on Sandia National Laboratories' Z accelerator[1], but reassigned to the mythical California Institute of Advanced Technology. This pinch creates an intense electromagnetic pulse (EMP) that violates energy conservation and blacks out the city's power grid long enough to complete the caper. A *Newsday* movie critic dubbed this kind of thing "technological minutiae."[2] But it is this sort of minutiae that we have been concerned with this week and that have occupied us, as early as the 18th century and at four previous international conferences on dense z-pinches (DZP). Experiments and theory have been conducted at many places in the U.S., including Los Alamos National Laboratory, Maxwell Laboratory (now Titan Pulse Sciences), Naval Research Laboratory (NRL), Phillips Laboratory (now the Air Force Research Laboratory), Physics International (PI), Sandia, UC Irvine, and at laboratories in England, France, Russia, and elsewhere. "I enjoyed the movie and 'the pinch' was an amusing twist but had little to do with science,"[3] said Jeff Quintenz, director of Sandia's Pulsed Power Sciences Center, explaining that Z is a rather poor EMP device except for its effect on sensitive diagnostics nearby. The *Ocean's 11* 'pinch' is thus a well-executed deception, but the history of z-pinch research suggests a real z pinch has also been a well-executed deception for us as researchers.

THE Z PINCH FOR CONTROLLED FUSION?

The z pinch was one of the earliest methods to try heating plasmas to thermonuclear temperatures.[4] A z-pinch reference from that era is to Malcolm Haines.[5] In the late 1950s, Oscar Anderson, Bill Baker, Stirling Colgate, Harold Furth, et al, observed copious neutrons (10^8/pulse) from a linear deuterium pinch imploded in 1.5 μs with a 20-kV bank,[6] but showed the neutrons had a nonthermal origin from deuterons accelerated in electric fields. The linear z pinch, in which current heats the plasma and creates a magnetic field confining the plasma, is beautiful in its simplicity, but is unstable. The neutrons were from the sausage (m = 0) instability. These disappointing results, however, begat magnetic fusion energy, which tries to heat a plasma by microwaves or neutral beams to 4 keV and an nτ of 10^{14} s/cm^3, and inertial fusion energy, which tries to use compressional heating to reach 4 keV and a ρr of 0.4 g/cm^2.

Other interpretations of the Anderson results suggested that, if the deuterium had been imploded on a faster time scale, before significant instability growth, fusion could be achieved. One of the first tests of this on a modern pulsed power generator was at Imperial College in 1978, using a 100-kA pinch within a 1-cm-diameter quartz tube. About 10^{12} neutrons/pulse were produced, T_e was 1 keV, and plasma density was 10^{18}/cm^3, but the ions did not get hot.[7] Again, the culprit was the m = 0 instability.

The subject of the 1st DZP conference in 1984 was fusion applications of the linear z pinch. Experiments and modeling of frozen deuterium fibers[8] and the possibility of controlled fusion by adding a stabilizing axial magnetic field[9] or by preventing radiative collapse were discussed at the 2nd DZP. By 1989, however, the scope of the conference had expanded to include the generation of radiation from gas puffs[10], imploding wire arrays, liners, and foils[11], the generation of intense fields by magnetic flux compression[12], and x-ray lithography and microscopy.[13]

THE EARLIEST HISTORY OF THE Z PINCH?

The Dutch chemist and physician, Martinus van Marum, vaporized metal wires with a z pinch more than two centuries ago, using the large electrostatic generator built in 1784 by the Englishman John Cuthbertson,[14] although the term "pinch" was apparently not used[15] until the 20th century. The 100-135 tinfoil-lined Leyden-jar capacitors, occupying 3.3 m^2, allowed a 60-kV charge and "could produce tongues of fire 60 cm long."[1] Assuming the exploding wires were in the center[16], the discharge time was ~0.5 μs and the peak current was 60 kA. Today, the most energetic successor to van Marum's device is Z. Although Z's capacitors are about the same size as van Marum's Leyden jars, the total capacitance is ~250 times larger and the capacitors are charged to about 90 kV. The 20-MA peak current on Z is more than 250 times larger, the pulse is shorter (100 ns), and the energy stored is 11.4 MJ instead of ~30 kJ. When the refurbishment of Z[17], called ZR, is completed in 2006, the peak current of 26-28 MA will be more than 400 times larger than in van Marum's device.

THE Z PINCH AS AN X-RAY SOURCE FOR EFFECTS TESTING

The z pinch is an efficient source of x rays. Following the discouraging results with beam-generated neutrons in the 1950s, and a decade of relative inactivity, U.S. z-pinch

---

[1]Van Marum's electrostatic generator is at the Teylers Museum in Haarlem, the Netherlands (http://www.teylersmuseum.nl/engls/ruimtaes/instrument/start.html).

research has focused for over 30 years on creating K-shell and L-shell x rays with energy >1 keV to test the response of materials and electronics to threats from nuclear weapons. The reemergence of U.S. interest in the z pinch followed the birth of modern pulsed power, in 1964, via Charlie Martin's experiments on SMOG at the Atomic Weapons Establishment.[18] From the mid 1960s to early 1980s, a major champion in the U.S. was Peter Haas of the Defense Nuclear Agency (DNA), now called the Defense Threat Reduction Agency (DTRA). As the DNA program manager, Haas, after whom a Pulsed Power Conference award is named, reinvigorated the research area by funding construction and operation of a number of devices—e.g., Aurora, Phoenix, Casino, and Gamble I/II. The Department of Energy also provided funding.

Although x rays can be created by exploding a single wire,[19] in the late 1970s researchers[20] found that a cylindrical array of a few wires offered better energy efficiency. Other load configurations have included a hollow gas puff, an annular gas jet, a thin foil formed into a cylinder, a low-density foam cylinder with or without a conductive coating, a discharge through a small capillary, a gas puff/wire array combination, or nested wire arrays. The electrical energy is converted into the kinetic energy of a magnetically confined, imploding plasma, with ~10% emitted at stagnation in soft (1–10 keV) x rays from localized "bright spots"[2] that may originate from the sausage instability. The remainder is mostly sub-keV radiation (~50%) and kinetic energy of the expanding plasma after stagnation. Optical radiation is generated near the start of the current pulse as material is ionized. Bremsstrahlung is also present,[21] generated by high-energy electrons from the strong electric field along the pinch axis.

At low current I, independent of the pulsed power device, z-pinch configuration, or element/alloy, the optimal x-ray yield (energy) $E \sim I^4$.[22] A slug model shows why this relation holds for conversion of electrical energy into kinetic energy of the imploding plasma.[23] At higher currents in experiments on four generations of Sandia accelerators and on devices at other institutions, $E \sim I^2$, essentially independent of pinch material, radius, and implosion velocity.[24] However, 2-D calculations suggest[25] additional energy can come from coupling the **J x B** force into the finite thickness sheath during stagnation, and a recent systematic examination of scaling of x-ray yield and power with current on Z[26] indicates optimizing pinch parameters for a specific device is more complicated. Many factors (wire heating, ablation, resistivity, Rayleigh-Taylor instability, pinch stagnation) can influence the x-ray production.

For radiation effects applications, the most important criteria are x-ray yield and long (15–50 ns) pulse widths. As a consequence, the effect of increasing wire number was ignored for nearly 20 years, since powers increased but yields stayed constant. In a 1988 review of x rays from z pinches,[27] Nino Pereira and Jack Davis referred to a 1981 paper[28] by Frank Felber (then at Maxwell), which concluded six or more wires are theoretically stable, and 1976 PI research[20] that found the x-ray power from four wires is indeed smaller than from six. However, according to Pereira and Davis,[27]

---

[2]Bright spots may not have hot temperatures compared to surrounding plasma, but the densities can be considerably higher—e.g., Clark, W., et al, p. 236, in *Proc. 5th Intl. Conf. High-Power Particle Beams*, 1983; Sanford, T.W.L., *Laser and Part. Beams* **19**, 539 (2001). But John Apruzese (*these proceedings, 2002*) suggests, for higher photon energies, temperatures, densities, or both can be much higher.

Experiments to increase the radiation by using 12 or even 24 wires fail[ed] to give a substantial improvement. Therefore, six wires [were] most common, because this minimize[d] the difficult handling of fragile wires....In practice the individual wires...are never identical or mounted symmetrically, and this asymmetry [can] spoil the implosion when the perturbation is unstable....Each wire can pinch by itself while it is accelerating toward the center. Because the instability of a single wire is fast, each individual wire may [pinch] before the wires meet on axis, resulting in an inhomogeneous plasma column. Consequently, wire implosions are often less uniform than gas puff implosions.

THE MODERN Z PINCH AND STOCKPILE STEWARDSHIP

In the early 1980s, z-pinch research continued, primarily with DNA funding (on Double Eagle at PI, Blackjack 5 at Maxwell, and Gamble II at NRL) to produce photon-pumped x ray lasers[29] and to generate x rays for radiation effects, although Sandia pursued a small effort on Proto II and Saturn. Sandia discontinued its research on generating z-pinch x rays to compress a capsule around that time because of the belief that ions[3] offered a more reasonable route to fusion.[30] In the meantime, an effort was pursued, mainly at NRL, Los Alamos, UC Irvine, and Imperial College, to develop a stable linear z pinch (sometimes called a "direct-drive z pinch") as a fusion source.[31] The hope was either that modern pulsed power devices could implode a deuterium pinch to fusion conditions faster than the instabilities could grow or that, if Bennett equilibrium[32] between joule heating and bremsstrahlung radiation losses could be achieved at the Pease-Braginskii current[33] (1.4 MA for hydrogen) by keeping the pinch at constant radius, the linear pinch would not become unstable.

Then, in the early 1990s, cessation of underground testing created the need to test predictions of advanced computational codes over a broad range of x-ray pulses and energies to ensure the safety and reliability of nuclear weapons. The 1994 "PBFA Z" project[34] to modify Sandia's then nine-year-old light ion accelerator, PBFA II, to allow z-pinch experiments in order to fill in gaps in radiation effects data at currents as large as 20 MA then led to a 1995 discovery[35] on the 7-MA Saturn that the soft x-ray power from an imploding z-pinch could be greatly enhanced by using an array of *many*, rather than a *few*, wires (200 to 400 being now typical for tungsten wires on Z, the renamed z-pinch modification of PBFA II). This breakthrough, which has resulted in an increase by an order of magnitude in the yields for radiation effects testing, was a result of several factors: a mature pulsed power technology capable of obtaining much higher currents, improvements in wire-array load design, and a better understanding of the behavior of z pinches obtained from modeling and diagnostics.

The flexible, efficient, well-characterized x-ray source that resulted has sparked renewed interest in two z-pinch fusion concepts—the double-ended vacuum hohlraum

---

[3]The ion approach began in the mid 1970s when Cornell, Sandia, and NRL developed efficient ways to produce intense ion beams after Graybill and Young's discovery in 1968 at Ion Physics Corp. that the collective effects of electrons could be used to accelerate a smaller number of ions to high energies.

and the "flying radiation case,"[36] a term Los Alamos has used and which Sandia calls the "dynamic hohlraum"[24]. The high-energy-density regimes accessible with modern pulsed power have also motivated interest in measuring material properties for stockpile stewardship and astrophysical applications and in enhancing the capabilities of Z. The intense soft x rays may also provide novel applications such as the ability to study dynamic features in the several hundred angstrom range in materials and in biological samples.

LESSONS LEARNED

What lessons have we learned from the last 40-odd years of z-pinch research?
- Despite its seeming simplicity, the z pinch is a complex, 3-D creature that requires sophisticated models and diagnostics (e.g., velocity interferometry, laser backlighting, and time-resolved spectroscopy) and careful treatment of the physics.
- Applying the z pinch to stockpile stewardship and fusion demands patience, skill, and innovation.
- Both newcomers and well-established researchers are still learning the intricacies of the z pinch.
- Open sharing of knowledge–nationally and internationally–is the best way to make progress in understanding this sometimes deceptive phenomena.
- Empirical discoveries happen, and we must be alert to understand what they tell us about the next step to take.
- Government funding of the research plays an important role in the U.S.

ACKNOWLEDGMENTS

Because of space limitations, this paper does not represent a comprehensive review of U.S. z-pinch research, and the list of references below includes only a relatively small number of the papers in the field. In particular, there was not enough space to discuss and reference the computations and diagnostics that have contributed to an improved understanding of z pinches. I would like to acknowledge discussions with Nino Pereira, Jack Davis, Sasha Velikovich, Malcolm Haines, Phil Spence, Bob Terry, David Mosher, my Sandia colleagues, and others at the Beams/DZP meeting in Albuquerque as well as a number of excellent review papers[15-16,22,24,27,31].

The U.S. Department of Energy (DOE) supported this work under Contract DE-AC04-94AL85000. Sandia Corporation, a Lockheed Martin Company, operates Sandia National Laboratories, a multiprogram laboratory, for the DOE.

REFERENCES

1. Spielman, R.B., Deeney, C., et al, p. 101, *Proc. 4th Intl. Conf. Dense Z-Pinches (DZP)*, American Institute of Physics (AIP), New York (NY), 1997.
2. Anderson, J., "Improving upon Rat Pack Mediocrity," *Newsday*, December 7, 2001 p. B-3.
3. Stein, B., "The Con-Artist Physics of 'Ocean's Eleven'," in *APS News* **11**, March 2002, p.5.

4. Bishop, A.S., *Project Sherwood: The U.S. Program in Controlled Fusion*, U.S. Atomic Energy Commission, Addison-Wesley Publishing Company (1958). These papers were presented at the 2nd United Nations Conference on Peaceful Uses of Atomic Energy in Geneva, Switzerland in 1958.
5. Haines, M.G., *Proc. Phys. Soc. London* **76**, 250 (1960).
6. Anderson, O. A., Baker, W.R., et al, *Phys. Rev.* **110**, 1375 (1958).
7. Choi, P., Dangor, A.E., et al, p. 69, in *Proc. 7th Intl. Conf. Plasma Physics and Controlled Thermonuclear Fusion Research*, 1978.
8. See papers by Sethian, J.D., Robson, A.E., et al, p. 308, Hammel, J., p. 303, and Lindemuth, I.R., p. 327, in *Proc. 2nd Intl. Conf. DZP*, AIP, NY, 1989.
9. See Golberg, S.M., Liberman, M.A., Velikovich, p. 345, and Rahman, H.U., Ney, P., et al, p. 351, in *Proc. 2nd Intl. Conf. DZP*, AIP, NY 1989.
10. See Spielman, R.B., Dukart, R.J., et al (on Saturn), p. 3, and Krishnan, M., et al (on Pithon), p. 17, in *Proc. 2nd Intl. Conf. DZP*, AIP, NY, 1989.
11. See Baksht, R.B., et al (on SNOP-3), p. 27, Deeney, C., et al (Double Eagle), p. 55, Degnan, et al (Shiva Star), p. 34, and Whitney, K.G., Thornhill, J.W., p. 143, in *Proc. 2nd Intl. Conf. DZP*, AIP, NY, 1989. Also, S. Stephanakis, et al (on Gamble II), *Appl. Phys. Lett.* **49**, 829 (1986).
12. See Felber, F.S., et al (on Proto II), p. 431, and Sorokin, S.A., Chaikovsky, S.A (SNOP-3), p. 438, in *Proc. 2nd Intl. Conf. DZP*, AIP, NY, 1989.
13. Qi, N., et al (on LION), p. 71, in *Proc. 2nd Intl. Conf. DZP*, AIP, NY, 1989.
14. Turner, G.L.E., Levere, T.H., *Martinus Van Marum Life and Work*, vol. 4, ed. by E. Lefebvre, J.G. De Bruijn, Noordhoff International Publishing, Leyden, 1973.
15. Ryutov, D.D., Derzon, M.S., Matzen, M.K., *Rev. Modern Physics* **72**, 167 (2000).
16. Spielman, R.B., DeGroot, J.S., *Laser and Particle Beams* **19**, 509 (2001).
17. Mazarakis, M.G., McDaniel, D.H., et al, "The ZR Refurbishment Project," *these proceedings*, 2002.
18. *J. C. Martin on Pulsed Power*, ed. by T.H. Martin, A.H. Guenther, M. Kristiansen, Plenum Press, NY, 1996.
19. Mosher, D., Stephanakis, S.J., et al, *Appl. Phys. Lett.* **23**, 429 (1973).
20. Stallings, C., Nielsen, K., Schneider, R., *Appl. Phys. Lett.* **29**, 404 (1976).
21. Putnam, S., Stallings, C., et al, *Bull Am Phys Soc.* **24**, 1078 (1979); Clark, W., Wilkinson, M., et al, *J. Appl. Phys.* **53**, 1426 (1982); Derzon, M.S., Nash, T., et al, *Rev. Sci. Instr.* **70**, 566 (1999).
22. Krishnan, M., Deeney, C., et al, p. 17, in *Proc. 2nd Intl. Conf. DZP*, AIP, NY, 1989; Thornhill, J.W., Whitney, K.G., Davis, J., *J. Quant. Spectros. Rad. Transfer* **44**, 251 (1990).
23. Mosher, D., NRL Report No. 3687 (1978).
24. Matzen, M.K., *Phys. Plasmas* **4**, 1519 (1997).
25. Peterson, D.L., Bowers, R.L., *Phys. Plasmas* **3**, 368 (1996).
26. Stygar, W.A., et al, submitted *Phys. Rev. Lett.* (2002).
27. Pereira, N.R., Davis, J., *J. Appl. Phys.* **64**, R1 (1988). Quote is from pp. R6 and R16.
28. Felber, F.S., Rostoker, N., *Phys. Fluids* **24**, 1049 (1981).
29. Dahlbaca, G., Gilman, C., et al, p. 245, *Proc. 4th Intl. Conf. High-Power Particle Beams*, Ecole Polytechnique, 1981; Maxon, et al, *J. Appl. Phys.* **57**, 971 (1985); Apruzese, et al, *Phys. Rev. A* **35**, 4896 (1987); Hussey, T.W., et al, p. 147, *Proc. 2nd Intl. Conf. DZP*, AIP, NY, 1989.
30. Yonas, G., Sweeney, M.A., p. 165, in *Proc. 12th Intl. Conf. High-Power Particle Beams*, Haifa, Israel, 1998, contains an abbreviated review of the quest for fusion using electrons or ions.
31. Sethian, J., p. 3, in *Proc. 4th Intl. Conf. DZP*, AIP, NY, 1997; also, z-pinch fusion papers in *Proc. 1st* (Naval Research Lab, Washington, D.C., 1984) and 2nd (AIP, NY, 1989) *Intl. Conf. DZP*.
32. Bennett, W.H., *Phys. Rev.* **45**, 890 (1934).
33. Pease, R.S., *Proc. Phys. Soc. London* **B70**, 11 (1957); Braginskii, S.I., *Sov. JETP* **6**, 494 (1957).
34. Spielman, R.B., Stygar, W.A., et al, p. 150, *Proc 11th Intl. Conf. High-Power Particle Beams*, 1996.
35. Sanford, T.W.L., Allshouse, G.O., et al, *Phys. Rev. Lett.* **77**, 5063 (1996). Tom Sanford credits Ken Whitney with encouraging him to increase the number of wires in the arrays and decrease their thickness. See also the theoretical model for a nested wire array z-pinch load by Davis, J., Gondarenko, N.A., Velikovich, A.L., *Appl. Phys. Lett.* **70**, 170 (1997).
36. Brownell, J., Bowers, R., *Bull. Amer. Phys. Soc.* **40**, 1848 (1995).

Southern Dense Pinch Discharges

Hernán Chuaqui

Departamento de Física, Pontificia Universidad Católica de Chile, Casilla 306, Santiago 22, Chile

Abstract. An overview of the work being carried out on dense pinch discharges at the Optics and Plasma Physics Laboratory is presented. Llampüdkeñ, a 250 ns rise time, 1 MA pulser is described. X-pinch work carried out on Llampüdkeñ is described, as well as a shell type gas embedded Z-pinch and a high current capillary discharge in GEPOPU.

INTRODUCTION

In this paper an overview is presented of the dense Z-pinch work carried out at the Plasma Physics and Optics Laboratory of our University. During the last few years a wide range of different dense discharges have been studied using GEPOPU, a 120 ns 200 kA pulser [1]. Recent results on a high current capillary discharge and on a hollow gas embedded laser initiated Z-pinch obtained on this generator will be presented. The main features of a new generator, Llampüdkeñ [2], a 250 ns rise time and up to 1 MA current, will be given, as well as preliminary results on X-pinch work carried out on this pulser.

LLAMPÜDKEÑ

After a number of years doing pinch work at our Laboratory it became apparent that a new generator was required. The two main design goals were a higher maximum current and the need to maintain a longer current ramp at higher values of dI/dt. Traditional designs usually result in higher impedance lines which cause problems due to a higher charging voltage. In these situations a substantial fraction of the line voltage is developed across the load in the first 30 ns of the discharge, or until main conduction begins. The need to withstand these high voltages usually requires the use of vacuum magnetic insulation, with its associated physical constraints. The present generator allows an MA current in the main conduction phase without the problem of high voltage at early times.

The generator LLAMPÜDKEÑ (meaning "butterfly" in Mapungdun, a Chilean indigenous language) is designed to reconcile conflicting requirements, at a very low cost. The generator is designed to have an impedance of 0.5 Ω in the main conduction phase while limiting the voltage across the electrodes at early times due to load mismatch. The design is based on the use of parallel plate water lines driven by two 28.8 kJ, 40 kV Marx generators. The novel features are the use of exponential transmission lines both for the transfer section as well as an auxiliary exponential line in parallel with the load section. The use of exponential lines allows propagation of waves between two points of different impedance, for certain limit pulse lengths, without reflection. The impedances seen by the source driving a line are not constant in time. This is important here, where the auxiliary line must present a low impedance at early times to divert excess voltage, when the plasma is resistive, but in the main discharge period must present a high impedance in order not to divert current. Furthermore, the energy diverted may be brought back to the load by leaving the end of the auxiliary line an open circuit. A current increase of about 30% is obtained in this generator. An electrical diagram of one-half of the generator is shown in figure 1.

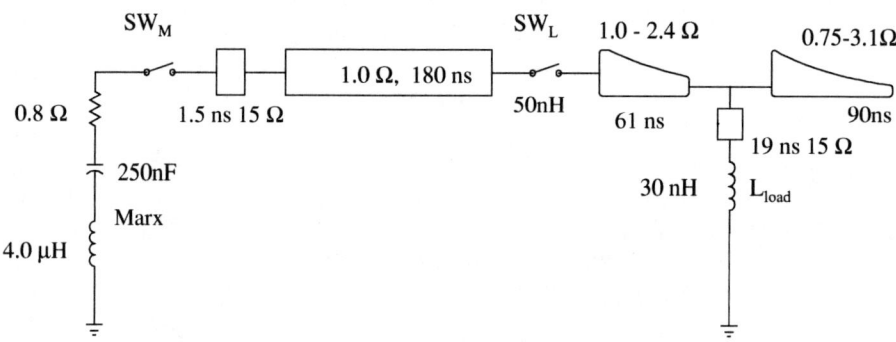

FIGURE 1. Schematic diagram of one arm of the generator (the other half is connected in parallel to the load). SW_M is the Marx switch, SW_L the line switch, L_{load} the load inductance. The values of transit time and corresponding impedance are given for the different lines.

Simulations of the generator using TopSpice were carried out using the parametersshown in figure 1 were carried out. Llampüdkeñ has been operated reliably at 50% of the maximum voltage. Figure 2 shows a direct comparison of experimental and simulated voltage and current traces for the case of a short circuit load with ±20kV charge voltage. In the experimental results the voltage traces are measured by the voltage probe at the output of each of the Marx banks and the current traces are those obtained by the integration of the output of a Rogowski coil placed inside the vacuum chamber. The simulation has been tailored to reproduce the Marx and switch jitters in this specific shot and the inductance of the switches has been adjusted to agree with the decay time of the voltage after the switches close. As can be seen there is very good agreement between the simulation and experimental measurements which both show a peak current of 385 kA with a risetime of 260 ns [3].

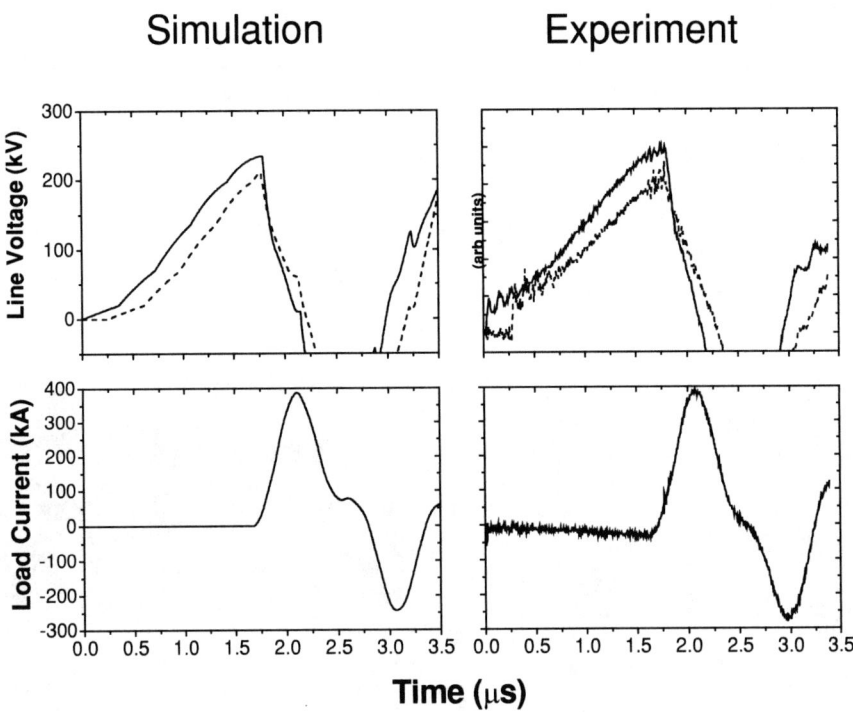

FIGURE 2. A comparison of a simulation and a real shot of the generator for 50% charge. The scales of the simulated and experimentally measured current traces are the same.

X-PINCH

A series of X pinch experiments were carried out on the Llampüdkeñ generator at a current level of ~400 kA. The x-ray emission from X pinch loads made from various numbers and diameters of Aluminium wire was studied, particularly in relation to the duration of the emission and the size of the hot spot source. The main diagnostics employed were a set of filtered PIN diodes and a time integrated multi filtered slit-wire camera [4]. Figure 3 shows images obtained through four individual slits of the slit-wire camera. The numbers on the figures indicate the diameters of the wires to which the shadows belong. A well defined shadow of the 25 μm wire indicates a source size of < 5 μm (the resolution of the diagnostic). Figure 4(a) and (b) respectively show images obtained for the 125 μm aluminum wire X pinch through a filter of 17 μm aluminized mylar and through a composite filter consisting of 3 μm silver and 24 μm mylar. The mylar filter transmits the aluminum K shell radiation whereas the composite filter does not but has a transmission window between 2 and 3.5 keV. Figure 3(c) and (d) show the image obtained with a X pinch of two and four 20 μm diameter aluminum wires respectively through the 17 μm aluminized mylar filter. In (b) the shadow of the 25 μm diameter wire can be seen indicating that for the 125 μm wire pinch the source size is < 5 μm for energies of 2 – 3.5 keV. This shadow is not seen in figure (a) however, indicating that the source is larger for photon energies down to ~1 keV. A source size of < 5 μm can be obtained for these lower photon energies when a 20 μm aluminium wire single or double X pinch is used as the shadow can be seen in (c) and (d). No image is obtained through the composite filter for the single 20 μm X pinch and with the 20 μm double X pinch only a very low intensity image is obtained, thus indicating that plasma is hotter in the case of the 125 μm wire pinch [5].

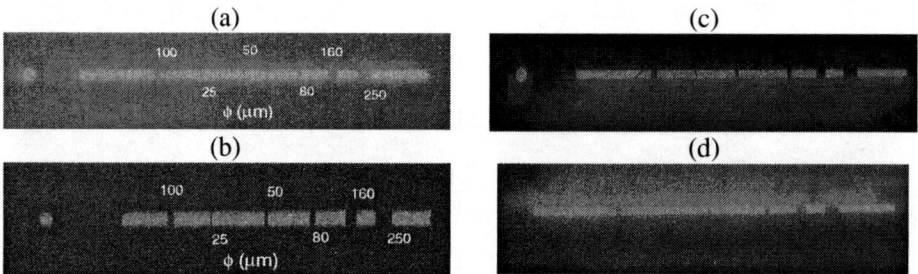

FIGURE 3. Images from the slit wire camera. The numbers on the images represent the diameters of the wires to which the shadows belong. The order of the wires is the same in all four figures. (a) 125 μm Aluminium X pinch with 40 μm Aluminized Mylar filter. (b) 125 μm Aluminium X pinch with composite filter of 3 μm silver and 17 μm mylar. (c) 20 μm Aluminium X pinch with 40 μm Aluminized Mylar filter (d) Double X pinch with 20 μm Aluminium wire with 40 μm Aluminized Mylar filter (this image does not have a pinhole picture).

HIGH CURRENT CAPILLARY DISCHARGE

Experiments carried out in GEPOPU using Argon in a capillary discharge to investigate the formation of a Z-pinch discharge were carried out. The main purpose of the experiments was to asses the extent to which transient hollow cathode (THCD) effects are relevant in these discharges. In these experiments operation is at a fairly good vacuum, but preionization is obtained with a laser generating a transient plasma in the hollow cathode (HC) volume, plus the addition of an auxiliary discharge [6]. The rate of rise of the main discharge current may be fixed at 1.5 or 3×10^{12} A/s. The THCD effects are found to extinguish if the preionizing current exceeds a threshold value, but e-beams are obtained with a different temporal behavior, characteristic of a diode, and the pinch column is markedly less homogenous at all times. Without the preionizing current no soft X-ray emission is obtained. The species observed by means of soft X-ray spectroscopy show that the wall material is the source of the pinch plasma and that the injected Ti plasma from the HC volume is, as expected, not significant. Results not reported yet shown that by redesigning the capillary discharge a totally different plasma can be achieved. Fig. 4 shows two spectra for the old and new capillary discharge, in which it is apparent that the plasma is essentially a Ti plasma. Further work is under progress.

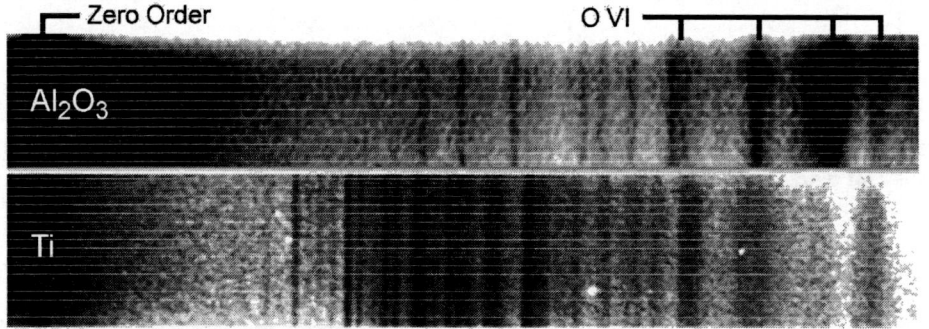

FIGURE 4. Spectra for simple capillary tube (top) and new design (bottom). It is apparent that in the new design the discharge is essentially in a Ti plasma.

SHELL Z-PINCH

The Z-pinch is the simplest geometry for confinement of plasmas, in which the plasma current generates its own magnetic field for confinement. It has been observed to be unstable to MHD instabilities, compressional pinches are unstable to $m=0$ instability and gas embedded pinches are unstable to $m=1$ instability. The present work puts forward a possible configuration to avoid both $m=0$ and $m=1$ instabilities. This is achieved by having a gas embedded Z-pinch, where the background gas suppresses the $m=0$ mode, and generating a central hollow discharge which has a gas or a high density inner core, which would prevent the growth of the $m=1$ instability.

One of the main purposes of the experiments carried out so far was to determine whether the proposed optical scheme would produce a hollow preionization. The results obtained are described in [7]. The plasma evolution was followed by means of double frame interferometry scheme using the second harmonic of the laser used to generate the preionizing plasma. The interferogram sequence shown in [7] provides a clear indication that a hollow plasma profile is established and compression is observed. At times of maximum compression and later it is not possible to observe fringes, probably due to a changing density which results in fringe motion during the laser pulse (~8 ns).

The maximum density observed is ~10^{-25} m^{-3}, indicating that there is compression of the initial gas. In the central region at and after maximum compression it would appear that an $m=1$ instability does develop, which agrees with the fact that at small radii the Kadomtsev criterion is not met. Diagnostics using a shorter laser pulse is planned. Future work requires the modification of the optical configuration in order to be able to introduce a plastic solid core, thus avoiding $m=1$ instabilities.

CONCLUDING REMARKS

Examples of the dense Z-pinch work being done at our Laboratory, a small modest installation, have been given. We feel that it is possible to carry out interesting and relevant work in a less than perfect funding situation.

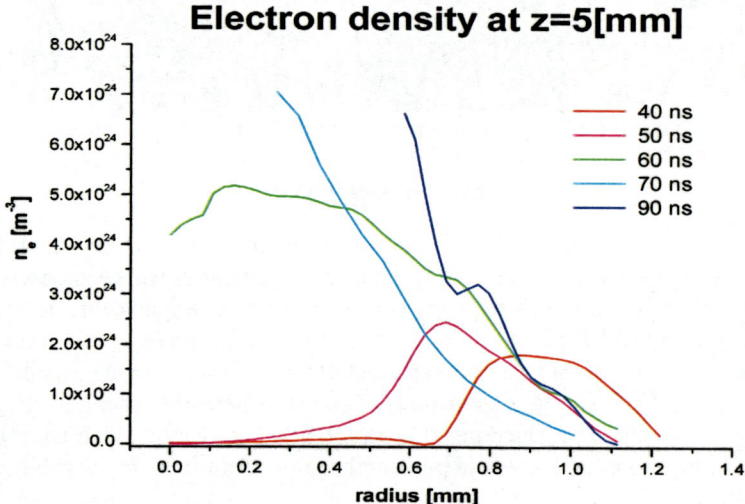

FIGURE 5. Electron density obtained from interferogram sequence for times after the beginning of the discharge current. At times earlier than those shown, the fringe displacement is too small to be able to extract a meaningful density. At small radii and at and after maximum compression fringes are washed out.

ACKNOWLEDGEMENTS

I am extremely grateful to my coworkers, Raúl Aliaga-Rossel, David Alvarez, Dirk de Jong, Mario Favre, Ian H. Mitchell, and Edmund S. Wyndham. This work was carried out under FONDECYT grants 7980023, 7980024, 1990435 and 8980011.

REFERENCES

1. Favre, M., *Small Plasma Physics Experiments II*, 170-186. Trieste (1989)
2. Chuaqui, H., Wyndham, E., Friedli, C., and Favre, M., *Laser and Particle Beams*, **15**, 241-248 (1997)
3. Chuaqui, H., Mitchell, I. H., Aliaga-Rossel, R., Favre, M., and Wyndham, E. S., "Performance of Llampüdkeñ with Short Circuit and Plasma Loads". This Proceeding
4. Choi P. Dumitrsescu C., Wyndham E, Favre M and Chuaqui H, *Rev Sci Inst* **73**, 2276-2281 (2002)
5. Mitchell, I. H., Aliaga Rossel, R, Gómez, J. A., Chuaqui, H., Favre, M. and Wyndham, E. S., "X Ray Emission From X Pinch Experiments on the Llampüdkeñ Generator", This Proceeding.
6. Wyndham, E., Chuaqui, H., Favre, M., Mitchell, I., Aliaga-Rossel, R., and Choi, P., "Experimental Observation in a High Current Capillary Discharge", This Proceding.
7. Pavez, C., Chuaqui, H., Aliaga-Rossel, R., Favre, M., Mitchell, I., and Wyndham, E., "Laser Initiated Hollow Gas-Embedded Z-Pinch", This Proceeding.

MULTI-MEGA-AMPERE
Z-PINCH PULSED
POWER DRIVES

The ZR Refurbishment Project

Dillon H. McDaniel, Michael G. Mazarakis, David E. Bliss,
Juan M. Elizondo, Henry C. Harjes, Harry C. Ives, III,
David L. Kitterman, John E. Maenchen, Timothy D. Pointon,
Stephen E. Rosenthal, David L. Smith, Kenneth W. Struve,
William A. Stygar, Edward A. Weinbrecht,[a]
David L. Johnson,[b] John P. Corley[c]

[a]*Sandia National Laboratories, P.O. Box 5800, Albuquerque, NM 87185, USA*[1]
[b]*Titan Pulse Sciences Division, 2700 Merced Street, San Leandro, CA 94577, USA*
[c]*Ktech Corporation, 2201 Buena Vista Drive SE, Albuquerque, NM 87106, USA*

Abstract. ZR is a refurbished (R) version of Z aiming to improve its overall performance, reliability, precision, pulse shape tailoring and reproducibility. Z, the largest pulsed power machine at Sandia, began in December 1985 as the Particle Beam Fusion Accelerator II (PBFA II). PBFA II was modified in 1996 to a z-pinch driver by incorporating a high-current (20-MA, 2.5-MV) configuration in the inner ~ 4.5 meter section. Following its remarkable success as z-pinch driver, PBFA II was renamed Z in 1997. Currently Z fires 170 to 180 shots a year with a peak load current of the order of 18-20 MA. The maximum z-pinch output achieved to date is 1.6-MJ, 170-TW radiated energy and power from a single 4-cm diameter, 2-cm tall array, and 215 eV temperature from a dynamic hohlraum. ZR in turn will, operating in double shift, enable 400 shots per year, deliver a peak current of 26 MA into a standard 4cm x 2cm Z-pinch load, and should provide a total radiated x-ray energy and power of 3 MJ and 350 TW, respectively, achieve a maximum hohlraum temperature of 260 eV, and include a pulse-shaping flexibility extending from 100ns to 300ns for equation of state and isentropic compression studies. To achieve this performance ZR will incorporate substantial modifications and upgrades to Marx generator, intermediate store capacitors, gas and water switches, water transmission lines and the laser triggering system. Test beds are already in place, and the new pulsed power components are undergoing extensive evaluation. The Z refurbishment (ZR) will be operational by 2006 and will cost approximately $60M.

---

[1] Sandia is a multiprogram laboratory operated by Sandia Corporation, a Lockheed Martin Company, for the United States Department of Energy under contract DE-AC04-94-AL85000.

INTRODUCTION

Z, the largest pulsed power machine at Sandia, began in December 1985 [1] as the Particle Beam Fusion Accelerator II (PBFAII, Fig.1) PBFA II was designed to produce high voltages (~15 MV) in a single-gap diode to accelerate and focus ions on small mm size fuel pellet and produce controlled fusion. PBFA II was subsequently modified in 1996 to provide high current rather than a high voltage in order to run a six-month set of scaling experiments for z-pinches, after which the machine was to return to light-ion research. As such, only the center vacuum stack, MITLs and water transmission lines and attachments were modified for this short campaign (Fig. 2). Because of the success of the six-month z-pinch campaign, the machine was never converted back, and was renamed Z in July 1997.

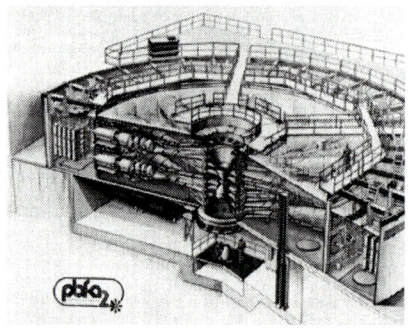

FIGURE 1. Particle Beam Fusion Accelerator II– Completed in 1985

FIGURE 2. PBFA II Center Modified to Z-Pinch Driver– Completed in 1996

Z now provides a unique capability to a number of DOE/NNSA, DTRA and basic science communities, and routinely produces x-ray power more than five times, and energy 50 times greater than any other non-pulsed power laboratory device. Z has now become a workhorse for HEDP physics and radiation effects research, Inertial Fusion Energy (IFE) studies, a range of basic scientific and university collaborations, and, more recently, material properties research. The bulk of the machine, however, is far from optimized for z-pinch operation, and most of the components are vintage 1985. Beginning in 1997, Z has evolved from a research machine to a user facility. Today about 270 shots per year could be possible from two shifts, although far fewer shots (173 in FY99 and 154 in FY00) have been realized because of operational funding constraints. The HEDP and ICF experimental community, however, clearly want Z to provide more shots, better precision and pulse shaping versatility, and more current. These factors are the impetus for refurbishing the accelerator as detailed in this paper.

PRESENT Z CONFIGURATION AND PERFORMANCE

The Z pulsed power design is based on the conventional Sandia pulsed power technology of Marx generators, water-pulsed-forming and transmission lines, vacuum Magnetically Insulated Transmission Lines, and post-hole convolutes [3]. The oil and water sections contain 36 modules with identical components. The prime power source of each module is a Marx generator with 60, 1.3 µF capacitors, which are usually charged to 90 kV. When the Marx erects, it transfers its energy to a water-dielectric coaxial capacitor, which reaches a peak voltage of 5 MV in 300 ns. In turn, the intermediate store capacitor discharges into a lower-inductance coaxial water capacitor through a laser-triggered gas switch in ~300 ns. From there and through self-breaking water switches the electrical energy is transferred first into a 4.32-Ω water transmission line and then through it into the water-vacuum-interface insulating stack and central vacuum section. The pulse at this point has a voltage of 2.5 MV and width of 105 ns FWHM. The total power generated by the accelerator is of the order of 60 TW. The pulses are then combined together in parallel into four equal number groups, 9 each, and feed four biconical constant impedance vacuum MITLs. The four pulses are then combined again via a double post-hole convolute section into a single ~20 MA, 2.5 MV pulse which finally drives the z-pinch load on axis. The 4m diameter vacuum MITLs are of constant impedance (2Ω the upper two and 2.75Ω for the lower two and are operating in quite a remarkable regime of 10 times higher fields than the cathode explosive emission threshold. The power transfer from the insulating stack to the load is extremely successful. The stack current and MITL currents are essentially the same. A small percentage of the total current (5% to 10%) is lost at the 12-post-hole convolute but this is to be expected because of the loss of self-magnetic insulation at the magnetic field nulls of the convolutes. The maximum voltage measured at the insulating stack exceeds the 3 MV and it routinely withstands peak fields of 100kV/cm without electrical breakdown. In additional test sets the insulating stack was operated open circuit at higher (up to 157kV/cm) fields without flashing over. The z driver successfully operates at the 60-TW, 5-MJ electrical design point and delivers routinely up to 20-MA currents to a variety of z-pinch and Isentropic Compression (ICE) loads. Figure 3 shows typical load current and radiated x-power waveforms for a 500-wire, 2-cm diameter, 1-cm high tungsten array

ZR GOALS AND OBJECTIVES

In the 1997 conversion of the PBFA II into Z only the center vacuum section was changed; the ZR refurbishment effort now aims to replace, redesign, improve and optimize the rest of the accelerator for z-pinch and ICE load. The outer tank, insulating stack and MITLs with convolutes will remain essentially the same with some modifications, especially for the insulating stack to withstand the increased electrical stresses.

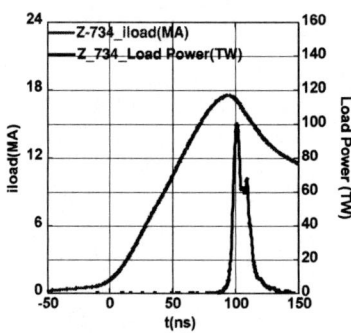

FIGURE 3. Typical load current and radiated x-power waveforms for a 500 wire, 2-cm diameter 1-cm high tungsten array

The ZR project will make Z capable of fulfilling the following objectives and goals.
- Enable the facility to routinely support a 400 shot per year program from two standard shifts. *[This capability is about 270 shots currently.]*
- Deliver a peak current of 26 MA into a standard 4cm x 2cm Z-pinch load, providing total radiated x-ray energy of 3 MJ. This will enable the program to achieve a peak x-ray power of 350 TW, and a primary vacuum hohlraum temperature of 170 eV.
- Enable a pulse length of 100 ns for a standard z-pinch load and a long pulse capability (250–300 ns) for equation of state and isentropic compression experiments, with the ability to routinely switch between configurations without extending the normal shot schedule.
- Provide independent triggering of all the 36 modules to facilitate pulse shaping. Ensure that a given peak current is repeatable over any 12-month period to within +/- 2% and over any sequential 30 shot series to within +/- 1% into a standard Z-pinch load. *[5% today]*
- During installation of new hardware and other upgrades, ensure Z is available for at least 150 experimental shots in any fiscal year. Provide compatible conceptual designs for single and double-sided drive configurations. Provide standardized access and supporting infrastructure for future diagnostics. *[Today, each new diagnostic requires a unique design for access and infrastructure support.]*

The achievement of the above goals will provide significantly more experiments per year, which will reduce the gap between the requested shots and the number of shots that can be fielded, will ensure higher quality experimental data from increased precision, better reproducibility, tighter tolerances on current levels, and increased flexibility of pulse profiles, and more energy and power to address future research needs.

ZR REFURBISHMENTS

Marx generators: The number of Marx generators will remain the same (36); however, they will be redesigned and the energy stored will be doubled. The present capacitors will be replaced with 2.6 µF capacitors, the resistors will be changed, and the triggering system will be made to withstand a larger number of shots before requiring maintenance. Each of the Marx generators will still have 5 rows of capacitors with a design option to extend into 6.

Intermediate store capacitors: New intermediate store capacitors must be developed to withstand maximum voltage of up to 7 MV. They will be longer (84 ns electrical length) and of larger diameter than the existing I.S. to accommodate the higher voltage and energy stored.

Laser triggered gas switches: Since the intermediate store capacitor peak voltage will exceed 6 MV, a new laser triggered gas switch is under development. Two options, one designed in Sandia and another developed in the HCEI in Tomsk, Russia or a combination of the two are currently under extensive testing. The gas switches will have as outer dielectric oil and, therefore, will be housed in a sealed oil module located inside the water tank unless the option to move the oil/water wall interface closer to the center is adopted. In that case the intermediate store capacitors and gas switches together with the Marxes will be all located in the same but larger oil tank.

Laser triggering system: In order to achieve the demanding pulse shaping requirements of the ICE experiments [4], 36 compact lasers will be utilized to individually trigger each of the gas switches. Pulse shaping will be accomplished mainly with staggered laser triggering. However, because of symmetric power flow requirements only groups of three modules will be fired at a particular time, and those must be symmetrically arranged around the machine.

PFL/Water switches: The Pulse Forming Lines (PFL) will be coaxial; however, the water switches will be linear, composed of a number of parallel rods. Thus, the transition from cylindrical to the triplate geometry of the downstream water transmission lines will occur at this point. The self-break water switches will operate at 4-5 MV and have been tested at the DTRA PITHON accelerator.

Vacuum insulating stack: The present vacuum insulating stack of Z and the vacuum MITLs are the most robust and conservatively designed components of the accelerator. It was felt, therefore, that it will be more cost and time effective to implement only modest changes such as changing the grading rings from aluminum to stainless steel and increasing the radius of the Rexolite^{TM} rings by 2.54 cm during the first years of ZR operation.

ZR PROJECT IMPLEMENTATION

Most of the component electrical parameters and geometric dimensions are already defined and engineering designs are under way. The entire ZR device is expected to become operational towards the end of FY 2005 (September 2005) and commissioned during the following six months. The refurbishment will occur in two major phases. In the first phase the energy storage section will be changed gradually as a rolling upgrade over about 12 months, by replacing old Marxes, power supplies, charging systems, etc. a few at a time with new ones during their normal maintenance cycle. The new units will be operated at lower level to match the output of the old ones until all 36 units are replaced. No additional downtime for the facility is expected to occur for this phase. The bulk of the hardware, other than the Marx banks, will be replaced during the second refurbishment phase. A 6-month shutdown period of the facility will take place to facilitate and expedite the changeover. Installation will not begin until all new hardware is on site, with as much pre-assembling as possible. Prototype hardware evaluation and system reliability programs will be implemented for the first time at Sandia with the ZR project. Those programs will be carried out at the pulsed power technologies development facility where the test beds are located. A $20°$ section of ZR is being built where two entire new modules will be tested and qualified, including new Marxes, intermediate store capacitors, gas switches, water switches, and triplate water transmission lines. In addition, another test bed is currently in operation where individual new components are being tested like gas switches, intermediate store capacitors, etc. The Mini Z accelerator was modified from APPRM and is being utilized for MFI studies, while the DTRA PITHON accelerator is being used for water switch development.

The system reliability program will continue past the ZR commissioning stage. It will operate at a much higher shot rate in parallel with the new Z and it will provide most valuable information on the lifetime, maintenance, and required replacement cycles of the various components.

SUMMARY

Refurbishment of Z using the existing building and tank structure will enhance Sandia's pulsed power capabilities in several key areas well into the next decade. Redesigning and upgrading the major sections of the machine will significantly improve maintenance, reliability, and robustness. These improvements should provide about 50% more shots from roughly the same size operational crew. Increased individual control of key timing components will allow a greater range of pulse widths and shapes and further study of equation of state and other material properties. In general, the Z accelerator, following the ZR refurbishment, will become practically a new device with a performance surpassing that of any other device.

REFERENCES

1. Turman, B.N. *et al.*, Proc. of the Fifth IEEE Pulsed Power Conf., Arlington, VA (1985), pp. 155.
2. Spielman, R.B. *et al.*, Proc. of the Seventh IEEE Pulsed Power Conf., Monterey, CA.
3. Spielman, R.B. *et al.*, Proc. of the Tenth IEEE Pulsed Power Conf., Albuquerque, NM (1995), pp. 396.
4. Struve, K.W. *et al.*, "Options for Pulse Tailoring on Z for Isentropic Compression Experiments," Proc. of the 29th IEEE International Conf. on Plasma Science, Banff, Canada (2002), p. 211.

Pulse Power System Development for Megajoule X-ray Facility BAIKAL

E.A. Azizov, V.V. Alexandrov, S.G. Alikhanov, V. H. Bachtin, V.I. Chetvertkov**, V.A. Glukhikh**, E.V. Grabovskii, A.N. Gribov, Yu.A. Hallimullin, V. A. Levashov, A.P. Lotocky, A.M. Zhitlukhin, E. P. Velikhov*, G.I. Dolgachev*, Ju.G. Kalinin*, A. S. Kingsep*, A. I. Kormilitcin***, V. G. Kouchinsky**, S.L. Nedoseev, O.P. Pechersky**, V.D. Pismenniy, G.P. Rikovanov***, V.P. Smirnov*.

SSC RF TRINITI, 142190. Troitsk, Moscow region, Russia
*RNC «Kurchatov Institute», Moscow,
** Efremov Institute, St.- Petersburg
***RFNC VNIITF, Snejinsk.

Abstract. TRINITI develops a project of multiterawatt generator «BAIKAL» to produce powerful pulses of soft X-rays, using electric pulse power 500 - 1000 TW. Parameters of proposed X-ray generator are: X-ray pulse energy - 10 MJ, X-ray pulse duration - 10 ns, load current amplitude - 50 MA. The methods of pulse power increasing proposed for Baikal project are studies and tested on 12 MJ inductive store installation MOL located at TRINITI Institute.

The implosion of fast liners is considered to-days as a possible source of generation of x-ray pulses of energy of MJ scale. TRINITI, Efremov Institute, Kurchatov Institute and VNIITF design a project «Baikal», in which an inductive stores are used and there as a result of several successive transformations an electric pulse is generated with parameters necessary for liners implosion.

Baikal facility must have the following parameters
- Radiation energy of SXR   10-20 MJ
- Radiation pulse duration   10 ns
- Generation method of SXR   plasma shells implosion
- Load current   50 MA
- Load current pulse duration   on compression phase   100 -300 ns.

The generator must have electric pulse power of 500 - 1000 TW. TRINITI has a unique complex of long pulse electric generators with inductive storage and switching system previously built for feeding tokamak T-14 magnets. This complex is located at TRINITI. It is proposed as a source of initial energy for "Baikal".

The general scheme of "Baikal" was presented in [1]. It consists from two consequently connected inductive stores, magnetic flux compression system and

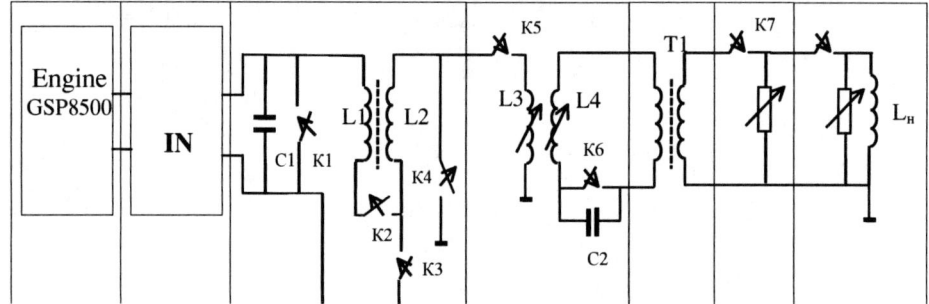

Fig. 1. The layout of "MOL". The names on items are placed at the text

POS. The installation "MOL" (fig. 1) is intended for investigation of two last stages of the generator "Baikal" and for electric pulse of megajoule range generation. It consists of inductive store IN, a capacitors battery C1, magnetic amplifier MA, magnetic compressor MC, voltage rising transformer T1 (VRT), capacitors battery C2 for magnetic flux generation, plasma opening switch (POS-1) and the load simulator. The load simulator ensuring rising impedance similar to one of the compressed liner. Load consists of a plasma opening switch POS-2 and an inductance.

Sectioned IN stores of 12,5 MJ energy at current of 50 kA and by using of 30 explosive open switches increases the output current due to parallel connection of its sections. Energy from IN transmits to the magnetic amplifier consisting of coupled inductive coils L1 и L2. After an electric opening switch (fuse) K2 openes the energy from L1 is passed to L2. After opening fuses K4 starts K3 и K5. In this case the current comes to the primary contour of MC, inductance of which during the current rise also increases. The energy in the primary contour of MC is 6 MJ. At the acceleration stage a magnetic flux is created in the second contour of MC by a discharge current of the capacitor C2. The switch K7 is intended to switch on POS 1 at the last stage of current increase in the second contour of MC. At the end of inductance decrease stage of L4 the second contour inductance is 50 nH.

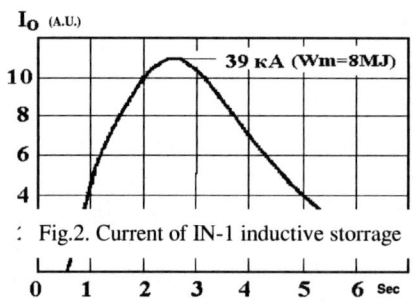

Fig.2. Current of IN-1 inductive storrage

At the same time current in the second contour of MC rises up to 11 MA at 1,8 µs. The transformer VRT is used to reduce the current trough the POS1. After opening of POS1 the current flows through load with rise time 150 ns. The load pulse characteristics are current amplitude is 1.8 MA, voltage amplitude is 5 MV and electric pulse energy is 0.77 MJ. The switch K8 is used to control parameters of POS1 at the initial stage of voltage rise. The simulator of of the liner load consist of POS2 having current rise time 150 ns and inductance.

Inductive store IN for MOL was a reconstructed and tested at 40 kA in 2000. A current profile at the test 2001Y is presented on Fig2.

Explosive opening switches BP 50/50 and BP50/1000 for 50 кА и 1 МА correspondingly designed in Efremov Institute are used IN. The magnetic amplifier is intended for increasing of power delivered to MC and fuses elaborated in VNIITF are used in it.

Fig. 3 Explosive opening switches for IN-1 RV50/50 –left, RV50/1000 right.

Fuse for Magnetic Amplifier

These fuse are working in condition of long current rise time. The time scale is considerably different from one occurring in existing installations. These devices were tested in TRINITI on a MA model and it was shown that voltage rise time is 30 µs, and electrical strength is - 1.5 кV/cm. The photos of these opening switches is presented on fig. 3. For fuse testing and elaboration of MA there was designed a simulator of two cascade store – Magnetic Amplifier taking into account usage of an electric explosive opening switches of current cutting off of magnetic store charging current at moment when it is close to its amplitude value. The inductive store consists of two windings tested at the current up to value of 100 кА. The scheme of MA simulator is presented in [1]. The calculated maximum current amplification (according the coefficient of winding magnetic coupling is equal to 1,72.

We have executed some number of experiments with a current up to 25 кА for difference substances where the conductor blew up. The coefficient of a current amplification depend both on a residual current in fuse and from magnitude of damping of a current in the output cascade of an inductance.

Maximum coefficient of current amplification was close to the calculated value and was equal 1,65-1,68. It was received for a wire destroyed in quartz sand at discharge current of 14-22 кА, charge pulse duration of 250-400 µs and current rise front duration in the output cascade $\tau \sim 50$ µs.

The magnetic field compression of MOL installation has to be performing by compressing it by two plain metallic fillets (liners) of 200 mm wide and 2 m length. For the investigation of the liner acceleration process there is under development a magnetic compressor that is fed from a capacitor battery «PUMA». In this magnetic compressor the current rise time, its amplitude, time profile would be close to one expected in MOL. The pushed unit is a winding in which acceleration of metallic fillets and magnetic flux compression are carried out. The winding is in a vacuum chamber of 1,5 m of diameter and 1,2 m length.

In the scheme of the installation «MOL» the voltage rising transformer (VRT) is situated between the magnetic compressor (MC) and the current plasma opening switch (POS). It is intended to match MC output current with current through POS and also to separate MC circuits from voltage appearing on the load at POS opening. To get current of 11- 12 MA in the secondary circuit of MC a circuit inductance must be not more 50 nH. This inductance is a MC residual inductance, an inductance of VRT conducting connections, an inductance of VRT dissipation and an inductance of VRT feeding cables which is considered to be closed up to the moment of current achieving the maximum value. From plenty of variants of VRT there was chosen a variant of a autotransformer with a transformation coefficient of 1:2. The design of VRT is coaxial (Fig5.). The output inductance at disconnected second winding - 250 nH. Dissipation inductance is 135 nH.

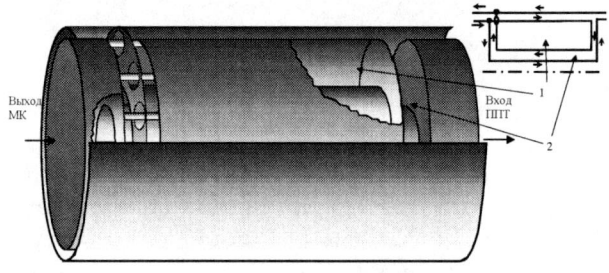

Fig.4 Design of VRT

In the program «Baikal» it is supposed to obtain a current pulse at a level of 50 MA of duration of 150 ns and voltage up to 10 MV. It is supposed to use current plasma opening switch POS as the output cascade. Maximum parameters of pulses obtained by POS on biggest installations do not exceed 3 MA and 2 MV, or 2 MA and 3 MV. It is supposed to enlarge indicated parameters on MOL installation (3.6 MA, 4-5 MV) and to determine ways of achieving parameters of the program «Baikal».

The development of POS is performed in Kurchatov Institute. Now experiments on the POS prototype "MOL" take place installation RS-20 for current 300-600 кА. On this installation by the activation analysis method the magnitude of voltage 3,2 MV was verified. This voltage is close to working value of "MOL" POS.

REFERENCES

1. Digest of tecnical Papers PPPS-2001, Pulse Power Conference, Las Vegas, Nevada, June 17-2 2001. p. 773-776.

Experiments Aimed at the "Baikal" Program

Alexander Kingsep [a], Yuri Bakshaev [a], Alexander Bartov [a], Petr Blinov [a], Andrey Chernenko [a], Sergey Danko [a], Georgy Dolgachev [a], Yuri Kalinin [a], Igor Kovalenko [b], Alexey Lobanov [b], Dmitri Maslennikov [a], Valery Mizhiritsky [a], Andrey Shashkov [a], Valentin Smirnov [a]

[a] *Russian Research Center "Kurchatov Institute", 1 Kurchatov Sq., 123182, Moscow, Russia*
[b] *Moscow Institute for Physics and Technology, 9 Institutskii Pereulok, 141700, Dolgoprudnyi, Russia*

Abstract. On the S-300 pulsed power generator (4.5 MA, 70 ns, 0.15 Ohm), within the frames of ICF program based on fast high-current Z-pinches, experiments are being carried out studying promising schemes of output units. In particular, a nanosecond-range plasma flow switch is being investigated aimed at sharpening the pulse. As a result, the switching rate as high as 2.5 MA / 2.5 ns has been achieved. The numerical simulation of such a device has been carried out. The results of experiments on the extrinsic magnetic field influence on high-impedance plasma opening switch (POS) operation are reported. It has been demonstrated that the POS output voltage could be increased by the factor of 1.5 compared to the POS free of an extrinsic field. Also the possibility of the expansion of POS conduction phase duration up to 40 µs has been presented. The development of more new laser and nuclear diagnostic methods will definitely contribute to attain new data which could help in understanding the POS operation physics.

INTRODUCTION

Fast compression of liners is under consideration as a possible approach to electric energy conversion into X-ray pulse at the energy scale of dozens of megajoules. Scientific cooperation including TRINITI, Kurchatov institute, Efremov institute, and VNIITF develops the "Baikal" project [1] to produce the electric pulse with parameters adequate for these purposes. One of its main features is the output cascade based on the powerful plasma opening switch (POS) instead of the pulse forming lines. The goal of our experiments is 1) to study the possibility of using POS in the multi-megaampere range of the current amplitudes and to construct the POS adequate for "Baikal", and 2) to study new kinds of output units those could provide the Hohlraum experiments at the current generation of high-current machines.

NANOSECOND PLASMA FLOW SWITCH

We investigated the output devices similar to the plasma flow switch but operating in the nanosecond range of pulse duration. The plasma bridge between the inner and outer cylinders is being accelerated along the axis by the current pulse of generator (Fig.1). Accelerated plasma bridge moves between the cylinders. When

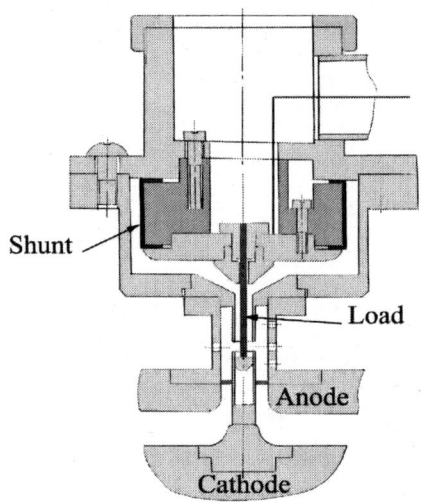

FIGURE 1. The sketch of output device

it is flying through the break of the inner cylinder, the magnetic flux enters the central cavity where the load is placed. Plasma bridge was created by means of the current-driven explosion of a thin foil in the beginning of current pulse. Diameters of inner and outer cylinders equal to 4 and 10 mm, respectively. The break of inner cylinder was varied between 1 to 2.6 mm. The diameter and length of central cavity equal to 3.6 mm and 10 mm, respectively. The current amplitude is close to 2.5–3.0 MA. The spatial homogeneity of breakdown of foils and the velocity of their sliding along the inner electrode are recorded by both frame and streak ICT photographs in visible range. The best results correspond to the aluminum-coated mylar films as thin as 1.2–1.5 µm. The maximal velocity of sliding recorded was up to 10^8 cm/s. In the first experiments, we used metallic wires or tubes as loads with the diameter of 0.5–2 mm. The current rise time on the load varied from 2.5 up to 10 ns. The current pulse duration (half-width) was recorded in the range of 7–20 ns. The shunt (see Fig.1) was made of steel foil as thin as it was necessary to provide the time resolution $\Delta\tau < 2$ ns. By means of the shunt measurements, the fact of the current switching onto the load has been proved, during the time of the order of 2.5 ns, by the amplitude 2.5 MA. Heretofore, such a switching rate (10^{15} A/s) has been achieved only in our experiments. In Fig. 2, on can see the oscilloscope traces of both input current and current switched onto the metallic tube of 1.5 mm in diameter that served as a load. The measurements of soft X-rays (SXR) were carried out by means of two vacuum X-ray diodes (XRD) with the *Ni* photo-cathodes, supplied by mylar filters with the mass thickness 0.34 mg/cm^2 and 0.67 mg/cm^2. The load in this series was the wire array of 8–16 tungsten wires of 5–6 µm in diameter situated at the radius of 1 mm. The geometry and layout of output device and diagnostics allowed XRD "to see" only the inner surface of the cavity while the direct radiation of the load being screened. The temperature of the inner wall thus determined was up to 50 eV. While the wall temperature being stable enough, the switching rate is much less stable because of its extreme sensitivity to the initial conditions which was confirmed by the series of numerical simulations. S-300 machine has been equipped by a new laser diagnostic set-up based on the YAG:Nd laser. To increase time and space resolution, master oscillator pulse was compressed by using the stimulated Brillouin scattering. As a result, five-frame shadow pictures of Z-pinch loads with exposure of 1ns during one shot have been obtained.

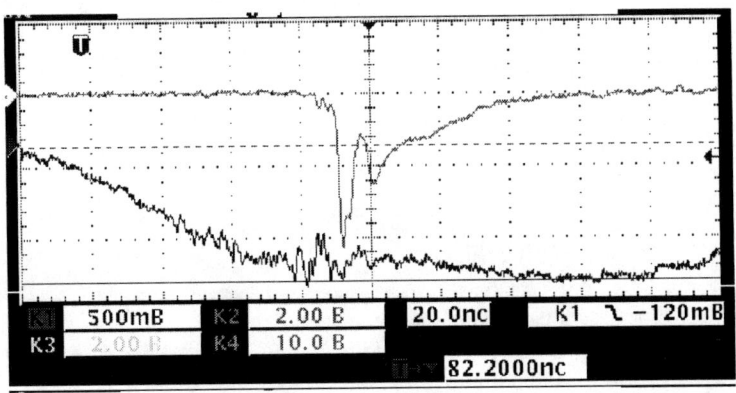

FIGURE 2. Traces of the current on a load (top) and total current (bottom)

POS EXPERIMENTS

To provide the necessary output parameters of "Baikal" facility (50 MA, 150 ns, 10 MV), the powerful system of POS has to be constructed. Our experiments are being carried out on the RS-20 generator by the same current flow density level as in future "Baikal" machine. The POS design includes 78 plasma guns, external solenoid and POS electrodes. The maximum linear charge density (along the circle of the outer POS electrode) is ~ 9 mC/cm. Diameters of electrodes are 10 and 18 cm. The basic electric parameters are as follows: 4 Marx cascades, capacitor charging voltage 32-48 kV, output Marx generator voltage 128-192 kV, drive current amplitude 200-320 kA, the switching ratio 2µs/150 ns. To increase the charge value passing through the POS, it was necessary to slow down the density decrease inside the gap. This problem has been solved by means of programmed POS gap fill by a plasma, the idea of this method based on some predictions of electron magnetohydrodynamics [3]. In this mode of operation we succeeded to transport all charge through the POS and in some shots (~ 80%) the switching voltage could increase up to 3-3.5 MV. These results were obtained only by using the extrinsic longitudinal magnetic field (see Fig.3). In this figure, the facilities with extrinsic magnetic field are marked by x-symbol, and facilities without extrinsic magnetic field marked by -symbol. The points corresponding to RS-20 machine are marked by 0-symbol: <u>1</u> – Marx voltage U = 160 kV and applied magnetic field B_Z = 1 T, <u>2</u> – 840 kV and 1.6 T, <u>3</u> – 840 kV and B_Z=0.

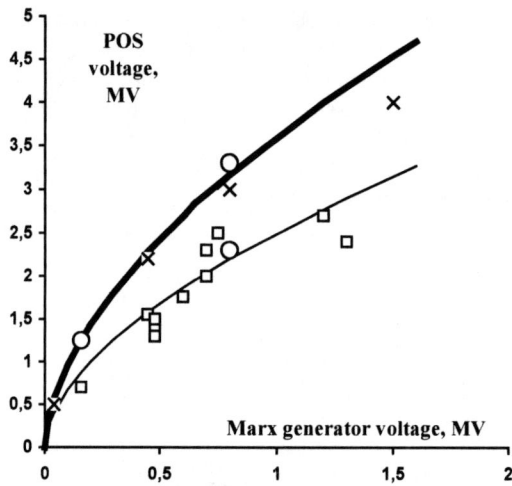

FIGURE 3. POS voltage as a function of Marx voltage with and without extrinsic magnetic field (bold line and hair-line, respectively).

One of the features of "Baikal" generator is 40 μs–long current prepulse which has to pass through the gap before switching. To slow down the decrease of plasma density, plasma guns were switched consequently 3 times by using separate capacitors with 5–10 μs time delay. As a result, the efficient POS operation delayed up to 40 μs after the current start has been demonstrated at the small "Taina-2" machine, and now we are working to reproduce this effect on RS-20 generator. It was extremely important from the standpoint of POS applications, to perform the detailed output voltage measurements. Direct measurements on the base of shunts and voltage dividers in the range of several MV are very difficult and not so reliable. In addition, the POS geometry prevents from proper operations. To overcome this obstacle, the high-energy limit of the electron Bremsstrahlung spectrum in a diode has been measured. The method of filters cannot be used because of the essentially non-monotonic dependence of the X-ray absorption upon the energy in the 2.5–3 MV range. Meanwhile, the cross-sections of the photo-nuclear reactions $Be^9(\gamma,n)Be^8$ and $D^2(\gamma,n)p$ with the thresholds 1.65 MeV and 2.25 MeV, respectively, are growing very sharp in this range with the photon energy. These features of reactions we have used provide very good quality of the voltage measurements. After all, we confirmed our result based on the conventional procedure, i.e., $U_{max} \approx 3.2$ MV.

The work was partially supported by the Russian Foundation for Basic Research, grants 00-15-96599 and 01-02-17359.

References

1. Alexandrov, V.V., Azizov, E.A., Branitsky, A.V., et al, *Proc. 13th Int. Conf. On High-Power Particle Beams,* Nagaoka, Japan, June 25-30, 2000, 147-150.
2. Smirnov, V.P., Chernenko, A.S., Chukbar, K.V., et al, *Proc.18-th IAEA Fusion Energy Conf.,* Sorrento, Italy, Oct. 2000, IFP/05.
3. Kingsep, A.S., Chukbar, K.V., and Yan'kov, V.V.,"Electron Magnetohydrodynamics", in *"Reviews of Plasma Physics"*, edited by B.B.Kadomtsev, **16**, Consultants Bureau, NY, 1990.

Study of the Plasma Focus as a Driver for the Magnetic Compression of Liners

Vladimir E. Fortov[a], Mikhail A. Karakin[b], Edil'girej Yu. Khautiev[b], Vyacheslav I. Krauz[b], Stanislav F. Medovschikov[c], Aleksandr N. Mokeev[b], Victor V. Myalton[b], Sergey L. Nedoseev[c], Valentin P. Smirnov[b], Valentin P. Vinogradov[b]

[a]*Institute for High Energy Densities of Associated Institute for High Temperatures RAS, Moscow, 127412, Russia*
[b]*RRC "Kurchatov Institute", Institute of Nuclear Fusion, Moscow, 123182, Russia*
[c]*TRINITI, Troitsk, Moscow Region, 142092, Russia*

Abstract. Possibilities of plasma focus – type facilities using in investigations on high density energy physics are discussed. The experimental results of studying wire arrays compression by PF-3 plasma focus facility current sheath performed in Kurchatov Institute are given. The technology of manufacturing the liners has been developed at TRINITI. Compression of the tungsten wire arrays having linear mass 0.33-0.6 mg/cm with velocity ~$(2\div3)\times10^6$ cm/s are shown. A new approach to forming liner load is proposed. Such load represent an ensemble of free fine-disperse particles of substance (dust). A given approach has some advantages related, first of all, with an opportunity to vary – on a wide scale – the mass, configuration and the element structure of a similar liner. In the first experiments at PF-3 facility an efficiency of interaction between the PF-sheath and the dust target has been shown that is manifested, in particular, in the enhanced MHD-pinch stability.

1. INTRODUCTION

The outstanding physical and quantitative experimental results in the recent years on the production of a soft X-ray radiation for the indirect way of irradiating a fusion target [1] allow one to consider the Z-pinch systems as the most probable candidate to the role of a driver in the circuit - diagrams of the pulsed fusion. The projects of the kiloterawatt drivers based on the Z-pinch are actively developed at present in the USA (X-1) [2] and in Russia ("Baikal") [3]. However, there is a wide variety of the problems related with the study of liner dynamics, X-ray radiation, emerging under compression of the liners, and a number of other physical laws the solution of which is not only possible but necessary at the experimental base at our disposal. The usage of the PF-type facilities as current generators in the solution of such tasks can be a rather promising direction that is related with an opportunity to produce an comparatively simple engineering laboratory devices for studies in the physics of high energy densities.

Recently the combined circuit-diagrams are discussed for creating laboratory soft X-ray radiation sources, where PF are used as inductive storage and the current sheath realizes the transport of its energy to the load placed at the system axis [4-6].

A given scheme of experiment allows one to perform an modeling studies of the compression of different liner loads using in the fast Z-pinch experiments.

At conducting liner disposition at the PF system axis current rise time on the load will determine by the current skin thickness, δ, and by the radial sheath velocity, V_r. At the real accessible parameters I ~ 3 MA, δ ~ 1 cm и V_r ~ 10^7 cm/c, it is possible to achieve a rate of a current increase on a load

$$\dot{I} \sim I (V_r / \delta) \geq 3 \cdot 10^{13} \text{ A/c}$$

In the current measurements done by N.Filippov [7] with the Rogowski coils assembled in the anode central zone current rise rate $\dot{I} \geq 10^{13}$ A/s was achieved at the place of possible liner disposition.

It is possible to underline two main direction of this works:
> ➤ plasma current sheath parameters improving and PF optimization as a generator of high current ($\geq$ 3 MA);
> ➤ search for optimal variant of liner loads allowing to perform an effective current transferring on the load.

2. SCHEME OF EXPERIMENT

The experiments were done on the PF-3-facility having Filippov-type electrode system. The total capacity of the power supply source is 9.2 mF, maximal charging voltage is 25 kV, maximal stored energy is 2.8 MJ. The minimal external inductance is 15÷40 nH. The main working gas is neon under pressure of 1 ÷ 5 Torr.

The low pressure spark gap switch is traditionally used as a switch for the facilities of such a type. The spark gap switch is a multisectional (24 sections) toroid with a large diameter, 1.2 m, and with a cross-section diameter 0.08 m. The spark gap switch is evacuated to the pressure of ~10^{-2} Torr.

Porcelain or glass ceramic insulators with the diameter of 0.9 m and 0.25 m high are used at PF-3-facility. The insulator size actually defines the anode diameter. The anode is a sectional copper disc with he total diameter of 0.92 m, 0.025 m thick. There is a changeable insertion in the central part of the anode. This allows one to considerably increase the life time of the anode, as well as to control the profile of the compressing sheath by changing the configuration of this insertion. At present, a funnel-like insertion with the diameter of 0.1 m in the upper part of the funnel and 0.07 m deep is used in the experiments. The vacuum frame made of carbon steel 2.6 m in diameter, 0.45 m high, serves as a cathode.

The calculated value of a short-circuiting current for the PF-3-facility, at the maximal charging voltage of 25 kV and at the minimal external inductance of 15 nH, is 20 MA. The actual magnitude of the discharge current is determined by plasma-current sheath dynamics and configuration. The experiments were done at W = 450-

750 kJ, that provided discharge current amplitude 2-3 MA. Current value at the stage of interaction with liner was some less (1.2-2.0 MA) because of inductive current drop at current sheath compression to the axis.

3. EXPERIMENTS WITH WIRE ARRAYS

A special supplying device with vacuum lock was developed for the experiments with foam liners and wire arrays (Fig.1, item 8). The device is located upon the upper cover (cathode) of the discharge chamber. All the construction elements are located outside the discharge volume and they do not introduce heterogeneity violating the PCS-dynamics. The liner (7) is suspended upon a thin (~ 60 μ) metallic filament (6) and descended to the necessary position at the system axis with the mobile rod located in the cylindrical column sluice. When the vacuum lock is closed, the sluice column volume can be autonomously evacuated to the pressure of ~ 10^{-2} Torr or filled with the atmospheric air for replacing the liner without violation of vacuum conditions in the discharge chamber. This allows one to perform a preliminary discharge chamber "training" for the PCS-parameters improving and to arrange the liner replacement without vacuum violation in the chamber. Usually, the central transversal liner cross-section was located 10-15 mm high over the anode level. The streak camera slit was oriented to the same altitude, parallel to the anode plane.

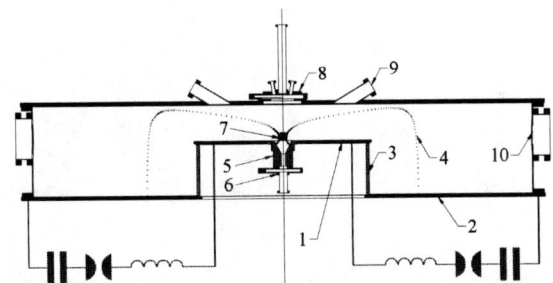

FIGURE 1. Diagram of the liner experiments at the PF-3-facility:
1 – anode; 2 – cathode; 3 – insulator; 4 – plasma current sheath; 5 – anode insertion;
6 – suspension ware; 7 – liner; 8 – loading unit with a vacuum lock; 9, 10 – diagnostics ports;
C_0 – capacitor bank; S – low pressure spark gap switch; L – external inductance.

Assemblies of 60 tungsten wires, 15 mm long, 6 μm (linear mass 330 μg) and 8 μm (linear mass 600 μg) in diameter, and pulled between two metallic discs along the diameter of 20 mm with a step of ~ 1 mm have been used in the experiments with wire arrays.

It has been shown that, after current sheath contact with the liner, an essential portion of a current continues to flow in the vicinity to the wires during the period necessary for transferring the current to the liner. But some portion of a current is "dropped" down to the axis, producing a pinch (Fig.2). The very wires remain to be immobile during ~ 150 ns for the light liner and during ~ 300 ns for the heavy one, then a rather fast compression of the liner to the axis occurs. The rate of this compression, estimated from the streak camera pictures in Fig. 3, is about 3.10^6 cm/s for the light liner, and $\sim 2.10^6$ cm/s for the heavy one. Later on, a comparatively-slow pinch expansion takes place at the rates approximately-lower than the compression ones by the order of magnitude.

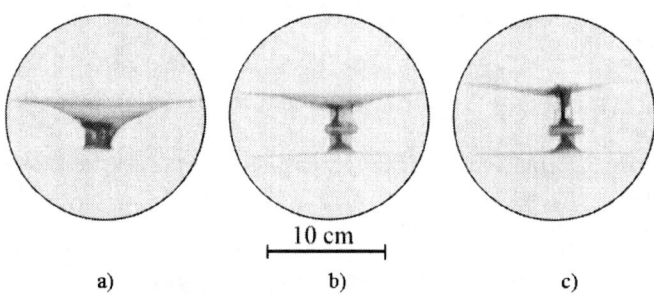

FIGURE 2. Frame camera pictures of the discharge with the wire array, m= 330 μg/cm. Frame exposure is 10 ns; a), b), c) different instants of the pinch formation with time shift of 150 ns

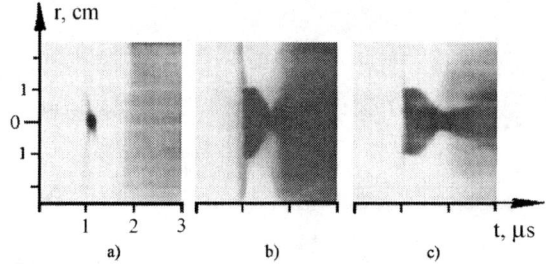

FIGURE 3. Streak camera pictures: a) discharge in neon without liner; b) wire array, 330 μg/cm; c) wire array, 600 μg/cm.

4. PF-DISCHARGE DYNAMICS IN THE PRESENCE OF A CONDENSED DISPERSE PHASE

In the beginning of the sixties, Ch.Maisonnier, L.G.Linhart and M.Haegi [8, 9] proposed to use a microparticle diode for the production of a plasma sheath in the experiments with a hollow dynamical pinch by means of a thin layer of microparticles.

The implementation of this technique seems us to be rather promising for producing the target as a dust cloud in the experiments by the liner program. A given approach has a number of the advantages related, first of all, with an opportunity to vary – on a wide scale – the mass, configuration and the element composition of a similar liner.

In the experiments performed on the PF-3 facility the dust target was produced at the system axis as a freely-falling flow of the fine-disperse (2 ÷ 10 μm) powder of Al_2O_3. Three modifications of a supplying device with different powder supply rates, allowing one to produce various dust column profiles, were used.

Under condition of a particle flow continuity, at an average particle diameter d ~ 4 μ, moving with an free-fall acceleration, the estimations of the following parameters in the expected compression zone over the anode surface are done for three modifications with the powder mass consumption equal to 80 mg/s, 200 mg/s, 500 mg/s correspondingly:

- linear mass of a dust target (mg/cm): 0.2 mg/cm; 0.5 mg/cm and 1.25 mg/cm;
- density of particles (n/cm^{-3}): $6.7 \cdot 10^5$; $1.7 \cdot 10^6$ and $1 \cdot 10^6$;
- mass density (g/cm^{-3}): $6.7 \cdot 10^{-5}$; $1.7 \cdot 10^{-4}$ and $1 \cdot 10^{-4}$;
- linear number of particles (n/cm): $2 \cdot 10^6$; $5 \cdot 10^6$ and $1.25 \cdot 10^7$;
- distance between particles (μ): 110; 80 and 100;
- total area of an interphase surface (cm^2): 1; 2.5 and 6.

Thus, with this supplying device one manages to produce the dust targets close to the foam liners and wire arrays used in the similar experiments respective to such important parameters, as, for example, an effective diameter and a linear mass. It is shown that more stable pinches, respective to the MHD-instabilities, are produced in the discharges with a dust target. In the discharge with pure neon, the pinch undergoes the development of instabilities already in Δt ~ 200 ns after an instant of maximal compression with the subsequent destruction (Fig. 4). At the same time, in the case of a dust target, a stable long-living plasma configuration at the axis is observed (Fig.5).

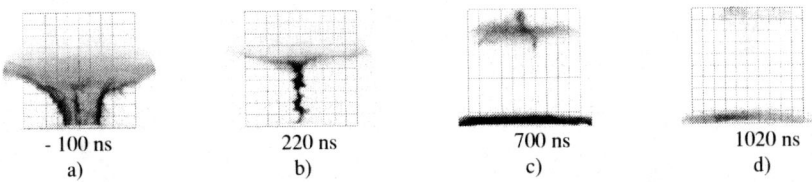

- 100 ns   220 ns   700 ns   1020 ns
a)   b)   c)   d)

FIGURE 4. Frame camera pictures of the discharges in pure neon obtained at the stage of pinch formation (a, b) and at stage of pinch destruction (c, d); exposition time is 10 ns

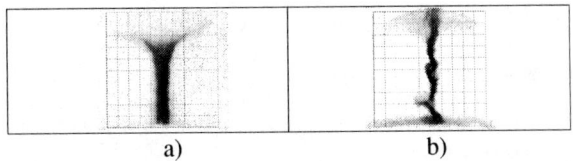

a)   b)

FIGURE 5. Frame camera pictures of the discharges with dust target: a) m= 1.2 mg/cm, Δt = 920 ns; a) m=0.2 mg/cm, Δt =1020 ns

In this case, with an increase with the linear mass, the pinch stability rises. However, even at the comparatively-low linear mass (200 μg/cm), in spite of the instability developments, the pinch configuration exists longer than 1 μs after the maximal compression.

In the discharges with a dust target, a soft X-ray pulse, is registered. The integral pinch photographs produced with the pin-hole camera oriented at the angle of 90° to the system axis, through a Be-filter, 10 μm thick, is given in Fig. 6. X-ray radiation time duration is ~ 30 ns long, Such a short time of radiation in comparison with the pinch existence duration is probably related with a shift of the radiation spectrum to a softer range not registered with our detectors.

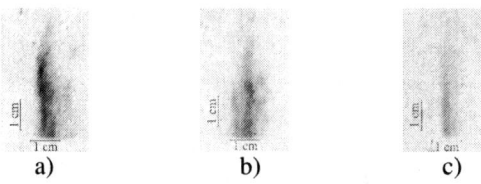

FIGURE 6. Pin-hole camera pictures of the discharge in pure neon (a) and in the presence of a dust target (b, c).

5. CONCLUSION

The results of above mentioned experiments show, that the PF-3 facility can be effectively used in the scientific program on the research of physical processes at the multicharge ions liner compression. The transportation of a current ~ 3 MA in a zone of the liner load disposition is executed. The time of the current switching on the wire array and velocity of its consequent compression to axis depend from linear mass of the liner. The new type of liner load as dust cloud of fine-disperse particles is offered. An essential result of the first experiments with the dust is the pinch MGD-stability improving.

REFERENCES

1. T.W.L.Sanford, R.E.Olson, R.C.Mock et.al. Physics of plasmas, v.7, N.11, November 2000, p.4669
2. J.P.Quintenz et al., Proc. Beams'98, Haifa, Israel, June 7-12, 1998, p.9
3. E.A.Azizov et al., PPPS-2001, June 17-22, 2001, Las-Vegas, Nevada, USA, rep. P1C31
4. M.Scholz et al., Phys.Lett., A 262 (1999), 453-456
5. V.V.Myalton, V.I.Krauz, E.Yu.Khautiev et al., Int. Symp. PLASMA-2001, Warsaw, 2001, O8-2,
6. V.I.Krauz, V.V.Myalton, E.Yu.Khautiev et al. Int. Symp. PLASMA-2001, Warsaw, 2001, I3-1
7. N.V.Filippov, T.I.Filippova, V.I.Krauz, et al, Current Trends in International Fusion Research - Proc. of the Second Symposium, Washington, 1997, (Editor – E.Panarella), 1999, NRC Research Press, Ottawa, Ontario, Canada, p.27-40.
8. Ch.Maisonnier, J.H.Linhart, M.Haegi, Nuclear Fusion: 1962 Supplement, Pt2, 727-732
9. Ch.Maisonnier, M.Haegi and J.H.Linhart, Proc. of Conf. on Plasma Phys and Contr. Fus. Res., Culham, 1965, V.II, IAEA, Vienna, 1966, p.345-365

Conceptual Design of Decade Half, a 15-MA, 300-ns PRS Driver

Phil Spence,[a] Pat Corcoran,[a] John Douglas,[a] Terry Tucker,[a] Bob Altes,[a] Kendall Childers,[a] Peter Sincerny,[a] Lavell Whitehead,[b] Van Kenyon,[b] Marc Babineau,[b] Tim Cotter[b], Pete Kurucz,[c] R. Davis[c]

[a]*Titan Pulse Sciences Division, 2700 Merced Street, San Leandro, CA 94577 USA*
[b]*Arnold Engineering Development Center, Arnold AFB, TN 37389-6400*
[c]*Defense Threat Reduction Agency*

Abstract. The Decade radiation effects simulator is located at AEDC and presently consists of four modules capable of operation as either a large area bremsstrahlung source (Decade Quad LAB)[1] or a plasma radiation source (Decade Quad PRS)[2]. During 2001 we investigated several conceptual system architectures for connecting the current of eight Decade modules to a single PRS load. This Decade Half radiation source will be used in combination with other synchronized radiation effects simulators to offer a combined-effects testing capability at AEDC.

We describe and compare the projected performance, risk, and relative cost of several rough conceptual designs, and discuss in more detail the "water bridge" concept that was selected for more complete conceptual design development. We present a self-consistent electrical and mechanical conceptual design for Decade Half, including an equivalent circuit model, details of the mechanical architecture (including operation, maintenance, and turnaround considerations), and electrical performance estimates of approximately 15 MA peak current delivered to a representative large radius, ~ 300 ns implosion time PRS load.

1. INTRODUCTION

The Decade radiation effects simulator was designed in the early 1990's to be a versatile, multi-module pulse generator configurable to drive both electron bremsstrahlung diode and fairly fast Z-pinch, plasma radiation source loads. The basic module building block is DM2, a 900-kV, ~ 0.5-ohm, 225-ns output pulse length (start-up to start-down) driver capable of delivering 365 kJ to a matched load. The DM2 driver circuit consists of a Marx generator, a novel, multi-tube, transfer capacitor, a set of six electrically-triggered output switches and a water dielectric output line. For convenience the Marx generators for four modules are housed in a single tank. This tank and the four modules connected to it form Decade Quad (DQ), one-quarter of the originally envisioned sixteen-module simulator.

For DQ bremsstrahlung mode the four modules are operated independently. Each one is fitted with a vacuum insulator stack, a vacuum inductive energy storage line, and a plasma opening switch/diode to produce a 40-ns (FWHM) radiation pulse.[1] The four diodes are close-packed to provide radiation uniformity over an approximately 60-cm diameter test area.

For DQ PRS mode the four output water lines are bused together in water to drive a common, two-sided, low-inductance vacuum insulator stack, radially converging MITLs and a post-hole-convolute, and a final, single-sided radial feed and fast Z-pinch. The Z-pinch axis on DQ is horizontal. Peak drive currents of 8 MA have been achieved with 300 ns implosion time loads.[2]

During 2001 we investigated several conceptual system architectures for busing the current of eight Decade modules together to drive a single PRS load. Such a Decade Half (DH) configuration will be required to drive the 14-15 MA PRS load planned for use in combination with other synchronized radiation effects simulators to provide a combined-effects testing capability at AEDC. Section 2 discusses four of the concepts we considered in a trade study and concludes with qualitative comparisons of their advantages and disadvantages.

Section 3 further describes the conceptual design of the architecture selected by the trade study – the 300 ns "water bridge" coupler concept. Summary and conclusions are given in Section 4.

2. CONCEPTUAL DESIGNS CONSIDERED FOR DECADE HALF (DH)

Figures 1 through 4 show the four different designs that were considered in an initial trade study. The Water Coupler design (Figure 1) is shown with two DQs having their output lines facing each other in an "opposing Quad" configuration. Each of the eight output lines is fitted with a coaxial to rectangular, or blade, transition. The blades are connected to a set of flat plate water transmission lines that extend upward and then horizontally across to connect the two quads in a "bridge" structure. The four upper modules, two from each DQ, drive the upper two horizontal plate output lines. Similarly, the four lower modules drive the lower pair of horizontal lines. A common, four-level, central, vacuum insulator stack is located in the center of the bridge. Four conical MITLs, a double post-hole-convolute (DPHC), and final single radial feed, all similar to their Saturn and Z counterparts,[3],[4] complete the power connection to the Z-pinch load. The vertical insulator stack axis is chosen to simplify MITL removal, insertion and alignment.

The Vacuum Coupler design shown in Figure 2 also uses an opposing Quad configuration. Here each of the eight output lines is fitted with an axial vacuum insulator stack and a long coaxial MITL. The coaxial MITLs converge toward a central coupler and undergo a coaxial to blade transition before they connect to a common set of four conical MITLs, a DPHC, final radial feed, and Z-pinch load. Both the Water Coupler and the Vacuum Coupler concepts were designed for ~ 300 ns Z-pinch implosion times.

Figure 3 shows the 150 ns PFL, with water coupler, design. This design has some similarity to the Water Coupler design, but with an added stage of power amplification accomplished by inserting a water dielectric pulse forming line stage between the transfer capacitor output switches and the water dielectric output lines. The PFLs are switched through ground planes with untriggered multi-site water switches to drive four parallel plate water output lines per module. The output lines are vertical triplates, 32 in total for the eight modules, and they converge and connect to a centrally-located, four-level, vacuum insulator stack in a configuration similar to the Saturn output section. The common water tank that contains the vertical plate output lines is configured as a bridge structure to allow test chamber access from under the bridge.

The final concept considered was the Mega-Quad shown in Figure 4. In this concept the Marx energy and individual module energy and power were doubled to allow DH to occupy less floor space within the Decade facility and to allow the existing DQ to remain intact and operational. The design shown is for a 300 ns implosion time. The architecture is very similar to the present DQ PRS driver but with larger transfer capacitors, higher voltage output switches, larger diameter water output lines and a larger water coupler. Both two- and four-level vacuum insulator stacks were considered. The insulator stack and Z-pinch axis is horizontal.

For each of the four concepts we first iterated the electrical and mechanical designs to achieve a viable mechanical concept (considering structural aspects, assembly and maintenance, and user chamber access). A summary of the four design concepts (peak current, total kinetic energy coupled to the load, relative cost, and overall figure of merit) is shown in Table 1. The figure of merit is the ratio between the performance (MJ coupled to the load) and the relative cost of the machine including risk reduction R&D. This analysis led to the selection of the water coupler as the baseline design.

3. DECADE HALF WATER COUPLER CONCEPT

The simplified circuit model for the front end of the chosen water coupler design is shown in Figure 5. The overall impedance of the water coupler is 0.0625 ohms. The water section drives a vacuum insulator and MITL section that is 7.3 nH. The load was a 13-cm diameter, 2.5-cm long plasma radiation source load with an initial inductance of 1.3 nH. The predicted output of the DH machine is shown in Figure 6. The new DH machine will be capable of delivering 14 MA into the load described above. One design issue that has been addressed is the symmetry of current delivered to the MITL. Two-dimensional analysis of the current flow from the two-sided output water lines into the circular vacuum section have demonstrated the ability to provide control of current symmetry (for magnetic insulation implosion symmetry) by cutting out sections of the water line. The cutouts force the current distribution to be much more symmetric at the input to the MITLs. An isometric sketch of the planned DH facility showing the location of the test object is shown in Figure 7.

4. SUMMARY

A trade study has been completed to determine the optimum (maximum energy delivered to the load per dollar) design concept for connecting two Decade Quad modules into a Decade Half machine. The water coupler concept that was chosen as the baseline will be capable of delivering 14 MA and 1.0 MJ of kinetic energy into a plasma radiation source load. The final design of this machine should be completed in FY-03 with IOC of the new facility at the end of FY-06.

5. FIGURES

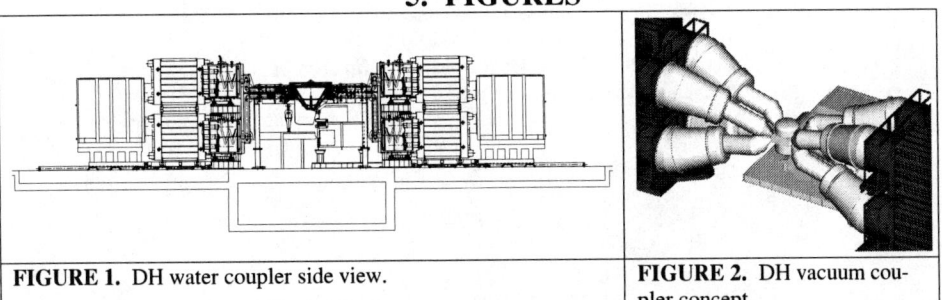

FIGURE 1. DH water coupler side view.

FIGURE 2. DH vacuum coupler concept.

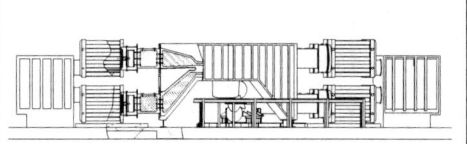

FIGURE 3. DH 150-ns concept side view.

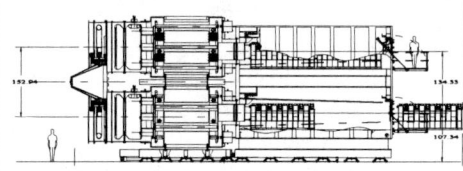

FIGURE 4. DH Super Quad concept side view.

Design Concept	Peak Current Into PRS Load	Total Energy Coupled to Load	Relative Program Cost, incl. Risk Reduction R&D	Risk Areas	"Figure of Merit"
Water Coupler, 300 ns	15.2 MA	1010 kJ	1.22	• PRS performance w/ 300-ns drive (applies to all 300-ns concepts)	.83
Water Coupler, 150 ns	13.69 MA	838 kJ	> 1.48	• Requires higher V triggered output (transfer) switch	< .57
Vacuum Coupler, 300 ns	13.1 MA	784 kJ	1.00	• Uncertain power loss in vacuum coupler – mandatory R&D • Difficulty in holding A-K gap alignment & tolerances	.78
Super Quad (4 layer), 300 ns	13.15 MA	719 kJ	2.14	• Requires higher voltage components	.34
Super Quad (2 layer), 300 ns	11.78 MA	839 kJ	1.86	• Output switches – prob. Laser triggered	.34

TABLE 1. Summary of DH trade study.

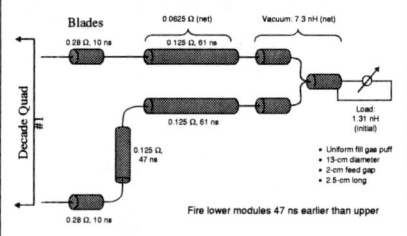

FIGURE 5. DH water coupler simplified circuit.

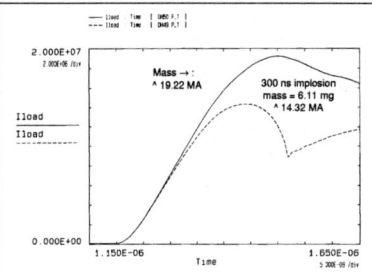

FIGURE 6. DH water coupler predicted performance.

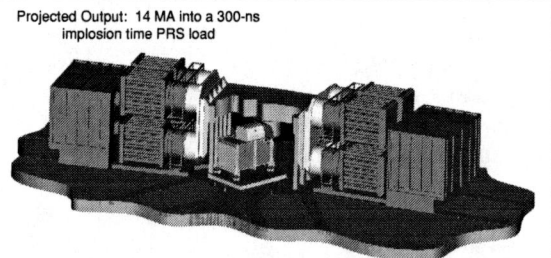

FIGURE 7. Down-select to water coupler option as the DH baseline design.

---

[1] P. Sincerny, K. Childers, D. Kortbawi, I. Roth, C. Stallings, J. Riordan, and B. Hoffman, 11th IEEE Pulsed Power Conference, Baltimore, Maryland, 1997.

[2] W. Rix, B. Altes, K. Childers, P. Corcoran, M. Danforth, P. Spence, P. Sincerny, T. Tucker, J. Douglas, M. Babineau, V. Kenyon, L. Christensen, G. L. Whitehead, P. Kurucz, K. Ware, J. K. Brandon, 13th IEEE Pulsed Power Conference, Las Vegas, NV, 2001.

[3] R. B. Spielman, P. Corcoran, J. Fockler, H. Kishi, and P. W. Spence, "A Double Post-hole Vacuum Convolute Diode for Z-pinch Experiments on Saturn," 7th IEEE Pulsed Power Conference, 1989.

[4] R. B. Spielman, W. A. Stygar, J. F. Seaman, F. Long, H. Ives, R. Garcia, T. Wagoner, K. W. Struve, M. Mostrom, I. Smith, P. Spence, and P. Corcoran, "Pulsed Power Performance of PBFA-Z," 11th IEEE International Pulsed Power Conference, 1997, p. 709.

New IES Scheme for Power Conditioning at Ultra-High Currents: from Concept to MHD Modeling and First Experiments

Alexandre S. Chuvatin,[a] Leonid I. Rudakov,[b]
Vladimir A. Kokshenev,[c] Leonid E. Aranchuk,[a] Dominique Huet,[d]
Vladimir A. Gasilov,[e] Alexandre Yu. Krukovskii,[e]
Nikolai E. Kurmaev,[c] Fiodor I. Fursov[c]

[a] *Laboratoire de Physique et Technologie des Plasmas, Ecole Polytechnique, 91128 Palaiseau, France*
[b] *Berkeley Scholars, Inc., Springfield, VA 22150, USA*
[c] *High Current Electronics Institute, 634055 Tomsk, Russia*
[d] *Centre d'Etudes de Gramat, 45600 Gramat, France*
[e] *Institute for Mathematical Modeling, Miusskaya sq., 125047 Moscow, Russia*

Abstract. This work introduces an inductive energy storage (IES) scheme which aims pulsed-power conditioning at multi- MJ energies. The key element of the scheme represents an additional plasma volume, where a magnetically accelerated wire array is used for inductive current switching. This plasma acceleration volume is connected in parallel to a microsecond capacitor bank and to a 100-ns current ruse-time useful load. Simple estimates suggest that optimized scheme parameters could be reachable even when operating at ultra-high currents. We describe first proof-of-principle experiments carried out on GIT12 generator [1] at the wire-array current level of 2 MA. The obtained confirmation of the concept consists in generation of a 200 kV voltage directly at an inductive load. This load voltage value can be already sufficient to transfer the available magnetic energy into kinetic energy of a liner at this current level. Two-dimensional modeling with the radiational MHD numerical tool *Marple* [2] confirms the development of inductive voltage in the system. However, the average voltage increase is accompanied by short-duration voltage drops due to interception of the current by the low-density upstream plasma. Upon our viewpoint, this instability of the current distribution represents the main physical limitation to the scheme performance.

CONFIGURATION AND BASIC SCALING LAWS

Currents of dozens of MA with microsecond rise-time are already achievable at magneto-cumulative generators or low-impedance capacitor banks [3, 4]. Further, ten-fold current pulse sharpening would allow the next generation of plasma radiation sources (PRS) to appear [5]. Current rise-times of dozens of nanosecond require a voltage of several MV to be generated across the load. It was suggested to use the voltage of inductive origin in nanosecond [6] or microsecond [7] regimes for efficient energy coupling into a Z-pinch type PRS. We introduce here a particular configuration, where staged plasma acceleration would allow 1 µs – 100 ns current sharpening, Fig. 1a. Two inductances, the load one (L_z) and that of the additional volume (L_0, L_1), are connected electrically in parallel (3) (i.e. we will conventionally call them LL scheme) and both are powered by a microsecond-discharge-time capacitor bank (1). During the discharge, when the switch (5) is open, the current first flows through a plasma conductor (4). This plasma is accelerated during ~10^{-6} s by the $j \times B$ force up to high velocity (> 10^7 cm/s). Electrodes geometry is chosen in order to provide $L_0 \ll L_g$, $dL_0/dt \ll \sqrt{L_g/C}$ at the first stage of the plasma motion and to ensure efficient current transfer from the low-impedance generator. When the plasma reaches the end of the acceleration region it starts to expand in a large vacuum cavity (6) and to be compressed onto the axis. As a result, the voltage $I \times dL/dt$ is generated across the interelectrode gap and after the closing of (5) the magnetic energy can be rapidly transferred into the load L_z.

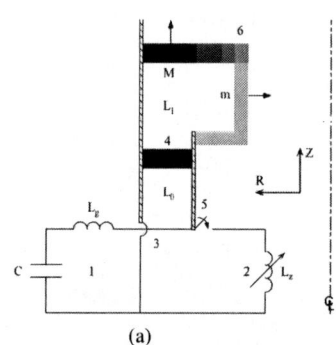 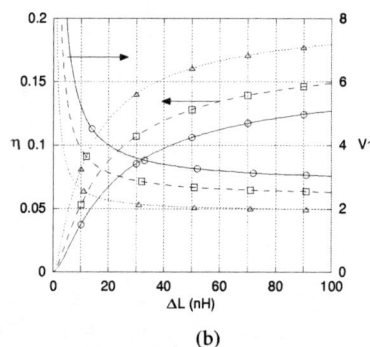

(a) (b)

FIGURE 1. (a) Plasma arrangement in electrical circuit of the LL-scheme: 1 – low impedance Marx generator, 2 – load (imploding liner), 3 – convolute, 4 – initial position of the plasma piston, 5 – closing switch, 6 – expanding plasma, L_0 and L_1 – inductances of the acceleration and expansion volumes accordingly. (b) Efficiency η and the value of $V\tau$ (in MV×μs) for L_0 = 2 nH (triangles), L_0 = 4 nH (squares) and L_0 = 6 nH (circles).

Neglecting the energy dissipation, from the flux conservation equation one can define the generator-to-load current transfer coefficient and the energy efficiency:

$$k \equiv I_z/I_0 = (L_1 - L_0)/(L_1 + L_z + L_1 L_z/L_g) \quad \eta \equiv E_z/E_0 = L_z k^2/(L_g + L_0) \quad (1)$$

Let us fix a constant inductance L_z = 7 nH representing some mean value for PRS and let us consider I_z = 60 MA [5]. Then the critical parameters would be L_1, V (capacitor bank erecting voltage) and $\tau \equiv \sqrt{(L_g + L_0)C}$.

$$L_g^{opt} = 4L_0 / \left(\sqrt{1 + 8(L_0/L_z + L_z/L_1)} - 1\right) \quad (2)$$

Maximum of efficiency, $\eta(L_g)$ occurs at some L_g^{opt}, Eq. (2). For $L_0 \to 0$ and $L_1 \to \infty$ we have η = 0.25 and k = 0.5. Minimum of the $V\tau$ value, Fig. 1b, for τ = 1 μs, L_0 = 4 nH and L_1 = 40 nH corresponds to the following primary generator: C = 28 μF, L_g = 10.5 nH, V = 2.8 MV, I_0 = 120 MA. These values could be achievable at existing level of the microsecond pulsed-power technology and provide the efficiency η > 10 %.

PROOF-OF-PRINCIPLE EXPERIMENT

The main objective of this series was demonstration of the load voltage generation at moderate current values (2-3 MA) when using the LL inductive switching scheme. The terminal part of GIT12 IES generator used in this experiment is shown in Fig. 2a. Tungsten wires (tighten horizontally in the figure) of the diameter 7.8 μ and 11 μ were mainly applied to form an array in the interelectrode gap of 3-5 cm with the acceleration distance up to 6 cm. The generator current, I_1, convolute current, I_{conv}, and wire array current, I_2, were measured with the help of dB/dt probes. The voltage was measured by the probe installed at the bottom of the 424 nH mechanical support [1] and then inductively corrected to the locations of the current monitors.

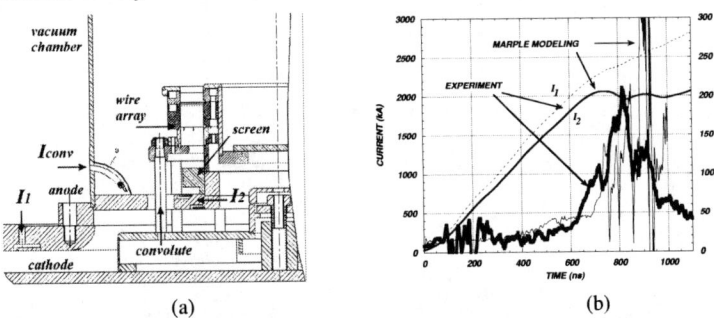

FIGURE 2. (a) LL scheme assembly on GIT12. (b) Current and voltage for the experiment #52 chosen for numerical modeling because of limited convolute current losses (I_1-I_2) in this shot. 12 tungsten wires of 7.8 μ diameter were used in the interelectrode gap of 3 cm with the distance between the initial array position and the output to the large vacuum volume of 6 cm. Experimental voltage (bold) is compared to that coming from 2D RMHD modeling performed under assumptions described below.

Wire plasma radiation during the shot lead to intense electron current losses at the convolute (> 1MA). The main part of electrode damage due to this leakage was located at the anode parts between the convolute rods, where the magnetic field had the minimum. Analysis of the experimental current waveforms allowed to conclude that the monitor I_{conv} was the most sensitive to these losses and could lead to overestimation of the generated voltage. Finally, the voltage estimated at the convolute level without taking into account this probe signals had the peak value of at least ~200 kV in several optimized configurations, the result from one of them is illustrated by Fig. 2b.

PLASMA DYNAMICS IN NUMERICAL MODELING

Numerical modeling was performed with the help of the *MARPLE* numerical tool for radiational magnetohydrodynamic plasma flows in plane or cylindrical two-dimensional geometries [2]. The wire array in the coaxial plasma region was initially positioned at Z = 10.5 cm (coordinate upon Fig. 3) as a 3 mm – thick tungsten plasma disk with the mass of 0.12 mg, the temperature of 1 eV and with a random initial density perturbations with the amplitude of ±5 %. 12 tungsten wires in the experiment

#52 (Fig. 2b) hardly seemed to form an azimuthally homogeneous plasma shell. Several additional experiments (not discussed here) revealed importance of the downstream plasma flow. The mass in simulations was taken 3 times smaller to match the moment of the voltage rise beginning to the experimental value (~ 730 ns). First *MARPLE* runs involving real generator electrical circuit without the convolute leakage current overestimated the LL voltage (1.5-2 times higher than that shown in Fig. 2b). However, when fitting the experimental plasma current waveform downstream the convolutes, I_2, the simulated voltage was close to the measured in the experiment. Fig. 3 illustrates the density and magnetic field contours appearing for the discussed shot at different time moments.

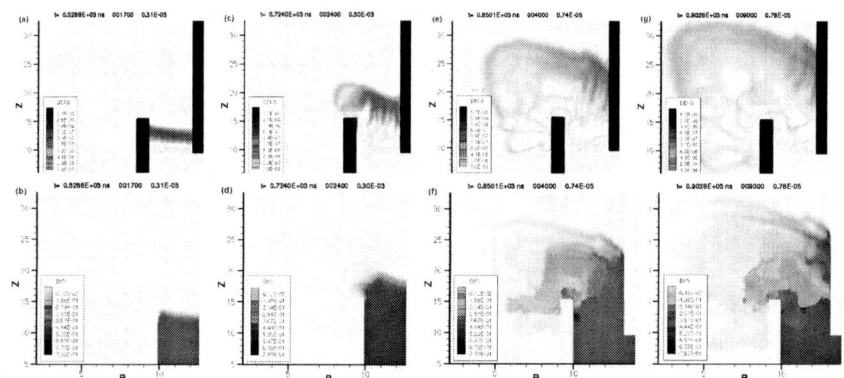

FIGURE 3. Results of 2D modeling: (a,b) acceleration phase, (c,d) RT instable plasma, starting expansion into the vacuum volume, (e,f) expansion and compression to the axis, voltage generation, (g,h) current interception by the residual upstream plasma, the voltage drops.

As soon as the electric field $E \propto [v \times B]$ is generated by the expanding plasma (c through f) and appears across the anode-cathode gap, the current can be redistributed to the residual plasma tail. Characteristic current loops are noticeable in Fig. 3f, 3h. This leads to abrupt voltage drops (see Fig. 2b). However, in its turn, the tail plasma starts to be accelerated by j×B force. The residual density is low and this plasma can be rapidly blown-away from the gap. The process of the current instability repeats further in time and represents the voltage amplitude limitation in the problem.

This work is supported by the contract DGA/CEG N° 00-25-043-00-470-46-51.

REFERENCES

1. A.A.Kim et al, *10th IEEE Pulsed-Power Conf.*, p.226, 1995.
2. V.A.Gasilov, et al, to appear in *Mathematical Modeling* (in Russian).
3. V.K.Chernyshev, *Proc. 7th Int. Megagauss Conf.*, v. 1, p. 41, 1996.
4. P.J.Turchi, *Proc. 12th IEEE Int. Pulsed-Power Conf.*, v.1, p. 3, 1999.
5. M.K.Matzen, *Phys. Plasmas* **4**, 1519 (1997).
6. L.I.Rudakov, *Proc. 12th IEEE Int. Pulsed Power Conf.*, Monterey, CA, 1999, v. 2, pp. 1102-1105.
7. A.S.Chuvatin et al, *Bull. Am. Phys. Soc.* **44**(7), 103 (1999).

ECF2 : A pulsed power generator based on magnetic flux compression for K-shell radiation production

P. L'Eplattenier[a], F. Lassalle[a], C. Mangeant[a], F. Hamann[a], M. Bavay[a],
F. Bayol[a], D. Huet[a], A. Morell[a], P. Monjaux[a], G. Avrillaud[b], B. Lalle[b]

[a]*Centre d'Etudes de Gramat, 46500 Gramat, France*
[b]*ITHPP, 46500 Thegra, France*

Abstract. The 3 MJ energy stored ECF2 generator is developed at Centre d'Etudes de Gramat, France, for K-shell radiation production. This generator is based on microsecond LTD stages as primary generators, and on the magnetic flux compression scheme for power amplification from the microsecond to the 100ns regime. This paper presents a general overview of the ECF2 generator. The flux compression stage, a key component, will be studied in details. We will present its advantages and drawbacks. We will then present the first experimental and numerical results which show the improvements that have already been made on this scheme.

INTRODUCTION

The 3MJ energy stored pulsed power generator ECF2 [1] is developed for Z-pinch K-shell radiation production at Centre d'Etudes de Gramat (CEG) within the context of the Sphinx project. This generator is based on the magnetic flux compression scheme [2] for power amplification from the microsecond to the 100ns regime. The flux compression scheme presents many advantages but also several drawbacks, especially for K-shell like Z-pinch loads. The present goal of the program at CEG is to better adapt this scheme to the requirements of K-shell radiation production by a Z-pinch implosion. This paper is organized as follows: in section I, we will present the ECF2 generator as well as the goals of the Sphinx project. In section II, we will present in more details the power amplification stage based on the magnetic flux compression scheme, its advantages and drawbacks. Section III will present the first experimental and numerical results obtained on ECF2.

I THE ECF2 GENERATOR

The ECF2 generator is based on 16 lines of 8 LTD_03 stages [1]. Each LTD stage is composed of two 4 µF capacitors, which can be charged up to 75 kV. Those LTD stages allow a rise time of the current of the order of 1 µs. A power amplification stage based on the magnetic flux compression is then used to shorten the pulse to about

100ns. Thus, the LTD technology coupled with the flux compression allows to avoid the pulse forming lines, the use of big oil and water tanks. The LTD switches are not pressurized and work with air. All this leads to an inexpensive pulsed power generator compared to the other ones in the world based on other technologies. It also is a very modular concept allowing fast changes in the configuration if necessary. In particular, the central part has been designed in order to be able to change the number of lines for the primary and for the secondary generators. The ECF2 generator represents a total of 3 MJ stored energy, and the goal is to be able to deliver a current of 10 to 15 MA rising in 100ns in a typical Z-pinch load. Among the 16 lines of the generator, 12 are usually used for the primary generator and 4 for the secondary generator. At this time, only 4 stages per lines have been assembled, the 4 other ones should be assembled by July 2002. Hence, all the results presented in this paper correspond to ECF2 with half the total stored energy.

The ECF2 generator will be used at CEG in the context of the Sphinx project for K-shell radiation production with Z-pinch loads. The first Z-pinches to be tested will be wire arrays ones, Aluminum first, then Titanium and Stainless Steel. Then Copper and gas puffs may also be tested. Up to now, the only loads that have been used are short circuit ones in order to test the flux compression stage.

II THE MAGNETIC FLUX COMPRESSION STAGE

The magnetic flux compression scheme is a power amplifier that can be used on HPP generators as an alternative to pulse forming water lines or Plasma Opening Switches [2]. This scheme can be used for many applications such as K-shell radiation production for weapon effect purposes, generation of high temperature hohlraums or generation of very high pressure loading in isentropic compression [3]. This scheme has been used on several experiments on the Z generator at Sandia National Laboratories (SNL) [3], [4], [6]. It was also used on the 640kJ energy stored ECF1 generator at CEG and is now used on the ECF2 generator. See [5] or [6] for a full description of the scheme.

The configurations used on the ECF2 generator were the following: the armature was 15cm long, with an initial radius in between 5cm and 10cm. It was initially made with 216 to 432 Aluminum wires for a total mass in between 7mg and 20mg. The stator was 1cm radius. Up to now, the only load used was a short circuit one with an inductance of 0.4 nH.

The flux compression as a power amplification stage presents some advantages. It first is an inexpensive power amplification. Moreover, this scheme depends on many parameters that can be changed very easily in the experiments, such as the mass, length, initial radius of the armature, the initial radius of the stator and even its shape, the location of the secondary injection gap,... Those many parameters leave many opportunities to pulse shape the current waveform obtained in the secondary after the amplification [5]. This is very interesting for applications like Isentropic Compression Experiments (ICE) for which the shape of the pressure waveform is very important. It is also interesting for Z-pinch loads for which those many parameters can be optimized together with the parameters of the pinch to increase the radiation output.

It also presents some drawbacks. There are 2 main ones. The first one is due to the 1D-2D behavior of the armature with a low density plasma flowing ahead of the main part of the mass and Rayleigh-Taylor (RT) like instabilities developing during its implosion. Those instabilities may present some flux losses at the time of impact of the armature on the stator. The second one is due to the presence of a current pre-pulse in the load, during the whole time of the injection of the secondary current, before the final amplification. This current pre-pulse is bad for Z-pinch loads, and especially for K-shell radiation production. Part of the present Sphinx program is to improve those drawbacks as will be presented in the following section.

III NUMERICAL AND EXPERIMENTAL RESULTS

Figure 1 presents the results of a typical shot on ECF2, with 4 stages per line, shot ECF418. The experimental primary and secondary currents measured by B-dots are presented as well as the corresponding simulated ones.

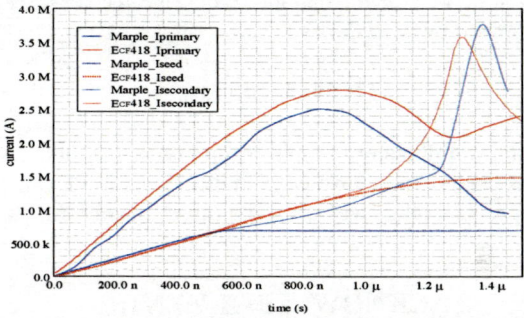

Figure 1: Experimental and numerical (simulation made with the 2D MHD code Marple) primary and secondary currents for shot ECF418

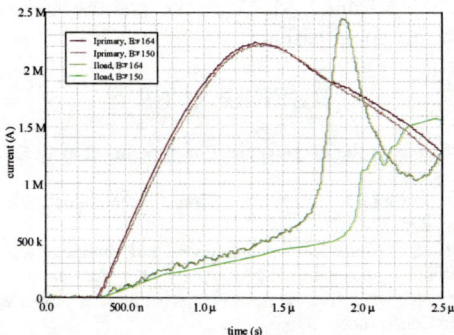

Figure 2: comparison of the secondary current with and without the insulator sheath on the central rod

On that shot, the liner was made with 428 aluminum wires for a total mass of 15mg. A 400 µm insulator sheath was placed on the central rod to prevent the instabilities in the liner to generate some flux losses as they implode on the stator. This insulator sheath has been used on most of the shots, and it is how we currently deal with the flux losses in the RT instabilities. To judge the efficiency of this thin insulator sheath, figure 2 shows a comparison of a shot with the sheath and a shot without the sheath.

Several solutions are now being studied to reduce the current pre-pulse in the load. As seen on figure 1, this pre-pulse has an amplitude of about 1.5 MA and lasts around 1 µs, compared to the main pulse that reaches 3.5 MA in 200 ns. The first idea that has been tested was to use a second gap inside the secondary loop. The current can then flow in the secondary only after this gap is crowbarded by the plasma precursor coming from the armature. This reduces the pre-pulse duration from 1µs to around 600 ns.

Other ideas are also being studied. They include the use of a plasma opening switch located at the injection gap to the secondary cavity, the use of a double armature with a secondary injection in between them, and more sophisticated schemes like the Current Doubler scheme proposed by L. Rudakov [7].

CONCLUSION

The ECF2 generator at full energy should be on line in July 2002. Several potential problems of the flux compression have been studied and solved. Other ones are being addressed now. The generator will then be used mainly for K-shell radiation production.

ACKNOWLEDGEMENTS

The authors are very indebted to J.F. Leon and the ECF1/ECF2 team at CEG, and also to R.B. Spielman, D. McDaniel, M. Mazarakis and the Z team at SNL for a very fruitful collaboration on flux compression.

REFERENCES

[1] Ph. Monjaux et al, "Status on the microsecond flux experiment at CEG", Pulsed Power Plasma Science, 2001, pp.310-313
[2] J.F. Leon et al, "Flux compression experiments on Z accelerator", IEEE International Pulsed Power Conference, 1999, pp. 275-278
[3] P. L'Eplattenier et al, "Numerical modelization and optimization of flux compression experiments", 12th APS on shock compression of condensed matter, Atlanta, june 2001
[4] M. Bavay et al, "The magnetic flux compression scheme as a power amplification and pulse shaping stage", ICOPS 2002 Conference
[5] P. L'Eplattenier et al, "0D numerical modelisation and optimization of flux compression experiments", Pulsed Power Plasma Science, 2001, pp.665-668
[6] M. Bavay, PhD thesis, "Compression de flux magnétique dans le régime sub-microseconde pour l'obtention de hautes pressions et de rayonnement X intense", to be published, Paris IX University.
[7] L. Rudakov et al, "Current doubler flux compression device for power amplification in vacuum", Pulsed Power Plasma Science, 2001, pp.1719-1722

Performance of Llampüdkeñ with Short Circuit and Plasma Loads.

Hernán Chuaqui, Ian H. Mitchell, Raúl Aliaga-Rossel, Mario Favre and Edmund S. Wyndham.

Departamento de Física, Pontificia Universidad Cátolica de Chile
,Casilla 306, Santiago 22, Chile.

Abstract. Llampüdkeñ [1] is a pulsed power generator designed to deliver a 1 MA, 250 ns risetime current pulse into a dense plasma load. The main novel feature of this generator is the two auxiliary transmission lines which transmit the energy not absorbed by the load, reflect it at the open end of the line and deliver it to the load when the energy from the main lines is decreasing. With the auxiliary lines an increase of 30% on the current as well as a decrease of the voltage at the load is obtained. To date Llampüdkeñ has been operated up to the 400 kA level, into both short circuit and plasma loads. Details of actual performance of the pulse power generator are presented and compared with simulations.

INTRODUCTION

This paper reports on the performance of, Llampüdkeñ[1] (1 MA, 250 ns), a new pulsed power generator designed for Z pinch experiments. The generator consists of two Marx banks, each one connected to a set of parallel plate transmission lines which transfer the energy to the load. The first transmission line in each set is a 1.0 Ω, 180 ns line is followed by a four channel, electrically triggered switch. After the switch there is an exponential transfer line which leads to a bifurcation. One arm of the bifurcation is connected to a short transfer line leading to the load and the other arm to an exponential auxiliary line. The purpose of the auxiliary line is to provide a path for energy which is not initially coupled to the load and then to reflect it back, thus significantly increasing the plasma current in the latter half of the discharge. The exponential nature of the line improves the coupling further and also limits the voltage at the load chamber feed for the case of high impedance (e.g. the early stage of a plasma shot).

SIMULATION

The generator is simulated by the commercial TopSpice package using the equivalent circuit shown in figure 1. In this particular simulation the Marx is treated as a simple capacitor of value equal to the erected Marx capacitance (250 nF) initially charged to the full erected voltage, in series with an 4.0 µH inductor and a 0.8 Ω resistor. The following short, high impedance section is the connection between the

---

[1] The name Llampüdkeñ comes from the native language of one of the indigenous groups in Chile. It means "butterfly".

Marx and the main transmission line. The line switch, located after the main transmission line is closed at a predetermined time in the simulation and includes an inductance of 50 nH. The two exponential lines are the transfer and auxiliary lines described above. Simulations previously reported [1] have shown that, under optimum conditions, the peak load current increases by 30% due to the reflection of uncoupled energy from the end of the line.

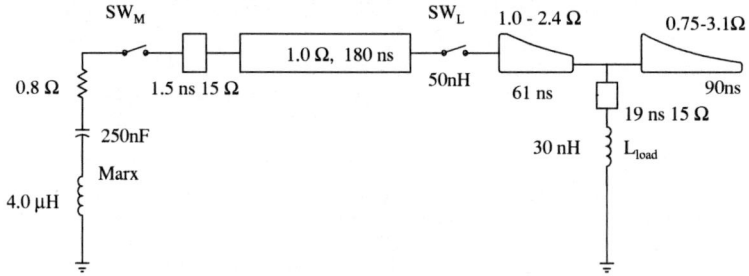

FIGURE1. An equivalent circuit of the Llampüdkeñ generator. The circuit only shows one of the two Marx banks and transmission line sets. The load is simply a 30nH inductor.

In the simulations of "short circuit" shots, the load is treated as a constant inductance. A value of 30 nH is used which includes the inductance of the load chamber. Figure 2 shows a simulation of the generator at full charge (480 kV Marx voltage) into such a load. The line switches are triggered when the lines are fully charged. The traces show the voltage at the output of the Marx bank and the current through the load. The load current reaches a value of 1.1 MA with a risetime of 200 ns.

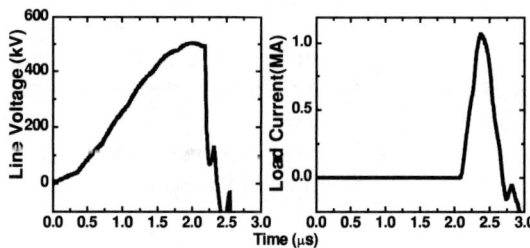

FIGURE2. A simulation of the generator at full charge, showing the line voltage and the load current.

EXPERIMENTAL RESULTS

Llampüdkeñ has been operated reliably at 50% of the maximum voltage. Figure 3 shows a direct comparison of experimental and simulated voltage and current traces for the case of a short circuit load with ±20kV charge voltage. In the experimental results the voltage traces are measured by the voltage probe at the output of each of the Marx banks and the current traces are those obtained by the integration of the output of a Rogowski coil placed inside the vacuum chamber. The Rogowski was previously calibrated using an RC discharge and a resistive probe. The simulation traces are the voltages and current obtained at the corresponding places in the equivalent circuit.

The simulation has been tailored to reproduce the Marx and switch jitters in this specific shot and the inductance of the switches has been adjusted to agree with the decay time of the voltage after the switches close. As can be seen there is very good agreement between the simulation and experimental measurements which both show a peak current of 385 kA with a risetime of 260 ns.

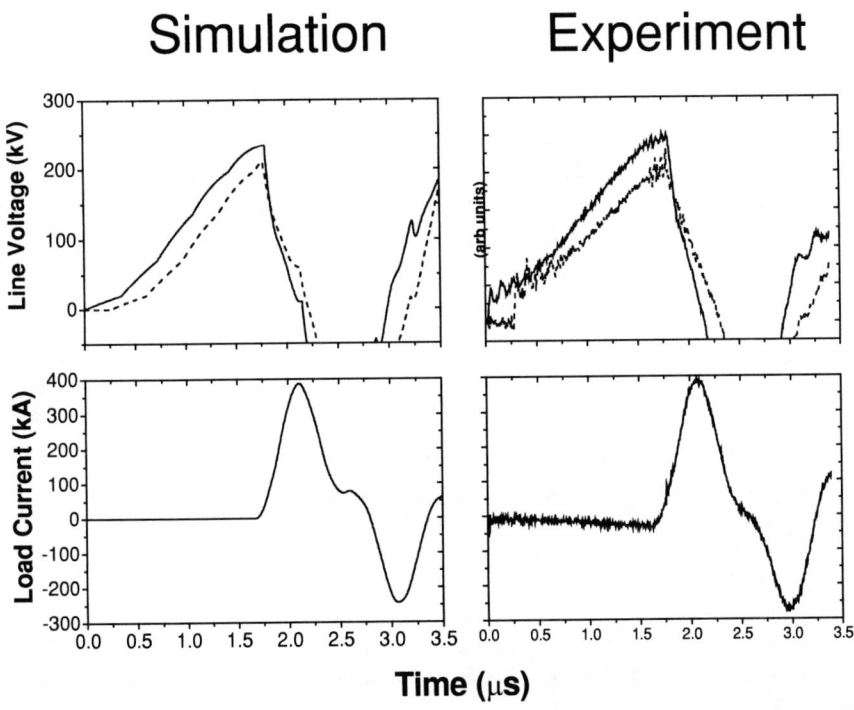

FIGURE 3. A comparison of a simulation and a real shot of the generator for 50% charge. The scales of the simulated and experimentally measured current traces are the same.

An important difference between the equivalent circuit for this shot and that in the optimised case shown in figure 2, is that the inductance of the switches had to be increased, one to 200 nH and the other to 300nH, in order to obtain the same fall time of the line voltage after switch closure. This is an indication that the line switches are not breaking down in the four channels but only in one or two. This situation may improve as the Marx voltage is increased.

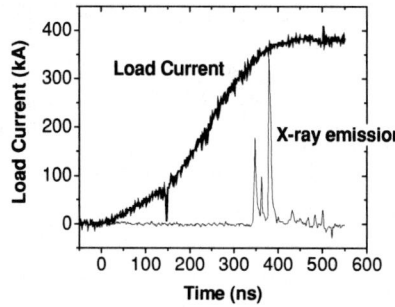

FIGURE 4. Experimental measurement of the load current for 50% charge with an Aluminium wire X pinch load. The x-ray emission from the plasma is also shown.

Experiments with X pinch loads have been carried out on the generator [2]. Figure 4 above shows the load current and X-ray emission from a typical shot. The current is measured by a Rogowski probe which takes the form of a groove located in the earth electrode. This probe was calibrated against the Rogowski coil used for the above measurements using short circuit shots. The X-rays are measured by a filtered PIN diode. In this particular shot the load was a X pinch made from two 125 µm diameter wires and x-rays were emitted from the plasma near peak current. The load current reaches a peak of 380 kA with a risetime of 260 ns.

CONCLUSION

The Llampüdkeñ generator has been commissioned at up to 50% of its maximum charge voltage. The experimental measurements are in very good agreement with the simulation of the generator. Into a short circuit load the experiment and tailored simulation both give a peak current of 380 kA with a rise time of 260 ns. Experiments with X pinch plasmas show that it is possible to reach this same peak current with a plasma load.
 Simulation of a realistic optimum setup demonstrates that with a 100% of charging voltage it should be possible to obtain slightly more than the target current of 1 MA.

ACKNOWLEDGMENTS

This work was carried out at the Pontificia Universidad Católica de Chile, supported by FONDECYT grant numbers 8980011 and 1990435.

REFERENCES

1. Chuaqui H., E. Wyndham, C. Friedli and M. Favre, *Laser and Part Beam* **15**, 241-248 (1997)
2. Mitchell I. H., Aliaga Rossel R. Gomez J.A., Chuaqui H., Favre M, and Wyndham E.S., *These proceedings*

Load Current Sharpening and Liner Implosions on a Capacitor Bank

Sergey A. Sorokin and Stanislav A. Chaikovsky

Institute of High Current Electronics, 4 Akademichesky Av., Tomsk 634055, Russia

Abstract. Experiments on sharpening the current through the liner load were performed on a capacitor bank ($I \sim 400$ kA, $T/4 \sim 1.1$ µs). Three load region configurations were tested. Each configuration includes two gas-puff shells. The current flows through the outer shell and switches to the inner shell at the current maximum. High efficiency of the inductive energy transfer to the liner and x-ray radiation (~50%) was demonstrated. Visible streak and x-ray pinhole cameras were used to observe the behavior of the shells and the formation of a compact pinch ($r_f < 1$ mm). The highest neon K-shell yield (up to 80 J) was obtained for the double shell liner configuration. The mechanism of the current switching to the inner shell is discussed.

INTRODUCTION

A variety of applications call for simple and low-cost soft-x-ray (SXR) sources providing high efficiency of electric-to-radiation energy conversion. The simplest way of realizing such a source is to produce magnetic compression of a gas-puff liner by the pulsed current of a capacitor bank. The absence of pulse compression units in such a scheme increases the efficiency of the energy transfer from the capacitor energy store to the liner. However, this scheme, along with the above advantages, has the disadvantage that the current rise time is too long ($T/4 \sim 1$ µs). Simple estimates show that the change-over from high-current pulse generators with a current rise time of ~100 ns to microsecond capacitor banks may be critical for the quality of the liner implosion and the performance of the pinch as a radiator in view of Rayleigh-Taylor instabilities [1]. It follows that, when using a capacitor bank, to attain comparatively stable liner implosions followed by the formation of a compact pinch, it is necessary to sharpen the current through the load. Such a sharpening unit (switch) must not be detrimental to the inherent advantages of the capacitor bank, that is, its use must not complicate the design and service of the load unit, and the energy losses in the switch must be as low as possible. In the experiment under consideration, three configurations, each including two gas puff shells, were tested for the load unit of the capacitor bank (20 µF, 34 kV). The capacitor bank current flows initially in the outer shell and is switched to the inner shell as the current approaches its peak value. To simplify the design and service of the system, the gas puff shells were both formed with the use of a single common gas valve. The gas mass in the shells was controlled by varying the plenum pressure and the nozzle throat widths. The dynamics of motion of the outer shell was simulated using a 0-dimensional model. For current measurements and for monitoring the dynamics of current-carrying shell Rogowski coils, magnetic probes, and a visible streak camera were used. X-ray images of the pinch were taken with a pinhole camera. The pulse-integrated soft x-radiation yield and power were measured with a foil bolometer, x-ray diodes, and *pin*-diodes. The normal operation of the Rogowski coils and magnetic probes placed beneath the initial position of the current sheath is disturbed due to the fact that, they are shielded by the plasma of the propagating current-carrying shell. Therefore, additional supporting evidence for the current sharpening and for the efficient energy transfer to the inner shell came from (a) the fact that the rate of implosion of the inner shell became greater than $1 \cdot 10^7$ cm/s; (b) the formation of a compact pinch as a result of comparatively stable implosion; (c) the SXR yield comparable to the energy of the magnetic field in the diode, and (d) the neon *K*-shell radiation yield W_k close to the estimate based on the results obtained on faster facilities and on the I^4 scaling of the *K*-yield [2].

Configuration 1

The first configuration is shown in fig. 1a. In this scheme, the outer and inner cylindrical shells are axially spaced to prevent the influence of the former on the latter. The conditions for switching the current to the inner shell are created as the outer shell arrives at the edge of one of the disk electrodes. Examination of the Rogowski coil and probe signals has shown that this configuration fails to provide efficient current switching to the inner shell. The Rogowski coil signal appears with a time delay relative to the arrival of the outer shell at the edge of the disk electrode. The coil signal amplitude and risetime vary from shot to shot. It seems that the radial motion of the current-carrying shell is replaced by its axial motion toward the inner shell region. Besides, the subsequent motion of the current-carrying shell is accompanied by reclosing the current from the coaxial transition region to the inner shell region.

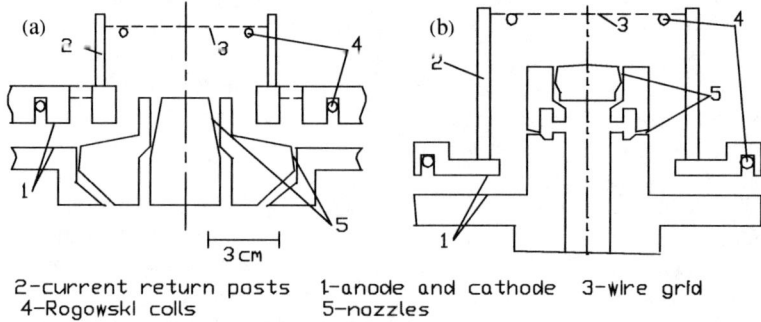

2-current return posts 1-anode and cathode 3-wire grid
4-Rogowski coils   5-nozzles

FIGURE 1. The load unit in configurations 1 (a) and 2 (b).

Configuration 2

The second tested scheme is a plasma flow switch (PFS) configuration (fig. 1b). In this scheme, the outer shell is produced with the use of a radial nozzle and is accelerated in the axial direction. By inclining the radial nozzle to the load, it is possible to prevent plasma breakup at the inner electrode. Upon some optimization of the geometry of the outer electrode and the distance between the radial and axial nozzles, stable current transfer to the inner shell was attained (fig. 2). In this case, by varying the throat width of the axial nozzle, the liner velocities in the range $(1-4) \cdot 10^7$ cm/s could be achieved. To demonstrate the use of this scheme of sharpening the current through the liner for the production of a compact pinch and efficient generation of SXR and K radiation, a series of shots was performed on a neon liner with the use of a complete set of the available diagnostic tools. The following neon liner parameters have been realized. The current through the liner increases during 100–200 ns. The K-radiation pulse FWHM and peak power are, respectively, ~30 ns and $(3-6) \cdot 10^8$ W, which corresponds to a K-radiation yield of about 10–20 J. According to bolometric measurements, the total radiation yield ranged between 500 and 700 J. The x-ray pinhole photographs showed a pinch with an average diameter of 1–2 mm. Figure 2 presents the waveforms of the R1 and R2 Rogowski coils and *pin-diode* currents for a typical shot on a neon liner. Let us estimate the efficiency of SXR generation as the ratio of the energy produced to the magnetic field energy at the onset of current passage through the liner. At the instant the current is switched into the load, the inductance of the electric circuit is close to its initial value, equal to about 26 nH, and the current is 320 kA. This corresponds to ~1.3 kJ of magnetic energy and ~50% efficiency. Extrapolation of the experimental data on the K-radiation yield of neon obtained on fast facilities to the current $I = 320$ kA by the scaling $W_k \propto I^4$ gives an estimate W_k ~10 J, which is in satisfactory agreement with the yield measured in this experiment and is indirect evidence that the current was switched into the load.

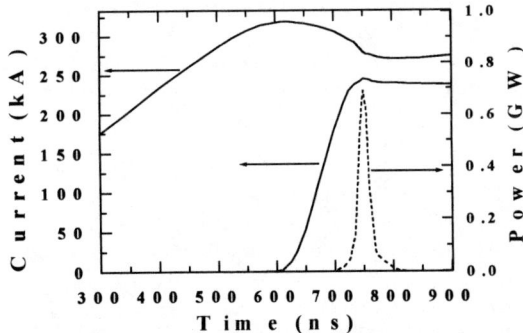

FIGURE 2. Waveforms of the R1 and R2 coils and pin-diode for a typical shot with a neon liner.

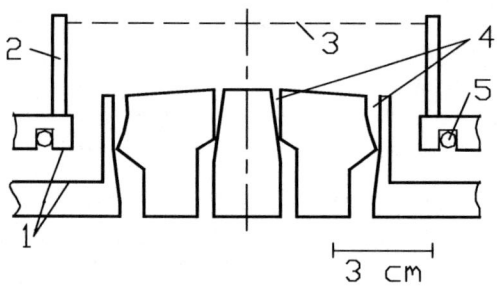

FIGURE 3. The double shell load configuration (1-anode and cathode, 2-current return posts, 3-wire grid, 4-nozzles, 5-Rogowski coil).

Configuration 3

During the implosion of a gas-puff liner, the liner plasma may break off one of the electrodes if the plasma velocity has a component normal to the electrode surface. For a double-shell liner, this may result in current switching to the inner shell. In this case, the time to the switching can be varied by inclining either the gas jet or the electrode with respect to the liner axis. Figure 3 shows the design of the load unit that was used to switch the current from the outer to the inner shell (at a near-maximum current in the outer shell, I ~ 380 kA). The dynamics of motion of the shells was monitored by a visible streak camera with the slit positioned both perpendicular and parallel to the liner axis. A detailed description of the experiment is given elsewhere [3]. Here, we present only the principal results. With the streak camera, the detachment of the outer shell from the cathode (nozzle) was observed which was followed, within a short time (100–150 ns), by acceleration of the inner shell to $v \sim 2 \cdot 10^7$ cm/s. In this case, the integrated pinhole picture (8-µm Al filter) showed a compact pinch of diameter 0.1–0.15 cm. For an optimum mass of the inner shell, the neon K-radiation yield was 50–80 J.

Thus, the operation of compact loads of a capacitor bank has been demonstrated which provide highly efficient conversion of the energy stored in the capacitor bank into x-ray radiation.

REFERENCES

1. Hussey, T. W., Roderick, N. F., and Kloc, D. A., *J. Appl. Phys.* **51,** 1452 (1980).
2. Krishnan, M., Deeney, C., Nash, T., et al., "Review of Z-pinch Research at Physics International Company," in *2nd Int. Conf. on Dense Z-Pinches,* edited by N. R. Pereira et al., AIP Conf. Proceed. 195, New York: American Institute of Physics, 1989, pp. 17-22.
3. Chaikovsky, S. A., and Sorokin, S. A., *Plasma Phys. Rep.* **27,** 947-952 (2001).

WIRE ARRAY EXPERIMENTS

Effect of Discrete Wires on The Implosion Dynamics of Wire Array Z-Pinches

S.V. Lebedev, J.P. Chittenden, S.N. Bland, D.J. Ampleford, C. Jennings and M.G. Haines

The Blackett Laboratory, Imperial College, London SW7 2BW, UK

Abstract. A review of recent experiments at the MAGPIE facility (1MA, 250ns) at Imperial College is presented. The experiments show that the core-corona structure of the plasma, combined with the 3-D topology of the magnetic field in wire array z-pinches, results in the implosion dynamics being significantly different from that of a thin plasma shell. During the first ~80% of the implosion time the interior of the array is gradually filled by the plasma ablated from the stationary wire cores. This phase ends with the formation of gaps in the wire cores, which occurs due to non-uniformity of the ablation rate along the wires. The final phase of the implosion, starting at this time, occurs as a rapid snowplough-like implosion of the radially distributed plasma, previously injected into the interior of the array. The density distribution of the precursor plasma being peaked on the array axis could be a key factor providing stability of wire array implosions.

THE FORMATION OF CORE –CORONA PLASMA STRUCTURE IN WIRE ARRAYS

It is now well established that the wires in a wire array Z-pinch are not completely converted into the plasma immediately after the start of a current pulse. Instead, a core-corona plasma structure is formed, which persists over a significant part of the implosion. Direct observation of dense wire cores in wire arrays was performed using x-ray radiography [1] with an X-pinch installed in the return current post. The characteristic sizes of the wire cores are different for different wire materials (~0.25mm for Al and ~ 0.1mm for W), but insensitive to the current per wire and to the initial wire diameter [1]. The radiographic measurements and other diagnostics (laser probing and optical streak photography) show that the wire cores are seen at their initial positions for the first ~80% of the implosion. The coronal plasma flows from the wire cores towards the array axis with characteristic velocity $V \sim 1.5 \times 10^7$ cm/s for aluminium and forms at stagnation a narrow, uniform and stable precursor plasma column, which is observed until the final implosion.

The side-on laser probing shows that the process of the gradual ablation of wire cores is not uniform along the wires. The instability seen in the coronal plasma has a well-defined, quasi-periodic structure with wavelength of ~0.5mm for Al and ~0.25mm for W, which does not change with time. This wavelength is 2 to 3 times the core size measured by radiography [1], which implies that the size of the region (~core size), where formation and acceleration of the coronal plasma takes place, determines the development of the instability. The observed axial non-uniformity of the formation and sweeping of the coronal plasma from the wires leaves an imprint on the axial distribution of mass in the wire cores, which becomes observable on radiographic images of Al wire arrays at ~80% of the implosion time (Fig.1). The wavelength of this structure is the same (~0.5mm) as seen in the coronal plasma from early time. Further growth of perturbations in the wire cores leads to the formation of breaks (gaps) in the cores, which are seen in laser probing (Fig.1) and in radiographic

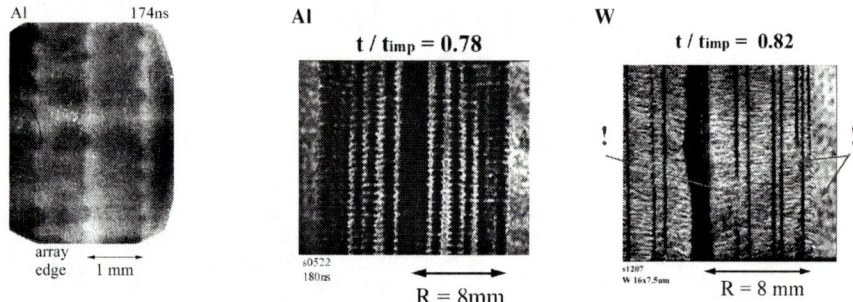

FIGURE 1. X-ray radiographic image (left) and laser probing images of Al and W wire arrays.

images[2]. Variations in the rate of ablation are apparently responsible for the non-simultaneous appearance of the gaps in the cores at the different axial positions. This could be related to the fact that the development of the global (correlated) m=0 mode of instability, which is seen from about this time, has a longer wavelength. For W wire arrays the disappearance of wire cores is also seen at the same 80% of the implosion time (Fig.1).

The existence of long-living wire cores in wire array z pinches leads to a significant deviation of the implosion trajectory [1] from a 0-D trajectory. The stationary position of the wire cores until their breakage imply that the current up to this time flows mainly in the low-density coronal plasma immediately around the cores, providing ablation of the cores and acceleration of the coronal plasma towards the array axis. For this situation a simple phenomenological analysis [2] provides an estimate of the ablation rate per unit length of an array:

$$\frac{dm}{dt} = -\frac{\mu_0 I^2}{4\pi R_0 V} \qquad (1)$$

The continuous ablation of wires leads to radial redistribution of mass and the above expression for the ablation rate allows calculate the density profile. For 16mm diameter, 32 (15µm) Al wire arrays in our experiments, at the start of the implosion (at ~80% of implosion time) about 17% of mass is in the precursor column on axis, ~23% is in-flight between the precursor and the initial array radius R_0, and the remaining ~60% is in the wire cores.

THE SNOWPLOW PHASE OF WIRE ARRAY IMPLOSION

The formation of gaps in the wire cores triggers a new phase in the dynamics of the wire array implosion, in which the imploding current sheath (piston) accretes the plasma previously injected into the interior of the array. The initial mass in the piston could be estimated from the measured implosion trajectory [2]. Fig.2 shows the trajectories of snowplow implosion of the radial mass distribution calculated using ablation rate (1) for the different initial masses in the piston, together with the

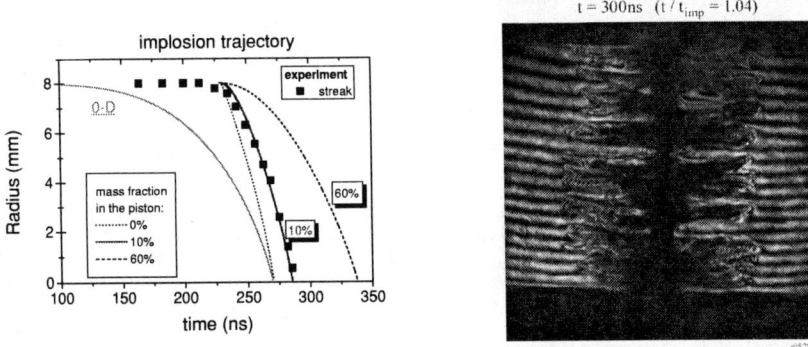

FIGURE 2. Implosion trajectory and laser probing image of 32 (15µm) Al wire array.

experimental data for a 32 (15µm) Al wire array. The starting time for the calculation corresponds to 80% of the experimentally measured implosion time. It is seen that the assumption that all the mass remaining in the wire cores (~60% of the initial array mass) participates in the implosion, does not agree with the experimental data, giving too late an implosion time (~335ns, instead of 285ns seen in the experiment). The best fit to the experimental data was obtained assuming that ~1/6 of the mass left in the cores (or ~10% of the initial array mass) participates in the implosion. This analysis suggests that a significant amount of the array mass (~30-50%) is left behind the

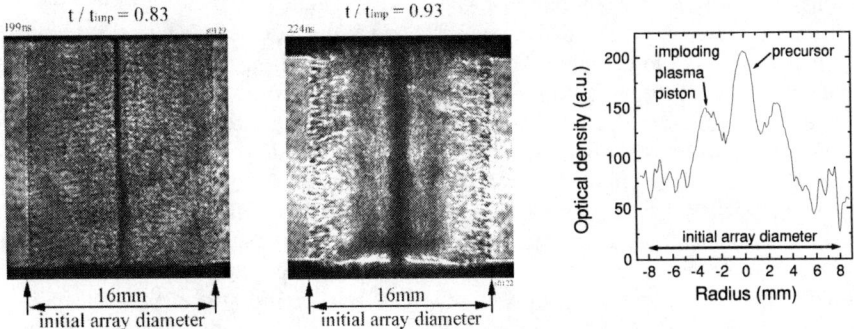

FIGURE 3. Laser probing images showing formation of imploding plasma sheath in W wire array.

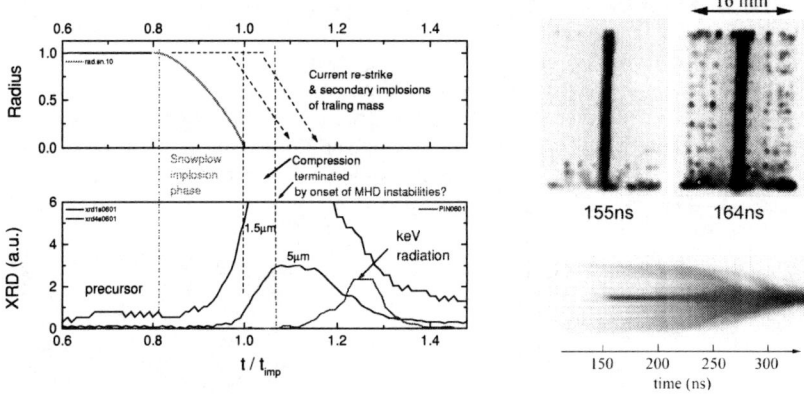

FIGURE 2. Relative timing of the implosion trajectory and the X-ray pulse, soft x-ray images at the time of gaps formation, and radial optical streak showing secondary implosions.

imploding piston. This is consistent with the results of laser probing, which shows (Fig.2) that the plasma remains at the initial diameter of the array even after the implosion.

For tungsten wire arrays the final snowplough phase of the implosion occurs similarly to that observed for aluminium. Formation of the imploding plasma piston in 32 (4µm) W wire array is seen in the side-on laser probing image (Fig.3). In addition to the precursor on axis, a shell-like structure is seen at ~3-4mm radius. The radial position of this shell-like structure is in good agreement with the implosion trajectory of radiating plasma measured by the optical radial streak photography. The image clearly shows that there is a significant difference between the plasma structures in the outer and the inner halves of the array radius. In the outer region, for r ≥ 4mm, the characteristic axial spatial scale of the non-uniformities is relatively large and the radially elongated features can be followed from the initial array radius (8mm) over the entire outer region. In contrast, the plasma in the inner region (r ≤ 4mm) has non-uniformities on a much smaller spatial scale and without detectable global structure.

It is instructive to compare the dynamics of the final phase of the array implosion with the shape of the x-ray pulse (Fig.4). At early time, from t/t_{imp} ~0.5 until ~0.8, the signal from the x-ray diode (XRD) detector filtered with 1.5µm polycarbonate shows emission from the precursor. The increase of the emission after t/t_{imp} ~0.8 coincides with the start of the final implosion phase and is related to the emission from the imploding piston. The final sharp rise of the soft x-ray emission, seen on XRD filtered with 5µm polycarbonate at t/t_{imp} ~0.98 apparently corresponds to the time when the piston reaches the radius of the precursor plasma column on the array axis. The rise-time of the x-ray pulse could thus be related to the compression of this plasma by the current sheath. The peak of the soft X-ray signal coincides with the start of the harder (>1.5keV) radiation pulse, measured using by *PIN* diode installed in a crystal (Mica) spectrometer. This could indicate that the compression phase terminates by the development of the MHD instabilities in the plasma column but more studies, both experimental and theoretical, are clearly needed here. Experimental data indicate that during the final implosion phase and the compression phase, the current re-strikes through the material left behind by the implosion. This leads to formation of the correlated "bright spots" seen on soft x-ray images [3], and also causes the secondary implosions seen in optical streak images (Fig. 4). These secondary implosions could act as a mechanism of additional energy deposition into the radiating plasma column, which increases the X-ray yield in excess of the 0-D kinetic energy associated with the first implosion. The stagnation/compression phase is also accompanied by the generation of an electron beam, which was measured by the collector inserted in the anode. For the implosion of 16mm diameter, 16 (15µm) Al wire array the e-beam current of >>2kA was measured behind 25µm thick Ti foil (cut-off energy ~70keV).

DISCUSSION

The experimental data show that due to the formation of a core-corona plasma, the implosion in a wire array z pinch occurs as a two-stage process. During the first stage the precursor flow of the coronal plasma from the wire cores provides a gradual redistribution of a significant fraction of the array mass. The wire cores remain stationary at their initial positions for ~80% of the implosion time, until the formation of gaps in the wire cores occurring due to the non-uniformity of ablation rate along the wires. From this moment the final stage of the implosion starts, which occurs as a fast snowplough-like implosion of the precursor plasma driven by a piston, which contains initially a relatively small fraction of the array mass. Stabilization of the snowplough implosion phase by the density profile peaked on axis [4,5] could be a key factor responsible for the remarkable performance of wire array z-pinches. The x-ray power generated during the snowplough implosion phase could be quite considerable (a few TW for Z conditions) and might account for the early preheat of the hohlraums observed e.g. in Ref. [6].

The mechanism determining the parameters of the main radiation pulse clearly requires further study. The present work suggests that the development of the R-T instability during the implosion phase could be less important than it is usually thought. Some other effects, such as non-simultaneous ablation of the wires (both axially and azimuthally) and the development of MHD instabilities during the plasma compression on axis could be at least of equal importance. Variations in the ablation rate of the wires, both statistical (ablation physics) and systematic (geometry of the magnetic field) could affect the symmetry of the plasma distribution in the array interior and the symmetry of the formed current sheath. Another important issue is a role of the material left behind by the imploding current sheath, which is seen at the initial array radius even at the time of stagnation. It is possible that some fraction of the current reconnects through the trailing mass, thus reducing a force available to compress the plasma and limiting the X-ray power. The secondary implosions of the trailing mass, driven by the current re-striking through this material, are indeed observed in the experiments. These secondary implosions could contribute to the total energy balance and might be responsible for the yield of the X-ray pulse generated in wire array z-pinch implosions being higher than the 0-D kinetic energy [7].

The conclusion that could be made from the presented experimental data and the analysis is that the 3-D effects and formation of precursor plasma flow needs to be taken into account in the optimisation of z-pinch loads. An extremely important outstanding issue, of course, is the scaling of the present analysis, based on the experimental data obtained at the 1MA current level, to larger currents. The recent experiments [8] at Z facility at SNL show a deviation of the implosion trajectory from the 0-D and an early arrival of the precursor plasma on array axis, which suggests a qualitative similarity to the implosion dynamics discussed in the present paper.

ACKNOWLEDGEMENTS

This work was supported by Sandia National Laboratories, Albuquerque, NM (Contract No. BF6405) and by the U.S. Department of Energy under Contract No. DF-FG03-98DP00217.

REFERENCES

1. S.V. Lebedev et al., Phys. Rev. Letters **85**, 98 (2000).
2. S.V. Lebedev et al, Phys. Plasmas **8**, 3734 (2001).
3. S.V. Lebedev et al, Phys. Rev. Letters **81**, 4152 (1998).
4. A.L. Velikovich, F.L. Cochran, J. Davis, Phys. Rev. Letters **77**, 853 (1996).
5. J.H. Hammer et al, Phys. Plasmas **3**, 2063 (1996)
6. M.E. Cuneo et al, Phys. Plasmas, **8**, 2257 (2001).
7. C. Deneey et al, Phys. Plasmas, **6**, 3576 (1999).
8. M.E. Cuneo et al, paper TU-O1-3I in this Proceedengs.

Ablation Rate of Wire Cores in Wire Array Z-Pinch Experiments

S.V. Lebedev, D.J. Ampleford, S.N. Bland, J.P. Chittenden, M.G. Haines

The Blackett Laboratory, Imperial College, London SW7 2BW, UK

Abstract. An effect of the global magnetic field on the ablation rate of wire cores in wire array Z-pinch experiments was studied using wire arrays with fixed wire number but different array diameter. The data suggest that the ablation rate is higher for the arrays with smaller diameter, when the global magnetic field is larger. A model for the scaling of the ablation rate with array parameters is also presented.

INTRODUCTION

For single wire z-pinch explosions both experiments [1,2] and cold-start simulations [3,4] clearly show the formation of a heterogeneous plasma structure with a dense cold wire core, surrounded by a low density coronal plasma. In wire arrays evidence of a similar core-corona structure has been detected by laser probing in experiments at different levels of current between 1MA and 7MA [5,6,7]. However, the formation of plasma in a wire array is expected to be significantly different from single wire z-pinches due to the net **JxB** force directed radially inwards, which blows the coronal plasma from the wires to the array axis. This may strongly affect the rate of ablation of the wire cores and the character of perturbations in the wires. The radial redistribution of mass by the precursor plasma flow during the early stages plays an important role in the subsequent dynamics of the implosion. The measurements of the implosion trajectories in wire array experiments show that the implosion starts with a delay in comparison with expected form the 0-D analysis. Experimental data [8] obtained in experiments at 1 MA level show that the implosion starts only when about half of the initial array mass has been ablated, after formation of gaps in wire cores occurring due to axial non-uniformity of the ablation.

A simple phenomenological analysis described in Ref. [8] provides an estimate of the ablation rate of the wires in an array. It assumes that the current is concentrated in the small region just around the wire cores, thus providing continuous ablation of material from the cores, conversion into plasma and acceleration of this plasma to the array axis by the **JxB** force. It also assumes that the plasma is leaving this region virtually without current, and moves towards the array axis force-free with a constant velocity V. This velocity is used as a parameter in the model. For the configuration with stationary wire cores and the flow of coronal plasma, the rate of mass removal from the cores (per unit length) could be written as a condition for the momentum balance:

$$V\frac{dm}{dt} = -\frac{\mu_0 I^2}{4\pi R_0} \qquad (1)$$

The left hand-side of equation (1) is the derivative of the momentum in the flow of plasma leaving the current-carrying layer, which is required to allow the existence of configuration with stationary wire cores. For our experiments velocity of the coronal plasma was measured by end-on laser probing [7], and the ablated rate calculated from this model is in agreement with results from 2-D (x-y plane) magneto-hydrodynamic (MHD) computer simulations, using the model described in Ref. [9]. To estimate the ablation rate in wire arrays at different experimental conditions using formula (1), the knowledge of the scaling of the effective velocity with array diameter and current is required.

EFFECT OF GLOBAL MAGNETIC FIELD ON THE ABLATION RATE

The experimental data and analysis presented in Ref. [8] show that the wire cores remain at the initial positions until the time when about half of the array mass is ablated from the cores. This observation was used to infer the scaling of the ablation rate with array radius, and thus with the magnitude of the global magnetic field. Experiments were performed with wire arrays of three different diameters (8mm, 16mm and 36mm), but with fixed wire number (32) and wire diameter (15µm Al). Due to the high impedance of the MAGPIE generator the current pulse-shape was the same for all three configurations.

Fig.1 shows the implosion trajectories measured from radial optical streak images. It is seen that the implosion of 8mm and 16mm diameter arrays starts at different time, despite the same current per wire, and no implosion occurs for 36mm diameter array. This is in qualitative agreement with formula (1), which suggests 1/r scaling of

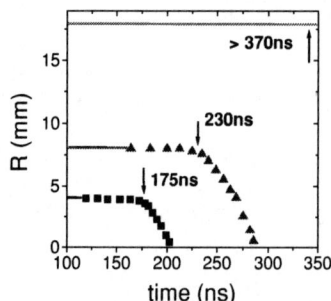
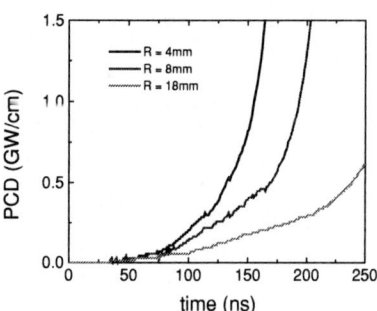

Figure 1. Implosion trajectories (left) and radiation power for 32 (15µm) Al wire arrays of different diameter measured by open PCD detectors.

ablation rate if velocity of the coronal plasma flow is independent on the array diameter and is constant in time. Assumption that the ablation rate is proportional to I^2 as in (1), and that at the start of implosion the same mass fraction was ablated in both 8mm and 16mm diameter arrays, indicates that the ablation rate is a factor of ~2.4 higher for the 8mm diameter array.

It is instructive to compare radiation from the arrays of different diameters, measured during the early stages of the discharge. Fig. 1 shows signals of open PCD detectors for the 8, 16 and 36mm diameter arrays with 32 (15µm) Al wires. It is seen that for all arrays the radiation starts at ~50ns after the current start. This time is the same as the "dwell" time [7], the time when the first coronal plasma is detected in our experiments by the laser probing. The radiation is higher for smaller diameter arrays, and the data suggest that the scaling of the total radiation power with radius is similar to the scaling of the ablation rate given by formula (1).

An estimate of the energy radiated per ablated ion was found from the ratio of the measured radiation power to the rate of ablation (1). For all three different array diameters this gave the same (and constant in time) energy per ablated ion, equal to ~200eV for Al (assuming ablation velocity $V=1.5 \times 10^7$ cm/s being constant in time and the same for the three array diameters). Similar measurements for 16mm diameter W wire arrays show that the total radiation power is a factor of ~3 higher than for Al arrays, which due to the higher atomic weight gives a higher estimate of the energy radiated per ablated ion, ~ 4keV. This result could be used to estimate the expected radiation power from wire arrays driven by a significantly higher current, e.g. on Z facility. If the scaling of the ablation rate (1) holds and the same energy is radiated per the ablated ion, the level of radiation from the ablating wires should be ~ 0.5TW at the time when the current through the 20mm diameter array is ~10MA.

The evidence that the ablation rate depends on the array diameter is also seen in experiments with conical wire arrays. Fig.2 shows laser probing image of 16 (20µm) Al conical wire array with 16mm smaller and 30mm larger diameters. It is seen that near the smaller diameter end of the array the wires are fully ablated over the length of ~8mm, while at a larger diameter the wires are still clearly seen at their initial positions.

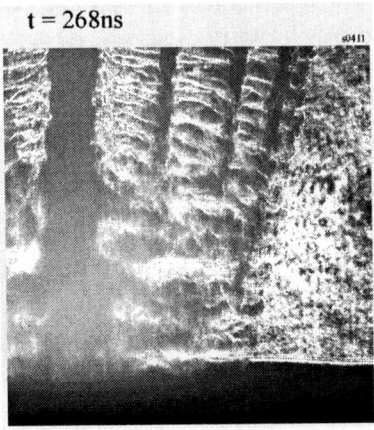

Figure 2. Laser probing of conical Al wire showing effect of array radius on ablation rate.

MODEL OF ABLATION OF WIRES IN A WIRE ARRAY

The expression for the ablation rate (1) requires velocity V of the coronal plasma flow as a parameter. Here we present an attempt to obtain a scaling for this effective velocity, and thus the ablation rate, with array diameter and current. We assume that the ablation is quasi-stationary. Let us consider array of radius R_0 with N wires and assume that all current is concentrated in the regions of radius δ around each wire core. Cold material is continuously ablated from the cores at a rate given by formula (1), and the ohmic heating (fraction $\alpha \sim 1$ of it) provides increase of the internal energy of this material to the equilibrium value [10] corresponding to temperature T (2). We will also assume that the size of the current-carying region is equal to the radiative cooling length (3), which is calculated using velocity V and cooling function L_Z taken from Ref. [11]. It is also assumed that at the boundary of the current-carrying region the magnetic Reynolds number Re_M is ~ 1 (4).

$$\alpha N \frac{(I/N)^2 \eta}{\pi \delta^2} = \frac{dm}{dt} E_{int}(T) \qquad (2)$$

$$\delta = V\tau \propto \frac{V(Z+1)T}{ZL_z(T)n_i} \qquad (3)$$

$$Re_M = \frac{V\delta}{D_M} \sim 1 \qquad (4)$$

Numerical solution of the system (1-4) allows finding an equilibrium temperature in the current-carrying region, which is a relatively weak function of current per wire. For 16mm diameter 32 Al wire array discussed previously, the temperature is in the range 15-20eV, $\delta \sim 1$mm and V$\sim 10^7$ cm/s, which is close to the experimentally observed. It is interesting to note that no solution exists for times less than $\sim$50ns, the time equal to the "dwell" time observed in the experiments. The model suggests the scaling for the ablation velocity $\propto (N/R_0)^{1/3}$, and for both the ablation rate and the total radiation power $\propto N^{-1/3}(R_0)^{-2/3}$. This scaling for the total radiation power (and the absolute value of it) is in close agreement with the experimental measurements of the radiation from Al wire arrays of the three different diameters discussed in the previous section. We should note, however, that despite the rather reasonable agreement of the model with the experimental data, justification of the assumptions used in this model is clearly required.

This work was supported by Sandia National Laboratories, Albuquerque, NM and by the U.S. Department of Energy under Contract No. DF-FG03-98DP00217.

REFERENCES

1 D. Kalantar and D. Hammer, Phys. Rev. Lett. **71** 3806 (1993).
2 S.A. Pikuz et al, Phys. Rev. Lett. **83**, 4313 (1999).
3 J.P. Chittenden et al, Phys. Rev. E **61**, 4370 (2000).
4 I.R. Lindemuth, Phys. Rev. Lett, **65**, 179 (1990).
5 I.K. Aivazov et al, Sov. J. Plasma Phys. **14**, 110 (1988).
6 C. Deeney et al, Rev. Sci. Instr. **68**, 653 (1997).
7 S.V. Lebedev et al, Phys. Plasmas **6**, 2016 (1999).
8 S.V. Lebedev et al, Phys. Plasmas 8, 3734 (2001).
9 J.P. Chittenden et al, Phys. Plasmas **8**, 2305 (2001).
10 D. Mosher et al, IEEE Trans. Plasma Sci, **26**, 1052 (1998).
11 D.E. Post et al, At. Data Nucl. Data Tables, **20**, 398 (1977).

Implosion Dynamics and X-ray Characteristics of Nested Wire Array Z-pinches

S.N.Bland, S.V.Lebedev, J.P.Chittenden & M.G.Haines

Blackett Labs, Imperial College, Prince Consort Rd, London, SW7 2BW, UK

Abstract. The results of nested wire array Z-pinch experiments dominated by a switching of current from the outer to the inner array are presented. The various stages of X-ray emission are related to the dynamics of the outer and inner arrays. The dependence of the X-ray pulse shape on the diameter of the inner array was also explored.

INTRODUCTION

One important advance in wire array Z-pinch research has been the use of 'nested' arrays. This produced a 40% increase in the maximum X-ray power obtained at implosion of an array, and a corresponding decrease in the rise-time of the X-ray pulse [1]. Plasma formation and dynamics in nested wire arrays, however, remains largely undiagnosed, especially on high current machines, where access to the pinch is usually restricted. Two different models of implosion have been suggested. In the first the wires of the outer and inner arrays both become plasma. During implosion the outer array collides with the inner array, and the resultant accumulation of mass reduces the level of R-T instabilities, which, in this model, determine the rise-time of the X-ray pulse. In the second model, the wires of the inner array remain as small discrete bodies, shielded from current by the inductance of the outer array. The outer array implodes through the gaps of the inner array, a fast inductive switch of current occurs from the outer to the inner, and the inner rapidly accelerates towards the axis [2-3].

Our previous experiments [4] produced implosions consistent with the current switching model by using an elongated inner array to suppress current flow. This paper presents further research on this type of nested array.

EXPERIMENTAL SET-UP

Experiments were carried out on the MAGPIE generator (1MA, 240ns 100% rise time). Diagnostics included side and end-on laser probing, an optical streak cameras set to sample across the radius of the array, side and end-on X-ray framing cameras, XUV and hard X-ray spectrometry, differentially filtered arrays of XRDs and diamond PCDs and an X-ray radiography system. Full details can be obtained using reference 5.

Nested array configurations typically used an outer array of 16 15µm Al wires on a 16mm diameter with a length of 23mm. This was identical to a single array configuration used in reference 5, providing comparison. Initially inner arrays consisted of 16 15µm Al wires on an 8mm diameter with a length of 83mm.

RESULTS AND DISCUSSION

Plasma formation

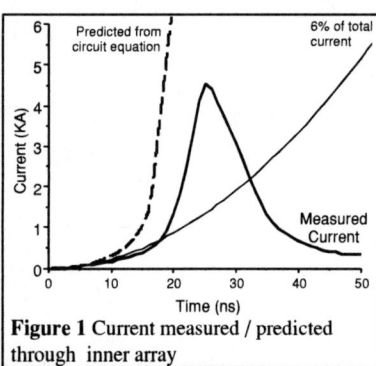

Figure 1 Current measured / predicted through inner array

The level of plasma formation at the wires of the outer and inner arrays is determined by the division of current between them. Fig. 1 shows the current that flowed through the inner array measured during the experiment by a pick up coil. Initially division of current was dominated by inductance and only ~6% flowed through the inner array. At ~15ns there was a sudden increase in the fraction of current flowing through the inner array. This corresponded to the time when the wires of the outer array went through solid-liquid and liquid-gas phase transitions, the substantial increase in resistivity shunting current to the inner array. At ~25ns current started to decrease, reaching zero by ~50ns. This was due to the wires of the outer array expanding and breaking down to form plasma, lowering the resistance of the outer array whilst increasing the inductive shielding of the inner array. The wires of the inner array, meanwhile, were expected to be ohmically heated to ~650K (below melting point).

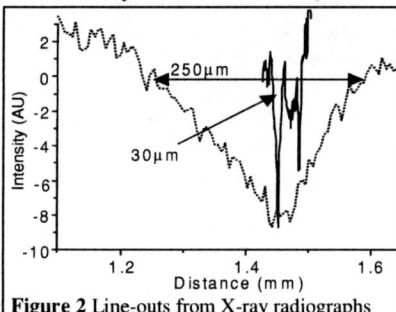

Figure 2 Line-outs from X-ray radiographs showing cores of outer (dotted) and inner array

Until ~80ns, current through the inner array could be monitored directly by the pick up coil and remained ~ zero. After 80ns, the presence of plasma between the outer and inner arrays made this measurement unreliable. Side-on laser probing and X-ray radiography (fig. 2) measurements showed that the wires of the inner array remained as stationary, small, discrete bodies of <30μm diameter until at least 221ns. Comparison of the effect of ohmic heating of a wire with the observed lack of expansion suggested that less than 1% of the current flowed through the inner array until this point.

The outer array, meanwhile, acted identically to a single array of the same wire number, mass and diameter. The wires of the outer array formed core-corona systems with cold, dense wire cores (size 250μm, the same as in a single array) being continually ablated to provide low density coronal plasma. The coronal streams were accelerated from around each core by the $\mathbf{J} \wedge \mathbf{B}$ force and then flowed in streams towards the axis of the array. As the streams from the outer array passed the position of the inner array, some interaction of the streams with the wires of the inner array was observable as emission on the side-on X-ray framing camera. However the effect of momentum transfer from the streams to the inner array was negligible due to high transparency of the inner array (estimated as ~98%). When the streams reached the axis of the array they collide forming a precursor plasma column at the same time as in a single array experiment (~120ns). Emission from the precursor was seen as a long, relatively low 'foot' on the 1.5μm polycarbonate filtered PCD channel.

Implosion of the Nested Array

All measurements of the implosion trajectories of the outer and inner arrays (fig. 3) were consistent with the current switching model. Initially the outer array continued to act identically to a single array. The onset of implosion was triggered by the formation of gaps in the wire cores of the outer array due to the ablation of coronal plasma not being axially symmetric. Current switched from acting close to the cores to the plasma that had pre-filled the array, accreting this in a snowplough implosion. It should be noted that the implosion trajectory of the outer was very different to that expected from the implosion of a shell containing 100% of the arrays mass being accelerated by 100% of the current. In particular, the outer array started to implode relatively late (195ns), but accelerated at a much higher rate. The implosion also left a large fraction of the arrays mass at the original wire positions – estimated as ~ 20% using interferometry. Between 195ns and 225ns, when the outer array reached the inner array, emission measured by the 1.5μm polycarbonate filtered PCD increased at the same rate as in single array experiments, consistent with the non-elastic accumulation of plasma expected during the snowplough.

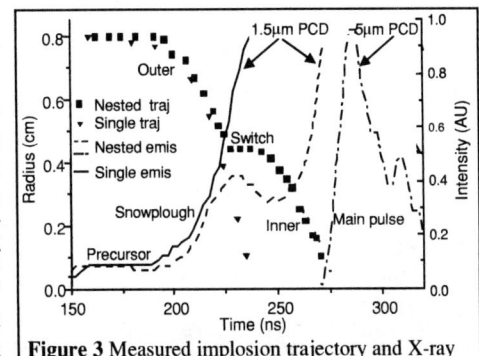

Figure 3 Measured implosion trajectory and X-ray emission from nested compared with a single array.

As the outer array imploded towards the inner, the wires of the inner array remained as small, discreet, stationary bodies. Immediately after the implosion of the outer array had reached the inner array, the inner array remained stationary, indicating that no momentum transfer had occurred. The inner array then accelerated along an implosion trajectory corresponding to a shell with 100% of the arrays mass being accelerated by 100% of the current. This was very different to the implosion trajectory observed for the outer array, where the pre-injection of mass dominated dynamics. Side-on laser probing images appeared to show little evidence of mass left at the radius of the inner array compared with the outer.

After the implosion of the outer had reached the inner array, emission measured by 1.5μm polycarbonate filtered PCD started to decrease. This was also consistent with current switching from the outer array to the inner array, which would remove the force accelerating the snowplough, leaving it to be decelerated by further mass accumulation. Emission started to increase again as the inner array accelerated towards the axis, and at 270ns, the main X-ray pulse was observed as the inner array stagnated.

The rise time of the main pulse was much shorter than in the case of a single array of the same configuration as the outer (14ns c.f. 33ns). The exact reason for this remains unclear. It could be related to the fast implosion of the inner array reducing the growth time of any instabilities, or to the different plasma density profiles that the inner array would encounter during implosion (due to the 'pre-implosion' of the outer) or possibly be related to the more symmetric delivery of 100% of the mass and current to axis. After peak the X-ray pulse decreased relatively slowly, likely due to un-imploded mass at the outer array being accelerated to the axis.

Inner Array Diameter and the Shape of the X-ray Pulse

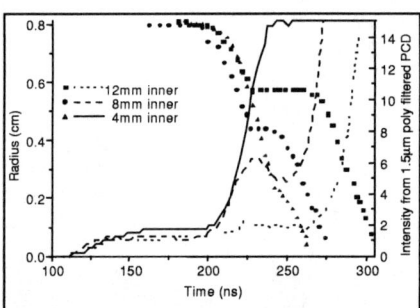

Figure 4 Implosion trajectories and X-ray emission for 4, 8 and 12mm inner arrays

Varying the parameters of the outer and inner arrays, we can attempt to control the relative timing of each implosion, adjusting the shape of the resultant X-ray pulse. Experiments were made using inner arrays of 4, 8 and 12mm diameter. The number of wire and mass of the inner array remained the same. In all cases the outer array again acted identically to a single array until switching current to the inner array. As the diameter of the inner array was reduced, the outer array imploded over a longer period before reaching the inner array and there was a corresponding increase in the level of snowplough emission (fig. 4). The time over which the inner array imploded also decreased with reduced diameter, shortening the time delay between the start of snowplough emission from the outer and the main X-ray pulse from 84ns to 63ns. Differentially filtered PCDs showed that the hardness of both the snowplough emission and the main X-ray pulse increased with decreasing inner diameter, indicating a possible increase in temperature.

SUMMARY

Suppressing the current to the inner array of a nested configuration resulted in implosions consistent with the current switching model. X-ray emission consisted of 3 parts: a long foot pulse associated with the formation of a precursor plasma on axis; a small peak due to emission from the snowplough implosion of the outer array; and the main pulse at stagnation of the inner array. Reducing the diameter of the inner array increased the length of time over which the snowplough of the outer array accelerated, increasing the corresponding emission. It also decreased the implosion time of the inner, reducing the time delay between the snowplough emission and the main pulse.

This work is supported by Sandia National Labs and the Department of Energy.

REFERENCES

1. C. Deeney et al, *Phys. Rev. Letters* **81**, 4883 (1999).
2. J. Davis et al, *Appl. Phys. Letters* **70**, 170 (1997).
3. R. E. Terry et al, *Phys. Rev. Letters* **83**, 4305 (1999).
4. S.V. Lebedev et al, *Phys. Rev. Letters* **84**, 1708 (2000).
5. S.V. Lebedev et al, Phys. Plasmas **8**, 3734 (2001) and references therein.

The Effect of Precursor Plasma Flow on Foam Targets in Wire Array Z-Pinch Experiments.

James B.A.Palmer[a], Sergey V.Lebedev[b], Simon N.Bland[b], Jeremy P.Chittenden[b], David J. Ampleford[b]

[a]*AWE Plc., Aldermaston, Berks, RG7 4PR, UK.*
[b]*The Blackett Laboratory, Imperial College, London, UK.*

Abstract. Previous experiments have demonstrated that the slow ablation rate of material from wire arrays results in the form[1]ation of a precursor plasma stream bombarding the axis [1]. This could have major repercussions for the centrally located foam targets used in dynamic and static walled hohlraum configurations on the Z facility at Sandia National Laboratory (SNL) [2]. Experiments to characterise the effect of precursor plasma flow on foam targets were carried out on the MAGPIE generator at Imperial College. The TPX foam used is similar in size and density to foam used in the experiments at SNL. Diagnostics included: x-pinch backlighter; x-ray framing cameras; diamond PCDs; laser shadowgraphy and interferometry; optical streak photography. Backlighter results suggested that the foam was compressed at a rate consistent with experimental estimates of the momentum of the bombarding plasma streams. Laser probing images, however, showed expansion of low density plasma from the foam surface that exhibited structure similar to an m=0 instability. Side-on XUV and x-ray imaging showed axially modulated emission from the foam.

INTRODUCTION

Foam targets are placed on the axis of wire array z-pinches on the Z machine pulse power facility, at SNL, to give the dynamic hohlraum (DH) configuration [2,3]. The detailed behaviour of this system is not well reproduced by simulation [4]. The effect of bombardment by precursor plasma streaming from the wire array onto the foam is not fully understood and is not included in the modeling of the DH performance. This paper describes results from experiments carried out on the MAGPIE generator at Imperial College to investigate how the precursor plasma affects on-axis foams.

EXPERIMENTAL SET-UP

Wire arrays consisted of 16 or 32 wires of tungsten on a 16mm diameter with a length of 23mm. All arrays were over-massed to prevent implosion and supply a constant stream of precursor plasma. The current pulse was the same in every experiment with a peak at 1 MA and a 240ns full rise time. The TPX (CH_2) foams were ~1.5mm diameter, ~20 mm long and 10-30 mg/cc (foams were supplied by SNL). The data discussed in this paper came from optical streak photography (1 streak/shot), X-pinch driven radiography (1 image/shot), laser shadowgraphy and interferometry (2 images/shot) and side on XUV and x-ray pinhole imaging (4 images/shot).

---

[1] ©British Crown Copyright 2002/MOD

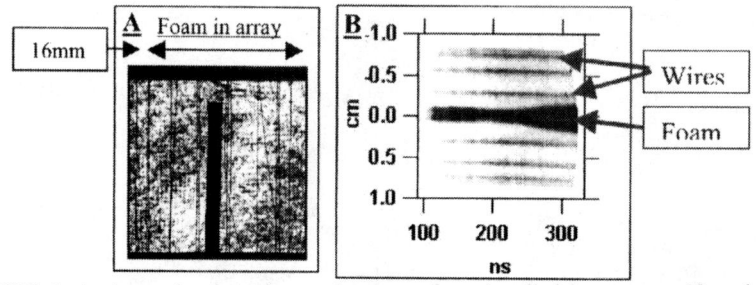

FIGURE 1. A. A pre-shot laser image showing a foam installed in an array. Note that the foam is shorter than the array and that a precursor plasma column formed in the space above it. **B.** Streaked optical image, time runs left to right. Emission is seen from the wires that remain stationary and the foam that is first compressed (from ~170 ns) and then expands (~240 ns).

OPTICAL STREAK PHOTOGRAPHY

A horizontal slit imaged a radial slice of the array onto a streak camera and so the radial evolution of the array could be recorded. Figure 1.B shows some typical data from the streak camera. The wires are seen to remain stationary, there was no implosion because the array was over-massed. Emission from the foam is also seen and shows compression then expansion of the foam.

X-PINCH RADIOGRAPHY

An Al x-pinch replaced one of the four current return posts and provided a point source of x-rays for radiography of the foam [5]. Images were recorded to DEF x-ray film behind a 12.5 µm Ti filter (transmission window 3-5 KeV). Over the 3 to 5 KeV window of the filter the absorption of the foam and tungsten were very similar and so any features on the radiographs could not be identified with certainty as being either foam or tungsten. However a cylindrical feature, which we assume is the foam, could be seen on axis and its diameter measured. The diameter of the foam vs. time is shown in Figure 2 where it is seen to decrease with time. This agrees reasonably well with the predicted compression of the foam given by Lebedevs precursor model [1]. Each data point is from a separate experiment. The radius at various positions along the foam axis, in a single radiograph, were measured and averaged. Low contrast in some radiographs led to large variations in the measured radius and this is indicated by the error bars. Some periodic axial structure is just visible in some radiographs and may be related to the axial periodic structure seen in the precursor streams.

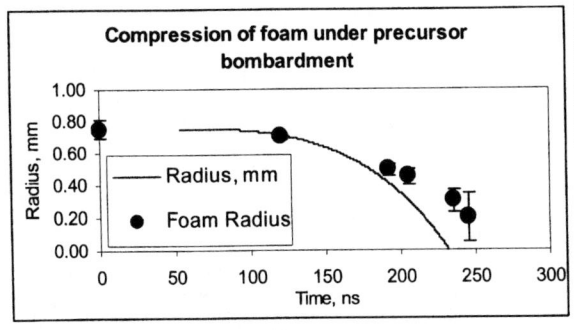 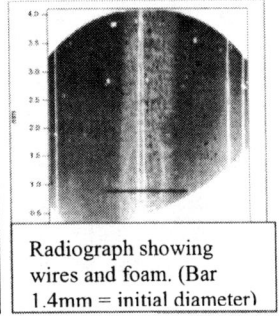

Radiograph showing wires and foam. (Bar 1.4mm = initial diameter)

FIGURE 2. Plot shows compression of foam seen in radiographs. Error bars indicate the uncertainty of measurements and the solid line shows predicted compression of foam according to the model in Ref. [1]. Radiograph shown was taken at 192ns.

LASER IMAGING

Laser imaging (shadowgraphy and interferometry) showed the location of low-density plasma. Pre-shot and shot images allowed the evolution of precursor streams, foam and wires to be recorded. The data showed uniform expansion of plasma from the foam surface at 129ns. By 175 ns the plasma had developed a structure similar to an m=0 instability and by 316ns this modulation was no longer seen. At all time the diameter of the plasma column measured by the laser probing is relatively small, presumably expansion is suppressed by the kinetic pressure of the incoming flow of the precursor plasma. The wavelengths of the ripples on the plasma surfaced were measured by taking the peak-to-peak and trough-to-trough distances. The mean wavelength vs. time is plotted in Figure 3 and is seen to increase. The mean wavelength of ~ 1.7 mm at later time is very similar to the wavelength of the global instability seen to form at late time in imploding arrays on MAGPIE [1]. A curious relationship is seen when the ratio of the mean wavelength and the *initial* foam diameter (which was found to vary by up to 240 microns between foams) is plotted, as this ratio increases linearly with time.

This data is in contrast to the radiographic data, which shows uniform compression of the foam but laser imaging shows non-uniform expansion of plasma from the foam during precursor bombardment. Though this is not necessarily inconsistent as the laser imaging is probing a much lower plasma density ($n_e \sim 10^{19}$ cm^{-3}). When no foam is present a precursor plasma column forms on the array axis and this column is not axially modulated. It is a uniform and narrow, ~2mm, column.

It is worth noting that the plasma seen to rise from the base of the array comes from the brass foam holder that occasionally projected up into the path of the precursor plasma streams. The brass ablated and moved axially up and was seen to have an effect in the side-on pinhole imaging, discussed next.

SIDE-ON XUV AND X-RAY PINHOLE IMAGING

A side-on four-frame pinhole-framing camera viewed a large portion of the array. With no filtering in place it imaged from XUV to x-ray and with filters of a few microns of Macrofoil only x-rays (~200 eV). In all images the emission from the foam is axially modulated. This is in contrast to the uniform emission seen from the precursor column formed in the absence of a central foam. The wavelength of this modulation was measured in the same way as with the laser images and was found to be similar to the modulation seen in the laser images.

A drop in emission was seen at the base of the foam when the laser imaging showed ablation of the brass foam holder, as discussed earlier. It is reasonably assumed that this plasma acted as a filter attenuating the foam emission, and also could effect the plasma flow impacting the foam.

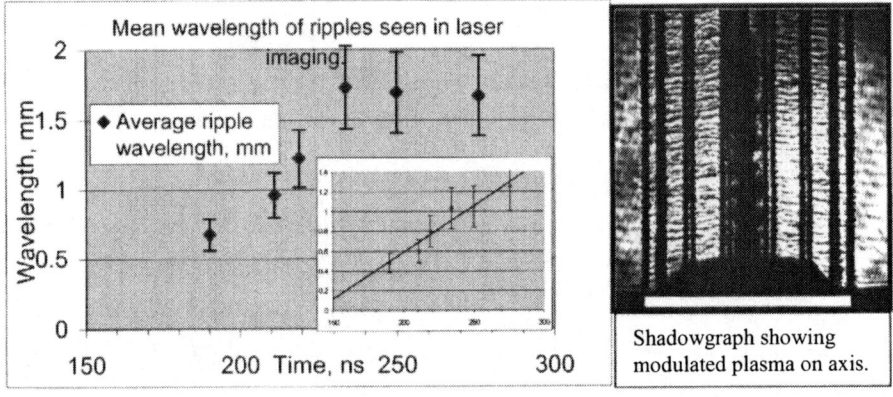

FIGURE 3. Axial modulation of low density plasma shown in shadowgraph on right, white horizontal bar is 16 mm. Main graph shows increase of modulation wavelength with time. Inset shows linear increase of ratio between the wavelength and the initial foam diameter.

SUMMARY

The experiments carried out on MAGPIE have shown that the precursor plasma does have a significant affect on foam targets placed on the array axis. Radiography showed compression of the foam by the plasma flow, in reasonable agreement with the model described in Ref. [1]. Laser imaging showed low density plasma expanding from the foam and developing axial modulation similar to an m=0 instability. XUV and x-ray imaging showed axial modulation of emission from the foam and the wavelength of this emission was similar to that seen by laser imaging. Further work is planned, to continue analysis of data collected and to perform more experiments.

ACKNOWLEDGMENTS

This work is supported by SNL, DoE. Also, Chittenden is supported by a William Penny Fellowship from AWE Plc. We thank SNL for supplying the foams.

REFERENCES
1. Lebedev, S.V., et al., *Phys. Plasmas,* **9**, 2293-2301 (2002).
2. Nash, T.J., et al., *Phys. Plasmas,* **6**, 2023-2029 (1999).
3. R.E. Olson, et al., *Fusion Technol.* **35**, 260 (1999).
4. Peterson, D.L., et al, this conference proceedings.
5. Lebedev, S.V., et al., *Rev. Sci. Inst.,* **Vol 72 No1**, 671-673 (2001)

The Effect of Array Configuration on Current Distribution in a Wire Array Z-Pinch

S.N.Bland, S.V.Lebedev, F.Beg, H.Kwek[a], J.P.Chittenden, & M.G.Haines

Blackett Laboratory, Imperial College, London, SW7 2BW, UK
(a) Physics Dept, University of Malaysia, Kuala Lumpur 50603, Malaysia

Abstract. The results of experiments that changed the current distribution inside the radius of an array, or between the wires of an array, are presented. Use of a low number of wires (4), Ni wires, or the introduction of a pre-pulse current to the array all resulted in the pre-cursor plasma on axis displaying MHD like instabilities. Implosion dynamics were affected in a consistent manner, with the arrays imploding later than expected. In arrays made from alternating Al and W wires, current was initially concentrated around the Al wires. When the Al wires started to implode, current was switched to the W wires, producing a 2 stage implosion.

INTRODUCTION

Recent experiments at various levels of drive current (1MA-20MA [1-4]) have shown that the dynamics of a wire array Z-pinch are dominated by 'discreet wire' effects. Experiments at Imperial College [1] demonstrated that the wires in an array initially form core-corona systems with stationary, cold dense wire cores being ablated into warm, low-density coronal plasma. The majority of current remains close to the wire cores, ablating them and accelerating the coronal plasma into streams that flow towards the axis of the array. At the axis the streams collide forming a narrow, stable precursor plasma column, which is probably inertially confined. Implosion of the array occurs when gaps form in the wire cores due to ablation occurring in an axially asymmetric way. Current then switches from acting close to the wire cores to the plasma pre-filling the array, accreting this in a snowplough implosion towards the axis. X-ray emission initially consists of a long duration, low intensity 'foot' associated with the formation of the precursor column. This is followed by a rise in emission due to the snowplough accreting mass in a non-elastic manner. Emission from the snowplough then merges with the 'main' X-ray pulse caused by the implosion stagnating on axis.

This paper describes experiments to explore the effect of array configuration on current distribution, inferred from the changes observed in the stability of the precursor plasma column on axis or the implosion dynamics.

EXPERIMENTAL SET UP

Experiments were carried out on the MAGPIE generator (1MA, 240ns 100% rise time). Diagnostics included side and end-on laser probing, an optical streak camera set to sample across the radius of the array, side and end-on X-ray framing cameras, differentially filtered XRDs and PCDs, XUV and hard X-ray spectrometers and an X-ray radiography system.

Arrays used consisted of 4-32 wires on a diameter of 16mm with a 23mm length.

RESULTS

Low Wire Number Arrays

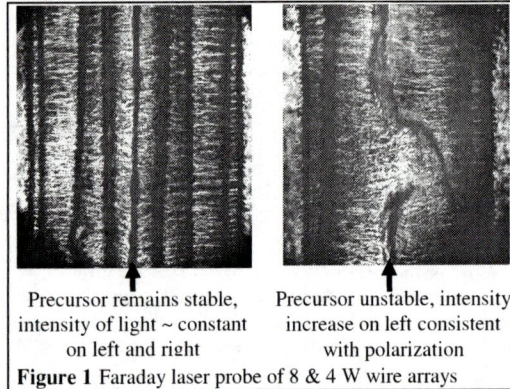

Precursor remains stable, intensity of light ~ constant on left and right

Precursor unstable, intensity increase on left consistent with polarization

Figure 1 Faraday laser probe of 8 & 4 W wire arrays

The precursor plasma in arrays in arrays made of 8 or more Al, Cu or W wires appeared to be a uniform column, which remained stable until implosion of the array. Estimates of the growth time of MHD instabilities [5] suggested that < 7% of the total current could flow through the precursor column of an 8 wire Al array (and likely much less). Faraday probing measurements of 8 wire W arrays estimated < 4% of the current flowed through the precursor column.

In arrays consisting of only 4 W wires, however, side-on laser probing images showed the precursor column be grossly m=1 unstable suggesting the presence of current. Faraday rotation implied >10% of the current would be necessary to obtain the change in polarization observed.

Ni Arrays

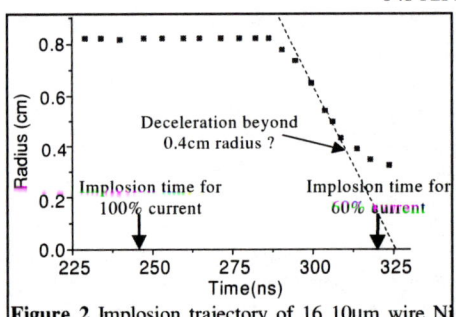

Figure 2 Implosion trajectory of 16 10μm wire Ni array and predictions of stagnation time based on Al/W/Cu array implosions

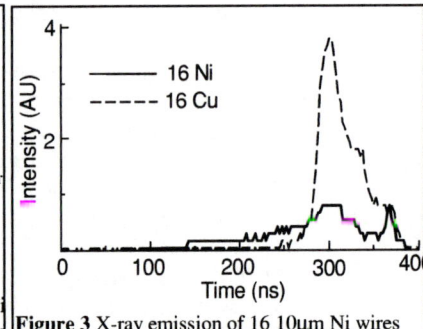

Figure 3 X-ray emission of 16 10μm Ni wires compared with 16 10μm Cu wires

In arrays made of 8 Ni wires the precursor plasma column formed was again seen to be unstable. The growth rate of the instability was 30-50ns, suggested ~20% of the current flowed through the precursor during the experiment.

The implosion trajectories of 8 and 16 wire Ni arrays were measured using the radial optical streak. The time at which the implosion of a Ni array started was significantly later than that expected for Al, W or Cu arrays of the same mass. This was consistent with some fraction of the current being injected into the precursor plasma, reducing that responsible for ablation of the wire cores. Estimates based on the observed trajectory of a 16 wire array (fig. 2) suggested that only 60% of the current was involved in the ablation and subsequent implosion process. The implosion trajectory of the array also appeared to slow as it approached the precursor column, possibly indicating the presence of a magnetic buffer region.

X-ray emission from Ni array was compared to that of a Cu array, which would be expected to have similar radiative properties ($\Delta Z = 1$) The initial emission from the precursor of a Ni array was higher, perhaps as a consequence of the presence of current. At stagnation, however, emission was relatively low (fig. 3). This poor performance was probably linked to the late implosion and low implosion velocity of the array, significantly reducing the available KE at stagnation.

Introduction of a Pre-Pulse Current

One further set of experiments where the distribution of current inside the array appeared to change involved the use of a 'pre-pulse' current, before the start of the main current drive. Arrays consisted of 32 15µm Al wires and the pre-pulse current was a 500ns long linear ramp with a peak of 18 or 35kA. The rate of rise of the main current drive was unaffected by use of the pre-pulse.

Laser probing immediately before the main drive showed no obvious expansion of the wires in the array. However, a very low density plasma column was observed on axis (electron line density ~5×10^{15} cm^{-1}). During the main pulse, the wires expanded to the same degree as in non-pre-pulsed experiments, and the formation of the precursor plasma occurred at about the same time on the radial streaks. From ~160ns, X-ray framing images showed the development of an instability along the precursor. The instability was accompanied by an increase in the X-ray emission measured by the 1.5µm polycarbonate filtered PCD. Estimates of the growth time of the instability suggested that > 20% of the current flowed through the precursor.

As observed in Ni experiments, the start of the implosion of the pre-pulsed array occurred relatively late, consistent with a reduction in the current close to the wires cores. The main X-ray pulse was also significantly reduced.

Mixed Arrays: Alternating Al and W Wires

The use of wires made from alternating materials was found to redistribute the current between the wires of the array, producing different ablation rates and sequential implosions. In arrays made from alternating Al and W wires, laser probing images showed the Al wires to expand and inject coronal plasma, whilst the W wires remained small and discreet. The precursor column was formed at the same time as in experiments with arrays made of only Al, significantly earlier than the time at which the precursor formed in arrays made from only W (120ns c.f. 180ns).

The implosion of a mixed array was a 2-stage process (fig. 4). The Al wires started to implode at the same time as in arrays made of only that number of Al wires. However the wires appeared to then follow a trajectory with little or no acceleration, resulting in a final velocity much lower then usual. The W wires started to implode relatively late compared with an array made of only W. They then accelerated at a much higher rate than usual (~1.4x). These observations are consistent with a switch in current from the aluminum to the W wires, once the Al has started to implode towards the arrays axis (100% switch at ~190ns).

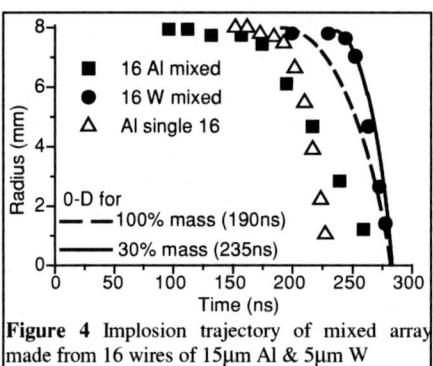

Figure 4 Implosion trajectory of mixed array made from 16 wires of 15μm Al & 5μm W

Discussion

The current distribution in an array is most likely determined by a combination of inductance and resistance. When the number of wires in the array was reduced to only 4, the resulting reduction in the inductive shielding of the precursor by the wires may explain the observed increase in current. In arrays made of Ni wires, the increase in current in the precursor was most likely related to the increase in resistivity of the wires (both at melting and boiling point) somehow affected the convection of current to the axis. The increase in current in the precursor observed during pre-pulsed experiments was probably due to the formation of the low density plasma column on axis, just before the start of the main current drive, which would have acted as a low resistance shunt. In mixed array experiments both inductance and resistance may have been important in suppressing current to the W wires, in a similar way to the current division between wires of the outer and inner arrays in a nested configuration [6].

SUMMARY

We have observed changes in the usual current distribution inside the radius of an array, or between the wires of an array. The use of low wire number arrays (4 wires), high resistivity wire (Ni) or the introduction of a pre-pulse current to the array were all seen to introduce current to the axis of the array long before implosion occurred. Alternately, using an array consisting of alternating wire materials, a 'mixed' array, resulted in the division of current between the wires of the array being changed. The change is current distribution was most likely caused by significantly differences in resistance and inductance compared with a typical array.

This work is supported by Sandia National Labs, the Department of Energy, and Maxwell Physics International.

REFERENCES

1. S. V. Lebedev et al, Phys. Plasmas **8**, 3734 (2001) and references therein.
2. V.P. Smirnov, Bull. Am. Phys. Soc. **45**, 61 (2000).
3. C. Deeney et al, Rev. Sci. Instrum. **68**, 653 (1997).
4. M.E.Cuneo, et al, Bull. Am. Phys. Soc. **46** P234 (2001).
5. N.R.Pereira et al, Jour. Appl. Phys. Vol55 P704 (1984).
6. S.N.Bland et al, DZP conference proceedings (2002).

Prolonged Plasma Production and Dynamics of Implosion of Multiwire Arrays

V. V. Alexandrov,[a] I. N. Frolov,[a] M. V. Fedulov,[a] E. V. Grabovskii,[a]
K. N. Mitrofanov,[a] S. L. Nedoseev,[a] G. M. Oleinik,[a] I. Yu. Porofeev,[a]
A. A. Samokhin,[a] P. V. Sasorov,[b,a] V. P. Smirnov,[a] G. S. Volkov,[a]
M. V. Zurin,[a] G. G. Zukakischvili,[a] K. Struve [c]

[a] *SSC RF TRINITI, Troitsk, Moscow reg., 142190 Russia*
[b] *Institute for Theoretical and Experimental Physics, Moscow, 117259 Russia*
[c] *Sandia National Laboratories, P.O.Box 5800 Albuquerque, NM 87185-1194 USA*

Abstract. Brief review of our recent results concerning prolonged plasma production (P^3) in multiwires arrays and its influence on their dynamics in pulse power facilities is presented. P^3 means that owing to overheating effects hot plasma with high conductivity is being produced from the relatively cold material almost during the whole period of current rising. P^3 results in formation of a plasma shell with a low aspect ratio and without any skin effect. It means that magnetic flux and electric current are distributed more or less uniformly along the shell and there is no magnetic piston. The latter fact is responsible for the more stable implosion of the plasma. P^3 is unavoidable effect for initially cold liners of all types in pulse power facilities, and it is useful for production of high power X-ray pulses with multiwires arrays because the process of P^3 leading to more stable implosion is well controlled in this case due to well defined periodic structure of the array. Theoretical evaluation of rate of plasma production is obtained, and it is compared with experiments. Simple simulation of plasma dynamics during implosion of multiwire arrays with P^3 is presented. Combining of such simulation with probe magnetic measurements inside cylindrical liners we obtain a certain information about dependence of the plasma production rate on time and about characteristics of depletion of plasma source. It appears that process of the depletion may be rather gradual, so that some amount of plasma is being placed at initial array position even till the moment of X-ray pulse maximum. It may lead to partial shunting of electric current and to some depression of X-ray pulse intensity. These results discover new opportunities to control plasma implosion and X-ray pulse parameters.

A lot of experimental data was obtained recently [1,2,3] in multiwire array implosion experiments at pulse power facilities Angara-5-1, MagPie and Z, which shows that production of well conducting plasma from dense relatively cold products of wire explosions lasts almost up to the moment of total current maximum when depletion of the cold material starts. This process is called as prolonged plasma production (P^3). P^3 was firstly discovered and considered in

Refs. [4,5]. However up to present times, important role of P³ in good implosion of multiwire arrays to the axis of pulse power machines was not understood properly.

The most complete investigation of multiwire array implosion taking into account P³ was published in Ref. [1]. The process of P³ leads to formation of inward plasma flow and then to plasma shell of a moderate aspect ratio. The flow and then the shell carry magnetic flux and electric current which are distributed more or less uniformly along the plasma. Thus there are no skin effect and no magnetic piston, that are explained by the fact that new plasma is formed just in the magnetic field. Plasma pressure in the inward flow is much less then the magnetic field pressure, so that plasma accelerates mainly by the Ampere force. Owing to such properties of plasma implosion and to moderate aspect ratio of the plasma shell formed by the P³ process, its compression becomes much more stable in comparison to hypothetical plasma liner [1].

The most important parameter of our model with the P3 process is a rate of plasma production, which can be measured in mass units per time unit and per unit of area of the lateral surface of the liner. The plasma production rate can be controlled by design of multiwire arrays, that opens a new way for optimizing of plasma implosion and hence X-ray pulse.

A simple model for the plasma production rate under condition, that depletion of cold dense material does not take place, was considered in Ref. [1]. A simple design of the multiwire array was assumed there, when wires are placed as elements of the cylinder uniformly along the generating circle. When the interwire gap is less than the boundary layer width, then structure of the boundary layer can be investigated in 1d slab geometry assuming steady state. The following dissipative processes were considered in Ref. [1]: Ohmic heating, thermal conductivity from the hot plasma toward the dense cold material, and diffusion of plasma across the magnetic field. The model investigated in Ref. [1] gives the following expression for the plasma production rate:

$$\dot{m} = k_{theor} \left(\frac{I_{MA}}{R_{L\,cm}}\right)^{1.8} \frac{\mu g}{cm^2\,ns}, \qquad (1)$$

where I is the total current through the liner, and R_L is liner radius. $k_{theor} = 0.2$ for tungsten [1]. Electron temperature, electron density and radial plasma velocity just out of the boundary layer are given by the following expressions:
$T_e = 15.7\left(I_{MA}/R_{L\,cm}\right)^{0.36}$ eV; $\quad n_e = 1.2 \cdot 10^{18} \left(I_{MA}/R_{L\,cm}\right)^{1.7}$ cm^{-3};
$V = 5.3 \cdot 10^6 \left(I_{MA}/R_{L\,cm}\right)^{0.22}$ cm/s, whereas the width of the boundary layer can be expressed as $d_s = 220 \left(I_{MA}/R_{L\,cm}\right)^{-0.85}$ µm. The model neglects azimuthal structure of the plasma flow. For larger interwire gaps, when the azimuthal structure

becomes important, the coefficient in Eq. [1] may depend on the interwire gap, diameter of cold dense cloud produced by initial wire explosion etc.

Plasma motion out of the boundary layer can be described neglecting the dissipative processes mentioned above but taking into account its nonstationary character and cylindrical geometry. Due to low plasma pressure in comparison with the magnetic one the former one can be neglected, and heat equation may be not considered at all. Such model was developed in Ref. [1].

We use it here for interpretation of magnetic probe measurements inside multiwire arrays performed at the Angara-5-1 facility [6]. The magnetic probes are placed at half radius of multiwire arrays (i.e. at $r = 0.5 \cdot R_L$) and at $0.85 \cdot R_L$ in different experiments. Such experiments appear to give an important information about the process of P^3 and about real $\dot{m}$. Four shots with magnetic probe measurements were simulated using the model mentioned above to determine $\dot{m}(t)$ and considering the latter function as a free parameter of the model.

Fig. 1 presents the typical shot (#3899) with the magnetic probe measurements inside the liner. Three other shots were processed by the same way with similar results. The multiwire arrays were 2 cm diameter for all experiments with the magnetic probe measurements. It is about two times more than in regular multiwire experiments at Angara-5-1. The array for the shot #3899 was made from 80 tungsten wires of 5 µm diameter, with nominal liner mass per unit length being equal to 320 µq/cm.

To parameterize $\dot{m}(t)$ we use its following representation: $\dot{m} = C[I(t)]^\mu$ for $t \leq t_q$, and $\dot{m} \approx C_1 \exp\left(-(t-t_q)/t_f\right)$ for $t > t_q$. These two branches were conjugated smoothly by a certain way at $t \approx t_q$. When t_q, t_f and μ are known then C and C_1 are determined by the conjugation condition and by the total liner mass. Thus our $\dot{m}(t)$ is parameterized actually by three parameters t_q, t_f and μ.

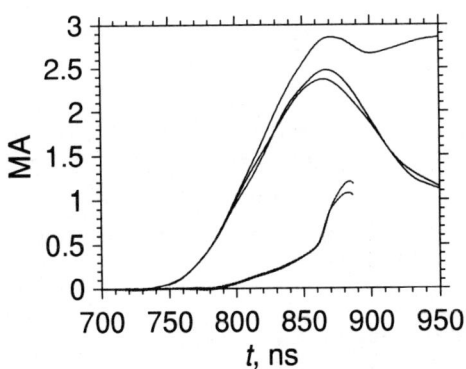

FIGURE 1. Experimental data for the shot #3899. The solid lines show the total current through the liner measured by different methods. The dashed lines show the current inside the half radius of the initial liner radius measured by different probes. The dotted lines show X-ray pulse measured by different detectors.

To reproduce the precursor in magnetic probe measurements starting at about 40-50 ns after beginning of current rise we chose $\mu = 2$. We can't distinguish $\mu = 2$ and $\mu = 1.8$ by this method but $\mu = 1.5$-1.6 is excluded. If the maximum of $\dot{m}(t)$ takes place at about 112 ns after beginning of current rise and $t_f \approx 16$ ns, then simulated electric current inside half liner radius coincides well with the measured one. See Fig. 2. Our choice of $\dot{m}(t)$ provides compression of main part of the liner mass inside the radius $0.1\,R_L$ by the experimental moment of X-ray pulse maximum (t ≈ 160 ns after beginning of current rise). It is very important that the process of P^3 should last up to the moment of implosion of main mass of the liner (though with relatively low rate) to provide a relatively low electric current inside the half liner radius.

Comparing of initial rise of $\dot{m}(t)$ obtained in the simulations with Eq. (1) we may conclude that the theoretical coefficient in Eq. (1) should be replaced by the following experimental value:

$$k_{exp} = 0.07 \div 0.13 \quad (2)$$

Difference between the theoretical value and the experimental one does not contradict declared accuracy of the theoretical estimation [1].

Analogous model was used for interpretation of results of laser probing for multiwire arrays at Z-machine. See Ref. [3]. The latter

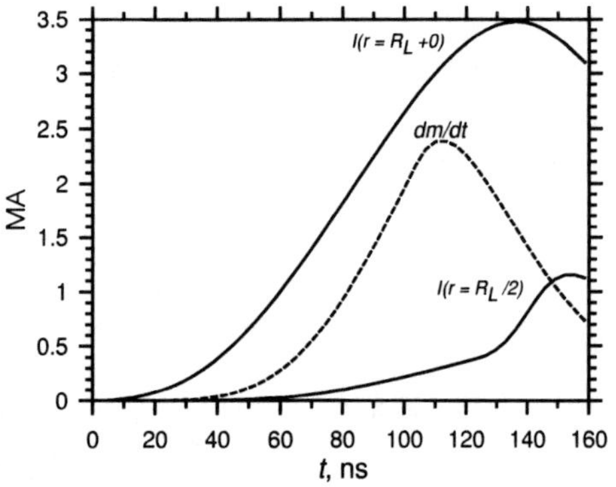

FIGURE 2. Comparison of simulated (solid line) and measured (dashed line) electric current inside the half radius of the liner for the shot #3899. They are labeled by $I(r=R_L/2)$. Total current through the liner in the simulation (solid line) and dm/dt (dashed line; in arbitrary units) are also shown.

results do not contradict to Eq. (1) with the coefficient (2) and with the exponent 1.8.

In summary, we conclude that magnetic probe measurements inside the liner give an important and unique information about the process of P^3 and, in particular, about the process of depletion of the plasma source at the initial position of wires. The form of $\dot{m}(t)$ may be controlled by using of more complicated design of multiwire arrays and by regime of initial implosion of wires. It opens a new way to control plasma implosion and X-ray pulse parameters.

REFERENCES

1. Alexandrov V. V. et al. *Plasma Physics Reports* **27**, 89 (2001).
2. Lebedev S. V. et al. *Physics of Plasmas* **8**, 3734 (2001).
3. Bliss D. E. et al. Poster report WE-P2-45 at BEAMS2002; see this volume.
4. Bekhtev M. V. et al. *Sov. Phys. JETP* **68**, 955 (1989); Ajvazov I. K. et al. *Sov. J. Plasma Phys.* **14**, 110 (1988).
5. Bobrova N. A. et al. *Sov. J. Plasma Phys.* **14**, 110 (1988).
6. Alexandrov V. V. et al. Poster report MO-P2-40 at BEAMS2002; see this volume.

Experimental Study Of Wire Array Implosion In Presence Of Prolonged Plasma Production On Angara-5-1 Facility

V.V. Alexandrov, M. V. Fedulov, I.N. Frolov, E.V. Grabovskii,
K.N. Mitrofanov, S.L. Nedoseev, G.M. Oleinik, I.Yu. Porofeev,
A.A. Samokhin, P.V. Sasorov[a], V.P. Smirnov, G.S. Volkov, M.V. Zurin,
G.G.Zukakischvili

SSC RF TRINITI, 142190, Troitsk, Moscow reg., Russia, [a]Institute for Theoretical and Experimental Physics, B.Cheremushkinskaya ul 25, Moscow, 117259 Russia

The study of the wire array implosion on installation Angara-5-1 is devoted. Magnetic probes and the x-ray backlighting technique were used to investigate a current and mass distribution during the implosion. The experimental data confirm the concept of prolonged plasma production: the long time of implosion the hot plasma flows from dense fixed wire cores and is accelerated by the Lorentz force to the array center together with frozen magnetic field.

INTRODUCTION

Cylindrical wire array implosion driven by a current is investigated widely in ICF now [1-3]. The outstanding parameters were obtained on Z installation [2].

It was shown [4,5], that in first several nanoseconds from the driven current start the plasma corona will be generated on the wires surface and the current switches from the wires to the plasma corona. The further heating of wire material is due to the heat transfer from plasma corona to wires. The plasma corona environs the dense wire cores. The cores remain at their initial positions during the significant part of the current pulse and are the fixed sources of the wires plasma. The permanently generated plasma is accelerated to the axis of the array by the Lorentz force. Originally empty internal region of wire array is filled by wire plasma with some part of the current. Mass and current distributions in the array during implosion were investigated on Angara-5-1 facility to understand the physics of implosion.

In present experiments the anode - cathode gap was 10mm, the array radius -6-10mm. Arrays had 40-120 tungsten wire. Diameter of wires - 5-10µm.

DISTRIBUTION OF THE CURRENT INSIDE ARRAY

Axial and azimuthal magnetic field were measured inside the array (⌀20mm) by magnetic probes. Diameter of probes was from 260µ to 880µ[6].

Temporal profile of the current measured inside cylinder (⌀10mm) has the following shape: signal appears in ~35ns after driven current start, slowly increases up to level 60-100kA (that is the current of precursor, ~3% of the total current) and then increases sharply due to the transfer of the main current shell inside the cylinder ⌀10mm. It is possible to estimate the velocity of current-carrying plasma as ~$2*10^7$cm/s. The estimation of current-carrying shell thickness gives more than 2mm.

The continuous plasma shell is generated due to the merging of plasma jets of all wires.

In special shots the external probing axial magnetic field was used to record the creation of the continuous plasma shell. Compression of such shell causes the compression of axial magnetic flux and the increasing of the axial magnetic field. The increasing of axial magnetic field was recorded in 30ns after start of driven current, when the total current was ~0.3MA. At this moment the continuous plasma shell (thicker than skin layer) was formed. The plasma shell passed to this time ~0.4cm from initial array radius.

BACKLIGHTING OF WIRE ARRAY DURING IMPLOSION

X-pinch was used as x-ray backlighting source [7]. One of 8 current return posts (at distance of 45mm from the liner axis) was replaced by X-pinch. X-pinch was produced by crossed molybdenum wires with diameter of 20μm. The crossing point of X-pinch flashes up brightly in 3-7keV region under action of current pulse of ~300kA (about 1/8 of total driving current). The flash of a X-pinch happened in 60-80ns after a beginning of a current through the liner, duration of flash was less than 2ns.

Shadow image was recorded on film. Films were protected from radiation of Z-pinch and a flux of microparticles by shielding system and titanium foil of 16μm thickness. Step attenuator from tungsten was situated in front of films for opportunity of transition from photometric density to density of substance.

To check spatial resolution of technique and to determine the size of x-ray source the test wires (diameter of 5μm) were set near the array. Image magnification for array wires was varied from 10 to 25; for test wires - from 5 to 10. The shadows of test wires and shadow of current driven array were obtained on same film simultaneously. From these measurements it was found, that the size of a X-pinch less 2μm. The spatial resolution of techniqui on the array ~1.7μm [8].

The shadow picture, obtained in x-ray probing of the array during implosion is presented in Fig.1 (wire diameter was 8μm, initial mass - 9.5μg/cm). Thin sloping line - the shadow of the test wire. Two dense wire cores are clearly visible. The experiments have shown, that in 60-80ns after current start the wires of array are augmented in a diameter in 3-4 times at level of current ~30-50kA/wire. The evaluation of expansion velocity of dense cores gives the value ~10^4cm/s. To the right of marked dense cores there are two poorly visible diffusion cores, which substance in a much more degree has passed in hot plasma. It testifies that the process of plasma production flows nonuniformly. The different luminescence of wires in the beginning of an implosion (including along their axis) was recorded in a visible light too.

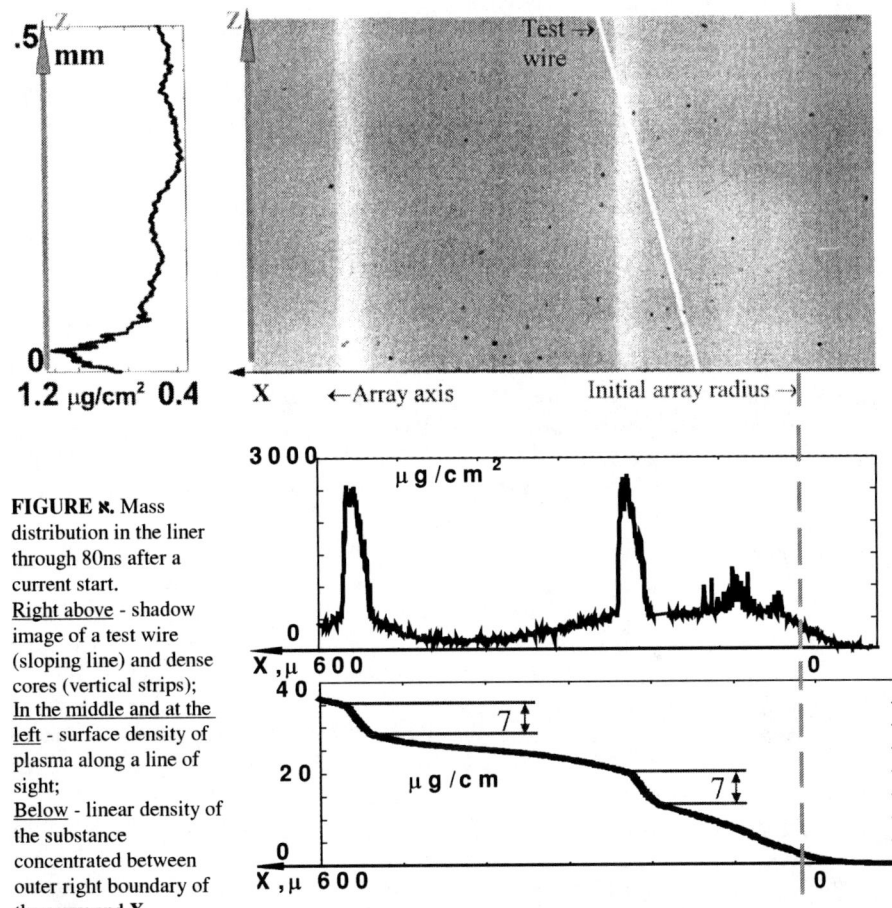

FIGURE א. Mass distribution in the liner through 80ns after a current start.
Right above - shadow image of a test wire (sloping line) and dense cores (vertical strips);
In the middle and at the left - surface density of plasma along a line of sight;
Below - linear density of the substance concentrated between outer right boundary of the array and **X**.

In the middle part of Fig.1 the surface density distribution of plasma along the line of sight at some height as the function of coordinate **X** is presented. Two dense core and two low dense cores are visible here also. Integral of this dependence as the function of **X** coordinate gives a linear density distribution of the substance concentrated between outer (right) boundary of the array and **X** (see bottom of Fig.1). The estimation of the mass in dense cores is ~70% of initial wire mass. The rest mass of wires (~30 % from initial mass) is spread inside ~200 microns on radius from initial array position. The small part of plasma is spread beyond the array periphery at distance ~100 microns.

In the left part of Fig.1 the surface density of plasma along the left wire as the function of **Z** coordinate is presented. The axial non-uniformity of the plasma density with a scale of 200μ is visible near core. The same size was recorded on the laser shadow and soft x-ray images previously.

At some images the axial spatial non-uniformity of dense core substance with spatial scale of 5-20μm is observed. This scale of the non-uniformity of dense cores does not coincide with the axial non-uniformity scale of surrounding low density plasma presented above.

CONCLUSIONS

The obtained data confirm the concept of prolonged plasma production: long time during an implosion hot plasma flows down from dense fixed cores and under action of the Lorentz force is accelerated to center together with frozen magnetic field. Briefly outcomes of this paper can be summarized as follows:

Current appearance inside a circle of a half of the initial radius takes place in 40ns after total current start. Precursor current is about 100kA. Internal boundary of the current - carrying shell carries up to 3% of the total current, has speed ~10^7cm/s and attains the liner axis at early stage of the discharge. This shell when passing through the half of initial array radius has a front width of more than 2mm, that is greater than skin-layer thickness.

A compression of axial magnetic flux by a current - carrying plasma shell was demonstrated. The axial magnetic flux is trapped by the precursor plasma. The merger of separate plasma streams in the continuous shell and the start of axial magnetic flux compression take place in ~30ns from the current start at the distance of 0.4cm from the initial liner radius. The axial magnetic field attains values not less than 300kG.

The mass of dense cores after 80ns from current start is ~70% from initial wire mass. The remaining mass of wires is spread on ~200 microns on radius inside initial array position. The diameter of wires was increased in 3-4 times during this time. Plasma of a small density are observed outside of the liner on a size ~100 microns. The axial stratification of a density of plasma with a size 200μm is visible in the vicinity of core. The wires lose its mass with different rate.

This work was supported in part by grand of RFBR # 01-02-17319.

REFERENCES

1. Sanford, G. O. , B. M. Marder, et al.,*Phys. Rev. Lett.* **77**, 5063 (1996).
2. Spielman, C. Deeney, G. A. Chandler, et al., *Phys.Plasmas* **5**, 2105 (1998).
3. S.V.Lebedev, F.N.Beg, S.N.Bland et.al. *Physics of Plasmas* **8**, 3734 (2001).
4. A.V.Branitskii E.V.Grabovskii et al. BEAMS'98. Haifa, Israel, June 7-12 1998. Proc, 599-602.
5. V.V.Aleksandrov, A.V.Branitskii, E.V.Grabovskii, et al. *Plasma Physics Reports*, **27**, 89–109, No.2 (2001). *Translated from Fizika Plazmy,* **27**, No.2, 99–120, (2001).
6. Euroconference On Advanced Diagnostics For Magnetic And Inertial Fusion (Varenna, Italy, September 3-7, 2001) Proc. 419-422.
7. Yu. Skobelev, S. A. Pikuz, A. Yu. Faenov et.al., JETP **81**, 692 (1995).
8. Euroconference On Advanced Diagnostics For Magnetic And Inertial Fusion (Varenna, Italy, September 3-7, 2001) Proc. 415-418.

Wire Array Z-pinch Experiment on Inductive Energy Storage Generator ASO-X

Yusuke Teramoto,[a,1] Hideyuki Urakami,[a] Susumu Kohno,[b]
Naoyuki Shimomura,[c] Sunao Katsuki,[a] and Hidenori Akiyama[a]

[a]*Graduate School of Science and Technology, Kumamoto University,
2-39-1 Kurokami, Kumamoto 860-8555, Japan*
[b]*Ariake National College of Technology, 150 Higashihagio-machi, Omuta, Fukuoka 836-8585, Japan*
[c]*Department of Electrical and Electronic Engineering, The University of Tokushima,
2-1 Minamijosanjima, Tokushima 770-8506, Japan*

Abstract. Wire-array z-pinch experiment was carried out on ASO-X generator. ASO-X is a 3-stage inductive voltage adder, and works as the inductive energy storage system with the plasma opening switch. The array diameter was 3 cm, and had 36 tungsten wires of 20 µm in diameter and 1 cm long. Taking time-resolved pictures of imploding plasma and observing x-ray signals, it was investigated that how the wire-array implodes for the different current rise times. For the x-ray diagnostics, the X-ray diode (XRD) and the diamond detector were used. When the current rise time is slow, the peaks of XRD and diamond detector were observed at the different time. In the time-resolved picture, the implosion of precursor plasma was observed followed by the main implosion. The strong kink instability was observed when the pinched plasma was completely created. When the current rise time is fast, no kink instability was observed. However, the implosion did not fully cover between the electrodes. The diamond detector did not detect the x-ray, although the XRD detected.

I. INTRODUCTION

X-ray radiated from z-pinches has recently been of much interests in applications as an inertial nuclear fusion driver [1]. The development of fast and high-energy pulsed power technology has allowed the use of a thin metal wire array as a z-pinch plasma source. A highly-symmetrical thin metal wire array can provide a symmetrical initial plasma, which then leads to a very uniform plasma implosion. The uniform implosion allows the production of an efficient x-ray radiation thus providing extremely high power and energy. The rise time of the drive current plays very important role in creating the uniform plasma implosion. In order to avoid the instability, the current rise time must be enough fast compared to the speed of the instability growth. The inductive energy storage system is superior to the capacitive storage system in terms of the simplicity and compactness of the generator system. In order to figure out the possibility of inductive system, and to investigate the behavior of imploding plasma, wire-array z-pinch implosion experiment was carried out on ASO-X generator [2] using the plasma opening switch.

---

[1] Present address: Ushio Inc., email: y.teramoto@ushio.co.jp

II. EXPERIMENTAL SETUP

Figure 1 shows the configuration of wire-array load. ASO-X generator is connected to the left end of the coaxial transmission line. The plasma opening switch (POS) having the eight cable plasma guns is placed between the ASO-X and the load. The wire-array is connected to the inner electrode and placed in the load camber. The diameter and length of the array can be changed. The anode side of the array has four separate arms to allow the individual current measurement of the wires. The X-ray diode (XRD), diamond detector, photodiode and x-ray pinhole camera were used. In order to make the x-ray detectors have the sensitivities for the different wavelength regions, the photodiode was filtered by a 3-µm Al filter. For both experiment, the high-speed framing camera was used. All the signals were recorded by the 4-channel digital oscilloscope (1.5 GHz, 4 GSa/s, Agilent Techology, 54845A). Charging voltage of ASO-X, total stored energy, typical peak current and rise time without POS were 30 kV, 8 kJ, 170 kA and 2 µs for the present experiment. The current of plasma gun and the time delay between the injection of the plasma and the firing ASO-X were adjusted to have the enough long conduction time of POS and the enough fast load current.

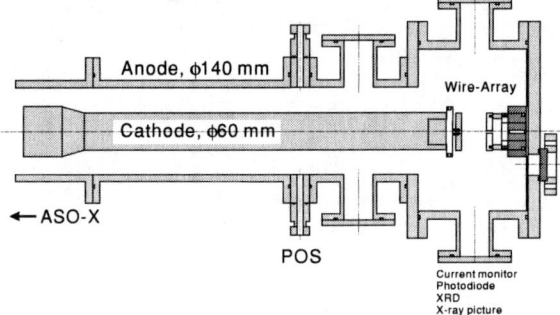

FIGURE 1. Wire-array load configuration on the ASO-X generator.

III. SLOW IMPLOSION EXPERIMENT

The experiment was carried out without POS to provide the slow current. Figure 2 shows the waveforms of the load current, XRD and diamond detector, with the timing of the framing camera. The current reached 170 kA at about 2 µs in this experiment. The signals of XRD and diamond detector started to increase at the beginning of the current. However, the XRD signal reached peak before the current peak, although the signal of the diamond detector had the peak at near the current peak. There was no signal for the Al-filtered photodiode. Therefore, the wavelength of the x-ray was not in the range of hard x-ray. The sensitivity of the diamond detector increases with decreasing wavelength in the range of sub-keV. Thus it can be said that the radiation of shorter wavelength occurred in the later time.

Figure 3 shows the time-resolved picture of the imploding plasma from 560 to 1860 ns. The time between each frame is 100 ns and the exposure time is 20ns. In the first two pictures, there were no clear images. At 560 ns, dark plasma column was created at central axis. This plasma column is thought to be created by the precursor plasma. The precursor plasma column became bright with increasing time, especially at the middle of the electrodes. At 1160 ns, when the XRD signal reached its peak, the firstly-imploded plasma is clearly seen. It is noteworthy that, around the time of the peak of XRD signal, the surrounding plasma starts to implode toward the center plasma column. And finally, the inner and the outer plasma shell became together and made a strong pinch. It took about 1900 ns for whole plasma to implode completely. Therefore, the kink instability is seen at the final picture that corresponds to the peak of the diamond detector signal. It means the hard x-ray radiated from the hot spot.

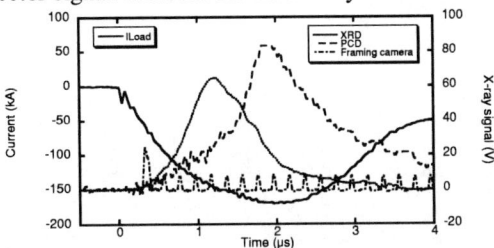

FIGURE 2. Waveforms of current, XRD, diamond detector (PCD) and timing of framing camera.

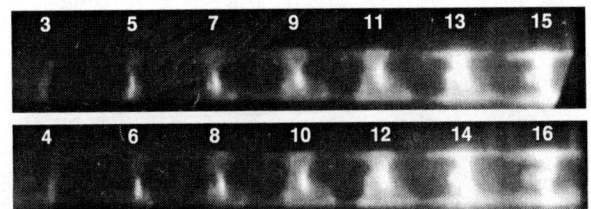

FIGURE 3. Time-resolved pictures of imploding plasma taking at from 560 to 1860 ns. Time between each picture is 100 ns.

IV. FAST IMPLOSION EXPERIMENT

With the same experimental setup used above, the fast implosion mode of the wire-array z-pinch was studied. In this experiment, the POS was used to decrease the current rise time and to increase the voltage by means of inductive energy storage scheme. Figure 4 shows the waveforms of the load current and XRD, with the timing of the framing camera. During approximately 1000 ns, the POS conducts the generator current storing the magnetic energy in the upstream. When POS opened, the load current rose within 50 ns. The peak current was approximately 110 kA. The reason of this lower current was the low density of plasma fill in the POS region. And it can be improved by increasing the charging voltage of the capacitor of the plasma gun.

In the fast implosion experiment, the implosion time was about 500 ns, that was four times faster than the slow implosion experiment. In the fast implosion experiment, the diamond detector did not detect the x-ray. Therefore it can be thought that only very soft x-ray was radiated. The peak amplitude of the XRD signal was higher than that for the slow implosion. The pulse width of the XRD signal was twice shorter than that for the slow implosion. Therefore it can be said that the implosion is more uniform than the slow implosion, and, then, the intensity of soft x-ray became stronger.

Figure 5 shows the time-resolved picture of the imploding plasma. At 1500 ns, that corresponds to the time of the peak of XRD signal, the pinch plasma is seen along the z-axis. However, the pinch plasma does not cover the whole inter-electrode region in this picture. After 1500 ns, the surrounding plasma seems to implode inward and become together with the inner part. Although the second implosion was seen in the slow implosion, there was no second strong pinch in the fast implosion.

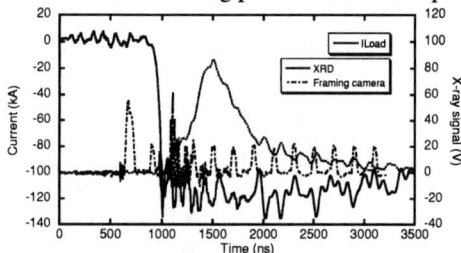

FIGURE 4. Waveforms of current, XRD, and timing of framing camera.

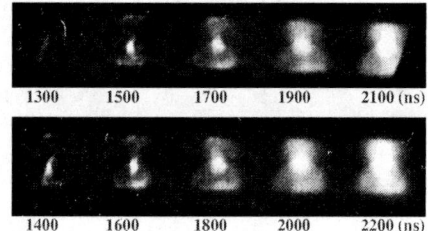

FIGURE 5. Time-resolved pictures of imploding plasma taking at from 1300 to 2300 ns. Time between each picture is 100 ns.

IV. SUMMARY

Tungsten wire-array z-pinch experiment was carried out on the inductive energy storage generator ASO-X. Both of slow and fast implosions were studied. The fast implosion mode seems to make more intense radiation in the soft x-ray region. In the slow implosion experiment, the twice implosions were observed. The weak implosion occurred by the pre-cursor plasma before the second strong implosion. Also, the kink instability was observed at the second pinch, which radiated the x-ray of the shorter wavelength.

REFERENCES

1   R. B. Spielman, et al, *Phys. Plasmas* **5**, 2105-2111 (1998).
2   S. Kohno, Y. Teramoto, I. V. Lisitsyn, S. Katsuki, and H. Akiyama, *Jpn. J. Appl. Phys.* **39**, 2829-2833 (2000).

GAS PUFFS EXPERIMENTS

Progress in Double-Shell Gas Puff Z-Pinches

H. Sze,[a] J. Banister,[a] B. H. Failor,[a] A. Fisher,[a] J. S. Levine,[a] Y. Song,[a]
J. P. Apruzese,[b] R. W. Clark,[b] J. Davis,[b] D. Mosher,[b] J. W. Thornhill,[b]
A. L. Velikovich,[b] B. V. Weber,[b] C. A. Coverdale,[c] C. Deeney,[c] T. L. Gilliland,[c]
J. McGurn,[c] R. B. Spielman,[c] K. W. Struve,[c] W. A. Stygar,[c] D. Bell,[d]
P. Coleman,[e] M. Babineau[f], C. Enis[f], V. Kenyon[f] and T. Worley[f]

[a] *Pulse Sciences Division, Titan Corp., San Leandro, CA 94577-0599*
[b] *Plasma Physics Division, Naval Research Laboratory, Washington, D. C. 20375*
[c] *Sandia National Laboratories, Albuquerque, New Mexico 87185*
[d] *Defense Threat Reduction Agency, Alexandria, VA 22310-3398*
[e] *Alameda Applied Sciences, San Leandro, CA 94577*
[f] *Sverdrup technology, TN 37389*

Abstract. An experimental study of high current (3-15 MA), high fidelity (multiple atomic number) and long implosion time (100-200 ns) gas puff loads using the 1-2-3-4 cm double-shell gas puff is in progress at Titan/PSD. Results of experiments conducted on Double-EAGLE, Saturn, Decade Quad and the Z accelerators will be analyzed and presented. The principal observations are: (1) The overall pinch quality and radiative characteristics of all the argon double shell z-pinches are quite satisfactory. The Ar K-shell yields varies from the expected I^4 scaling in the inefficient regime for 3 to 7 MA to I^2 scaling in the efficient regime from 7 to 15 MA. (2) On all experiments from 3 – 15 MA, selective seeding of the shells demonstrates that the hottest mass of the pinch originates from the inner shell. This suggests that mixing between the two plasma shells during their collision and final implosion is limited. (3) On the 15 MA Sandia Z accelerator, with a load mass of 0.8 mg/cm, the K-shell x-ray output reached 275 kJ in a 15 TW peak power, 12 ns pulse. The analyzed ion and electron densities reach 5×10^{19} and 1.0×10^{21} /cc and the highest electron temperature observed is up to 2.2 keV with a 2.0 keV continuum

INTRODUCTION

During the last three years, we developed a 1-2-3-4 cm double shell gas puff and successfully demonstrated it on Double-EAGLE, Saturn, Z, and Decade Quad. The design characteristics of the nozzle are described in Reference 1. The double-shell gas puff as shown in Fig. 1 has a double plenum with an inner shell mean radius of 1.5 cm and an outer shell mean radius of 3.5 cm. The shell widths are 1 cm. The argon gas density profiles[2] are shown in Fig. 2.

 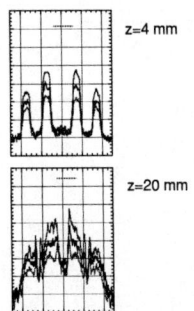

FIGURE 1. Double-shell nozzle.   FIGURE 2. Double-shell nozzle.

RADIATIVE CHARACTERISTICS ON DOUBLE-EAGLE, SATURN, Z, & DECADE QUAD

The radiative characteristics of the nozzle on Double-EAGLE, Saturn, Z, and Decade Quad are shown in Table 1. This table lists the implosion time, peak current, pinch lengths used, optimized K-shell yield, K-shell pulse width, ion density, electron temperature, and deduced K-shell radiating mass. Because the load in all of the experiments is identical, and because the results are obtained with an identical nozzle, this data excludes scaling effects from different initialization and multi-dimensional effects.

TABLE 1.. Summary of 1-2-3-4 cm Double-shells on DE, Saturn, Z, and DQ.

Gas Puff	Double-EAGLE	Saturn	Z	Decade Quad
Implosion time, ns	187	156	112	213
Peak Current, MA	3.7	6	>15	4.6
Pinch Length, cm	3.8	2.0	2.4	3.8
K-shell Yield, kJ/cm	3.8	7.5	110	6.5
K-shell pulsewidth, ns	10	11	12	11
K-shell diameter, mm FWHM	2.0	1.6	2.5	2.5
Ion Density, 10^{19} cm^{-3}	1.0	--	11	2.3
Electron Temperature, keV	1.4	1.0	2.0	1.5
K-shell Radiating Mass, µg/cm	27	--	280	70
Shot #	4428	2736	663	498

Data obtained with same diagnostic suite

CURRENT AND IMPLOSION TIME SCALING

For low mass implosion, the K-shell yield should scale as I^4 in the so-called "inefficient" regime. As the radiated yield becomes comparable to the available implosion energy, energy conservation limits the yield and then the yield scales as I^2. Fig. 3 plots peak argon K-shell yield versus peak current for a variety of pulsed power machines. Note that most of the data points are taken with different nozzle configurations and diagnostics in the 100 ns implosion time experiments. In spite of these potential, significant differences between machine, nozzle, and diagnostics characteristics, there is a remarkable consistency in the data and the clear radiation trend transition from the I^4 to I^2 scaling. In addition to the 100 ns implosion time data, Fig. 4 shows the K-shell yield of a list of ~ 200 ns implosion time experiments. The data seem to suggest a K-shell yield decline with longer implosion time in the inefficient regime.

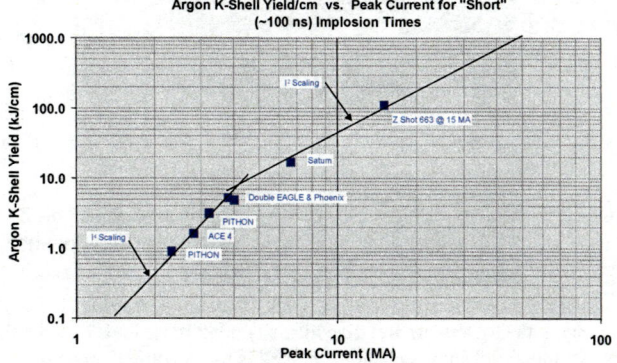

Fig. 3. Yield/length consistent with I^2 scaling.

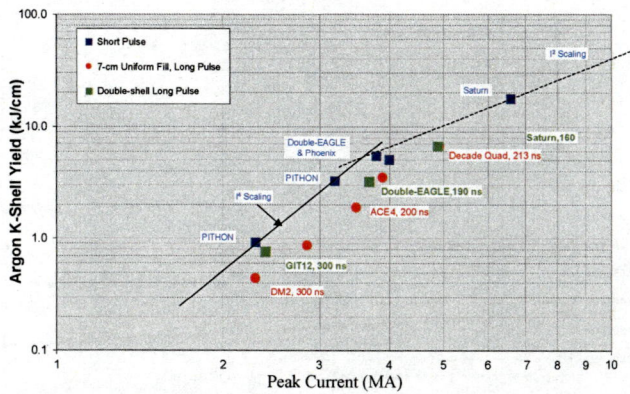

FIGURE 4. Empirical database for argon shows the implosion time penalty.

IMPLOSION DYNAMICS

Using a chlorine tracer, x-ray streaking spectroscopy suggests that: (1) the mass that starts nearest the axis gets the hottest and (2) the two shells do not mix or interpenetrate significantly during the implosion. Note this implosion characteristics is observed in the Double-EAGLE, Saturn, Z, and Decade Quad with identical hardware and diagnostics.

Fig.5 shows the time-integrated but radially-resolved spectra for two Z tests. On test #662 with the chlorine tracer in the inner shell, the K-lines are quite evident relative to the argon lines. With the tracer moved to the outer shell on shots #663, the chlorine emission is weaker by at least a factor of 4. The streaked x-ray spectra in Fig. 6 show that the inner-shell emission turns on and off more rapidly than the outer-shell emission and the Lyman-like to helium-like ratio is higher when the chlorine is in the inner shell, which implies again that the inner mass gets hotter.

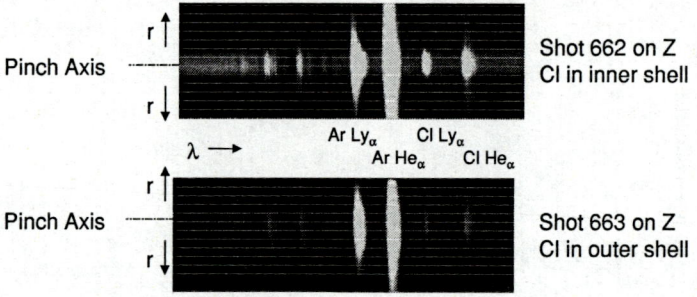

FIGURE 5. Spectroscopy with Cl Dopant. Relative intensity of Cl to Ar lines indicate that shells do not mix. Cl line ratios indicate that inner shell is much hotter than outer.

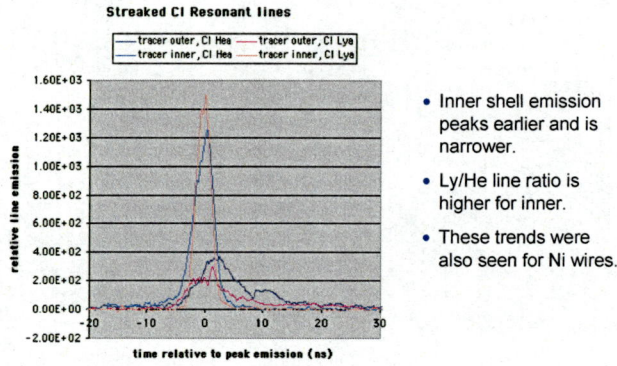

FIGURE. 6. Cl Line Ratio & Pulse Width Vary With Tracer Location

CONCLUSION

We have presented the radiative characteristics of the 1-2-3-4 cm double-shell gas puff on Double EAGLE, Saturn, Z, and Decade Quad. The data presented is collected with a same set of diagnostics. In spite of all the significant machine differences, the data suggest: (1) an I^4 to I^2 current scaling trends for short "100 ns" implosion times and (2) a K-shell yield penalty in the low mass regime for "200 ns" implosion times. X-ray streaking spectroscopy also suggests there is poor outer- and inner-shell mixing during implosion and the mass that starts nearest the axis gets the hottest.

ACKNOWLEDGEMENTS

This work is supported by the Defense Threat Reduction Agency.

REFERENCES

1. Song, Y., et al., Rev. Sci. Instrum, 71, 3080 (2000).
2. Failor, B. submitted to Review of Scientific Instruments.
3. Sze, H., et al., Phys. Plasmas, 7, 4223 (2000).
4. Sze, H., et al., Phys. Plasmas, 8, 3135 (2001).
5. Levine, J ,et al., this conference.

K-shell Yield Performance Assessment of Argon Gas Puff Loads Imploded on Double-Eagle, Z, and Decade Quad

J. W. Thornhill,[a] J. P. Apruzese,[a] J. Davis,[a] A. L. Velikovich,[a] K. G. Whitney,[b] H. Sze,[c] J. S. Levine,[c] B. H. Failor,[c] C. Deeney,[d] C. A. Coverdale,[d] and D. Bell[e]

[a]*Radiation Hydrodynamics Branch, Plasma Physics Division, Naval Research Laboratory, Washington, DC 20375 USA*
[b]*Berkeley Scholars, Inc. Springfield, VA 22150 USA*
[c]*Titan Systems Corp., Pulse Sciences Division, 2700 Merced Street, San Leandro, CA 94577*
[d]*Sandia National Laboratories, Albuquerque, NM 87185 USA*
[e]*Defense Threat Reduction Agency, Alexandria, VA 22310-3398*

Abstract. The K-shell yield performance of Titan double-shell argon gas puff loads imploded on the Z machine and Decade Quad (DQ) is compared to the performance of Double-Eagle loads employing the same nozzle configuration. Specifically, the K-shell yields obtained on the Z and DQ machines are compared with I^4, I^2, and I^0 current scaling projections made from Double-Eagle yields, where I is the peak load current. This analysis allows initialization and multidimensional effects to be factored out of the scaling behavior. This projection analysis is useful for evaluating argon gas-puff load performance because it relates I^4, I^2, and I^0 K-shell yield scaling transitions that a load, of a given specific energy, is predicted to undergo as the total load mass increases. Our analysis shows that the maximum K-shell yields were near optimum for the nozzle configuration used and the energy coupled to the load in the Z and Decade Quad experiments. At present it is not understood why an additional 22 nH of inductance for DQ, 6 nH for the Z machine, and 8 nH for Double-Eagle is required in their respective equivalent circuit models to reasonably match measured peak currents and implosion times. Note, we do not think that there is extra inductance in the real circuit; it is included in the modeled equivalent circuit as a method to account for the possible anomalous effects that are reducing the current from its expected value. These effects may include: multidimensional effects, current losses, and anomalous resistivity.

INTRODUCTION AND REVIEW OF I^2 AND I^4 YIELD SCALING

In this work we examine the performance of Titan 4-3-2-1 nozzle argon loads on the Z machine and Decade Quad (DQ) relative to their performance on Double-Eagle (DE). This nozzle is a double puff design with gas flowing from two concentric annuli (outer annulus located between 4 cm and 3 cm radius, inner annulus between 2 cm and 1 cm radius).[1] The K-shell yields obtained on Z and DQ are compared with I^0, I^2, and I^4 scaling projections from DE K-shell yields for implosions having the same implosion velocities, radial mass distribution, and cathode recession. Because the loads are so similar this scaling analysis automatically excludes most scaling effects resulting from different initialization and multidimensional effects. This allows load performance to be evaluated almost entirely in terms of its current scalings from its performance on DE. The DQ and Z experiments are analyzed in the context of three physics based scaling trends, which are: (1) I^4 scaling, which is valid in the low mass regime because radiation rates there increase at their optically thin density squared rates, (2) since energy inputs increase as I^2, at large enough load mass energy conservation requires that K-shell yields asymptote to an $I^{\leq 2}$ scaling, and (3) total radiation rates often increase with mass faster than ionization and thermalization rates such that radiation losses cool the plasma faster than it can be heated and ionized resulting in either a fall off or in I^0 saturation of K-shell yield. Trend (3) especially occurs for low specific energy ($\eta^* < 2$) implosions because they barely have enough energy to ionize into and radiate from the K-shell.[2] η^* is defined as the JxB coupled energy-per-ion normalized by $E_{min} \cong 1.012\ Z^{3.662}$ eV/ion, the minimum energy needed to instantly heat and ionize into the K-shell.[3] Conversion efficiencies of 20-30 percent of JxB energy into K-shell emission are calculated in the I^2 regime for high specific energy ($\eta^* > 4$) implosions. [3] These efficiencies have been achieved experimentally for plasmas with atomic numbers as high as Z=18.

The DE experimental yields from which scaling projections to DQ and Z are performed are shown as a function of experimental implosion time (τ) which depend on the load mass that the measured peak values of current in Fig. 1 can implode. These results are taken from Ref. [4]. The I_{peak} curve shown in Fig. 1 required that an additional 8 nH of inductance be included in the circuit model in order to obtain a reasonable fit to the measured peak current values as a function of τ. The inferred η^* and mass curves of Fig. 1 are based on this circuit model. The inferred masses are close to the values listed in Ref. [4]. The DE K-shell yields are projected for DQ along I^α scaling curves using,

$$Y^{I^\alpha}(\eta^*) = Y_{DE}(\eta^*) \times \left(\frac{I_{DQ}(\eta^*)}{I_{DE}(\eta^*)}\right)^\alpha \qquad \text{for fixed } \eta^* \text{ and } \alpha = 0, 2, 4 \qquad (1)$$

An I^0 yield projection implies that the K-shell yield is saturated at the DE K-shell yield. Likewise, an I^2 {I^4} projection implies that the yield is increasing in proportion to the ratio of the JxB {(JxB)^2} energy from the DQ pinch to that from DE.

ANALYSIS OF DQ AND Z EXPERIMENTS

Figs. 2 and 3 show the results for the DQ experiments in which the voltage was reduced by the factor 0.7/0.85. Fig. 2 demonstrates that adding 22 nH to the DQ circuit produces a reasonable match with the measured peak current values as a function of τ. The masses and η^* values that are inferred from this circuit as well as the calculated peak current, without any additional inductance, are also shown in Fig. 2. This figure indicates, for a fixed τ, that the measured peak current values are about 1 MA below their expected values (no extra 22 nH). This represents approximately 30 percent less energy coupled to the load than expected. The I^α K-shell yield scaling projections are shown in Fig. 3. This figure illustrates that the 244 ns implosion was in a saturated, I^0, scaling regime. In other words, at this low level of η^* ($\eta^* \cong 1.7$), as the mass is increased from 150 µg/cm on DE to 260 µg/cm on DQ, the total radiation rates have increased with mass and cooled the plasma to the extent that it is difficult to ionize to and radiate from the K shell. Thus, increasing the mass beyond 260 µg/cm at $\eta^* \leq 1.7$ will not produce any additional K-shell emission. As the mass is reduced and correspondingly η^* is increased (230 ns shot), the DQ K-shell yield approaches an I^2 scaling projection in which the K-shell yield is increasing in proportion to the coupled energy. The conversion efficiency of coupled energy to K-shell emission is about 12 percent for this shot, which is reasonable for this relatively low $\eta^* \cong 2$ implosion. Fig. 3 demonstrates that it is futile to attempt to optimize K-shell emission in future experiments by operating with this nozzle configuration with mass loads in excess of 200 µg/cm having η^* values less than 2. However, an encouraging consequence of the I^2 scaling achieved for the 230 ns experiment (mass $\cong$ 200 µg/cm) is that loads of less than 200 µg/cm that are imploded at η^* values greater than 2 can be expected to be in an $I^{>2}$ scaling regime. This consequence was verified by the DQ shot at full voltage. The 28 kJ of K-shell emission achieved by this shot (not shown) lie between I^2 and I^4 scaling projections. It is about 18 percent of the J×B energy.

The Z machine results for 2.4-cm length experiments are displayed in Fig. 4. The experimental results shown in this figure are taken from Ref. [5]. An additional 6.0 nH in the circuit model was required to reasonably match the measured peak currents with their respective implosion times (not shown). The inferred masses (Fig. 4) are also in agreement with those in Ref. [5]. The peak current calculated with extra inductance is 2-3 MA lower than that calculated without extra inductance. This represents about a 30 percent reduction in J×B energy for a given τ. Neither the DQ nor this current reduction is presently understood. Figure 4 illustrates how the yield behavior changes from nearly saturated, I^0, at the higher mass and lower η^* values of the 135 ns shot, to yields that are clearly in excess of I^2 projections from DE as the mass is reduced and the η^* values correspondingly increased. The K-shell yield for the lowest mass shot (800 µg/cm, η^*=5.5, τ=112 ns) is so large that it begins to approach typical theoretical values of 30 percent of the J×B coupled energy.

CONCLUSIONS

This analysis shows that the Titan nozzle's K-shell yield performance on Z and DQ is in accord with established current scaling trends from its performance on DE. It also shows that the efficiency of converting J×B energy to K-shell emission increases with implosion velocity for this configuration, i.e. the maximum conversion efficiency for an $\eta^* \leq 2$ load is about 12 percent whereas for an $\eta^* \cong 5.5$ it is at least 27 percent. At present it is not understood why DQ, Z and DE require additional inductance in their respective equivalent circuits (22 nH for DQ) to obtain a reasonable match between measured peak currents and implosion times. In addition to understanding and improving power flow, another option for achieving higher argon K-shell yields on DQ is to implode loads from larger radius, which will allow more energy to couple to the load at higher specific energy.

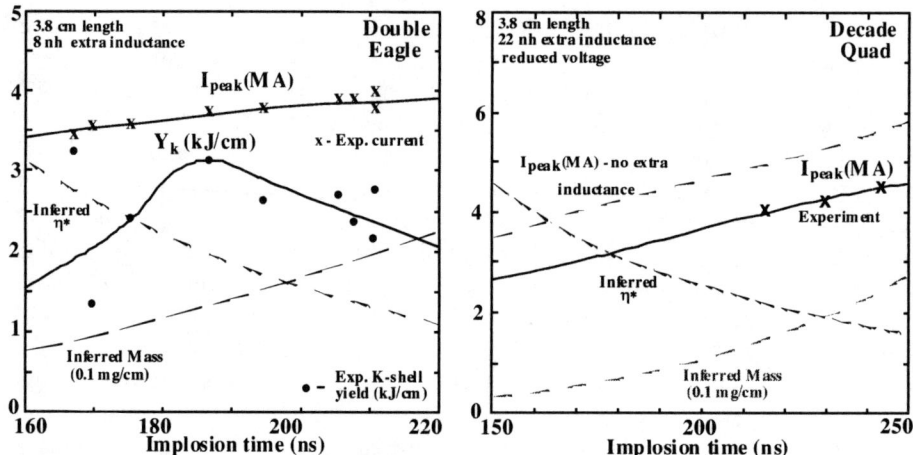

Figures 1 and **2.** Double-Eagle and Decade Quad experimental peak currents as a function of experimental implosion time. Also shown are inferred mass loads, η^* values, and calculated peak currents. In **Fig.** 1 the experimental K-shell yields are shown as a function of implosion time.

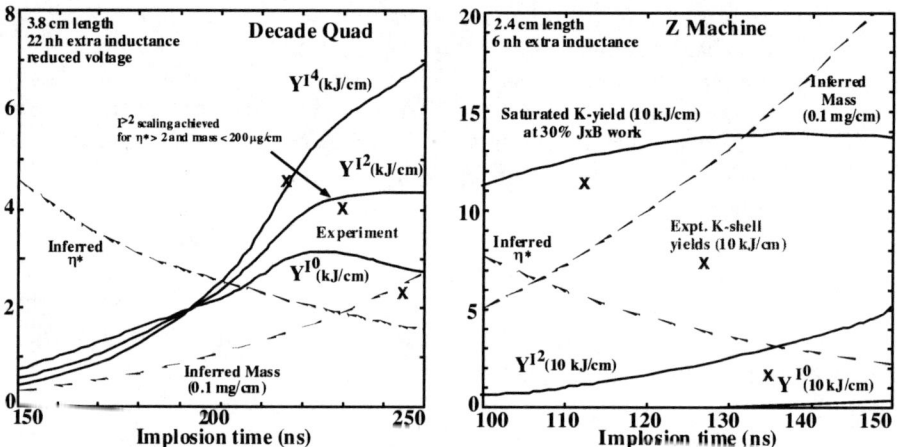

Figures 3 and **4.** Comparison of Decade Quad and Z machine experimental K-shell yields as a function of implosion time with I^0, I^2, and I^4 scaling projections from Double-Eagle K-shell yields. Also shown are inferred mass loads and η^* values.

ACKNOWLEDGMENTS

This work supported by DTRA and Sandia National Laboratories

REFERENCE

1. Y. Song, P. Coleman, B. H. Failor, et al., Rev. Sci. Instrum., **71**, 3080 (2000).
2. J. W. Thornhill, K. G. Whitney, J. Davis, and J. P. Apruzese, *J. Appl. Physics* **80**, 710 (1996).
3. K. G. Whitney, J. W. Thornhill, J. P. Apruzese, and J. Davis, *J. Appl. Phys.* **67**, 1725 (1990).
4. H. Sze, P. L. Coleman, B.H. Failor, A. Fisher, J. S. Levine, et al., *Phys. of Plasmas* **7**, 1 (2000).
5. H. Sze, P. L. Coleman, J. Banister, B. H. Failor, A. Fisher, et al., *Phys. of Plasmas* **8**, 3135 (2001).

Z-Pinch Experiments on Decade Quad Using a Double-Shell Gas Puff

J. S. Levine,[a] M. A. Babineau,[b] J. Banister,[a] D. Bell,[c] C. Enis,[b]
B. H. Failor[a], V. Kenyon,[b] H. M. Sze,[a] and T. Worley[b]

[a]*Titan Systems Corporation/Pulse Sciences Division 2700 Merced Street, San Leandro, CA 94577*
[b]*Sverdup Technology, Tullahoma, TN 37389*
[c]*Defense Threat Reduction Agency, Alexandria, VA 22310*

Abstract. An initial capability for producing argon K-shell radiation was demonstrated at Decade Quad. Using a double-shell nozzle gas puff, designed for 200 ns implosions, 17.7 and 28.7 kJ of K-shell radiation at 4.06 and 4.64 MA, respectively, were radiated. Although the load current was 20% lower than circuit model calculations predict, the radiation produced was consistent with previous experiments performed at Double-EAGLE at comparable load current.

INTRODUCTION

Decade Quad (DQ) was originally built as four independent modules for producing Bremsstrahlung radiation. It was subsequently modified with the addition of a convolute section that joins the modules together into one high-current pulse power driver for z-pinch research and testing. Because DQ takes 300 ns to reach peak current (into a short circuit) it requires long-implosion time z-pinch loads.

In a short, "introductory," experiment to demonstrate an initial argon K-shell radiation capability, we fielded our double-shell nozzle [1] which has been used previously on Double-EAGLE and Z, where we produced 19 and 274 kJ of argon K-shell radiation, respectively [2,3]. Since the nozzle was designed for 200 ns implosions, it was not anticipated that this test would illustrate the full potential of DQ. The early implosion time, relative to the Decade pulse width, meant that the voltage spike produced at implosion would be added to a substantial voltage remaining from the drive pulse. It was feared that this might cause arcing that could damage the machine. Our plan, therefore, called for operating at a Marx charge voltage reduced from the standard level of 85 kV to 70 kV. (As reported below, the final two shots were taken at full charge without damage to DQ.)

MODELING

To predict the behavior of our nozzle in Decade, we used a circuit model for DQ (open circuit voltage waveform with a fixed resistance and inductance in series with the z-pinch) driving a snowplow model of the z-pinch with the following parameters: uniform thick shells from r = 1.0 to 2.0 cm and from r = 3.0 to 4.0 cm, with equal mass in each shell, return current radius 6.8 cm, pinch length 3.8 cm, compression ratio (initial outer radius/final radius) 15. The z-pinch appears as a variable inductance to the circuit, with its value determined self-consistently from the snowplow dynamics. The peak current and the η values [4] (the ratio of the ion kinetic energy to the energy required for ionization into the emitting shell) are calculated. The K-shell yield is then calculated based on two models: the Thornhill/Whitney/Giuliani [5] (Yield$_{Whitney}$) and Mosher/Qi/Krishnan [6] (Yield$_{Mosher}$) models. Some of the results of the calculations are summarized in FIGURE 1. While there are differences between the two model predictions, an implosion time of 200 ns should be near ideal. The parameters of this target configuration are: mass 305 μg/cm, peak current 5.1 MA, η 1.7, yield 6.5 kJ/cm or 24.7 kJ total.

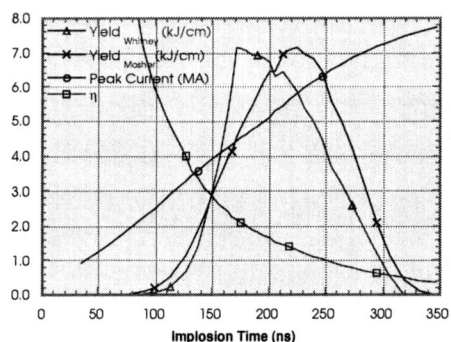

FIGURE 1 Results of circuit model simulation of argon double-shell gas puff on Decade.

FIGURE 2 The complete double-shell gas puff assembly installed in Decade.

DESCRIPTION OF EXPERIMENT

We used our standard double-shell nozzle, with a 1 cm recess, on all of the shots at Decade. FIGURE 2 shows the nozzle as assembled into Decade.

The electrical and gas connections were made to the nozzle through a 57 cm tall, 8 μH transit time isolator that was designed for the physical constraints of the Decade facility. A ring preionizer [7], that generates UV radiation to ionize the outer surface of the argon before the application of high voltage, was mounted on the front vacuum door and is therefore not visible in the pictures.

To measure the load current, we designed an anode "top hat" that had four B-dot coils (see FIGURE 2) and held the current return posts. The coils protruded into an annular groove, instead of being shielded in a hole, to avoid frequency-dependent skin effects. All the anode top hats (one for each shot), with the B-dots installed, were calibrated at the Z facility at Sandia National Laboratories.

The timing of the implosion and the width (FWHM) of the x-ray pulse were determine using XRDs and PCDs. The K-shell x-ray yield was measured with a set of tantalum calorimeters. A CCD-based X-ray pinhole camera and an axially-resolved CCD-based spectrograph imaged the implosion [8].

The philosophy behind the selection of diagnostics was to provide a minimally complete set of measurements using only the diagnostics previously used at Double-EAGLE. The Decade diagnostics could then be cross-calibrated and used to enhance the DE diagnostics.

EXPERIMENTAL RESULTS

The basic parameters of the shots are shown below in Table 1 and the parameters of the radiating plasma in Table 2. The plasma parameters were estimated using the NRLTENI code, based on the measured argon spectrum, the radiated power and the pinhole camera image. [9]

Table 1. Basic parameters of the shots on Decade.

Shot Number	Marx Charge (kV)	Mass @2cm (µg/cm)	Peak Current (MA)	Implosion Time (ns)	FWHM (XRDs) (ns)	FWHM (PCDs) (ns)	Yield (BBCals) (kJ)
493	70	325	4.50	244		10.0	8.8
494	70	260	4.25	230	10.6		15.5
495	70	208	4.06	216	12.6	15.3	17.7
497	70	156	3.56	213		14.1	9.5
498	85	286	4.64	214	11.2	14.4	28.7
499	85	260	4.73	213	10.5	12.7	28.4

Table 2. Plasma parameters of the radiating pinch.

Shot Number	K-Shell Diameter (mm)	Electron Temperature (eV)	Ion Density (10^{19} ions/cm^3)	Emitting Mass (µg/cm)	(%)
493	2.17	1291	1.50	36.84	11.3
494	2.66	1424	1.47	54.14	20.8
495	2.60	1498	1.41	49.71	23.9
497	2.66	1493	1.00	36.95	23.7
498	3.07	1563	1.49	73.40	25.7
499	2.43	1400	2.29	70.59	27.2

It is immediately obvious that the observed peak currents are significantly lower than those predicted by the circuit model (see FIGURE 1). [We confirmed that this was not a current diagnostic problem by calculating the implosion time based on the measured current waveform and the initial mass distribution and verifying that it was consistent with the observed implosion time.] The cause of this discrepancy has not been determined and is being investigated.

With that caveat, the z-pinch behaved and radiated as expected. Shot 495, for example, has typical Double-EAGLE parameters: 4 MA producing 18 kJ of radiation is a very good, but not extraordinary, result.

The first four shots, at 70 kV Marx charge, were to define the optimum implosion time. Since the last of these shots suffered an early diverter closure, it is not completely clear where the optimum occurs, but Shot 495 should be near that point.

Current scaling was investigated in the final two shots by increasing the Marx charge voltage to 85 kV (full charge for DQ). The results, 28.7 kJ at 4.64 MA on Shot 498, give a current scaling of $I^{3.6}$. This is consistent with the theoretical transition from inefficient (I^4) to efficient (I^2) scaling occurring at about 5 MA. [6,10]

CONCLUSION

We succeeded in demonstrating an initial argon K-shell x-ray capability on Decade Quad. This was a non-optimized experiment that used a double-shell nozzle sized for 200 ns implosions. At 4 MA load current, 18 kJ were produced, completely consistent with results at Double-EAGLE. At the higher current of 4.6 MA, 29 kJ were produced, scaling as $I^{3.6}$, again consistent with expectations.

The current delivered to the z-pinch load was significantly lower (20%) than predicted by the circuit model. There were no indications of current loss in the vacuum region and the Marxes and waterlines appeared to be operating properly. This is thus a mystery that needs to be resolved.

The next step will investigate larger diameter nozzle (12-15 cm diameter) that are designed for the full 300 ns pulse width to access the full current available. This step is certainly non-trivial and will require multiple rounds of testing and optimizing.

ACKNOWLEDGEMENTS

This work was supported by the Defense Threat Reduction Agency.

REFERENCES

1. Song, Y., Coleman, P., Failor, B.H., Fisher, A., Ingermanson, R., Levine, J.S., Sze, H., Waisman, E., Commisso, R.J., Cochran, T., Davis, J., Moosman, B., Velikovich, A.L., Weber, B.V., Bell, D., and Schneider, R., *Rev. Sci. Instrum.*, **71**, 3080-3084 (2000).
2. Sze, H., Coleman, P.L., Failor, B.H., Fisher, A., Levine, J.S., Song, Y., Waisman, E.M., Apruzese, J.P., Chong, Y.K., Davis, J., Cochran, F.L., Thornhill, J.W., Velikovich, A.L., Weber, B.V., Deeney, C., Coverdale, C.A., and Schneider, R., *Phys. Plasmas*, **7**, 4223-4226 (2000).
3. Sze, H., Coleman, P.L., Banister, J., Failor, B.H., Fisher, A., Levine, J.S., Song, Y., Waisman, E.M., Apruzese, J.P., Clark, R.W., Davis, J., Mosher, D., Thornhill, J.W., Velikovich, A.L., Weber, B.V., Coverdale, C.A., Deeney, C., Gilliland, T.L., McGurn, J., Spielman, R.B., Struve, K.W., Stygar, W.A., and Bell, D., *Phys. Plasmas*, **8**, 3135-3138 (2001).
4. Whitney, K.G., Thornhill, J.W., Apruzese, J.P., and Davis, J., *J. Appl. Phys.* **67**, 1725-1735 (1990).
5. Thornhill, J.W., Whitney, K.G., Davis, J., and Apruzese, J.P., *J. Appl. Phys.* **80**, 710-718 (1996).
6. Mosher, D., Qi, N., and Krishnan, M., *IEEE Trans. Plasma Science* **26**, 1052-1061 (1998).
7. Song, Y., Coleman, P.L., Failor, B.H., Kortbawi, D., Levine, J.S., Riordan, J.C., Sze, H.M., Thompson, J.R., Commisso, R.J., Moosman, B., Stephanakis, S.J., Weber, B.V., Fisher, A., and Schneider, R.F., 1999 IEEE International Conference on Plasma Science, June 20-24, Monterey, California, 1999, p. 298.
8. Failor, B.H., Coleman, P.L., Levine, J.S., Song, Y., Sze, H.M., LePell, P.D., Coverdale, C.A., Deeney, C., Pressley, L., and Schneider, R., *Rev. Sci. Instrum.* **72**, 2023-2031 (2001).
9. Failor, B.H., Coleman, P.L., Levine, J.S., Song, Y., Sze, H.M., Chong, Y.K., and Apruzese, J.P., *Rev. Sci. Instrum.* **72**, 1232-1235 (2001).
10. Thornhill, J.W., Cochran, F.L., Davis, J., Apruzese, J.P., and Whitney, K.G., in *Dense Z Pinches*, edited by N.R. Pereira, J. Davis, and P. Pulsifer, American Institute of Physics, Woodbury, 1997, pp. 193-197.

Gas Puff Radiation Performance As a Function of Radial Mass Distribution

Philip L. Coleman[a], Mahadevan Krishnan[a], Rahul Prasad[a], Niansheng Qi[a], Eduardo Waisman[a], B. H. Failor[b], J. S. Levine[b], and H. Sze[b]

[a]*Alameda Applied Sciences Corp., 2235 Polvorosa Ave., Suite 210, San Leandro, CA 94577 USA*
[b]*Pulse Sciences Division, Titan Corp., San Leandro, CA 94577 USA*

Abstract. The basic concept of a z-pinch, that JxB forces implode a shell of mass, creating a hot dense plasma on-axis, is coming under closer scrutiny. Wire arrays may start with an initial cold mass in a near "ideal" shell, but in fact they appear to develop complex radial mass distributions well before the final x-ray output [1,2]. We consider here the situation for gas puff z-pinches. While the ideal of a gas "shell" has been the nominal objective for many years, detailed measurements of gas flow show that nozzles used for plasma radiation sources (PRS) also have complex radial distributions. In particular, there are significant data [3] showing that the best x-ray yield comes from the least shell-like distributions. Recent experiments on the Double Eagle generator with argon have further enhanced this view [4]. For those tests with a double "shell" nozzle, there was a factor of almost 4 increase in yield when the relative mass (outer:inner) in the two shells was changed from 2:1 to less than 1:1. We suggest the following explanation. A configuration with most of its mass at large radii is subject to severe disruption by instabilities during the implosion. A more continuous radial mass distribution with $d\rho/dr < 0$ may mitigate instability development (via the "snowplow stabilization" [5] mechanism) and thus enhance the thermalization of the kinetic energy of the imploding mass. In addition, the appropriate balance of outer to inner mass maximizes the formation of a strong shock in the core of the pinch that heats the plasma and leads to x-ray emission.

INTRODUCTION

For over 20 years, the basic notion of a z-pinch has been encapsulated in the ideal of the implosion of a thin shell. But recent experiments with both wire arrays [1,2] and gas puffs [3] suggest that the most successful pinches work because the radial mass distribution is in fact *not* shell-like, but rather continuously distributed in radius. In the case of gas puffs, the evidence has been circumstantial. That is, a nozzle that was known to have a very shell-like distribution showed lower K-shell x-ray output than a nozzle with a "filled-in" radial distribution, when both designs were tested on the same pulsed power generator.

In experiments described here, we used a single nozzle to explore the transition from shell-like to filled-in. By varying the mass ratio between the two nominal "shells" of this double shell nozzle, we could vary the net gas distribution from more shell-like to more continuous as a function of radius. (In realty, because of the natural divergence of the gas flow, each "shell" has a significant extent, >1 cm, in radius.)

REVIEW OF EXPERIMENTAL RESULTS

Data from a few "typical" shots will be compared. For test #5079, the outer shell (nozzle exit radii from 3 to 4 cm) had about twice the mass (per unit axial length) of the inner shell (nozzle exit radii from 1 to 2 cm). This more "shell-like" case is compared with test #5080 where the outer shell mass was about 80% of the inner shell mass. Overall mass was adjusted so that both tests had comparable implosion times and peak currents. (Shell masses are controlled by changes in the plenum pressures. Note also that this nozzle employed a recess: the nozzle face was set back 1.6 cm from a sparse wire grid that served as the cathode of the pinch.)

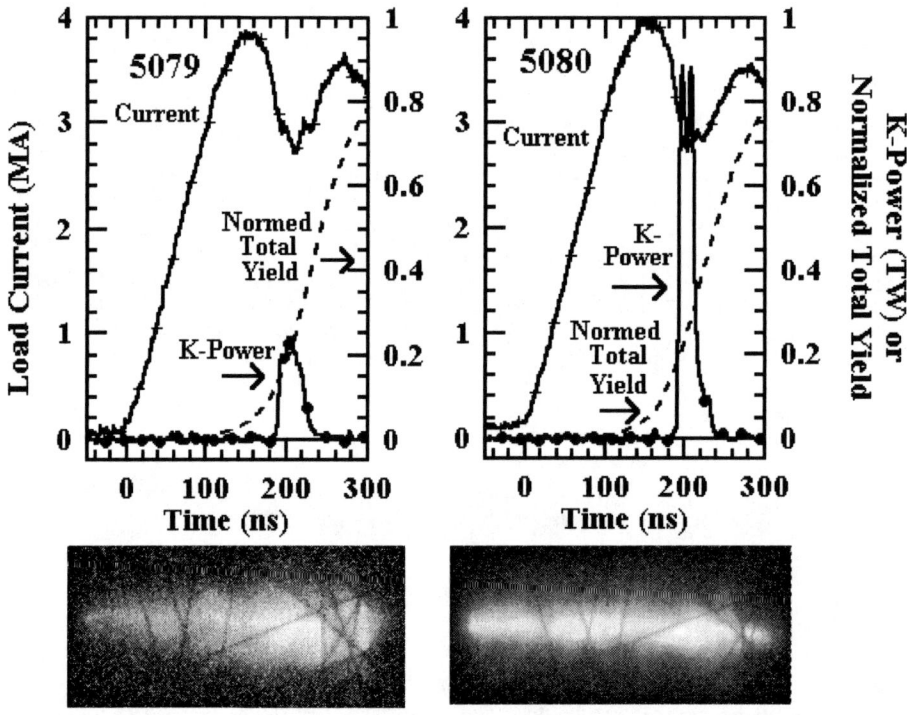

FIGURE 1. (Top) Current and power waveforms for the two tests; 5080 had low outer::inner mass ratio. The dotted trace is normalized cumulative total yield. (Bottom) Time-integrated pinhole images of K-shell emission. The anode end is to the right for each image. Pinch length was 3.8 cm.

Figure 1 compares key features of the two tests. With the more filled-in distribution of test 5080, K-power was over 4 times higher, K-yield was over 3 times larger (17.7 ± 0.6 vs. 5.3 ± 0.3 kJ), and pinch diameter was significantly reduced (2.6 vs 5.0 mm). (A one-dimensional imaging ["zipper"] diagnostic indicated that more than half of the apparent pulse width [18 ns] was due to zippering; the intrinsic pulse width was less than 7 ns FWHM for test #5080.) On the other hand, a bare aluminum cathode XRD used as a measure of total radiated output, showed no significant difference between the tests. (Note that K-shell output was complete before 50% of the total output has appeared.)

On all tests with low mass ratios, one of the gas plenums contained argon (Z=18) gas with 2% Freon ($C_2Cl_2F_2$) which supplied chlorine atoms (Z=17) as a tracer. As reported [6] with other double shell gas puffs, the intensity of chlorine x-ray emission relative to argon, as seen with a spectrograph, was reduced when the tracer was located in the outer plenum. This suggests that it is the mass that is initially nearest the axis that will participate best in the final hot plasma. However, even with the tracer in the outer shell, chlorine emission was readily seen, hence there must be significant mixing of the two shells either during the initial gas flow or subsequent implosion. From a streak spectrograph covering the dominant K-shell lines of argon and chlorine, the timings of onset of K-shell emission of the two species matched to within a nanosecond. Limited data do suggest that when the chlorine was in the inner shell, its emission lasted longer than the argon emission by up to ~5 ns.

DISCUSSION AND SUMMARY

Figure 2 summarizes the K-shell yield as a function of the mass ratio for five roughly comparable tests. The dotted curve is only a phenomenological fit. Above a mass ratio of 1, the curve was chosen to vary as about the square of the ratio. Ideally, the shape of this curve should be the result of detailed 2D MHD calculations that have been benchmarked against the measurements.

In the absence of such calculations, we consider the following simple picture. The role of the mass at largest radius is simply to accumulate energy from the generator during the rise of the current pulse. The stability of the implosion (or lack thereof) of this outer mass is relatively unimportant. (However, if there is little mass on axis, the imploded mass is severely disrupted and radiates poorly.) The kinetic energy of the outer mass is transferred to the mass at smaller radius via a shock. Convergence of the shock on the axis leads to plasma heating and radiation. The mass that can radiate is the mass nearest the axis during the final stages of the implosion.

At the 4 megamp level of Double Eagle, argon is still "inefficient" (yield scales as the fourth power of current), and we might expect the K-yield to vary as the square of the mass of the inner "shell" at least for small inner mass. (On a generator like Z where argon is "efficient", the yield would vary linearly with inner mass and the dotted curve of Figure 2 might be a straight line for mass ratios above 1.) When the implosion time is fixed, the available energy is independent of the inner mass. Thus, at sufficiently high inner mass there will be too many radiator atoms available, temperature will be limited and K-yield should decrease. Hence the curve of Figure 2 is assumed to fall to near zero for very low mass ratios. The benefit of good modeling of this configuration would be to predict the optimum mass ratio and guide future experiments.

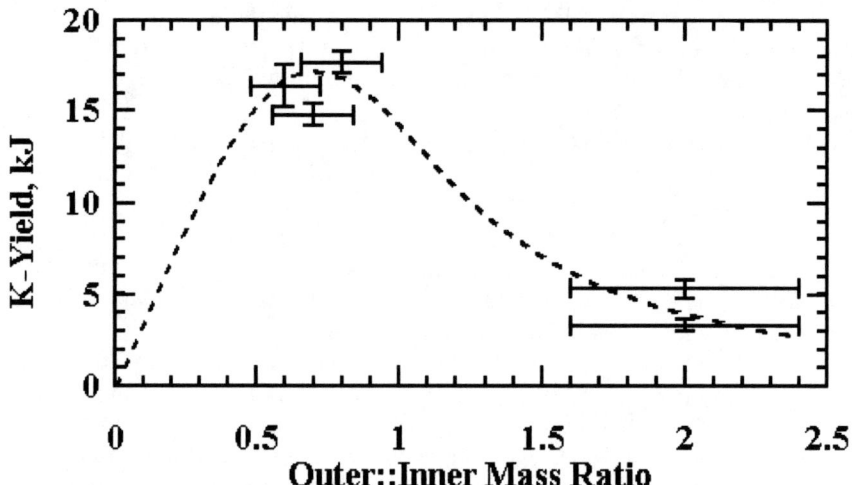

FIGURE 2. Argon K-shell yield as a function of the outer to inner mass ratio with implosion time held approximately constant near 190 ns. The error bars in mass ratio are due to the machining tolerances on the nozzles. A large mass ratio corresponds to a more shell-like, large-radius overall distribution of mass.

ACKNOWLEDGMENTS

This work was supported by the U.S. Defense Threat Reduction Agency. We greatly appreciate the efforts of the Double Eagle crew in making this test series a success.

REFERENCES

1. Lebedev, S. V., et.al., *28th IEEE International Conference on Plasma Science*, (2001).
2. Cuneo, M., et.al. *43rd Annual Meeting of the APS Division of Plasma Physics*, (2001).
3. Coleman, P.L., et.al., *Laser and Particle Beams*, **19**, 409 (2001).
4. Coleman, P.L., et.al., *29th IEEE International Conference on Plasma Science*, (2002).
5. Golberg, S.M and A.L. Velikovich, *Phys. Fluids B*, **5**, 1164 (1993).
6. Sze, H., et.al., *Phys. Plasmas*, **7**, 4223 (2000) and **8**, 3135 (2001).

Double Gas Puff Z-Pinch with Axial Magnetic Field for K-Shell Radiation Production

Alexander V. Shishlov, Rina B. Baksht, Stanislav A. Chaikovsky,
Aleksey Yu. Labetsky, Vladimir I. Oreshkin, Alexander G. Rousskikh,
Anatoly V. Fedunin

High Current Electronics Institute, 4 Academichesky Ave., Tomsk, 634055, Russia

Abstract. A double gas puff with a solid fill inner shell and an annular outer shell with axial magnetic field is proposed as a possible load configuration for a plasma radiation source for K-shell radiation production of high-Z materials. This load configuration is investigated in the experiments with neon gas puffs on the IMRI-5 generator (400 kA, 430 ns) and with argon gas puffs on the GIT-12 generator (2.5 MA, 300 ns). Influence of the axial magnetic field on z-pinch stabilization and K-shell radiation yield is studied.

INTRODUCTION

Creation of a plasma radiation source (PRS) with the quantum energy higher than 10 keV presents a considerable challenge. High atomic number materials should be used as a working medium, for example, krypton, whose K-lines are in the spectral range from 12 keV to 17 keV. A multi mega-ampere generator is required in order to obtain plasma, which efficiently radiates in this region. However, when high-Z materials are used, the kinetic processes such as relaxation of ionization states, interchange of energy between ions and electrons, radiation in lines, etc are brought to the forefront. In [1], Kr K-shell radiation yield from a z-pinch implosion on a pulsed power generator with peak currents between 60 MA to 100 MA was investigated. The authors showed that the efficiencey of K-shell radiation generation is very low, because Kr has a high radiative ability in M- and L-lines. Therefore, achieving the conditions required to generate K-shell radiation is difficult, since the energy deposited in the pinch is emitted in a softer spectral range.

Another obstacle which appears in the experiments with z-pinch PRS is instabilities that develop inevitably during a z-pinch implosion. It is especially critical for a PRS which radiates in hard spectral range, since in this case the need to use high-Z materials dictates the necessity to start an implosion from a large initial radius in order to obtain a final implosion velocity of 10^8 cm/s. Indeed, according to [2], the initial gas puff radius of 3 cm is optimal for generation of Kr K-shell radiation using a pulsed power generator with a peak current of 60 MA and a current rise time of 100 ns, and the initial radius should be increased proportionally if the current rise time increases. It was shown in a row of experimental works (see, for example [3,4,5]) that z-pinch implosions from large initial radii are very unstable and, therefore, they are inefficient radiators.

In this paper, we investigate a double gas puff with an axial magnetic field as a possible load design that could help to overcome the problems stated above.

DOUBLE GAS PUFF WITH AN AXIAL MAGNETIC FIELD

A schematic drawing of a double gas puff with an axial magnetic field is shown in Fig.1. In this load design, the inner shell is a solid gas column (solid fill), and the outer shell is an annular gas puff with an axial magnetic field inside.

Employing such load configuration has the following advantages. Two stabilization mechanisms are present: stabilization by an axial magnetic field [6, 7] and snow plow stabilization [8]. This will allow long time implosions and implosions from large initial radii providing better coupling between the load and a generator. It is necessary to note that the magnetic field provides stabilization of the outer shell and should have small influence on the implosion dynamics of the inner shell. The main mechanism responsible for the inner shell stabilization in this load design is the mechanism of snow plow stabilization.

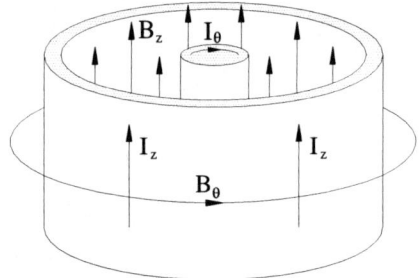

FIGURE 1. Schematic drawing of a double gas puff with an axial magnetic field

When a high-Z material is used as a working medium, the role of the magnetic field is not limited only to implosion stabilization. The magnetic field prevents a z-pinch from excessive compression that leads to large radiation losses in the soft spectral range and does not allow obtaining the conditions required for K-shell radiation generation. In the presence of the axial magnetic field, heating of the inner shell matter up to high temperatures by the action of a strong shock wave occurs at comparatively weak compression [9], and the life-time of the high temperature plasma capable of K-shell radiation generation is prolonged. The 1D RMHD simulation carried out for this load configuration showed that efficient generation of Kr K-shell radiation can be achieved using this load design [10].

It is necessary to note, however, that application of the axial magnetic field can present a pitfall for a K-shell PRS. As it was reported in [11], in spite of better implosion stability and better quality of the final pinch, application of an axial magnetic field resulted in reducing the K-shell power and yield. The reason for this behavior seems to be the energy losses on axial magnetic field compression that leads to reducing the energy delivered to plasma. Therefore, the proposed load should be tested experimentally to answer the question if it is possible to find optimal load parameters that provide z-pinch stabilization without significant reduction in the efficiency of K-shell emission. Such experiments were carried out with neon and argon B_z double gas puff on the IMRI-5 and the GIT-12 generators, respectively.

EXPERIMENTAL RESULTS

Prior to the experiments, 1D RMHD simulations were carried out in order to determine a region of the optimal load parameters with respect to the K-shell radiation yield. The simulation results for the conditions of the IMRI-5 generator and Ne double gas puff with an axial magnetic field are shown in Fig.2. The initial value of the axial magnetic field in the simulation was chosen in accordance with the stabilization criterion $B_0(kG) = (10 \div 30) I_{max}(MA)/R_0(cm)$ from [6] (for the case of double shell gas puff, R_0 is the outer shell initial radius). The simulation showed that the K-shell yield has a maximum at the following double gas puff parameters: $d_{in}/d_{out} = 0.2 \div 0.25$; $0.5 \leq m_{in}/m_{out} < 1$.

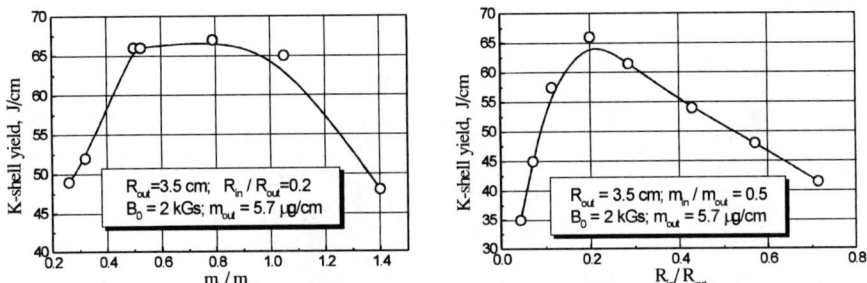

Figure 2. Simulation results of the Ne K-shell radiation yield as a function of the shell mass ratio and the shell radius ratio of a B_z double gas puff on the IMRI-5 generator.

The first set of experiments with a B_z double gas puff was carried out on the IMRI-5 generator (400 kA, 430 ns). Four nozzles configurations with different radius ratio of the outer and inner shells were tested in shots both with and without an axial magnetic field. Details of these experiments are given in [12]. Here, we would like to remind the main results. The maximum K-shell yield was observed in the nozzle configuration with the outer shell diameter of 60 mm and the inner shell diameter of 12 mm both in the shots with and without the axial magnetic field. At the optimal double gas puff parameters, the maximum K-shell yield was 25% lower in the experiments with the axial magnetic field, however the K-shell radiation power was twice higher. Apparently this increase in the K-shell radiation power occurs due to a better implosion stability observed in the experiments when the axial magnetic field was applied. The initial value of the axial magnetic field, at which z-pinch stabilization was observed without significant reduction in K-shell yield, is $B_0 \approx 0.5 \cdot B_{st}$, where $B_{st}(kG) = 10 \cdot I_{max}(MA)/R_0(cm)$.

The second set of the experiments was conducted on the GIT-12 generator (2.5 MA, 300 ns) with an argon gas puff. A nozzle configuration with the outer shell diameter of 80 mm and the inner diameter of 16 mm were tested. The outer shell mass was chosen so that the implosion times of 250-300 ns and the load peak current of 2.3-2.5 MA were ensured. In most of the shots, the outer shell mass was 60±10 µg/cm. The mass of the inner shell was varied to find the optimum in respect to the K-shell radiation yield.

The Ar K-shell radiation yields and powers obtained in the experiments are shown in Fig.3. It is necessary to note that in the experiments without the axial magnetic field, the double gas puff with the solid fill inner shell showed a better performance; the maximum Ar K-shell yield increased 1.5 times compared to our previous experiments [13]. However, in the shots with the axial magnetic field, the K-shell radiation yield decreased and did not exceed 400 J/cm. A very interesting feature of the K-shell yield dependency on the inner shell mass in the experiments without magnetic field is a pronounced maximum at the inner shell masses of 40-60 μg/cm and a sharp decrease in the K-shell yield at higher masses.

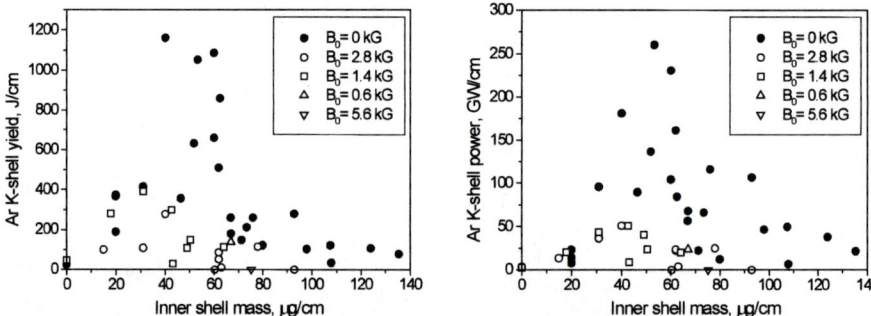

Figure 3. Ar K-shell radiation yield and power as a function of the inner shell mass for the shots with and without axial magnetic field.

Implosion stabilization was observed at the initial value of the axial magnetic field of 2.8 kG (0.5 B_{st}) and 1.4 kG (0.25 B_{st}). Further decrease of the initial magnetic field up to 0.6 kG results in the same final pinch quality as it was in the shots without the axial magnetic field. The influence of the magnetic field on z-pinch stabilization is illustrated in Fig.4. A general tendency is that the final pinch has a smaller diameter when the axial magnetic field was applied. However, analysis of the experimental data shows that this is not associated with a higher degree of pinch compression, but it is a result of lower temperatures and lower densities achieved in the implosions. For example, for the shots presented in Fig.4 the electron temperatures were 1.5 keV and 1.1 keV, and the ion densities were $8.5 \cdot 10^{18}$ cm^{-3} and $6 \cdot 10^{18}$ cm^{-3},

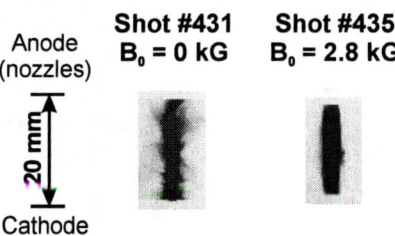

Figure 4. Time-integrated pinhole camera pictures of Ar double gas puff implosion with and without axial magnetic field (in Ar K-shell).

correspondingly for shot #431 and shot #435, whereas the outer and the inner shell masses were the same. Reduction in the final plasma parameters (temperatures and densities) is also confirmed by the fact that in the experiments with the axial magnetic field radiation yield in the soft spectral range measured by XRD is, on the average, two times lower.

The data obtained with the help of a visible light streak camera indicates changes in implosion dynamics (see Fig.5). First of all, the inner shell velocity becomes lower in the shots with the axial magnetic field. In the range of inner shell masses from 40 μg/cm to 60 μg/cm, only in one shot the inner shell velocity was higher than $5 \cdot 10^7$ cm/s (in all shots without magnetic field, which produced a kilo-joule K-shell yield, the inner shell implosion velocities were $(5 \div 6) \cdot 10^7$ cm/s). Second, the streak camera pictures show that implosion stops at larger radius, the pinch bounces back and then implodes again. Such behavior is not usually observed in the shots without the axial magnetic field.

Comparison of dI/dt traces also shows that the negative dip associated with the z-pinch implosion is significantly less in the shots with the axial magnetic field. This and the fact that the inner shell implosion velocities are lower in the experiments with the axial magnetic field allow us to conclude that energy deposition into the inner shell is less in this case.

The reason for this is not clear. It is reasonable to suggest that the axial magnetic field becomes too high in the

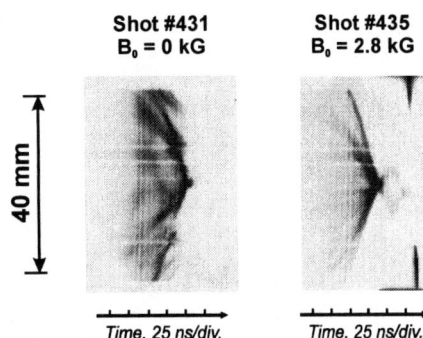

Figure 5. Streak camera pictures show changes in implosion dynamics.

final stage of implosion and stops the implosion, and, accordingly, too much energy is spent on compression of the magnetic field. Simple estimations show that this could not be the reason. Really, let us assume that the inner shell has a very low conductivity, all the magnetic flux is preserved (though 1D simulations show that at least 30% of the magnetic flux should be lost due to diffusion), and the outer shell compresses the magnetic field of 2.8 kG from the radius of 40 mm to the final radius of 1 mm. In this case, at the final radius of 1 mm and the peak load current of 2.5 MA, azimuth and axial magnetic fields become comparable (but implosion should still continue, since the imploding shell has a rather high velocity by this time), and the energy spent on B_z field compression is 2.5-3 kJ. Such energy losses could result in 50% reduction in the K-shell yield. However, if we consider the initial magnetic field of 1.4 kG, the energy losses is less than 1 kJ, and they can not be responsible for the K-shell yield reduction observed in the experiments. Therefore, the role of the axial magnetic field and its influence on the z-pinch implosion dynamics is not well understood and requires further theoretical and experimental investigations.

SUMMARY

Experiments with a double shell gas puff with an annular outer shell and a solid fill inner shell were carried out on the IMRI-5 and the GIT-12 generators. The influence of the axial magnetic field on the implosion stability and K-shell radiation yield was investigated. In the experiments on the IMRI-5 generator the optimal parameters of the B_z double gas puff were found that provide implosion stabilization without a significant reduction of K-shell radiation yield and a twofold increase in K-shell radiation power. In the experiments on the GIT-12 generator, a double gas puff with a solid fill inner shell showed a better performance with respect to K-shell radiation emission. The Ar K-shell radiation yield of 1.1 kJ/cm was registered in 300-ns implosion at the current level of 2.3-2.5 MA, that is 1.5 times higher compared to our previous experiments. However, in the shots with the axial magnetic field, the K-shell radiation does not exceed 400 J/cm. It was shown experimentally that the value of the initial axial magnetic field required for z-pinch stabilization in this load configuration can be determined as $B_0(kG) = 2.5 \cdot I_{max}(MA)/R_0(cm)$. Such values of the magnetic field should not alter the z-pinch implosion dynamics, and the energy losses associated with the axial magnetic field compression can not lead to the K-shell yield reduction observed in the experiments. The influence of the axial magnetic field on the z-pinch dynamics is not well understood and requires additional theoretical and experimental work.

ACKNOWLEDGMENTS

The authors wish to express their gratitude to Dr. N.Ratakhin and Dr. A.Velikovich for stimulating discussions of this work. The excellent performance of the GIT-12 operation crew is gratefully acknowledged. The work was supported by DTRA.

REFERENCES

1. Davis J., Giuliani J., Rogerson J., and Thornhill J., in *Proc. of 11th Int. Conf. on High Power Particle Beams, Prague, Czech Republic*, 1996, pp.709-712.
2. Mosher, D., Qi, N., and Krishnan, M., *IEEE Trans. Plasma. Sci.* **26**, 1052-1061 (1999).
3. Baksht, R.B., Datsko, I.M., Fedunin, A.V., Kim, A.A., Labetsky, A.Yu., Loginov, S.V., Shishlov, A.V., and Oreshkin, V.I., *Plasma Phys. Rep.* **21**, 907-913 (1995).
4. Coleman, P., Rauch, J., Rix, W., Thompson, J., and Wilcon, R., in *Proc. of 4th Int. Conf. on Dense Z-pinches, Vancouver*, 1997, AIP Conf. Proc. 409, p.119-123.
5. Commisso, R.J., Apruzese, J.P., Black, D.C., Boller, J.R., Moosman, B., Mosher, D., Stepanakis, S.J., Weber, B.V., and Young, F.C., *IEEE Trans. Plasma Sci.* **26**, 1068-1085 (1998).
6. Budko, A.B., Felber, F.S., Kleev, A.I., Liberman, M.A., and Velikovich, A.L., *Physics Fliuds B* **1**, 598 (1989).
7. Rudakov, L.I., in *Proc. of 2nd Int. Conf. on Dense Z-pinches, Laguna Beach*, 1989, AIP Conf. Proc. 195, p.290.
8. Golberg, S.M., and Velikovich, A.L., in *Proc. of 3rd Int. Conf. on Dense Z-pinches, London*, 1993, AIP Conf. Proc. 299, pp.42-50.
9. Oreshkin, V.I., *Russian Physics Journal* **38**, 1203 (1995).
10. Oreshkin, V., Rudakov, L., Davis, J., Clark, R., Velikovich, A., The 28th IEEE Int. Conf. on Plasma Science, Las Vegas, USA, 2001, *Book of Abstracts*, p.201.
11. Sorokin, S., and Chaikovsky, S., *Proc. 4th Int. Conf. on Dense Z-pinches, Vancouver*, 1997, AIP Conf. Proc. 409, pp. 593-596.
12. Chaikovsky, S.A, et al., "Effect of an Axial Magnetic Field on K-Shell radiation of a Neon Double Gas Puff," these proceedings.
13. Shishlov, A.V., Baksht, R.B., Fedunin, A.V., Fursov, F.I., Kovalchuk, B.M., Kokshenev, V.A., Kurmaev, N.E., Labetsky, A.Yu., Oreshkin, V.I., and Rousskikh, A.G., *Phys. Plasmas* **7**, 1252-1262 (2000).

Effect of an Axial Magnetic Field on the K-shell Radiation of a Neon Double Gas Puff

Stanislav A. Chaikovsky, Aleksey Yu. Labetsky, Alexander V. Shishlov, Anatoly V. Fedunin, Vladimir I. Oreshkin, Rina B. Baksht, Alexander G. Rousskikh

Institute of High Current Electronics, 4, Akademichesky Ave, 634055, Tomsk, Russia

Abstract. Optimization of the neon K-shell radiation yield and power was performed in double shell gas puff experiments with and without axial magnetic field on the IMRI-5 generator at a current level of 400 kA and a current risetime of 430 ns. Optimum parameters of a double gas puff with an axial magnetic field have been found that provide both implosion stability and a K-shell yield comparable to the expected K-shell yield for a generator with a risetime of 100 ns.

INTRODUCTION

Powerful laboratory-scale radiation sources operating in a spectral range of 10–20 keV are of great interest for material studies. A possible approach to produce such sources is utilizing the z-pinch of a material whose K-shell radiation falls in this spectral range. The main obstacles to efficient production of 10–20 keV K-shell x-radiation with a z-pinch are the plasma implosion instabilities and enormous radiation losses in softer x-rays [1]. By 1-D MRHD simulations it has been demonstrated that a double shell z-pinch with an axial magnetic field and solid fill inner shell can provide efficient production of Kr K-shell (12–17 keV) x-rays on a multimegaampere generator [2]. The axial magnetic field and double shell z-pinch structure can provide stability of the implosion [3]. According to the simulations [2], the magnetic field also prevents excessive plasma compression during the implosion which results in reduced soft radiation losses and more efficient plasma heating.

Since experiments on large machines are very expensive, we believe that step-by-step testing is an appropriate way of studying this promising load configuration. The experiments described here were our first step to study this scheme of a radiation source. We used a small 400-kA, 430-ns generator to implode double-shell neon z-pinches. The main goal of the experiments was to find out a double gas-puff configuration with a B_z field, which could provide implosion stability at a reasonably high energy transferred to the plasma, and, hence, sufficiently high K-shell power and yield.

EXPERIMENTAL

The experiments were carried out on the IMRI-5 generator, which provides a short-circuit peak current of 470 kA with a risetime of 430 ns. Preliminary calculations using the 1-D MRHD HCEI [4] code have shown that the inner-to-outer shell diameter ratio $D_{in}/D_{out} = 0.2$–0.25 is an optimum value for the Ne K-shell radiation production on the IMRI-5 generator. The optimum outer shell diameter has been found to be 6-7 cm and the optimum inner-to-outer shell mass ratio is in the range $0.5 < m_{in}/m_{out} < 1$.

In these experiments, we used four sets of nozzles with mean outer shell diameters of 4.4, 6.0, 8.0, and 10 cm. The inner shell was a solid fill with the nozzle exit diameter that made up 1/5 of the mean outer shell diameter. A schematic drawing of the IMRI-5 generator load unit is given elsewhere [5].

For each configuration, the inner shell mass optimum for the generation of K-shell radiation for a double shell gas puff with and without an axial magnetic field was found. The scan involved adjusting the outer shell mass to ensure a 450-500-ns implosion time. The initial magnetic field B_0 was varied in the range 0.25-2.5 kG.

Additional shots were performed for each nozzle configuration both with the outer shell only and with the inner shell only. The shell masses were than estimated by comparing snowplough calculations and these experimental results.

TABLE 1. K-shell yield Y_k and power P_k for shots where a maximum K-shell yield at $B_o=0$ and $B_o \approx 0.5 B_{st}$ was obtained

D_{out}, cm	D_{in}, cm	B_0, kG	Y_k, J/cm	P_k, GW/cm
4.4	1.0	0	12	0.4
		1.1	1.5	0.1
6.0	1.2	0	63	2
		0.66	48	4
8.0	1.6	0	38	2.8
		0.55	29	1.7
10.0	2.0	0	32	0.7
		0.49	4	0.2

The Ne K-shell radiation power and yield were measured by two vacuum x-ray diodes (XRD) with aluminum cathodes and the following filters: aluminum 10 μm + Mylar 3 μm (XRD1) and aluminum 0.6 μm + Kimfoil 6 μm (XRD2). Two pinhole cameras (with an 8-μm -thick aluminum filter and a 0.2 μm aluminum + Kimfoil 2 μm filter) were fielded to record a time-integrated image of the pinch. The implosion dynamics was observed using a visible light streak camera.

RESULTS AND DISCUSSION

The optimum configuration that provides the maximum K-shell radiation yield in the experiments both with an axial magnetic field and without magnetic field was found to be a double gas puff with an outer shell diameter of 6 cm and an inner shell diameter of 1.2 cm (Table 1) and with respective masses m_{out} = 10-12 μg/cm and

m_{in} = 5-7.5 µg/cm. The optimum outer shell radius and mass ratio m_{in}/m_{out} are in good agreement with 1-D RHMD simulations.

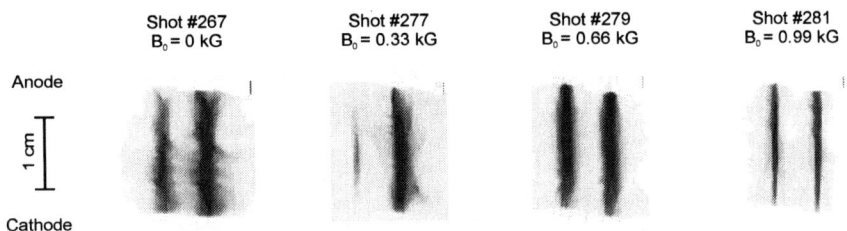

FIGURE 1. Time-integrated pinhole images obtained for different values of the initial axial magnetic field (in the pinhole camera pictures: the pinch image in Ne K-lines are on the left and the pinch image in soft x rays is on the right).

The value of the axial magnetic field sufficient to stabilize implosions is $B_0 \geq 0.66$ kG (Fig. 1), which is half the value determined by the criterion $B_0 \geq B_{st}$ [kG] = $10 I_{max}$ [MA]/R_0 [cm] [6]. Here, R_0 is the initial radius of a single-shell z-pinch or the initial radius of the outer shell of the double-shell z-pinch, I_{max} – the maximum generator current. The Table 1 presents K-shell yield Y_k and power P_k for shots where a maximum K-shell yield at B_0=0 and $B_0 \approx 0.5\ B_{st}$ was obtained.

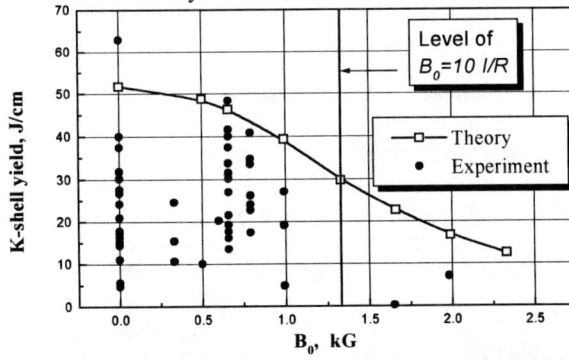

FIGURE 2. Ne K-shell radiation yield as a function of the initial axial magnetic field for outer shell initial diameter of 6 cm. Experimental and 1-D RMHD simulation results.

The maximum neon K-shell yield was measured to be 63 ± 12 J/cm without a magnetic field and 48 ± 10 J/cm with B_0 = 0.66 kG. The K-shell yield was estimated to be 74 J/cm for a single-shell implosion on a generator with the same current level (400 kA) and a current risetime of 100 ns. A two-level model [7] was used and a fifteen-fold radial compression was assumed for the estimation.

Experimental and 1-D MRHD calculated dependencies of the Ne K-shell yield and power on the initial axial magnetic field for the optimum nozzle configuration are shown in Fig. 2 and Fig.3. For the experimental values of the K-shell radiation power, there is a pronounced maximum for the initial axial magnetic field of 0.66 kG.

An increase in initial magnetic field results in a decrease in shot-to-shot variation of the K-shell radiation yield (see Fig. 2). With the average values of the K-shell yield practically identical in the shots with $B_0 = 0$ and $B_0 = 0.66$ kG, the spread in experimental values about an average value in the latter case is half that in the former.

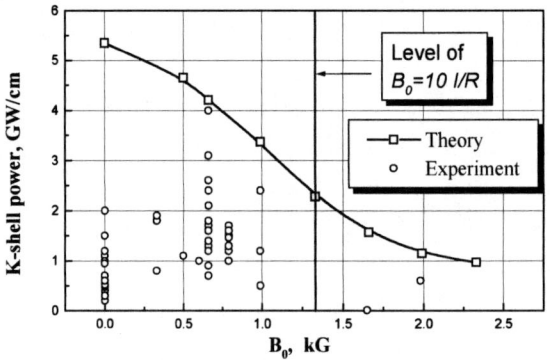

FIGURE 3. Ne K-shell radiation power as a function of the initial axial magnetic field for outer shell initial diameter of 6 cm. Experimental and 1-D RMHD simulation results

An additional set of experiments was carried out with a double shell gas puff with an outer shell diameter of 6 cm and an annular inner shell. The mean diameter of the inner shell was 1.1 cm and its outer diameter was approximately equal to that of the solid fill. The maximum K-shell radiation yield and power in shots with an annular inner shell were 1.5-2.5 times lower than those obtained in shots with a double gas puff with a solid fill inner shell.

SUMMARY

Optimization of the Ne K-shell radiation yield and power was performed in the double shell gas puff experiments with and without an axial magnetic field on the IMRI-5 generator at a current level of 400 kA and a current risetime of 430 ns.

Optimum parameters of a double gas puff with an axial magnetic field and solid fill inner shell have been found which provide implosion stability at implosion times of 450–500 ns and a K-shell yield comparable to that expected for a single-shell z-pinch produced on the generator with a current risetime of 100 ns. The optimum parameters and K-shell yield are in good agreement with the results of 1-D RMHD calculations.

Application of an axial magnetic field makes it possible to increase the K-shell radiation power and reduce shot-to-shot variations of the K-shell yield.

*) Work is supported by Contract DTRA01-01-P-0102.

REFERENCES

1. Davis J., Giuliani J., Rogerson J., and Thornhill J., *Proc. 11th Intern. Conf. on High Power Particle Beams, Prague, Czech Republic*, 1996, p.709.
2. Oreshkin V., Rudakov L., Davis J., Clark R., and Velikovich A., *The 28th IEEE Intern. Conf. on Plasma Science, Las Vegas, USA, 2001*, Book of Abstracts, p.201.
3. Sorokin S. and Chaikovsky S., *Proc. 3th Int. Conf. on Dense Z-pinches, London*, 1993, p.83.
4. Oreshkin V., *Preprint HCEI SB RAS №4* (1994).
5. Chaikovsky S. and Labetsky A., *see a companion paper in this volume*.
6. Bud'ko A., Felber F., Kleev A., Liberman M., and Velikovich A., *Phys. Fluids*, B, 1, 598 (1989).
7. Mosher D., Qi N., and Krishnan M., *IEEE Trans. Plasma Sci.* 26, 1052 (1999).

Soft X-ray Production and Spectrum Measurements in Imploded Krypton Liners

Sergey A. Sorokin and Stanislav A. Chaikovsky

Institute of High Current Electronics, 4 Akademichesky Ave., Tomsk 634055, Russia

Abstract. Experiments on efficient conversion of the pulse generator energy to soft x rays were carried out on the MIG facility (I ~1.9 MA, τ ~80 ns). Krypton gas-puff liners were used to produce a pulse of N-shell and M-shell radiation (0.1–1 keV). A radiation yield of 50±5 kJ in ~30 ns was attained, which made up ~60% of the energy in the water transmission line. The x-ray power and time resolved spectrum was measured with an x-ray diode array. Both the deconvolution and edge filter techniques were used to obtain the spectra. Gas-puff filters were used to extend the spectral range of the XRD's to <150 eV. It has been demonstrated that x-ray spectra can be adjusted by varying of the initial radius of the liner.

INTRODUCTION

Magnetic compression of cylindrical gas-puff liners is widely used for the production of high-power pulsed soft x rays (SXR) [1, 2]. The efficiency of the conversion of the electric energy of the generator to x-ray radiation, determined by the energy transferred to the vacuum diode, may be over 50 %. The spectral constitution of the radiation depends in large measure on the atomic number of the liner material and on the velocity of the imploding plasma. For measuring the radiation power and pulse-integrated yield, vacuum x-ray diodes (XRD's) [3] and foil bolometers [4] are most commonly employed. In general, a signal from one recording channel gives no information either about the spectrum-integrated radiation power or about the radiation energy distribution over the spectrum. The use of a set of probes with various, partially overlapping spectral response functions permits one, in principle, to restore the form of the radiation spectrum [5, 6, 7]. The accuracy of reconstruction of the spectrum depends in many respects on the choice of the combination of the photoelectric cathode and filter materials and on the accuracy to which the mass thickness of the filters is estimated. The films of thickness < 1 μm that would have a high transmissivity in the range of energies below 150 eV are not produced commercially. The use of gas filters substantially extends the choice of combinations of the cathode and filter materials. A gas-filter-producing system, once manufactured and adjusted, allows one to do away with the procedures of thin film (0.1–1 μm) preparation, measuring their thickness, and providing their survival in the process of operation. The mass thickness of a gas-puff filter can readily be varied and, in fact, can be made as small as desired; that is, the sensitivity of the XRD can be extended toward the region of low energies. In addition, gas filters can be used to measure the SXR spectrum by the edge filter technique since for the majority of gases the absorption K-edges are in the SXR range.

This paper presents experiments with imploding krypton liners, performed on the MIG generator ($I \sim 1.9$ MA, $\tau \sim 80$ ns) [8] with the aim to produce soft N-shell and M-shell x radiation. The liner kinetic energy per atom, K, is determined by the liner velocity, and it was assumed that $K \sim 2E_{min}$ ($E_{min} = 13.6$ keV is the sum of the ionization energies and the ion and electron thermal energies to reach the ionization state Z = 19), which corresponds to the liner velocity $2.5 \cdot 10^7$ cm/s. In the course of the experiment, the radiation spectral distribution was adjusted by varying the liner velocity. Since for the generator current $I \sim 1.9$ MA the cooling time of the radiating krypton plasma, τ_r, is shorter than the time of plasma confinement in the compressed state, $\tau_i \sim r_f/v_f$, the efficiency of the generation of x rays is determined in the main by the efficiency of the energy transfer from the generator to the liner. The efficiency of the generator-to-liner energy transfer was optimized by 0-dimensional computations.

To measure the pulse-integrated yield, power, and spectral distribution of SXR in the range 70–1500 eV, a copper-foil bolometer and a set of five XRD's were used. Along with thin-film filters, gas-puff filters were also employed which were produced by feeding a gas into the diagnostic channel tube in a pulsed manner with the help of a fast electrodynamic valve [9]. The original liner of initial diameter 14–20 mm and length 30–40 mm was produced using a fast valve and a Laval nozzle.

DIAGNOSTICS

For resistive elements of bolometers copper strips of length 14 mm, width 1 mm, and thickness 2 (on a dielectric support) and 15 μm were used. A reference current of amplitude 100–200 A and duration about 10 μs was passed through the resistive element. The requirement that the XRD should depend linearly on radiation flux poses a restriction on the radiation flux onto the surface of the photoelectric cathode: $W < 10^5$ W/cm^2. When operating an isotropic SXR source of power $P > 10^{12}$ W, it is necessary to place the XRD away from the source at $r > (P/4\pi W)^{0.5} \approx 10^3$ cm. The use of radiation flux attenuators reduces the length of the measuring channels. In the experiment under consideration, the radiation flux onto the photoelectric cathode was attenuated by a slit placed in the measurement channel tube normal to the pinch axis. The flux attenuation coefficient is defined as $(1+ a/b)$ Δ/L, where L is the pinch length, a and b are the respective distances from the slit to the pinch and to the photocathode, and Δ is the slit width. The diagnostic line-of-sight with a gas filter is shown schematically in Fig. 1. The fast valve is centrally located. The time of gas supply into the tube (<500 μs) should be shorter than the time it takes for the gas to expand in the tube to reach the XRD and the tube edge. In this case, the mass thickness of the filter can readily be determined from the known gas mass in the tube and the tube cross-

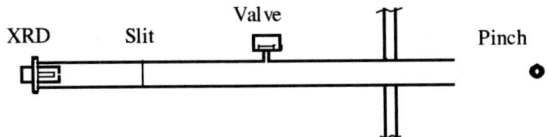

FIGURE 1. Schematic of the diagnostic line-of-sight with a gas-puff filter.

sectional area. To reconstruct the spectrum by the iteration method, four aluminum-cathode XRD's were used [filters: oxygen, $1.2 \cdot 10^{-4}$ g/cm^2 (1); 2.5 μm Mylar (2); 2.5 μm Mylar + 0.4 μm aluminum (3), and 8 μm aluminum (4)] and one graphite-cathode XRD (filter: 0.3 polystyrene μm). To calculate the response functions of XRD's, data on the quantum efficiency of photocathodes [10] and on the filter transmissions [11] were used. XRD signals were recorded with a Tetronix 640A digital oscilloscope. The graphite-cathode XRD with a 0.4-μm polystyrene filter has a slightly varying response function in the range 70–900 eV and allows radiation power measurements accurate to about 30%. It should be noted that the use of gas filters makes it possible to roughly construct a spectrum by XRD signals with quasi-constant response functions in certain spectral ranges (edge filter method). In the experiment under consideration, to measure the radiation spectrum in the range 125–690 eV, an array of four aluminum-cathode XRD's was used with the following filters: (1) acetylene, $2.3 \cdot 10^{-4}$ g/cm^2; (2) nitrogen, $1.4 \cdot 10^{-4}$ g/cm^2; (3) oxygen, $9 \cdot 10^{-5}$ g/cm^2, and (4) teflon, 5 μm. The radiation powers measured in proper spectral ranges can be used as a zero approximation in an iterative algorithm of spectrum reconstruction.

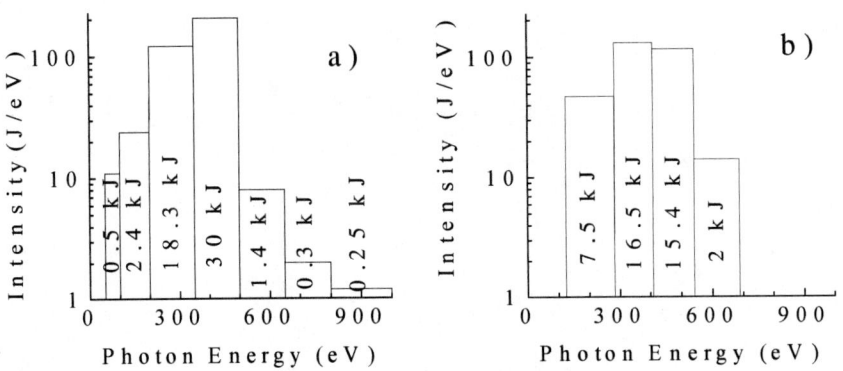

FIGURE 2. Histograms of time-averaged spectra reconstructed by an iterative algorithm (shot 27) (*a*) and measured by the edge filter technique (shot 37) (*b*).

RESULTS

Figure 2 presents histograms of time-averaged spectra for two shots with close parameters performed on a krypton liner. The spectrum in Fig. 2a (shot 27) has been reconstructed by an iterative algorithm and that in Fig. 2b (shot 37) has been measured by the edge filter method. The radiation yield integrated over the measured spectral range is, respectively, 54 and 42 kJ. The integrated yield for shot 27, measured with the graphite-cathode XRD is estimated (with the spectral sensitivity averaged over the range 70–700 eV) to be ~57 kJ, which makes up ~70% of the energy in the water transmission line.

The yield measured by a bolometer (2 μm) is ~20% lower than that measured by an array of five XRD's. This, perhaps, is related to the heat removal by the support.

REFERENCES

1. Turchi, P. J., and Baker, W. J., *J. Appl. Phys.* **44,** 4936 (1973).
2. Pereira, N. R., and Davis J., *J. Appl. Phys.* **64,** R1-R27 (1988).
3. Kornblum, H. N., and Slivinsky, U. W., *Rev. Sci. Instrum.* **49,** 1204-1206 (1978).
4. Degnan, J. H., *Rev. Sci. Instrum.* **50,** 1223-1225 (1979).
5. Burns, E. J. T., *Adv. X-Ray Anal.* **18,** 117 (1974).
6. Bailey, J., Fisher, A., and Rostoker, N., *J. Appl. Phys.* **60,** 1939 (1986).
7. Basov, N. G., Zakharenkov, Yu. A., Rupasov, A. A., et al, *Dense Plasma Diagnostics*, Moscow: Nauka, 1989, pp. 126-132 (in Russian).
8. Luchinskii, A. V., Ratakhin, N. A., Fedushchak, V. F., et al., *Izv. Vyssh. Uchebn. Zaved.* **38,** 58-66 (1995) (in Russian).
9. Sorokin, S. A., *Pribory i Tekhnika Eksperimenta* **N 3,** 136-138 (1999) (in Russian).
10. Day, R. H., Lee, P., and Saloman, E.B. et al., *J. Appl. Phys.* **52,** 6965-6973 (1981).
11. Henke, B. L., Gullikson, E. M., and Davis, J. C., *Atomic Data Nucl. Tables.* **54,** 181 (1993).

Spatial Structure of X-ray Emission of a Gas-puff Z-pinch Plasma

Keiichi Takasugi, Satoru Narisawa[a1] and Hisashi Akiyama[a2]

Institute of Quantum Science, Nihon University
[a]*College of Science and Technology, Nihon University*
1-8 Kanda-surugadai, Chiyoda-ku, Tokyo 101-8308, JAPAN

Abstract. Spatial and spectral characteristics of Ar gas-puff z-pinch plasma were investigated using a convex crystal spectrograph. Radial distribution of He-like Ar line corresponded to hot spots. H-like Ar line was observed in Cu electrode experiment. K_α lines of metal atoms had a wide distribution, which corresponded to cloud structure of x-ray image. Fe K_α spectrum was again observed in Al electrode experiment.

INTRODUCTION

As gas-puff z-pinch plasma is reproducible and easy in handling, it is noticed as a repetitive radiation source of x-ray. Hot spots generated in the pinch column are strong source of radiation. Spatial structure of hot spots was analyzed, and it was found to be well related to the spatial mode of MHD instability.[1-4] Cloud structure of x-ray image has been observed in gas-puff z-pinch experiment,[5] but the characteristics and mechanism of generation have not been made clear.

Spectra of x-ray radiation emitted from gas-puff z-pinch plasma have been measured, and K-shell radiations of Ar ions have been observed.[6] In order to observe wide spectral range of x-ray radiation, a convex crystal spectrograph was employed. As the spectrograph has no focusing spatial image of selected spectrum can be obtained.

In this research spectral and spatial characteristics of x-ray radiation emitted from Ar gas-puff z-pinch plasma was studied using a convex crystal spectrograph. By attaching a vertical slit axial distribution was also obtained. Spatial structures of K-shell radiations of Ar ions were investigated, and spectra for cloud structure were obtained.

EXPERIMENTAL SETUP

The experiment was carried out on the SHOTGUN gas-puff z-pinch plasma. The energy storage section consists of 24 µF high-speed capacitors, which is charged up to

---

[1] Present address, ULVAC, Inc.
[2] Present address, Nihon Engineering System Co., Ltd.

25 kV (7.5 kJ). Gas is injected through a hollow nozzle mounted on the anode using a high-speed electromagnetic gas valve. Ar gas is used in the experiment. The distance between the electrodes is 30 mm. The cathode has many holes in order not to prevent the gas flow. The electrode material is Cu or Al.

A convex crystal spectrograph, which is small and mobile, is employed for the observation of wide spectral range of x-ray emission. (Fig. 1) As the spectrograph has no focusing, it can take spatial information simultaneously. A mica crystal ($2d = 19.884$ Å) is used in this device. This crystal has the characteristic that reflectance is high for higher order reflection with odd number. The radius of curvature of the crystal is 19.5mm, and the radius of x-ray film is 75 mm. The x-ray films used here was Kodak DEF-392.

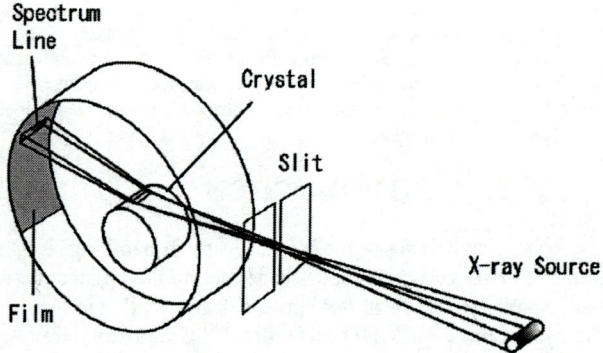

FIGURE 1. A convex crystal spectrograph with a vertical slit. A mica crystal (2d=19.884 A) is used.

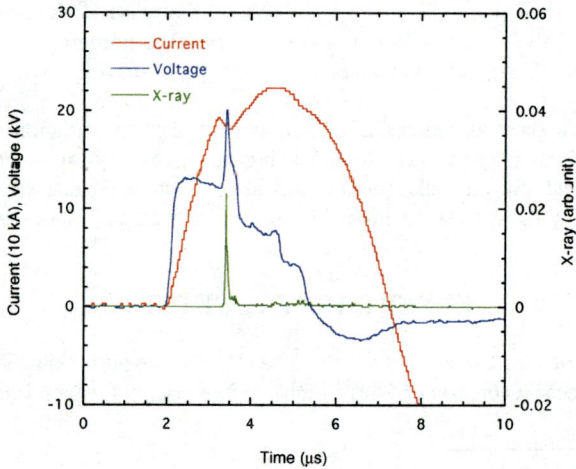

FIGURE 2. Typical plasma current, voltage and x-ray signal.

EXPERIMENTAL RESULTS

Typical plasma current, voltage between the electrodes and x-ray signal detected by a scintillation probe are shown in Fig. 2. Plasma current at the maximum pinch is about 200 kA. High voltage appears at this moment. The voltage was nearly doubled due to rapid increase of plasma inductance. Simultaneously pulsed x-ray is emitted. The data shows that all the x-rays were emitted simultaneously.

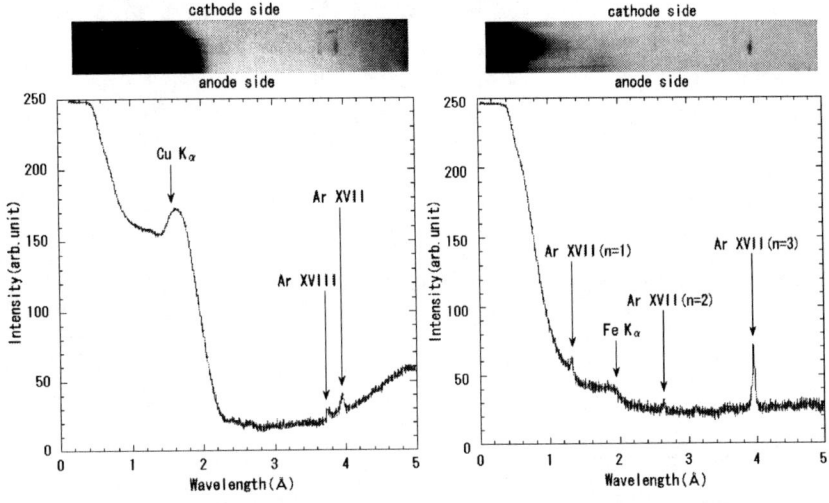

(a) Cu electrode (b) Al electrode

FIGURE 3. X-ray spectra taken by the convex crystal spectrograph. Ar spectra were sharp, but K_α spectra of metals were widely spread..

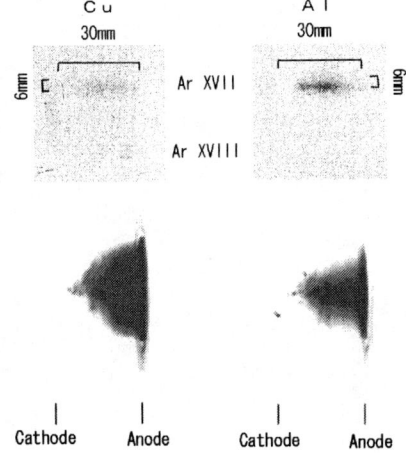

FIGURE 4. Spectrally resolved x-ray images Ar ions (top) and x-ray pinhole photographs (bottom).

Figure 3 shows x-ray photographs taken by the spectrograph for (a) Cu electrode and (b) Al electrodes and their intensity traces. Both spectrograms were exposed over 200 shots. The wavelength is plotted for third order reflection. In the Cu electrode experiment both Ar XVII (He-like) and Ar XVIII (H-like) lines were observed. (Fig. 3 (a)) The H-like Ar image was made up of three spots locating close to anode. Among the darkened film a widely spread image of Cu K_α line (1.542 A) was observed. In the Al electrode experiment intense Ar XVII (He-like) line was observed. Ar XVIII (H-like) line was not observed in this case. (Fig. 3 (b)) A widely spread image of Fe K_α line (1.938A) was observed. Al K_α line (0.834 A) was not in the observation range.

The spread of these lines are compared with pinhole photographs in Fig. 4. The spreads of Ar lines are about 6 mm, which correspond to the distribution of hot spots. The spreads of Cu and Fe lines are about 30 mm, which corresponds to the cloud structure.

SUMMARY

Spectral characteristics of x-ray radiation from Ar gas-puff z-pinch plasma have been investigated using a convex spectrograph. Cloud structure of x-ray was observed around hot spots, and all the x-rays were emitted simultaneously at the maximum pinch. Voltage spike was observed at this moment, which occurred due to rapid increase of plasma inductance. This voltage could be the source of electron beam generation.

Radial spread of He-like Ar line was about 6 mm, which corresponds to the distribution of hot spots. The H-like Ar image was observed only in Cu electrode experiment.

K_α line of Cu atom was observed in the Cu electrode experiment, and K_α line of Fe atom was observed in the Al electrode experiment. Both widths were about 30 mm, which corresponds to the cloud structure of x-ray image. As no other lines were found in the spectra, it is concluded that the cloud structure was formed by K_α line of metal atoms.

H-like Ar line was observed only in the Cu electrode experiment. This indicates that the electron temperature was higher than that of Al electrode experiment. If high-Z material is mixed into the plasma, it will release energy by radiation, and the pinch proceeds further. This is the so-called radiative collapse, and finally the pinch reaches high temperature.

REFERENCES

1. K. Takasugi, T. Miyamoto, K. Moriyama and H. Suzuki, *AIP Conf. Proc.* **299**, 251 (1993).
2. K. Moriyama, K. Takasugi, H. Suzuki and T. Miyamoto, *NIFS-PROC*-14, 43 (1993).
3. K. Moriyama, K. Takasugi, T. Miyamoto and K. Sato, *NIFS-PROC*-18, 90 (1994).
4. R. Muto, K. Takasugi and T. Miyamoto, *NIFS-PROC*-50, 139 (2001).
5. K. Takasugi, A. Takeuchi, H. Takada and T. Miyamoto, *Jpn. J. Appl. Phys.* **31**, 1874 (1992).
6. E. O. Baronova, K. Takasugi, V. V. Vikhrev and T. Miyamoto, *Proc. 13th Int. Conf. High Power Particle Beams 784* (2001).

Puff-gas Z-Pinch Experiment on "Yang" Accelerator

Jianjun Deng[a,b], Libin Yang, Yuancao Gu, Zhenghong Li, Xianbing Huang, Zhenghua Yang, Ning Ding, Cheng Ning, Bonan Ding, Xianjue Peng

[a] *Engineering Physics Dept., Tsinghua University*
[b] *China Academy of Engineering Physics, P.O.Box 919-150, Mianyang, Sichuan, 621900, PRC*

Abstract. This paper describes the puff-gas Z-pinch experiment undergoing at Institute of Fluid physics, China Academy of Engineering Physics. The introduction of the accelerator and the layout of the experiment are presented. Preliminary experimental results are also given

INTRODUCTION

Z-pinch can produce high power soft x-ray therefore is a potential technique approach for ICF. Recently the technologies and physics of Z-pinch have attracted many scientists' attention. This paper describes the puff-gas Z-pinch studies undergoing at Institute of Fluid Physics, China Academy of Engineering Physics. The experiment facility and preliminary results are introduced in the paper.

DESCRIPTION OF THE ACCELERATOR

The accelerator consists of Marx generator, Blumlein pulse forming line (PFL), switches, transmission line, magnetic insulating transmission line (MITL), and load (as shown in Fig.1). Both of the impedances of Blumlein PFL and Transmission line are 2 ohms.

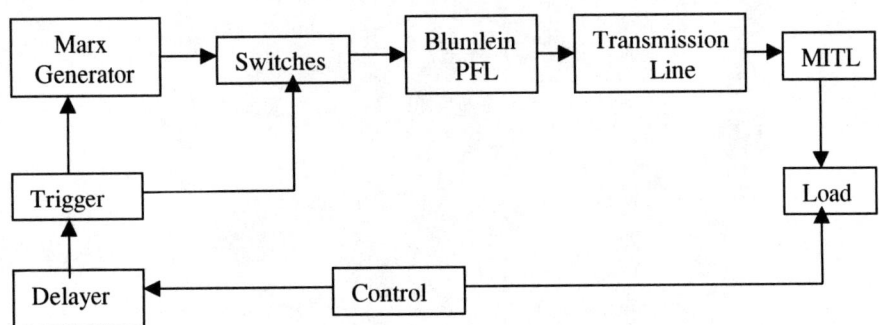

FIGURE 1. The schematic of the accelerator

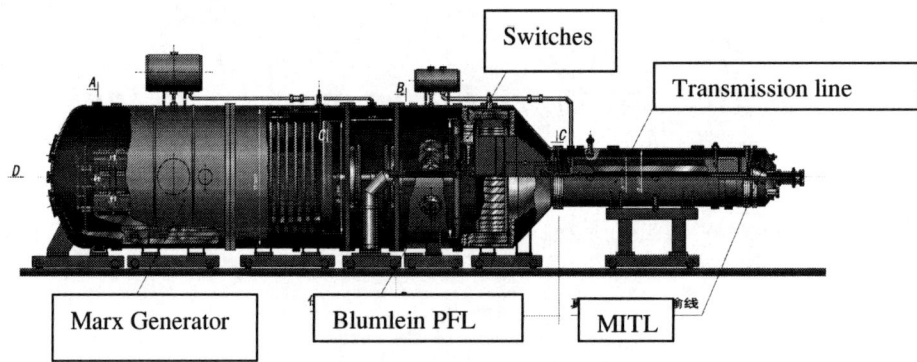

FIGURE 2. The drawing of the accelerator

In order to increase the output current on the load, upgrade to the accelerator has been carried out by replacing the transmission line with a transformer, which reduced the impedance from 2 ohms to 0.5 ohm. The current could be increased up to 1MA.

SETUP OF PUFF-GAS EXPERIMENT

The puff-gas experimental setup is shown in Fig. 3. The nozzle was put on the anode which was grounded. The gap between anode and cathode is about 15 mm.

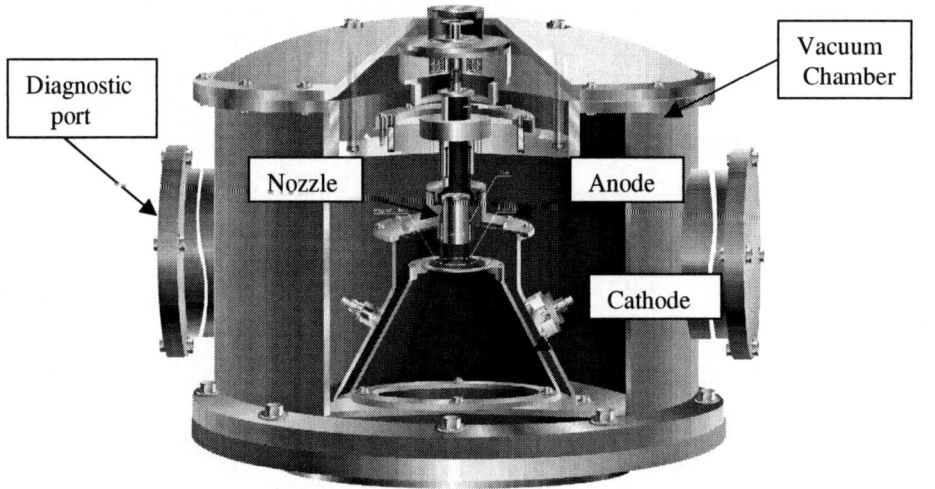

FIGURE 3. The layout of the load area

After the electro-magnetic switch of the nozzle is turned on, the Marx generator and the switches of the Blumlein PFL are triggered in turn after preset time delay so that the output current could be synchronized with the gas load at proper time.

EXPERIMENTAL RESULTS

The output x-ray was measured with PIN probe and photo-electron-multiplier. And the Z-pinch process was recorded with visible light stream camera. The preliminary results are given in Fig. 4 to Fig. 6. The upper one in Fig.4 is the current waveform. Fig. 5 is the x-ray output signal (Lower); the image of the stream-camera is shown in Fig. 6. The experiment just started and was still undergoing. There are some problems need to be solved.

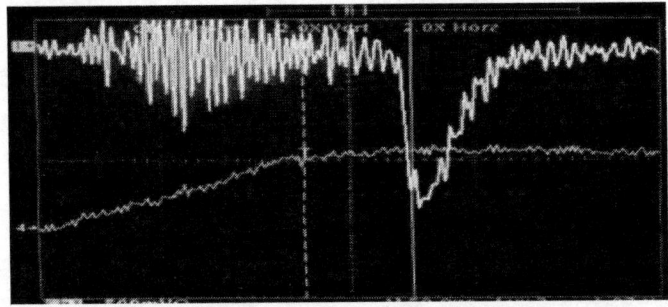

FIGURE 4. Current waveform (upper) on the load

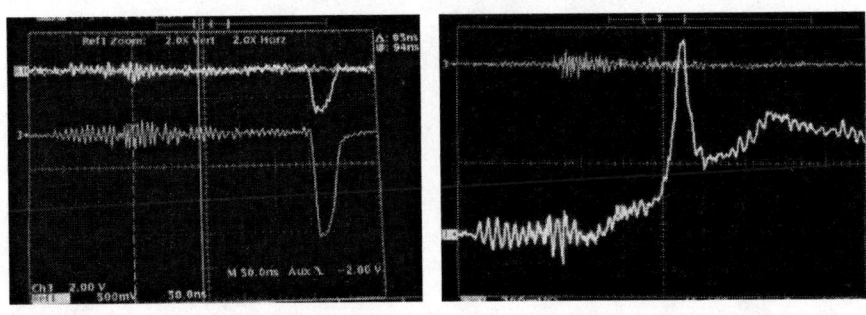

a) Waveforms of the PIN Probe with different sensitivities

b) Waveform of photo-electronic multiplier (Lower)

FIGURE 5. The x-ray output.

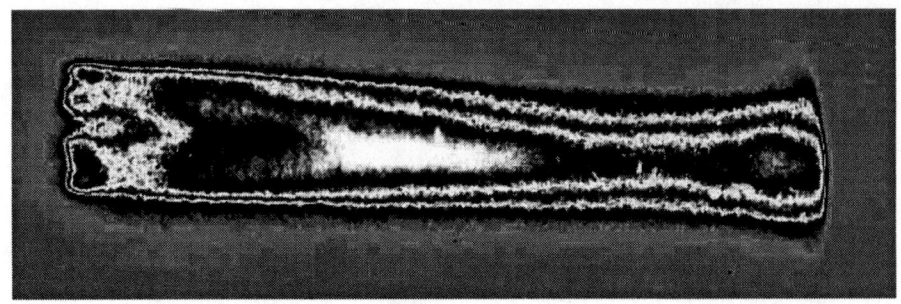

FIGURE 6. The visible light image recorded with stream-camera

REFERENCES

1. Gu, Y. C., Design of the transformer, Technique report, Institute of Fluid Physics, 2001.
2. Chang, L.H., et al, Technique report, Institute of Fluid Physics, 2002

Z-PINCH, X-PINCH, FOCUS AND CAPILLARY DISCHARGE PLASMA DYNAMICS AND DEVICES

The X Pinch as a Point Source for Point-Projection X-ray Radiography

T. A. Shelkovenko,[a,b] S. A. Pikuz,[a,b] D. B. Sinars,[a,c] K. M. Chandler,[a] M. D. Mitchell[a] and D. A. Hammer[a]

[a]*Laboratory of Plasma Studies, Cornell University, Ithaca, NY 14853, USA*
[b]*Permanent address: P. N. Lebedev Physical Institute, Moscow 119991, Russia*
[c]*Present address: Sandia National Laboratories, P.O. Box 5800, Albuquerque, NM 87185-1193, USA*

Abstract. Point-projection radiography of dense plasmas and other objects using an X pinch as the x-ray source is discussed. An X pinch, generated using two or more wires that cross and touch, in the form of an X, as the load of a pulsed power generator, reliably produces a small, bright x-ray radiation source very near the original cross point of the wires. Taking advantage of the small source size (in some cases <2 µm at energies above a few keV), radiographs of objects can be obtained without using pinholes. In addition, the short x-ray emission duration (<20 ps in some cases for energies above a few keV) of the X pinch radiation enables sharp radiographs of rapidly changing dense plasmas. Detailed measurements of the emission characteristics of X pinches made using different wire materials and in different energy ranges using a set of x-ray diagnostics with high temporal and spatial resolution are presented. A comparison of the radiated energy and the imaging properties of two and up to seven wire X pinches are discussed. The application of the X pinch x-ray source for phase-contrast x-ray radiography of low absorption objects is demonstrated.

An X pinch is generated using two or more fine metal wires that cross and touch in a single point, in the form of an X, as the load of a pulsed power generator [1, 3]. Very near the original cross point of the wires, a very small, bright, sub-nanosecond x-ray source in the 1-10 keV energy range is reliably formed. Recent experiments [3-8] have shown that x-ray radiography of high density plasmas using an X pinch as the x-ray source is an effective diagnostic method for studying exploding wires, imploding wire arrays or other X pinches. Two methods of radiography, monochromatic (using a Bragg crystal reflector as the wavelength selector) and point projection (i.e., no pinhole) x-ray radiography, were designed and used to investigate wire or wire array explosions on pulsed power generators. In recent years we have concentrated on the point-projection radiography method and its application to imaging dense plasmas and other objects. With this method it is possible to project a high-resolution, magnified image of an object directly onto x-ray sensitive film [1, 3].

For point projection x-ray radiography the size and duration of the x-ray source are important because they directly determine the spatial and temporal resolution of the imaging system. In previous articles [1, 3-6], we showed that an X pinch is an ideal x-ray source for point-projection radiography, and we described measurements of temporal and spatial parameters for the bright spots ("micropinches") formed in X pinches generated using two wires. Many of those experiments were carried out using two 2-wire X pinches as the load of our 450 kA pulsed-power generator. We have found, for example, that 2-wire Mo X pinches driven by 200-225 kA in this double X pinch configuration are very good point sources for point-projection imaging of single wire explosion or for studying the X pinches themselves, providing 1-3 µm spatial resolution and as good as 10 ps time resolution [1-6]. However, these X pinches do not have enough radiation > 5 keV to study wire array z-pinch implosions or > 1 mm size biological objects. For such studies, or to image such objects as gas puff z-pinches, x-ray sources are needed with different parameters from those we have used most often up to now. That means X pinches from different materials, wire sizes, and number of wires, and also different films and filters, will have to be used [1, 3].

In order to increase the radiation intensity, especially above 5 keV, in an X pinch x-ray burst we have found it necessary to increase both the current and the mass of wire material involved in pinch formation. However, simply using a single X pinch with larger diameter wires in a 2-wire X pinch does not lead to improved imaging because of an increased probability of multiple bright spot formation. In recent experiments with 4-7 wires in a single X pinch, we have found that multispot structure can be avoided [7]. In this article, we compare the temporal, spatial and energy parameters of multi-wire X pinches and compare imaging properties of X pinches in double and single configurations made from two and many wires.

Four-wire X pinches were initially used to increase the mass involved in the pinching process in order to increase the x-ray burst timing relative to the start of the current pulse in experiments with an x-ray streak camera. Very interesting radiography results and energy parameters were obtained. Therefore, multi-wire X pinches were studied using filtered PCDs, time-resolved spectroscopy, pinhole cameras and x-ray imaging. A detailed description of the diagnostics has been presented elsewhere [1, 3, 6]. Experiments have been carried out with X pinches made from 2, 3, 4, 6, and 7 Ti wires and 2 and 4 Al, Mo, NiCr, manganin and W wire have been studied. Note that both the mass and the symmetry of the X pinches change with more than two wires. The experimental results we report here were obtained using the XP facility [470 kA peak current, 100 ns full-width at half-maximum (FWHM)]. The experimental arrangement including several of the diagnostics is shown in Fig. 1.

From previous experiments [3,6,7], we know that only continuum radiation from micropinches gives high-resolution images. Evidently the line radiation from low and intermediate Z materials is radiated from a ≥ 10 µm scale-size volume, and the radiation time scale can be up to ~1 ns. The new results show that in all multi-wire X pinches studied, only the first burst, or two very close bursts, have intense continuum radiation (see Figs. 2 and 3). For example, in Fig. 2c time resolved spectra are presented together with PCD signals (Figs. 2a and b) and a high quality point-projection image (Fig. 2d) for the case of a 7-wire Ti X pinch. We can see in the PCD signals that three burst have comparable amplitude in all three energy ranges studied (see Figs. 2a and b) but only the first burst has the intense continuum radiation used for obtaining the image of the small beetle with high spatial resolution (Fig. 2d). It is important that this was obtained with a Cu filter, which is almost opaque to the Ti K-shell line radiation.

We have not yet studied the size of the radiation source and its duration in different energy ranges as was done for double 2-wire X pinches [1,3,5,6]. However, streaked spectra obtained with ~ 10 ps time resolution show that the duration of the Ti continuum radiation used for imaging is 30-50 ps, whereas the duration of the K-shell line radiation is more than 1 ns in a sequence of bursts [8]. With Mo-wire X pinches, Ne-like Mo spectra obtained with ~ 10 ps time resolution show that the continuum radiation duration is 10-30 ps while the line radiation lasts up to 0.5 ns, as shown in Fig. 3d.

Pinhole images obtained with 5 µm resolution and E > 3 keV show a single bright spot structure for multi-wire X pinches. For example, in Fig. 3e, one bright spot < 6 µm in size was recorded with a 4-wire Mo X pinch. The high spatial resolution of the resulting images (Fig. 3a-c) implies a very small radiation source. Thus, the multi-Mo-wire X pinches have temporal and spatial characteristics comparable to double 2-wire X pinches.

Finally, we point out that the X pinch x-ray source is ideal for "phase contrast imaging" [9], or more precisely, edge-diffraction and refraction sharpening of images [10]. If the source of radiation is spatially coherent (such as because it is essentially a point source) and the film is placed far enough behind the object being radiographed, wave optics, rather than just the usual x-ray optics, must be taken into account in the radiographic image analysis. At discontinuities in the object, the image shows a fringe pattern on the film (or other detector) that results from the combination of diffracted, refracted and directly transmitted x-rays. These fringes, together with the absorption by structures in a sample, enhance the edges of sample features on the film. A detailed description of the method using an X pinch was presented in [11]. Example images of different biological objects with different density and internal structures are presented in Figs. 1d, 2d and 3a, b. It is noteworthy that in the case of Fig. 3c, a thin plastic bag containing the object in a small amount of liquid was placed in an air chamber inside the vacuum chamber. Even in this case very high contrast and spatial resolution is observed in the image.

This research was supported by DOE Grants Nos. DE-FG02-98ER54496 and DE-FG03-98DP00217, and Sandia Contract BD-9356.

References.
1. Shelkovenko, T. A., Sinars, D. B., Pikuz, S.A., Hammer, D.A., Phys. Plasmas **8**,1305-1318 (2001).
2. Pikuz, S. A., Shelkovenko, T. A., Sinars, D.B., Greenly, J. B., Dimant, Y. S., Hammer, D. A., Phys. Rev. Lett. **83**, 4313-4316 (1999).
3. Shelkovenko, T. A., Sinars, D. B., Pikuz, S. A., Chandler, K. M., and Hammer, D. A., Rev. Sci. Instr. **72**, 667-670 (2001).
4. Lebedev, S. V., Bland, S. N., Beg, F. N., Chittenden, J. P., Dangor, A. E., Haines, M. G., Pikuz, S. A., Shelkovenko, T. A., Rev. Sci. Instr. **72**, 671-673 (2001).
5. Shelkovenko, T. A., Pikuz, S. A., Sinars, D. B., Chandler, K. M., Hammer, D. A., in Proc. of SPIE **4**, 180 - 187, 2001, G. A. Kyrala and J-C Gauthier editors.
6. Sinars, D. B., Pikuz, S. A., Shelkovenko, T. A., Chandler, K. M., Hammer, D.A., Rev. Sci. Instr. **73**, 2948-2956 (2001).
7. Shelkovenko, T.A., Pikuz, S. A., Sinars, D. B., Chandler, K. M. and Hammer, D. A., Phys. Plasmas **9**, 2165-2172 (2002).

8. Pikuz, S. A., Shelkovenko, T. A, Sinars, D. B., Skobelev, I. Yu., Chandler, K. M., Mitchell, M. D., and Hammer, D. A., paper ZYX in these proceedings.
9. Fitzgerald, R., "Phase-Sensitive X-ray Imaging," Physics Today, July 2000, pp. 23-26.
10. Margaritando, G., and Tromba, G., J. Appl. Phys. **85**, 3406-3408 (1999).
11. Pikuz, S. A., Shelkovenko, T. A, Sinars, D. B., Chandler, K. M., Hammer, D. A., Proc. of SPIE **4**, 234–238, 2001, G. A. Kyrala and J-C Gauthier editors.

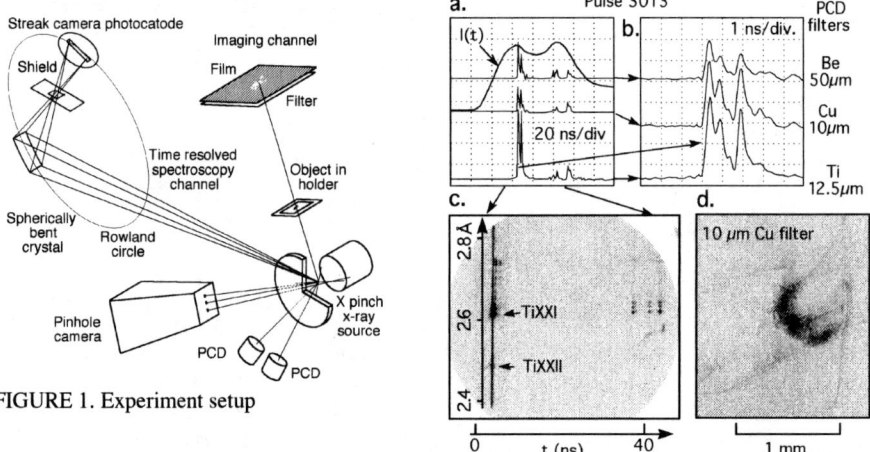

FIGURE 1. Experiment setup

FIGURE 2. Time resolved spectra (c) together with PCD signals (a, b) and a high quality point-projection image of pea aphid (d) for 7-wire Ti X pinch.

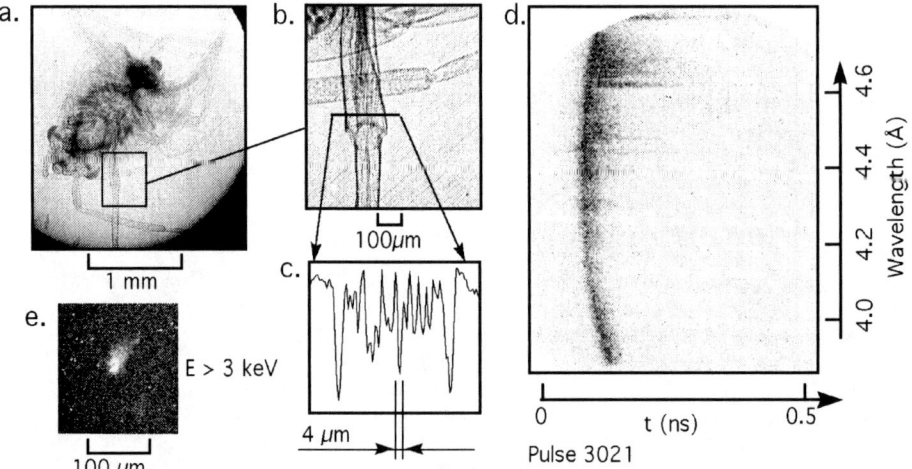

FIGURE 3. High quality point-projection image of tarnished plant bug (a), enlarged part (b) with outline (c) of the image showing 4 μm spatial resolution; Ne-like Mo spectrum obtained with high temporal resolution (d); pinhole image obtained with 5 μm resolution and E > 3 keV (e).

X Ray Emission From X Pinch Experiments on the Llampüdkeñ Generator.

Ian H. Mitchell, Raúl Aliaga-Rossel, Jorge A. Gómez, Hernán Chuaqui, Mario Favre and Edmund S. Wyndham

*Departamento de Física, Pontificia Universidad Cátolica de Chile
,Casilla 306, Santiago 22, Chile.*

Abstract. The results from the first plasma physics experiments on the Llampüdkeñ Generator (1MA, 250 ns) are presented. X Pinch experiments have been undertaken at current levels of 400 kA with a rise time of ~250 ns. X pinches were produced mainly from aluminium wires of different diameters and with varying numbers of wires. Results from X-ray diagnostics characterising the emitted radiation are presented. The diagnostics include filtered PIN diodes and a pinhole and slit-wire camera. Radiation of energy greater then 2.5 keV was emitted from hot spots in timescales of a few nanoseconds. Using the results from the slit-wire camera, the diameter of the hot spots is shown to be less than 5 µm.

INTRODUCTION

This paper reports on X pinch experiments carried out on the Llampüdkeñ [1][2] generator at a current level of ~400 kA. The x-ray emission from X pinch loads made from various numbers and diameters of Aluminium wire was studied, particularly in relation to the duration of the emission and the size of the hot spot source. The main diagnostics employed were a set of filtered PIN diodes and a time integrated multi filtered slit-wire camera[3].

The X pinches were 22 mm long with an angle of 65º between the limbs. The PIN diodes used were BPX photodiodes with the glass window removed. A bias of 120 V was applied to the diodes in order to improve the response time. The PINS were placed at a distance of 85 cm from the X pinch.

The slit-wire camera used had six slits, each slit containing six wires ranging in diameter from 250 µm to 25 µm. There was also a pinhole image (100 µm diameter pinhole) associated with each slit. Each slit and pinhole pair was covered by a different filter thus providing spectral resolution. The mount which held the slits and wires was placed at 25 cm from the source and the film plane of the camera was at 8 cm, resulting in a magnification of the shadows of the wires of 1.3. HP5 film was used to record the images.

RESULTS

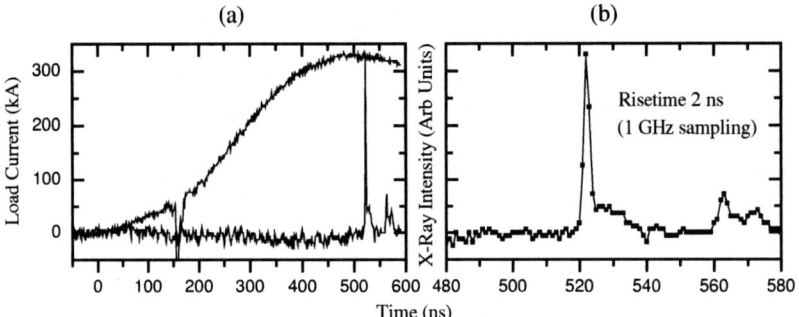

FIGURE 1 (a) Load current and X ray emission from a 125 μm Aluminium X pinch. (b) Detail of x-ray pulse shown in (a) showing the risetime of the pulse.

Figure 1 shows results obtained from an X pinch made from two 125 μm diameter Aluminium wires. Part (a) of the figure shows the load current along with the signals from a PIN diode filtered with 40 μm of Aluminised Mylar. The x ray emission for this type of X pinch was typically around peak current. Part (b) contains the PIN signal on a faster time scale showing the risetime of the pulse to be around 2 ns. Figure 2 shows the timing of the x ray emission for various types of X pinch and indicates that the 125 μm wire X pinch presents the best matching to the generator. The emission from the other types of X pinch had similar risetimes but less intensity than that of the matched load.

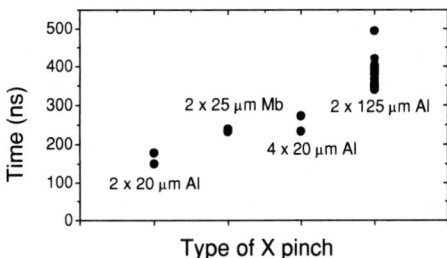

FIGURE 2 Graph showing the timing of the principle X ray pulse obtained from different types of X pinch. The times are relative to the start of the current rise. Peak current is typically around 450 ns.

Figure 3(a) shows the traces obtained on three filtered PIN diodes for a 125 μm X pinch load. This shot shows a common feature; two X ray pulses separated by about 20 ns. Such double pulses are a common feature and have been commented on by other authors[4]. The risetime of the signals is 1.6 ns. The transmission of the filters is shown in figure 3(b). The signals suggest that the hotspots emit K shell radiation plus a significant amount of harder radiation at energies above 2.5 keV. The spectrum of the two peaks appears to be similar given that there is no significant change in the ratio of the signals detected by the different filters from one pulse to the next.

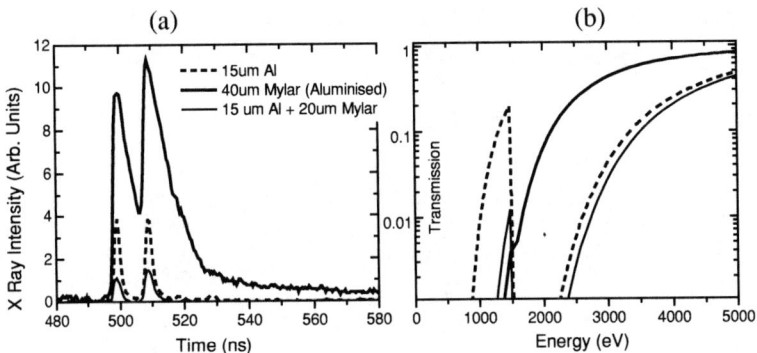

FIGURE 3. (a) Traces from set of three filtered PN diodes for a 125 μm Aluminium wire X pinch. (b) The transmission of the filters of the PINs in (a). The legend corresponds to both graphs. Aluminium K shell radiation falls between 1.6 and 2.0 keV.

Figure 4 shows images obtained through four individual slits of the slit-wire camera. The numbers on the figures indicate the diameters of the wires to which the shadows belong. A well defined shadow of the 25 μm wire indicates a source size of < 5 μm (the resolution of the diagnostic). Figure 4(a) and (b) respectively show images obtained for the 125 μm aluminum wire X pinch through a filter of 17 μm aluminized mylar and through a composite filter consisting of 3 μm silver and 24 μm mylar. The mylar filter transmits the aluminum K shell radiation whereas the composite filter does not but has a transmission window between 2 and 3.5 keV. Figure 4(c) and (d) show the image obtained with a X pinch of two and four 20 μm diameter aluminum wires respectively through the 17 μm aluminized mylar filter. In (b) the shadow of the 25 μm diameter wire can be seen indicating that for the 125 μm wire pinch the source size is < 5 μm for energies of 2 − 3.5 keV. This shadow is not seen in figure (a) however, indicating that the source is larger for photon energies down to ~1 keV. A source size of < 5 μm can be obtained for these lower photon energies when a 20 μm aluminium wire single or double X pinch is used as the shadow can be seen in (c) and (d). No image is obtained through the composite filter for the single 20 μm X pinch and with the 20 μm double X pinch only a very low intensity image is obtained, thus indicating that plasma is hotter in the case of the 125 μm wire pinch.

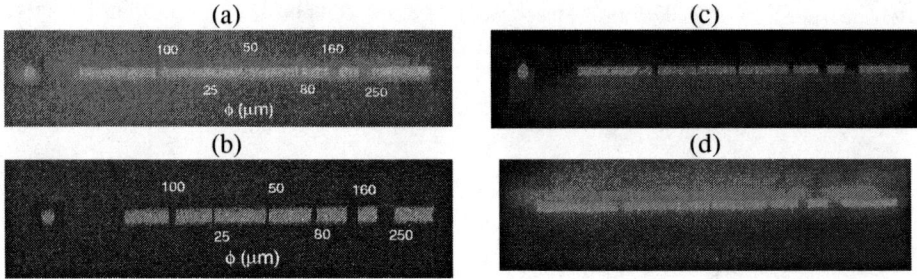

FIGURE 4. Images from the slit wire camera. The numbers on the images represent the diameters of the wires to which the shadows belong. The order of the wires is the same in all four figures. (a) 125 µm Aluminium X pinch with 40 µm Aluminized Mylar filter. (b) 125 µm Aluminium X pinch with composite filter of 3 µm silver and 17 µm mylar. (c) 20 µm Aluminium X pinch with 40 µm Aluminized Mylar filter (d) Double X pinch with 20 µm Aluminium wire with 40 µm Aluminized Mylar filter (this image does not have a pinhole picture).

CONCLUSIONS

X pinch experiments on the Llampüdkeñ generator produce small x-ray sources of diameter < 5 µm with pulse risetimes of < 2 ns. In the case of a matched load, an X pinch made from 125 µm Aluminium wires, the radiation includes the K shell emission and a significant proportion of higher energy photons.

ACKNOWLEDGMENTS

This work was carried out at the Pontificia Universidad Católica de Chile, supported by FONDECYT, grant numbers 1020835, 7980023 and 8980011.

REFERENCES

1. Chuaqui H , F. Wyndham, C. Friedli and M. Favre, *Laser and Part Beam* **15**, 241-248 (1997)
2. Chuaqui H., Mitchell I.H., Aliaga Rossel R, Favre M and Wyndham E.S. *These proceedings.*
3. Choi P. Dumitrsescu C., Wyndham E, Favre M and Chuaqui H, *Rev Sci Inst* **73**, 2276-2281 (2002)
4. Pikuz S.A et al, *Phys Plasmas*, **6**, 4272 (1999)

Regimes of Energy Input in the Pseudospark Discharge in the Sources of EUV Radiation

Yu. D. Korolev, O. B. Frants, V. G. Geyman, R. V. Ivashov,
N. V. Landl, I. A. Shemyakin

Institute of High Current Electronics, 4 Akademichesky Ave., 634055 Tomsk, Russia

Abstract. The paper describes the regimes of energy input in the pseudospark discharge in typical conditions of EUV source operation. It is shown that in a mode of superdense glow discharge the electron temperature in the discharge plasma reaches rather high values without self-magnetic compression of the plasma channel.

INTRODUCTION

One of the new approaches to generation of EUV radiation with a high pulse repetition rate is a use of a pseudospark discharge [1]. Distinctive features of the discharge conditions here are a rather fast current rise and short pulse duration. When interpreting a mechanism of EUV radiation, most of investigators proceeds from the concept of self-magnetic compression of the plasma column in the main gap. Correspondingly, the process of increasing the plasma temperature is treated as a transformation of the kinetic energy of the accelerated plasma into the inner energy of the plasma at the discharge axis.

However, the conditions of the pseudospark discharge differ dramatically from that for traditional Z-pinch with a high stored energy. This paper demonstrates that in typical EUV sources based on the pseudospark discharge the magnetic compression of the plasma is negligibly small. The discharge with hollow cathode and hollow anode burns in a mode of superdense glow with extremely high current density. Correspondingly, the main reason for enhancing the plasma temperature is the energy input in the discharge plasma from the electron beam accelerating in the cathode voltage drop region.

DISCHARGE AT AN ENCHANCED PRESSURE

Typical experimental conditions for nitrogen and oxygen correspond to pressure 0.5 Torr and less. The whole range of pressure can be subdivided for two regions. When the pressure is higher than 0.02 Torr the current waveform is smooth and the so-called current quenching phenomenon is not observed. Typical current waveform and corresponding discharge image taken by CCD camera is shown in Fig. 1.

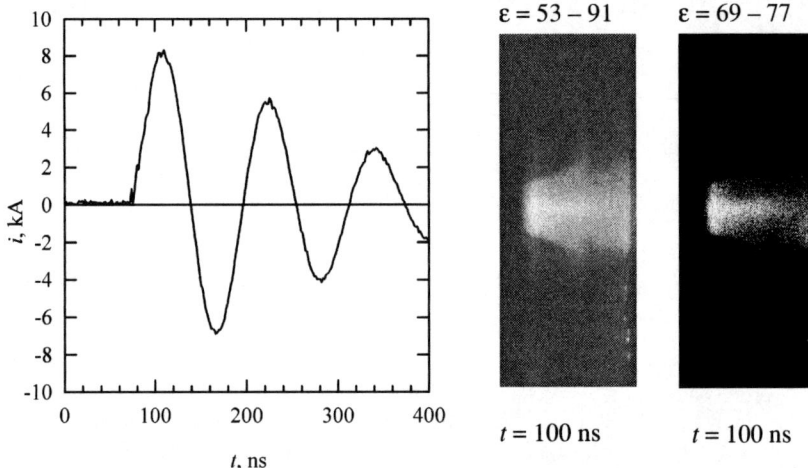

FIGURE 1. Current waveform and discharge photograph in oxygen ($V_0 = 12$ kV, $p = 0.5$ Torr, $C_0 = 16$ nF, $L_0 = 20.8$ nH, gap distance $d = 0.4$ cm, cathode is at the left side).

The exposition time for CCD camera was 100 ns, and an instant of the camera shut-off is pointed out below the photographs. The first of them corresponds to maximum interval for intensity ε of the discharge image (in arbitrary units from 53 to 91). To distinct in more detail the most bright discharge regions the second photograph is given in the interval of intensities from 69 to 77.

The gap luminosity has a rather intricate structure characteristic of a glow type of discharge. Most bright regions, which can be referred to as a negative glow [2], are localized inside the cathode and anode bore holes. The plasma channel diameter in the main gap is comparable with the bore hole diameter. Decrease in the gas pressure or increase in the initial voltage V_0 leads to increasing the plasma diameter in the main gap.

For the conditions of Fig. 1, an average current density in the plasma channel at an instant of the current maximum corresponds to $j = 2.5 \times 10^5$ A/cm². In order that the plasma channel be able to pass a certain value of current, the following condition for plasma density n_e in the channel must hold true:

$$n_e > 4j/ev_e, \tag{1}$$

where v_e is the average thermal velocity of the electrons in plasma.

As applied to the experiment under discussion we obtain $n_e > 6 \times 10^{16}$ 1/cm³. On the other hand, the gas density is only 1.75×10^{16} 1/cm³, *i.e.* we deal with a fully ionized plasma in which the multi-charged ions are definitely available. Then the radiation of the EUV can be expected from the hollow cathode plasma under similar conditions.

One of the important problems is interpretation of an extremely high current density in the superdense glow discharge. It had been demonstrated [2] that the current in this stage closes to inner surface of the bore hole on which a numerous cathode spots are available. The glow mode in this case is determined by a uniform distribution of the spots over the cathode surface.

On the other hand, there exist a double electric layer between the near cathode metal plasma and the negative glow plasma [3]. This layer conceptually plays a part of a cathode voltage drop for a specific gap in which the surface of near cathode metal plasma serves as an efficient emitter of electron beam. Due to the hollow cathode effect the energy of electrons accelerated in the cathode voltage drop region is dissipated in the negative glow region thereby ensuring a high electron temperature in the negative glow plasma and a high current density.

In some cases more bright filament forms at the discharge axis (Fig. 1). We suppose that an electron beam, which arises in the initial stages at the gap axis, is responsible for formation of the filament channel and the physical reason for appearing the filament is not connected with the discharge pinching. The process of the filament formation resembles that for the process of constriction of a high-pressure gas discharge. It means that the filament appears due to enhanced ionization rate at the discharge axis as compared to ionization rate at the periphery regions.

The case in which there is no filament at the discharge axis is shown in Fig. 2.

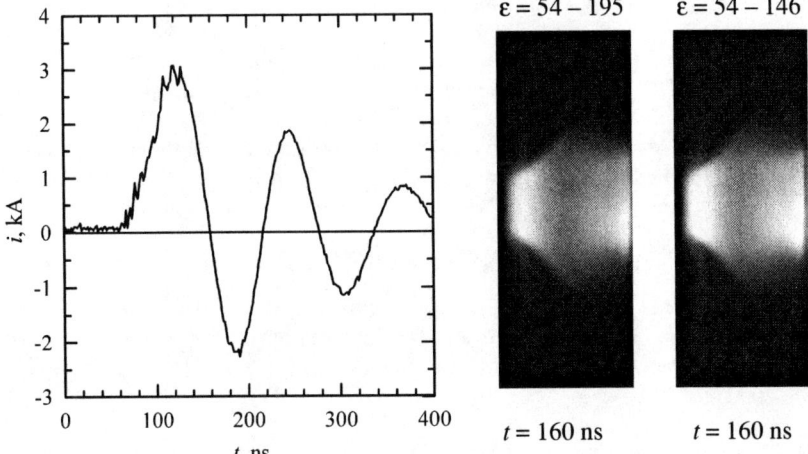

FIGURE 2. Current waveform and discharge photograph in oxygen ($V_0 = 6$ kV, $p = 0.5$ Torr, $C_0 = 16$ nF, $L_0 = 20.8$ nH, gap distance $d = 0.4$ cm, cathode is at the left side).

During the first half period the discharge still persists in a form of the glow discharge. It is seen that the bright negative glow plasma partly penetrates from the cathode bore hole in the main gap. Beside that, the bright plasma region is available in the anode bore hole and there is a dark region between the hollow cathode and hollow anode plasmas.

Decreasing the pressure to a value less than 0.02 Torr results in appearing the current quenching phenomenon (Fig. 3).

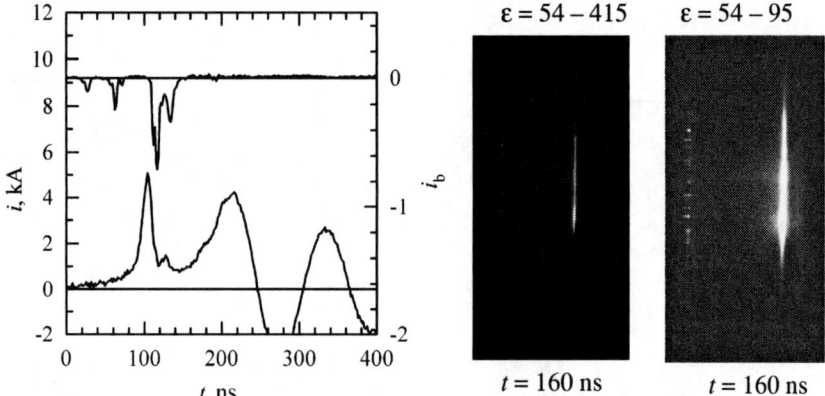

FIGURE 3. Waveforms of the discharge current and electron beam together with the discharge image (the beam at the discharge axis). $V_0 = 20$ kV, $p = 0.015$ Torr. $C_0 = 16$ nF, $L_0 = 20.8$ nH, gap distance $d = 0.4$ cm, cathode is at the left side).

Due to a fast increase in the gap resistance the sharp quenching in the current is observed at a current level of about 5 kA. Correspondingly, an inductive high-voltage kick is applied to the gap during the quenching and the fast electron beam arises at the discharge axis. The brightest region of the gap luminosity corresponds to edge of the anode bore hole, since this edge is subjected to bombardment by the electron beam so that the anode spots appear. However, when the most fraction of the beam forms at the gap axis the brightness of the anode spots is not extremely high. The brightness of the anode spots is increased for the pulses, in with the beam is randomly shifted from axis.

ACKNOWLEDGEMENT

The research is sponsored by NATO's Scientific Affairs Division in the framework of the Science for Peace Program and by the Program of the Ministry of Industry, Science, and Technology of Russian Federation (Project No 40.030.11.1125 of 01.02.2002).

REFERENCES

1. Bergmann K., Schriever G., Rosier O., Muller M., Neff W., Lebert R., "Highly repetitive, extreme-ultraviolet radiation source based on a gas-discharge plasma", *Appl. Optics*, **65**, 5413 – 5417 (1999).
2. Korolev Yu. D., Frank K., "Discharge formation process and glow-to-arc transition in pseudospark switch," *IEEE Trans. Plasma Sci.*, **27**, 1525 – 1537 (1999).
3. Kondrat'eva N. P., Koval' N. N., Korolev Yu. D., Schanin P. M., "A spectroscopic investigation of the near-cathode regions in a low-pressure arc", *J. Phys. D: Appl. Phys.*, **32**, 699 – 705 (1999).

Optical Observation of a Pseudospark Discharge Development in a Source of EUV Radiation

Yu. D. Korolev, I. M. Datsko, O. B. Frants, V. G. Geyman, R. V. Ivashov,
N. V. Landl, N. A. Ratachin, I. A. Shemyakin

Institute of High Current Electronics, 4 Akademichesky Ave., 634055 Tomsk, Russia

Abstract. The results of pseudospark discharge observation with spatial and temporal resolution together with the measurements of EUV radiation are presented. It is shown that in a mode of superdense glow discharge the electron temperature in the discharge plasma is high enough to provide for generation of EUV radiation without magnetic compression of the plasma channel.

INTRODUCTION

Considerable interest has recently been generated in development of the EUV sources with a high pulse repetition rate and a small energy in a single pulse. One of the promising approaches is a use of electrode system of the pseudospark discharge in such installations. Typical regimes of the energy input here are characterized by short pulse duration (less that 100 ns) and moderate energy stored in a primary capacitor bank (1 to 10 J).

Interpretation of the nature of appearing EUV radiation in the pseudospark discharge in the first publications [1] was based on the concept of magnetic compression of the plasma column. The present paper presents the results of parallel measurement of EUV radiation and observation of the discharge luminosity. It is demonstrated that a typical output of radiation with efficiency 0.1 per cent is provided in the regimes of the superdense glow discharge without constriction of the plasma column. The physical reason for generation of EUV is discussed.

MEASUREMENTS FOR DISCHARGE IN OXYGEN

Temporal development of the discharge luminosity has been investigated with a use of CCD camera by means of side-on observation. The signal of EUV radiation was recorded with fast vacuum photodiode located in the anode cavity at the discharge axis. In order to cut out the radiation with the wavelength larger than 13.5 nm, a mylar film of thickness 1 μm was used.

Characteristic feature of the pseudospark discharge is that in the prebreakdown stage an electron beam forms at the discharge axis [2]. This electron beam can reach the photodiode and produce a signal of negative polarity. The signal of EUV radiation in the experimental conditions under discussion has a positive polarity. To suppress the signal of electron beam, a permanent magnetic field which deflects the beam from the discharge axis has been used.

Typical current waveforms and side-on photograph of the gap for discharge in oxygen are shown in Fig. 1.

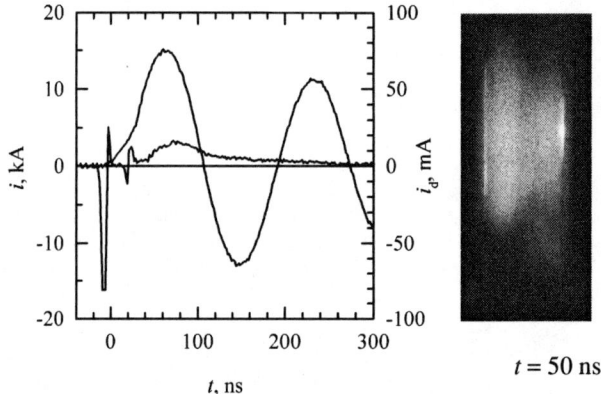

FIGURE 1. Waveforms of the discharge current i and the current of the vacuum diode i_d, together with the discharge image. Oxygen, $p = 0.07$ Torr, $C_0 = 32$ nF, $V_0 = 16$ kV, cathode is at the left side of the photograph.

The discharge column has a typical structure characteristic of a glow type discharge. There is no a constricted filament at the discharge axis. However, the discharge plasma generates EUV radiation. It means that in some discharge regions we deal with the fully ionized plasma in which the multi-charged ions are available.

Most likely these are the regions of the cathode and anode bore holes. It is known that the pseudospark discharge in its temporal development passes through the stages of dense glow discharge, superdense glow discharge, and constricted arc with distinctively expressed cathode spot [3]. The photograph in Fig. 1 corresponds to the stage of the superdense glow discharge. The current here closes to the inner surface of the cathode bore hole. Due to a high current density, the numerous microexplosions arise at the cathode surface. Each act of microexplosion can be considered as an initiation of a cathode spot. Nevertheless, when a large number of the microexplosions are uniformly distributed over the cathode, there is no preferential attachment of the current to a single spot. The total discharge current, which flows across the hollow cathode plasma, is also distributed uniformly thereby ensuring the glow type of discharge.

The cathode surface turns out to be covered by a sheath of high-density metal plasma, whose outer boundary plays a part of effective plasma cathode. The cathode voltage drop is concentrated between the metal plasma and the hollow cathode plasma. Correspondingly, mechanism of the energy input in the hollow cathode plasma for the case under consideration is the same as that for the classical low-pressure glow discharge with hollow cathode. The electrons are accelerated in the cathode voltage drop region and their energy is dissipated in the hollow cathode plasma.

The estimates show that the electron emission current density from the metal plasma can reach about 10 per cent of the total discharge current. Then with a typical cathode voltage drop of about 500 to 1000 V, the power introduced from the electron beam to the hollow cathode plasma is sufficient to generate the fully ionized plasma with multi-charged ions.

The conditions in Fig. 1 correspond to optimal pressure from the viewpoint of EUV output. An increase in pressure results in a decrease in intensity of EUV radiation. This case is shown in Fig. 2.

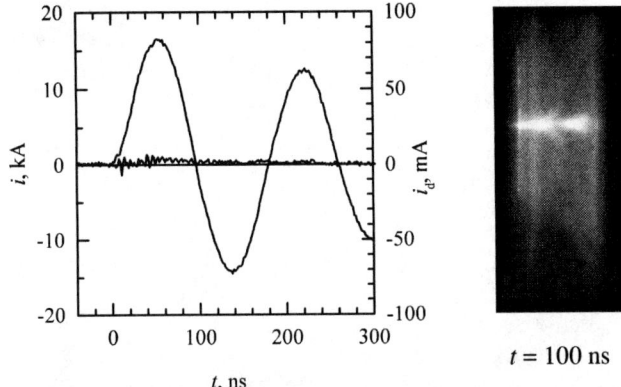

$t = 100$ ns

FIGURE 2. Waveforms of the discharge current i and the current of the vacuum diode i_d, together with the discharge image. Oxygen, $p = 0.2$ Torr, $C_0 = 32$ nF, $V_0 = 16$ kV, cathode is at the left side of the photograph.

The reason for decreasing the intensity of EUV radiation is the process of glow-to-arc transition. In the stage of superdense glow discharge the microexplosions arise not only at the inner surface of the bore hole but at the edge of the bore hole as well. In some shorts appearing the cathode spot initiates the transition from the superdense glow discharge into metal vapor arc during the first half period of the discharge current. Photograph in Fig. 2 shows this case, i. e. the arc channel on the background of glow stage. For this discharge mode the intensity of EUV radiation is sharply decreased.

MEASUREMENTS FOR DISCHARGE IN XENON

The main regularities for the discharge in xenon are the same as for the discharge in oxygen. The maximum output of EUV radiation corresponds to the superdense glow stage. However, the optimal gas pressure here is lower and the output energy of radiation is higher as compared to oxygen. The total energy of radiation in the range of 13.5 nm and less has been estimated as 0.1 per cent of the energy stored in the primary capacitor bank C_0.

Typical waveforms and the discharge photograph are shown in Fig. 3.

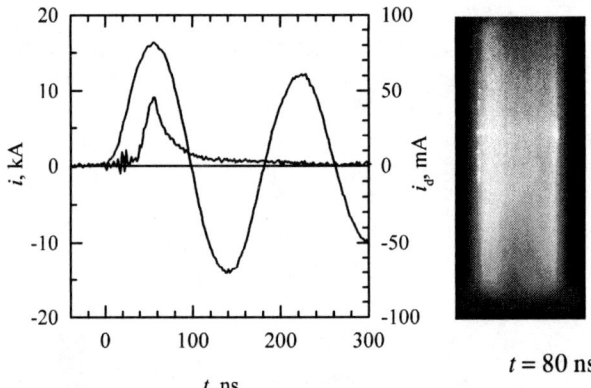

FIGURE 3. Waveforms of the discharge current i and the current of the vacuum diode i_d, together with the discharge image. Xenon, $p = 0.02$ Torr, $C_0 = 32$ nF, $V_0 = 16$ kV, cathode is at the left side of the photograph.

It is seen that the luminosity of the main gap corresponds to the superdense glow discharge. The cathode spots in a stage of their initiation and the anode spots are observed at the edges of the cathode and anode bore holes. However, the glow-to-arc transition process does not develop instantly, especially if we deal with a nanosecond time scale. For the conditions in Fig. 3 the process had no enough time to be developed and the discharge during the first half period had the glow form.

ACKNOWLEDGEMENT

The research is sponsored by NATO's Scientific Affairs Division in the framework of the Science for Peace Program and by the Program of the Ministry of Industry, Science, and Technology of Russian Federation (Project No 40.030.11.1125 of 01.02.2002).

REFERENCES

1. Bergmann K., Schriever G., Rosier O., Muller M., Neff W., Lebert R., "Highly repetitive, extreme-ultraviolet radiation source based on a gas-discharge plasma", *Appl. Optics*, **65**, 5413 – 5417 (1999).
2. K. Frank and J. Christiansen, "The fundamentals of the pseudospark and its applications", *IEEE Trans. Plasma Sci.*, **17**, 748-753, (1989).
3. Korolev Yu. D., Frank K., "Discharge formation process and glow-to-arc transition in pseudospark switch," *IEEE Trans. Plasma Sci.*, **27**, 1525 – 1537 (1999).

Experimental Setup for Investigation of the Pseudospark Discharge as Applied to Generation of EUV Radiation

Yu. D. Korolev, I. M. Datsko, O. B. Frants, V. G. Geyman,
R. V. Ivashov, N. V. Landl, I. A. Shemyakin

Institute of High Current Electronics, 4 Akademichesky Ave., 634055 Tomsk, Russia

Abstract. The paper describes the main features of the experimental installation for investigation of a high current pulsed gas discharge, which burns in typical conditions of EUV source operation.

INTRODUCTION

The problem of generation of EUV and soft X-ray radiation is traditionally solved with using a plasma focus method, a capillary discharge or a gas puff Z pinch with a rather high energy stored in a primary capacitor bank (larger than 1 kJ). On the other hand, in connection with EUV lithography at a wavelength of about 13 nm, considerable interest has recently been displayed in the installations with a moderate energy in a single pulse that would be capable of working with a high pulse repetition rate. One of the new approaches is application of a pseudospark electrode system, or strictly to say, the high-current hollow-cathode discharge [1] for obtaining the EUV radiation. The promising results in this direction for the discharge in oxygen and xenon at extremely low stored energy (about 1 J) have been demonstrated in [2]. In particular, for radiation of beryllium and lithium like ions in a range from 10 to 18 nm an efficiency of 0.1 % has been obtained at a pulse repetition rate up to 150 Hz.

The present paper describes the experimental installation, which has been intentionally developed for investigation of the pseudospark discharge in the EUV sources.

EXPERIMENTAL CHAMBER AND ELECTRIC CIRCUIT

Experimental arrangement is shown schematically in Fig. 1. The hollow electrodes *2* and *3* are fixed on the upper and lower flanges of ceramic chamber *1*, which measures 70 mm in inner diameter and 95 mm in height. The gap between the flat parts of the electrodes, the bore hole diameters, and the thickness of the flat part of the electrodes are equal to each other and measure 4 mm. The chamber is equipped with windows *4* for the side-on observation of the discharge image by means of CCD camera. Intermediate flange *6* is used for gas evacuating. A variety of assembles for diagnostics (measuring a fast electron beam, recording of EUV radiation, discharge observation with temporal resolution, and the like) can be mounted on flange *8*. When measuring the EUV radiation, a thin Be foil or mylar film is placed at the plate *9* in order to filter a spectral range less than 14 nm.

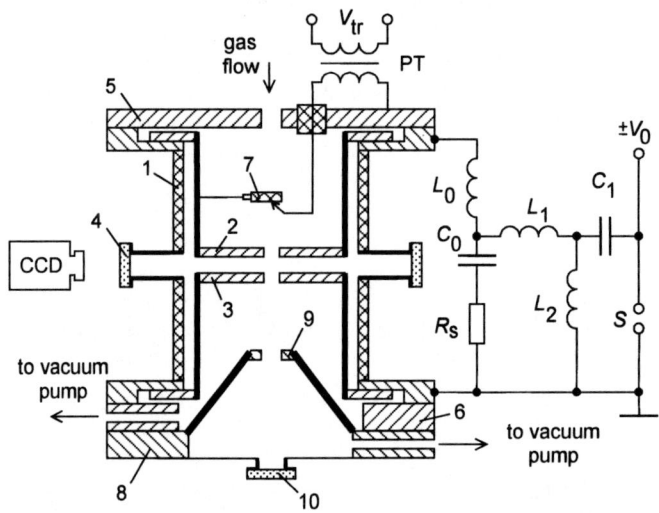

FIGURE 1. Schematic of the experimental arrangement ($C_0 = 16 - 32$ nF, $L_0 = 20.8$ nH)

The discharge is powered by means of ceramic capacitor banks TDK which are arranged around the chamber thereby insuring a low inductance circuit. A pulsed voltage at the main gap appears as result of resonant charging the capacitor C_0. When the voltage at the gap reaches its maximum value, the discharge is triggered externally by means of the trigger unit located in the cavity of the potential electrode. The unit is based on the surface breakdown over semiconductor cylinder *7* [3] and offers a possibility to trigger the gap both with positive and negative polarity of the potential electrode. The electric circuit and trigger unit allows the installation to operate with a pulse repetition rate up to 100 Hz.

The current waveforms are recorded by taking the voltage signal to oscilloscope from the current shunt R_s.

EXPERIMENTAL CONDITIONS

A characteristic feature of the experimental arrangement is the pulsed resonant charging of the main capacitor bank C_0 with a charging time of about 1.5 μs. As a result, a high overvoltage is readily achieved at the main gap. As far as the main discharge is triggered externally, the experiments can be carried out in a wide range of pressure. This situation is illustrated in Figs. 2 and 3.

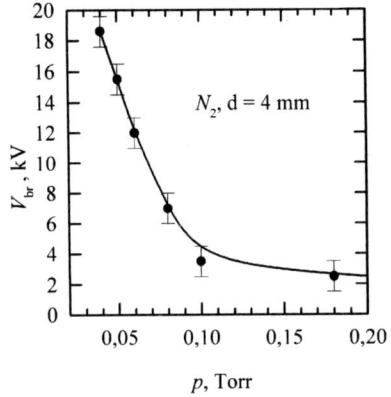
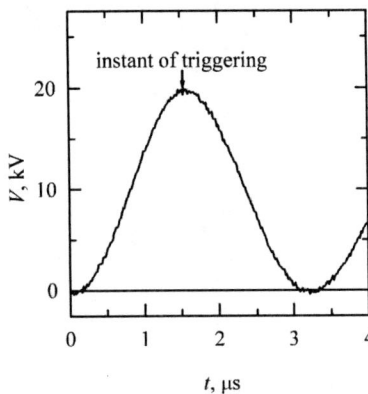

FIGURE 2. Static breakdown voltage for discharge chamber versus the gas pressure (O_2, $d = 0.4$ cm).

FIGURE 3. Voltage across the gap under the resonant charging ($C_0 = C_1 = 16$ nF, $L_0 = 20.8$ nH, $p = 0.2$ Torr).

For example, with a pressure $p = 0.2$ Torr the static breakdown voltage for oxygen and air corresponds to approximately 3 kV, while using the pulsed charging allows us to work with the voltage up to 20 kV. Even with a pressure $p = 0.5$ Torr, when the static breakdown voltage is 500 V, a pulsed voltage up to 12 kV is achieved at the gap due to fast charging.

Typical current waveforms for different pressures are shown in Fig. 4. This case corresponds to comparatively high pressures when the current waveform is smooth and so-called current quenching phenomenon is absent. The quenching in current for discharge in oxygen and nitrogen appears for a pressure less than 0.2 Torr.

The waveforms in Fig. 4 show that in an initial stage of the discharge formation the delay in the current rise is observed. Current increases slower than it would be increased in the case of short circuit experiment. It means that a characteristic time of the ionization rate in the gap is less than a characteristic time of the current rise in LC circuit, $\pi(LC)^{1/2}/2$. The gap resistance for this stage is larger that the impedance of the electric circuit. This stage becomes expressed more distinctively with decreasing the initial voltage and decreasing the pressure.

After the stage of the delay in the current rise the gap conductivity sharply increases which is reflected in increasing the slope of the current waveform. In the conditions of experiments such a transition occurs at a current level of about 2 kA independently on the gas pressure. The gap resistance becomes lower than the impedance of the electric circuit and the discharge burning voltage decreases. At an instant of current maximum the discharge burning voltage corresponds to 0.5 – 1 kV. Typical energy input in the gas discharge plasma to the instant of current maximum is about 20 to 30 per cent of a total energy $C_0 V_0^2 / 2$ stored in the capacitor bank.

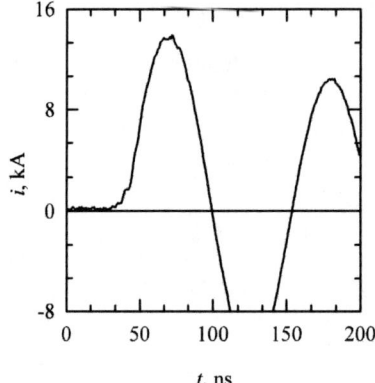 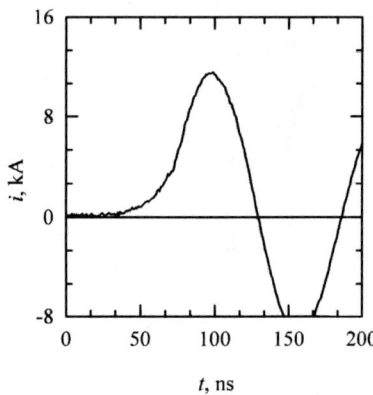

FIGURE 4. Current waveforms through the gap for oxygen at initial voltage $V_0 = 20$ kV for different pressures: $p = 0.2$ Torr (left) $p = 0.02$ Torr (right).

The discharge luminosity has a rather intricate structure, which is characteristic of a high current superdense glow discharge [1]. The brightest regions of the discharge plasma are the areas inside the cathode and anode bore holes. The plasma channel diameter in the main gap is equal to the bore hole diameter or exceeds it.

In some cases more bright filament forms at the discharge axis. However, the physical reason for appearing the filament is not connected with the discharge pinching due to self-magnetic field. The process of the filament formation resembles that for the process of constriction of a high-pressure gas discharge. It means that the filament appears due to enhanced ionization rate at the discharge axis as compared to ionization rate at the periphery regions.

ACKNOWLEDGEMENT

The research is sponsored by NATO's Scientific Affairs Division in the framework of the Science for Peace Program and by the Program of the Ministry of Industry, Science, and Technology of Russian Federation (Project No 40.030.11.1125 of 01.02.2002).

REFERENCES

1. Korolev Yu. D., Frank K., "Discharge formation process and glow-to-arc transition in pseudospark switch," *IEEE Trans. Plasma Sci.*, **27**, 1525 – 1537 (1999).
2. Bergmann K., Schriever G., Rosier O., Muller M., Neff W., Lebert R., "Highly repetitive, extreme-ultraviolet radiation source based on a gas-discharge plasma", *Appl. Optics*, **65**, 5413 – 5417 (1999).
3. V. D. Bochkov, V. M. Dyagilev, V. G. Ushich, O. B. Frants, Yu. D. Korolev, I. A. Shemyakin, "Sealed-off pseudospark switches for pulsed power applications (current status and prospects)", *IEEE Trans. Plasma Sci.*, **29**, 802 – 808, (1995).

Fast Capillary Discharge Experiments in a Small Device.
Spectra in the VUV Region

Leopoldo Soto[1], Andrey Nazarenko[2], Cristian Pavez[1,3], Patricio Silva[1], José Moreno[1], Karla Aubel[4], and Tomislav Vucina[4]

1 Comisión Chilena de Energía Nuclear, Casilla 188 D, Santiago, Chile
2 Institute of Spectroscopy, RAS, Troitsk, Moscow Region, 142090 Russia
3 Universidad de Concepción, Chile
4 Universidad de Santiago de Chile

Abstract. Experiments in a small and fast capillary discharge were performed. Millimeters capillaries with centimeters length were used. The discharge was operated in argon, with a cathode pressure of 300-900 mtorr. The system works with differential vacuum, thus the anode pressure is around five to ten times less than the cathode pressure. Experiments in capillaries with 0.8 and 1.5mm internal diameter, and 8 to 40mm were performed. For an applied voltage of 10 to 15kV, a peak current of 3 to 5 kA with a rise time of 2 to 5 ns is obtained ($\sim 10^{12}$ to 5×10^{11} A/s). In addition to usual electrical diagnostics, time resolved spectra in the region of 30 to 100 nm have been obtained. The Ar-IX (46.9nm) line was detected.

INTRODUCTION

In order to produce high brightness radiation with an electrical discharge, high energy densities should be provided to the plasma, $\sim 10^{12}$ J/m^3 (i.e. some MJ into a plasma column of a centimeter in length and radius). This is usually done by means of huge facilities such as the Sandia National Laboratory in the USA. Intermediate devices have been used to produce vacuum ultraviolet (VUV) to soft X ray radiation [1-4] and VUV lasing has been also reported [1]. Interestingly, 10^{12} J/m^3 can also be provided using small devices. In effect, by confining the plasma to submilimetric volumes, less than 1 J is required to reach high energy densities (as an extreme example 0.1 J is enough for a 60 µm diameter sphere). The key point to produce those conditions in small devices is the current rate, which should reach about 10^{13} A/s [5]. Consequently, the inductance should be very low.

To allow studies on the physical mechanisms related to population inversion and high brightness emission in the VUV to soft X-ray region in a table-top device, a fast and small capillary discharge was constructed at the *Comisión Chilena de Energía Nuclear*, CCHEN [6, 7]. Our device follows the design reported by P. Choi and M. Favre (0.1 - 1J, $dI/dt \sim 10^{12}$ A/s) [8].

EXPERIMENTAL SETUP AND DIAGNOSTICS

A pair of 90mm diameter brass electrodes forms the anode and the cathode of the discharge, as well as a parallel plate capacitor. The insulator (Mylar and polyvinylidenefluoride, PVDF) between the electrodes is also the capacitor dielectric. A 1 - 1.5 mm inner diameter capillary tube, 0.8, 2.5, 5.0 cm long, is placed at the axis. The discharge is produced inside the capillary. A primary 7 nF capacitor charges the storage capacitor by means of a pulse. The length of the cable connections between the spark gap and the capacitor of the main discharge is calibrated to obtain the required timing between the discharge and the diagnostics. The ionization in the interelectrode space is assisted by a high-energy electron beam, which is originated in plasma inside the hollow cathode region. The discharge operates in Argon with a cathode pressure of 0.1-1torr. The anode pressure is around five to ten times less than the cathode pressure. Current rates about 5×10^{11}-10^{12} A/s were obtained applying voltages of -10 to -15kV.

The voltage in the capillary capacitor, the current and the current derivative were measured in our experiments. Also VUV diagnostics have been developed and applied in previous experiments [6, 7]. A coaxial cable was biased at -200V and located at the discharge axis, behind the anode, in order to detect electrons (negative signal) or photons with energy above 4.3 eV (positive signal, like a X ray diode). A multipinhole camera with a magnification of m=3, attached to a microchannel plate (MCP) sensitive to X rays and VUV radiation is used to register 1 frame every 4 ns. VUV spectra of the capillary discharge have been obtained by means of a grazing incidence spectrograph GISVUV1-3S attached to a MCP. A gold-coated replica grating of 300 lines/mm, with 1m curvature radius, was used. The grazing angle was 4º. Results obtained in our previous experiments [6, 7] shown negative signals in the diode detector before breakdown indicating electron beams. About 2-3 ns after the peak current there were indications of photon radiation impacting on the diode, which produced positive signals. Pinhole images taken during a discharge along an 8mm length capillary and 500mtorr suggested the existence of a homogeneous plasma column at early time and after a possible compression in the plasma column. Time-integrated spectrum over ten discharges along an 8-mm length capillary (a single shot did not show enough intensity under these set up conditions to register a spectrum). Several lines are attributed to ArVIII, ArIX and ArX ions (second and higher orders were identified: ArIX-2, ArIX-3, ArIX-5). The ion temperature was roughly estimated at about 40-60 eV.

The research also has been complemented with numerical simulations. A one-dimensional model derived from the magnetohydrodynamics (MHD) equations was applied by A. Esaulov to study the plasma dynamics and temperature evolution during the discharges [9]. The calculations predict a final radius of 100-200 microns and electron densities of 2×10^{24}m^{-3} occurring three nanoseconds after the peak current. The corresponding electron and ion temperatures are 80 and 40 eV. Under these conditions, theoretically soft X-ray lasing can occur at 46.9 nm, due to population inversion by collisional electron excitation of Ne-like ArIX ions.

The VUV spectroscopy diagnostic has been improved in order to obtain time resolved spectra. In this work we report spectra at different stages of the discharge.

RESULTS AND DISCUSSION

Experiments for capillary lengths of 8, 28 and 40 mm, with a internal diameter of 1.5 mm, were performed in Ar with a cathode pressure of 300, 450, 600, 900 mtorr. Voltage of charge was –13kV. Interesting spectra have been obtained in discharges with capillary length of 28mm. At these conditions a peak current of ~3kA, with a half period of 12 ns is observed. Figure 1 shows the corresponding current trace and spectra at different stages of the discharge. The spectra were obtained from different discharges and the gating pulse in the MCP was 5 ns.

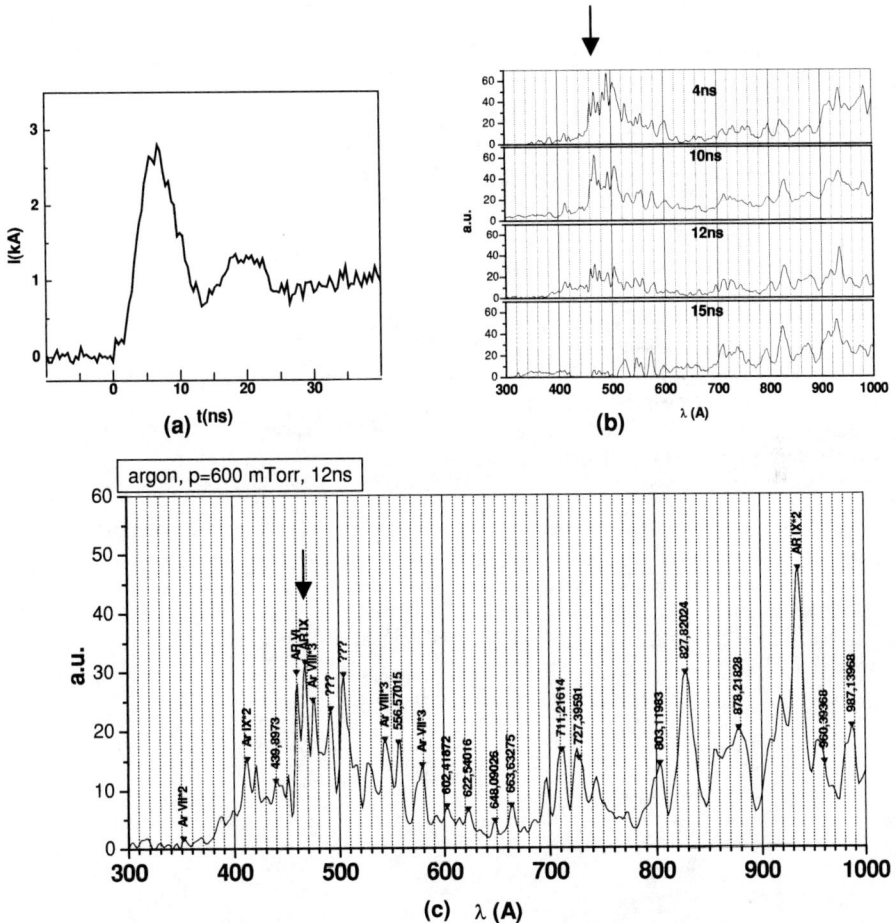

Figure 1. Current trace and spectra for discharges operating in Ar at 600mtorr. a) Current trace, b) spectra at 4, 10, 12 and 14 ns. c) Some lines have been identified. In b) and c) the arrow indicates the Ar-IX (46.9nm) line.

The Ar-IX (46.9nm) line was detected. The maximum emission of this line occurs 2-4 ns after the peak current, 7-10 ns after the peak current the line was not observed. These results are in agreement with predictions from theoretical simulations performed by A. Esaulov [9]. The next step is obtaining several spectra at different times in the same discharge; 3 frames separated 4 ns each will be implemented. Improvements in the device also are being implemented in order to work with longer capillaries, keeping the rise time and the value of the peak current.

ACKNOWLEDGEMENT

This work has been funded by a *Cátedra Presidencial en Ciencias* awarded to L. Soto by the Chilean Government and by the grant CCHEN 563. The authors are thankful to K.N. Koshelev and P. Antsiferov (ISAN, Troitsk, Russia) and to H. J. Kunze (Ruhr Universitat, Germany), for fruitful discussions and comments.

REFERENCES

1. J.J. Rocca, *Rev. Sci. Instrum.* **70**, 3799 (1999).
2. J.J. Rocca, v. Schlyaptsev, F.G. Tomasel et al, *Phys. Rev. Lett.* **73**,2192 (1994).
3. C.H. Moreno, M.C. Marconi, V.N. Shlyaptsev, B.R. Benware, C.D. Macchietto, J.L.A. Chilla, and J.J. Rocca, *Phys. Rev. A* **58**, 1509 (1998).
4. P.S. Antsiferov, L.A. Dorokhin, E.Yu. Khautiev, Yu.V.Sidelnikov, D.A. Glushkov, I.V. Lugovenko, and K.N.Koshelev, *J. Phys. D: Appl. Phys.* **31**, 2013 (1998).
5. P. Choi, J. G. Lunney, A. Engel, C. Dumitrescu-Zoita, T. N. Hansen, I. Krisch, J. Larour, J. Rous, Proceedings IEEE Pulse Power Conference, p. 287 (Piscataway, NJ, USA, 1999)
6. L. Soto, A. Esaulov, J. Moreno, P. Silva, G. Sylvester, M. Zambra, A. Nazarenko, and A. Clausse, Physics of Plasma **8**, 2572 (2001).
7. L. Soto, A. Esaulov, P. Silva, G. Sylvester, J. Moreno, M. Zambra and A. Nazarenko, IX Latin American Workshop on Plasma Physics, La Serena, Chile, 2000,. AIP Conf. Proc. **563**, p. 264, American Institute of Physics (2001).
8. P. Choi and M. Favre, Rev. Sci. Instrum. 69, 3118, (1998).
9. A. Esaulov, P. Sasorov, L. Soto, M. Zambra and J. Sakai, Plasma Physics and Controlled Fusion, **43**, 571 (2001).

Soft X-Ray Radiation of Fast-Capillary-Discharge CAPEX 2

Jiri Schmidt[a], Karel Kolacek[a], Vladislav Bohacek[a], Milan Ripa[a],
Olexandr Frolov[b], Pavel Vrba[a], Alexandr Jancarek[c], Miroslava Vrbova[c]

[a]*Institute of Plasma Physics, Academy of Sciences of the Czech Republic, Za Slovankou 3, P.O.Box 17, 182 21 Prague 8, Czech Republic*
[b]*Faculty of Mathematics and Physics, Charles University in Prague, Ke Karlovu 3, 121 16 Prague 2, Czech Republic*
[c]*Faculty of Nuclear Sciences and Physical Engineering, Czech Technical University, Brehova 7, 115 19 Prague 1, Czech Republic*

Abstract. The paper reports on technological modifications of the capillary discharge experiment CAPEX, especially in capillary region. A temporal evolution of axial soft X-ray radiation and a spectroscopic measurement in the soft X-ray region on the modified capillary are presented as well. The strong spectral line at the wavelength of laser transition (Ne-like Ar, $\lambda=46.9$ nm) was observed.

INTRODUCTION

Since their first realization in 1984 [1], the soft X-ray and extreme ultraviolet lasers based on the collisional excitation of Ne-like ions have been pumped by laser drivers. Alternatively, Rocca's experiment demonstrated large amplification at 46.9 nm in a discharge-pumped argon plasma [2], opening a new route to the development of simpler and compact ultrashort wavelength lasers based on excitation by capillary discharge. In comparison with laser-driven X-ray laser devices, this device is inexpensive, compact in size and has high conversion efficiency from electric energy to laser energy. In many applications of soft X-ray lasers, the output energy of the laser is required to be sufficiently high. In practical operation of the system, it is possible to increase the laser energy by increasing the gain-length product (up to saturation), the gain volume and the laser pulse width.

The capillary discharge is a multi-parametric system. Not only driver characteristics (capillary current amplitude, rise rate), but also conditions (initial gas pressure, pre-ionization current, geometry of capillary, wall material) influence plasma dynamics, compression and heating.

In previous works [3,4] we investigated capillary discharges in plastic capillaries (polyamide, polyamide/polyimide). These were accompanied by a large soot production causing that the whole apparatus (inclusive detection system) was covered by carbon. Therefore, we rebuilt capillary part and designed fully-ceramic capillary.

The modified apparatus CAPEX 2 and the results of last measurements of the soft X-ray radiation from alumina (Al_2O_3) capillary are presented in this paper.

EXPERIMENTAL SETUP

The apparatus consists of five main parts: a Marx generator, a coupling section (spacer), a fast cylindrical capacitor (pulse forming line), a main spark gap and a capillary. Schematic drawing of the apparatus is shown in Fig.1.

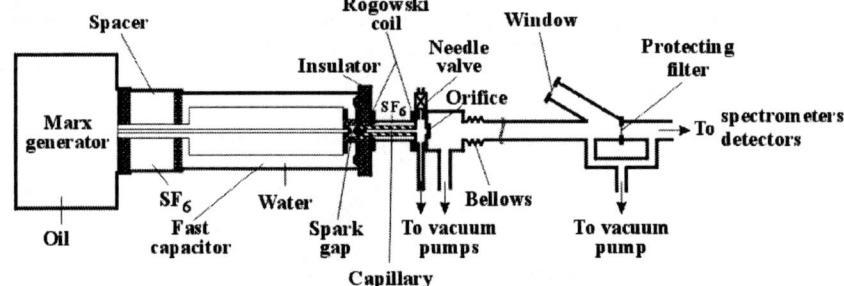

FIGURE 1. Schematic drawing of the apparatus CAPEX 2

The fully-screened oil-insulated Marx generator is used as a power supply for the capillary discharge experiment. At present the Marx generator has 8 stages and its erected voltage is typically up to 450 kV (for charging voltage of 28 kV). The coupling section (spacer) is a short coaxial cylindrical line which is filled with SF_6 gas. It functions as a safety interface between the oil insulated Marx generator an the water filled fast capacitor. A pulse forming line (fast capacitor) is a coaxial cylindrical line as well. As a dielectric we use de-ionized or destilled water. At the end of the line the main spark gap is placed. The inner part of the spark gap is filled with SF_6 gas, while its outer part is filled with water. The quantity of water in the outer part of the spark gap can be changed by a suitable insertion piece. In this way the spark gap capacity and its conductance are changed, which have a significant influence on the pre-breakdown current. A more detail description of the apparatus could be found in [5].

A significant changes have been made in the capillary part of the apparatus. The 180 mm long capillary is directly attached to the main spark gap (having one common electrode). The gas filling and pumping part is attached to the outer end of the capillary. We substituted plastic (polyamide or polyamide/polyimide) capillary for ceramics one. The spectroscopic measurement was performed in the alumina (95% AL_2O_3) capillary 3.2 mm in inner diameter, while its outer diameter is 40 mm. The capillary is placed in a shielding metallic cylinder of 80 mm in diameter. The volume between the metallic cylinder and the ceramic capillary is filled with pressurised (3 atm) SF_6 gas for an increase of electrical strength. The generated radiation is brought out through ϕ 3.5 mm orifice in the outer electrode. The diagnostic part is separated by another ϕ 0.8 mm orifice (~55 mm apart of the grounded capillary end).

The capillary is filled with argon by a needle valve. Typical values of argon pressure in the capillary prior breakdown are in the range from 40 Pa to 200 Pa.

SOFT X-RAY RADIATION MEASUREMENTS

We have performed measurement of the time evolution soft X-ray radiation as well as the soft X-ray spectroscopic measurements of the discharge in ceramic capillary.

Temporal evolution of soft X-ray radiation

First, we observed temporal evolution of soft X-ray radiation. We used PIN diode SPPD 11-40 powered by 140 V. The PIN diode was covered by 0.75 µm aluminium filter [5]. HP 54542C oscilloscope (0-500 MHz analog bandwidth, 700 ps rise-time, 2 GSa/s maximum sampling rate in each of four channels) was used for recording signals. The dependence of soft X-ray pulse shape in the correlation with the capillary current is shown in Fig. 2. The full width at half maximum (FWHM) of the large short intense spike of the soft X-ray signal is aproximately 4 ns and it appears 29 ns prior the capillary current maximum. However, this intense spike detected by PIN diode need not necessarily mean lasing, but efficient pinching only. In this case a 3.2 mm inner diameter, 180 mm long alumina capillary filled with 100 Pa of argon gas, was used. The amplitude and period of the capillary current pulse were approximately 44 kA and 238 ns. The main capillary current pulse was preceded by a pre-ionization current pulse of the amplitude ~40 A and of the duration several hundreds of nanoseconds to produce a uniform preionization.

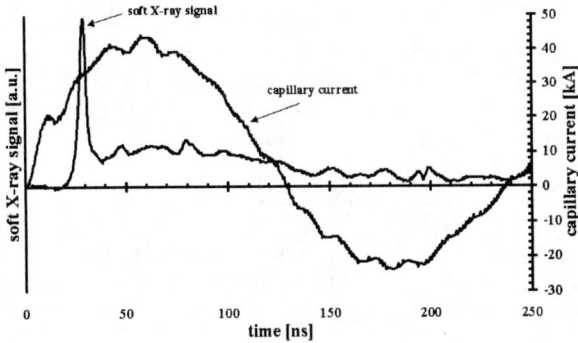

FIGURE 2. Temporal evolution soft X-ray radiation in the correlation with the capillary current.

Spectroscopic measurements

The soft X-ray emission in the axial direction was measured with a home-made soft X-ray flat field spectrometer having average groove density 450 gr/mm (Jobin Yvon grating with non-equidistantly spaced grooves). The spectral range of used flat field spectrometer is 10-100 nm. The soft X-ray spectrum was registered with the help of home-made microchannel plate detector. It has two MCPs in tandem arrangement and a "phosphor" screen. The 12-bit trigerable digital cooled CCD camera (PCO SensiCam, 1280x1024 pixels) was used to record the soft X-ray spectrum. The more detail description of the spectrograph arrangement is in [5].

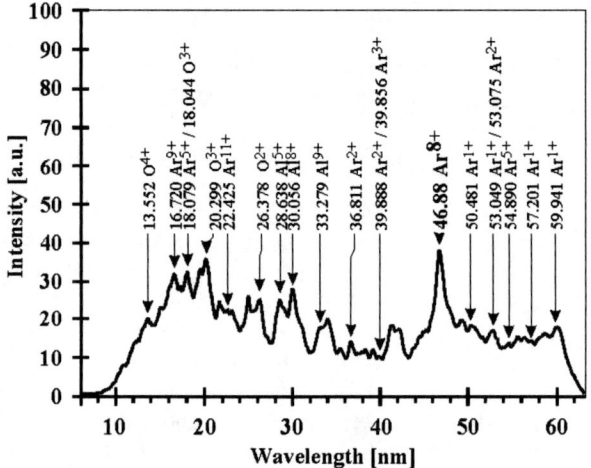

FIGURE 3. The soft X-ray spectrum of a pulse discharge burning in a 3.2-mm-diam capillary filled with approximately 100 Pa of argon.

Axial soft X-ray emission from the capillary plasma was observed through a hole in the grounded electrode. A typical spectrum from these discharges in optimal conditions (argon pressure, pre-ionizition current) is shown in Fig. 3. MCP detector was gated by a 50 ns pulse. Due to long exposition many spectral lines belong to ions of low ionization stages. Nevertheless, the strong amplification of the J=0-1 line in Ne-like Ar (λ=46.88 nm) was found. Measurements with a shorter MCP gated pulse is desirable and it will be performed in near future.

ACKNOWLEDGMENTS

This work has been performed under auspices and with the support of both the Ministry of Education Youth and Sports of the Czech Republic under Contract INGO LA055 and the Academy of Sciences of the Czech Republic under Contract AVOZ 2043910.

REFERENCES

1. Matthews, D.L., et al, *Phys. Rev. Letters* **54**, 110 (1985).
2. Rocca, J.J., et al, *Phys. Rev. Letters* **73**, 2192 (1994).
3. Kolacek, K., et al, *Journal of Technical Physics, Special Suppl.* **39**, 161-165 (1998).
4. Kolacek, K., et al, *Proc 13th Int. Conf. On High-Power Particle Beams (Beams2000)*, Nagaoka, Japan, **73**, 151-154 (June 25-30, 2000).
5. Kolacek, K., et al, in *Proc 20th SPPT; Czechoslovak Journal of Physics*, Prague, Czech Republic, Supplement D, **52**, 199-204 (2002).

Plasma Dynamics and Lasing Condition of Fast Capillary Discharges

Nobuhiro Sakamoto, Gohta Niimi, Yasushi Hayashi, Majid Masnavi,
Mitsuo Nakajima, Eiki Hotta, Kazuhiko Horioka

*Department of Energy Sciences, Tokyo Institute of Technology,
Nagatsuta 4259 Midori-ku Yokohama, 226-8502, Japan*

Abstract. The dependence of the lasing condition of fast capillary discharge plasmas on the initial gas pressure and rising rate of the load current; dI/dt has been experimentally investigated. The results indicate that there is an optimum condition at suitable values of initial gas pressure and dI/dt. Based on the experimental results and a simplified scaling model for the compression process, the correlation between the plasma dynamics and the lasing condition in capillary pinch plasmas is discussed.

INTRODUCTION

Fast capillary Z-discharges has been paid attention as promising laser sources at soft x-ray region from when J.J.Rocca *et al.* at Colorado State University succeeded a lasing at 3p-3s transition (~469Å) of Ne-like Ar [1]. The discharge pumping system has a possibility to accomplish practical lasing sources compared with the high-power laser system, from the view point of energy efficiency, the size of devices, and so on.

For the lasing by electron-collisional excitation scheme, the electron temperature and electron density of plasma have to be increased up to appropriate values. Capillary Z-discharges realize the lasing condition by fast pinching effects. The fast compression process inevitably accompanies the energy dissipation. In general, the faster compression process makes the hotter and lower density plasmas. The interaction between the current sheet and the shock wave makes an inner structure with temperature, density and velocity distribution, in the plasma column [2]. All of these parameters affect the gain distribution, refraction of stimulated emission and opacity of the resonance levels. Therefore, the lasing condition of compressed plasma is expected to be strongly dependent on the pinching process. However, the relations between those features and the lasing process have not been made clear so far. We have experimentally investigated the lasing condition of Ne-like Ar, as a function of the initial Ar gas pressure in the capillary; P_{Ar}, and the rising rate of driving current; dI/dt. Based on the results and a simplified scaling, we discuss the correlation between the plasma dynamics and the lasing condition in fast capillary discharge plasmas.

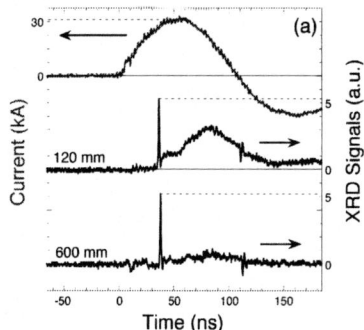 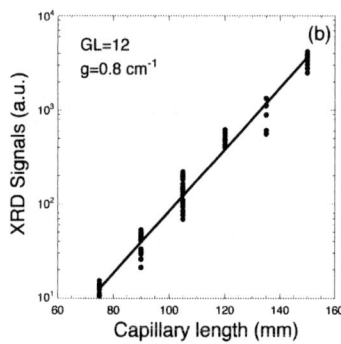

FIGURE 1. Experimental results of measurement for directivity (a), and amplification of X-ray signals (b).

EXPERIMENTAL RESULTS AND DISCUSSIONS

For fast capillary discharge experiments, we use two pulse-power devices that can drive maximum peak current of 200 kA [2] and 50 kA [3].

At first, lasing signals were confirmed by measurements of directivity of X-ray signals and gain coefficient as showed in Fig.1(a) and (b), respectively. Fig.1(a) shows XRD signals that the distance from the exit of the capillary to the XRD are 120 mm and 600 mm when the peak current is 31 kA. The attenuation of the spike-like waveform is quite less than other parts of the waveform when the distance is increased. This means that the spike-like waveform has a strong directivity. Fig.1(b) shows the peak values of the spike-like XRD signals as a function of the plasma length. The spike-like XRD signal values have an exponential enhancement when the capillary length was increased. From this result, we can estimate the gain coefficient to be 0.8 cm^{-1}.

We have investigated the peak values of the spike-like XRD lasing signals as a function of P_{Ar} and dI/dt. Fig.2(a) shows peak values of the spike-like XRD signals as a function of P_{Ar} with dI/dt as a parameter. There appeared the distributions that had a peak place. Fig.2(b) shows a schematic n_i(ion density)-T_e(electron temperature) diagram of the expected compression processes of the capillary discharge plasmas and the lasing parameter. The solid line in Fig.2(b) corresponds to an isentropic process from an initial condition expressed by $n_i = AT_e^{1/(1-\gamma)} = AT_e^{3/2}$ where A is a constant that is determined by the initial condition and γ is the adiabatic coefficient for mono-atomic gas. The dot-dashed line and dotted lines schematically show the expected compression process for the formation of lasing plasmas. Although these lines on the n_i-T_e diagram are drawn only for references, they give us some insights into the pinching process for achieving the lasing condition. For example when dI/dt was 7~8 × 10^{11} A/s, we got the maximum spike-like XRD signals at about P_{Ar} = 550 mTorr. Then, the compression process is considered to correspond to the dot-dashed line because the core of the estimated Ne-like Ar lasing domain is intersected by the dot-dashed line that originates from the initial pressure of 550 mTorr. When P_{Ar} is decreased to 400 mTorr, the compression route is expected to shift to the lower dotted line and when P_{Ar} is increased to 700 mTorr, the process is thought to shift to the upper line.

On the other hand, Fig.3(a) shows the change of peak values of the spike-like XRD signals versus dI/dt when P_{Ar} was fixed. The spike-like XRD signals also have the distributions against dI/dt. A dot-dashed line and dotted lines in Fig.3(b) also give the anticipated compression processes depending on the change of dI/dt. As shown in Fig.3(a), at P_{Ar} = 300 mTorr, we obtained the maximum spike-like XRD signals when dI/dt is 5.7 x 10^{11} A/s. Then, the compression route is considered to be on the dot-dashed line. In this case, when the dI/dt is increased, the compression process is expected to shift to the lower dotted line with smaller inclination shown in Fig.3(b), because stronger compression is considered to be more dissipative during the compression processes.

From the relationship between the plasma parameters in the lasing region and the initial conditions from which we could get lasing, we can derive an effective adiabatic coefficient γ'. For example, the derived γ' is to be about 2.57 when the initial Ar gas pressure and temperature are assumed to be 300 mTorr and 300 K, respectively. Under these initial conditions, we could get the peak XRD signals when the dI/dt value is 5.7 x 10^{11} A/s. We think inspecting the correlation between dI/dt values at peak gain and values of γ' over a wide experimental condition in detail, gives an effective scaling equation that can express a degree of dissipation in the pinching plasma. As the pinch effect of plasma itself is considered to be similar, if we can get a scaling formula including the dissipative effect, we can extrapolate it for shorter wavelength lasers together with a spectroscopic measurement of the pinching plasma [4].

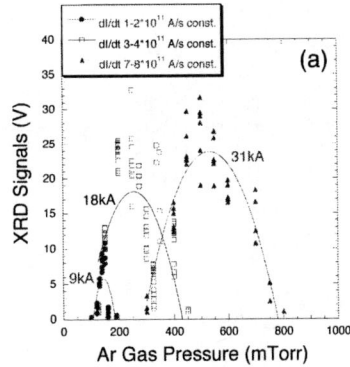
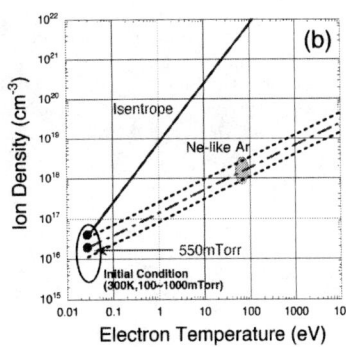

FIGURE 2. The change of peak values of spike-like XRD signals versus P_{Ar} when dI/dt was fixed (a), and a schematic n_i-T_e diagram of the expected compression processes of the capillary discharge plasmas and the lasing parameter (b).

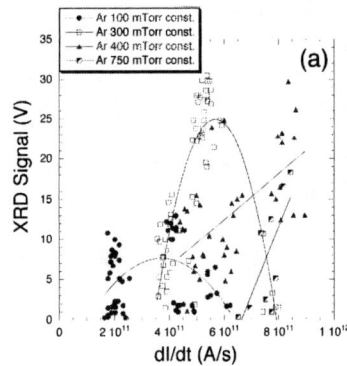 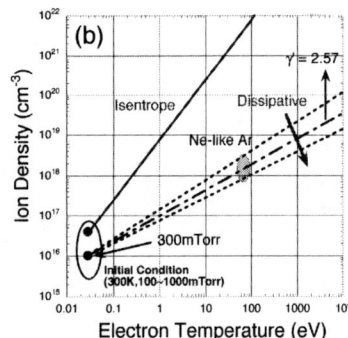

FIGURE 3. The change of peak values of spike-like XRD signals versus dI/dt when P$_{Ar}$ was fixed (a), and a schematic n_i-T_e diagram of the expected compression processes of the capillary discharge plasmas and the lasing parameter (b).

CONCLUDING REMARKS

Lasing in a fast capillary Ar plasma was confirmed by measurements of directionality and amplification scaling of output signals. The lasing signals were investigated over a wide range of parameters as a function of initial filling pressure P$_{Ar}$ and the rise rate of discharge current dI/dt. The consideration of the pinching process using the experimental results and a simplified compression model based on a concept of effective adiabatic constant indicates the possibility of the scaling formula for shorter wavelength regions.

REFERENCES

1. J.J.Rocca, V.Shlyaptsev, F.G.Tomasel, O.D.Cortazar, D.Hartshorn, and J.L.A.Chilla, Phys. Rev. Lett. **73**, 2192 (1994).
2. T.Hosokai, M.Nakajima, T.Aoki, M.Ogawa and K.Horioka, Jpn. J. Appl. Phys. **36**, 2327 (1997).
3. G.Niimi, Y.Hayashi, N.Sakamoto, M.Nakajima, A.Okino, M.Watanabe, K.Horioka and E.Hotta, IEEE Trans. Plasma Sci., **30**, No.2, (2002).
4. M.Masnavi, M.Nakajima and K.Horioka, Jpn.J_Appl. Phys. **41**(8), (2002). (in print)

X-ray Spectroscopic Studies of X-pinch Plasma Micropinches with ~ 10 ps Resolution

S. A. Pikuz,[a,b] T. A. Shelkovenko,[a,b] D. B. Sinars,[a,c] I. Yu. Skobelev,[d] K. M. Chandler,[a] M. D. Mitchell[a] and D. A. Hammer[a]

[a]*Laboratory of Plasma Studies, Cornell University, Ithaca, NY 14853, USA*
[b]*Permanent address: P.N. Lebedev Physical Institute, Moscow119991, Russia.*
[c]*Present address: Sandia National Laboratories, P.O. Box 5800, Albuquerque, NM 87185*
[d]*Permanent address: VNIIFTRI, Mendeleevo, Moscow Region 141570, Russia.*

Abstract. X pinches have been successfully developed as point sources of soft x-rays for radiography, but the parameters of the micropinch plasmas that produce the x-rays are not well established. To determine micropinch conditions, time-resolved x-ray spectra produced by 2- and multi- wire X pinches have been collected using an x-ray streak camera with <10 ps time resolution. Together with a spherically-bent mica crystal spectrograph, the streak camera recorded 1-10 keV radiation emitted from X pinches made from different wire materials. Spectral features varied on time scales ranging from 10 to 300 ps, depending on the wire material. From the spectra, we infer that some micropinches produced within exploding wire X pinches are near solid density and have an electron temperature above 1 keV. For example, an electron density up to about $3 \times 10^{23}/cm^3$ and an electron temperature up to about 2.5 keV were inferred from the He-like and H-like lines from a Ti X pinch. Results demonstrate that x-ray spectra must be recorded with sub-ns time-resolution to obtain plasma conditions in micropinches.

Many years of applying x-ray spectroscopy to hot plasmas has demonstrated the value of this method of determining the conditions in plasmas, especially those that are bright and dense [1 - 6]. Crystal spectrographs with appropriate geometry enable investigation of the plasmas with high spatial resolution as well as spectral resolution. This is especially important for exploding wire z-pinch experiments, in which small, bright x-ray sources called micropinches are generated. Single exploding wire z-pinches produce several such sources at random locations along the wire, thereby making an experimental determination of the plasma conditions extremely difficult. Spatially resolved spectroscopy can reveal not only the micropinch plasma conditions, but also the position and structure of the x-ray source.

A variation on the single-wire z-pinch, the X-pinch [7, 8], enables the use of very high spatial and temporal resolution diagnostics to study the conditions in these x-ray sources because the micropinch location and time can be predicted in advance. An X pinch is composed of two or more fine metal wires (typically 10-50 µm diameter) that cross and touch at a single point, in the form of an X, that are exploded to form plasma by a 200-400 kA, 100 ns FWHM current pulse. Near the original cross point of the wires a z pinch plasma forms and micropinches that are reproducible in space to within ±150 µm and in time to within ±1-3 ns [8] develop.

In earlier work, we determined that X pinches contain micropinches that evolve on time scales of order 0.25-1 ns or less [8,9]. As a result of the short time scale and small size, ~ 1-10 µm, of these x-ray sources, the best information obtained on the properties of micropinches came from time-integrated, spatially resolved x-ray spectroscopy [8]. These spectra often implied plasma densities near solid density and plasma temperatures of ~1 keV. By using a Kentech x-ray streak camera capable of 10 ps time resolution as the detector for a spherically bent mica crystal spectrograph, recent experiments [10] to be discussed here show that time resolution of ≤ 10 ps can be needed to accurately determine micropinch plasma conditions.

Time-resolved spectra from two- and multi- wire X pinches have been studied using many wire materials, including Al, Ti, Mo and W. Here we restrict ourselves to results using Ti wires. By using different streak camera time windows ranging from about 1 ns to about 40 ns, individual x-ray bursts or the long-time emission characteristics of a micropinch could be investigated. In Fig. 1, a Ti spectrum obtained using a slow sweep rate and a 3-wire Ti X pinch is presented together with photoconducting diode (PCD) signals. The experimental arrangement is that shown in Fig. 1, Ref. 11, which is also published in this volume. The streaked spectrum shows much more information than can be obtained from filtered PCDs: intense continuum in the first burst, intense H- and He- like Ti line radiation in the second burst and intense characteristic K-shell line radiation are all visible (see Fig. 1). Thus, a relatively slow sweep rate shows the complete behavior of the X pinch, but does so with ~0.2 ns time resolution.

To determine plasma parameters, spectra with 10 ps temporal resolution have been used. An example of a Ti spectrum obtained from a 7-wire Ti X pinch using a fast sweep is shown in Fig. 2a. In Fig. 2b we show the time-dependence of X pinch radiation bands containing: 1. mainly the resonance $He_{\alpha 1}$ line of Ti XXI (curve 3, 2.615-2.624 Å); 2. the intercombination $He_{\alpha 2}$ line of Ti XXI (curve 2, 2.629-2.638 Å);

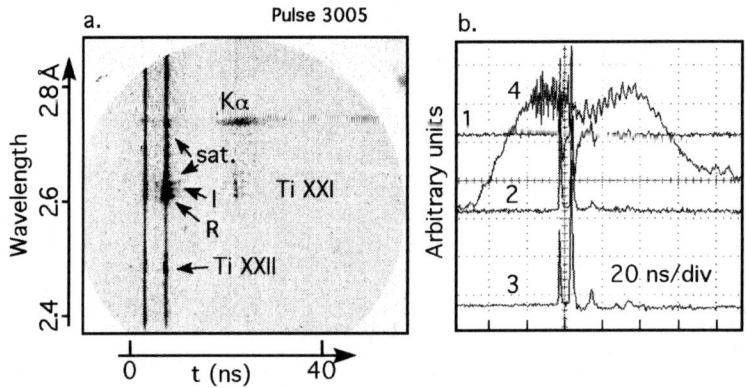

FIGURE 1. (a) The spectrum obtained using a slow sweep rate and a 3-wire Ti X pinch; (b) PCD signals for the same pulse (curve 1: 50 µm Be filter, curve 2: 10 µm Cu filter, curve 3: 12.5 µm Ti filter) together with the load current (curve 4). The peak current was about 400 kA.

3. the k, j satellite lines (i.e., $1s2p^2\ ^2D - 1s^22p\ ^2P$ transitions in Ti XX ions, curve 1,

2.642-2.651 Å); 4. the resonance Ly_α line of Ti XXII (curve 5, 2.489-2.498 Å); and 5. transitions in the continuum (curve 4, 2.500-2.509 Å). In Fig. 2 it is clear that in the time interval 0 - 600 ps, the brightness of the plasma has three maxima. It is natural to connect these with the formation of three micropinches. The shortest x-ray burst, ~50 ps, is radiated by the first micropinch. The duration of the second burst is about 50 ps for the Ly_α line and about 80 ps for the $He_{\alpha 1,2}$ lines. The third micropinch radiates the $He_{\alpha 1,2}$ lines for ~110 ps while the Ly_α and k, j satellite line emission lasts only ~ 50 ps.

The differences in duration of plasma radiation in various lines is caused by the difference in excitation channels of resonance and satellite lines of H- and He- like ions. The resonant levels 2p and 1s2p can be occupied as a result of excitation by electron impact from the ground states, 1s and $1s^2$, respectively, of the H- and He- like ions, or by three-body or radiative recombination of the bare Ti nucleus or H- like ions. However, the doubly-excited states $1s2p^2$ are populated mainly by dielectronic capture from the basic $1s^2$ state of the He-like ion.

The time-depended spectra shown here can be used to determine micropinch plasma parameters as a function of time. For this purpose the emission spectra of the He-like Ti XXI ions, including dielectronic satellites caused by transitions in Li-like Ti XX were analyzed within the framework of a standard quasi-stationary radiative-collisional kinetic model [3, 5]. When the temperature of the plasma is high, the model spectrum is function only of two plasma parameters, the electron density, n_e, and electron temperature, T_e, which can, therefore, be determined by adjusting those parameters until the model spectrum matches the observed one. Note, that the ratio of intensities of the $He_{\alpha 1}$ and $He_{\alpha 2}$ lines was the most sensitive to plasma density, and the relative intensities of the k, j satellites to $He_{\alpha 1}$ was most sensitive to plasma temperature. The experimental spectrum for the time t = 494 ps is shown in Fig. 3 together with a computed spectrum for $n_e = 10^{23}$ cm^{-3} and $T_e = 2400$ eV. Plasma

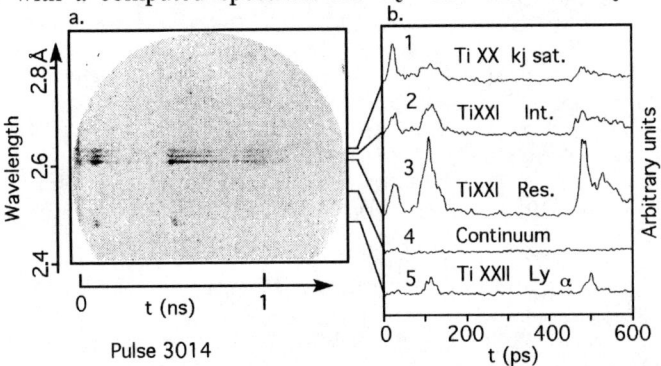

FIGURE 2. (a) Spectrum obtained with the streak camera from a 7-wire Ti X pinch using a fast sweep; (b) the time-dependence of the indicated x-ray radiation bands (wavelengths given in the main text).

parameters as a function of time determined in this manner are presented in Table 1. The tabulated conditions indicate that in the first micropinch, the plasma temperature did not exceed 1250 eV, and a peak value of electron density 3×10^{23} cm^{-3} was reached. By contrast, the second micropinch plasma reached 2100 eV with a density

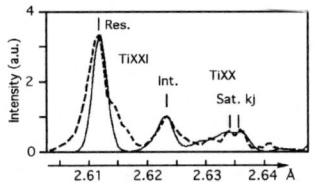

FIGURE 3. A comparison of the experimental spectrum at the time t = 494 ps and a computed spectrum for n_e = 10^{23} cm^{-3} and T_e = 2400 eV.

TABLE 1. X pinch plasma parameters as a function of time

Time (ps)	T_e (eV)	n_e (10^{23}cm^{-3})	Time (ps)	T_e (eV)	n_e (10^{23}cm^{-3})
32	1250	3.0	480	1500	0.5
88	1700	1.0	487	2000	0.6
95	1750	0.8	494	2400	1.0
102	2100	0.8	501	2500	1.0
109	2100	0.6	508	2400	0.9
116	1650	0.2			
123	1400	<0.1			
130	1150	<0.1			

≤ 1x10^{23}cm^{-3}. The peak temperature of the plasma, 2500 eV, is achieved in the third micropinch at an electron density of 1x10^{23} cm^{-3}.

The experimental results and calculations presented in this paper clearly demonstrate the necessity to have spectroscopic data with ≤ 10 ps time resolution to obtain plasma conditions from Ti micropinch x-ray spectra. The complete behavior of the X pinch requires time-resolved data over a long time interval.

This research was supported by DOE grant DEF02-98ER54496.

References.
1. Bhalla, C. P., Gabriel, A. H., and Presnyakov, L. P., *Mon. Not. R. Astron. Soc.* **172**, 359 (1975).
2. Vinogradov, A. V., Skobelev, I.Yu, and Yukov, E. A., *Sov. J. Quantum Electron.* **5**, 630 (1975).
3. Boiko, V. A., Vinogradov, A. V., Pikuz, S. A., Skobelev, I. Yu., and Faenov, A. Ya., *J. Sov. Laser Res.* **6**, 85 (1985).
4. Abdallah Jr., J., Faenov, A. Ya., Hammer, D. A., Pikuz, S. A., Csanak, G., and Clark, R. E. H., *Phys. Scr.* **53**, 705 (1996).
5. Pikuz, S. A., Shelkovenko, T. A., Sinars, D. B., Hammer, D. A., Lebedev, S. V., Bland, S. N., Skobelev, I. Yu., Abdallah Jr., J., Fontes, C. J., and Zhang H. L., *J. Quant. Spectrosc. Radiat. Transf.* **71**, 581 (2001).
6. Skobelev I.Yu., Faenov A. Ya., Bryunetkin B. A., Dyakin V. M., Pikuz T. A., Pikuz S. A., Shelkovenko T. A., Romanova V. M., and Mingaleev A. R., *JETP* **81**, 692 (1995).
7. Zakharov, S. M., Ivanenkov, G. V., Kolomenskii, A. A., Pikuz, S. A., Samokhin, A. I., and Ulshmid, J., *Sov. Tech. Phys. Lett.* **8**, 456 (1982).
8. Shelkovenko, T. A., Sinars, D. B., Pikuz, S. A., and Hammer, D. A., *Phys. Plasmas* **8**, 1305 (2001).
9. Shelkovenko, T. A., Sinars, D. B., Pikuz, S. A., Chandler, K. M., and Hammer, D. A, *Rev. Sci. Instrum.* **72**, 667 (2001).
10. Sinars, D. B., Ph.D. dissertation, Cornell University, 2001.
11. Shelkovenko, T. A., Pikuz, S. A., Sinars, D. B., Chandler, K. M., Mitchell, M. D., and Hammer, D. A., paper XYZ in these proceedings.
12. Shelkovenko, T. A., Pikuz, S. A., Sinars, D. B., Chandler, K. M., and. Hammer, D. A., Phys. Plasmas **9**, 2165 (2002).

Spectroscopic Analysis of 1MA X-pinch Implosions at the Nevada Terawatt Facility

Alla S. Shlyaptseva,[a] Stephanie B. Hansen,[a] Victor L. Kantsyrev,[a] Dmitry A. Fedin,[a] Nicholas D. Ouart,[a] Kevin B. Fournier,[b] Ulyana I. Safronova[c]

[a] *Physics Department/220, University of Nevada, Reno, NV 89557 USA*
[b] *Lawrence Livermore National Laboratory, P.O. Box 808, Livermore, CA 94550 USA*
[c] *Department of Physics, University of Notre Dame, Notre Dame, IN 46566 USA*

Abstract. An overview of the detailed spectroscopic analysis of more than fifty z-pinch shots at the Nevada Terawatt Facility is presented. Experimental x-ray spectra generated by Ti, Fe, Mo, W, and Pt z-pinches have been collected in wire implosions driven by a 0.7-1 MA current. Different configurations and masses of z-pinch loads are studied. In x-pinch experiments, pinhole images and damage to the anodes after shots evidence the presence of electron beams. Non-LTE collisional-radiative atomic kinetics models that include hot electrons have been developed and applied to interpret experimental x-ray spectra. In particular, diagnostics for K-shell spectra of Ti ions and L-shell spectra of Mo ions are presented.

INTRODUCTION

The Nevada Terawatt Facility (NTF) has produced a number of Ti, Fe, Mo, W, and Pt z-pinches driven by a 0.7-1 MA current from which x-ray spectra have been collected. Ti wire explosions are being intensively studied at Sandia National Laboratories. Studies of titanium wire-array Z-pinch experiments and modeling have been published elsewhere [1,2]. Recent wire-array experiments on the Z accelerator at SNL have used Mo wires in imploding arrays to produce hot plasmas and generate multi-keV energy x-rays [3]. The purpose of the NTF work is to study the radiation physics of z-pinches for a wide variety of HED materials on a 1MA-level pulsed-power device and to complement SNL study of wire array implosions. The SNL and NTF results stimulated the development of detailed time-dependent atomic kinetics models to diagnose the electron density (n_e), temperature (T_e), and electron beam characteristics of z-pinch plasmas [4]. Moreover, the development of K-shell spectropolarimetry [4,5] requires detailed plasma modeling with inclusion of hot electrons. This paper focuses on the NTF Ti and Mo results. Time-integrated K-shell spectra have been collected from Ti plasmas produced from a wide variety of pinches including single wires, x-pinches, and conical and cylindrical arrays with Al wires. Time-integrated and time-resolved L-shell spectra have been collected from Mo plasmas produced from x-pinches. The experimental details are presented in [6,7]. The kinetic models, diagnostic techniques, and Mo L-shell and Ti K-shell results are discussed below.

SPECTROSCOPIC MODELS

Two atomic kinetics models have been developed to diagnose experimental x-pinch spectra from the NTF. Both the L-shell Mo and K-shell Ti models include the ground states of all ionization stages from the bare ion to the neutral atom. Levels within these ionization stages are coupled by radiative recombination and collisional ionization [8, 9], Auger decay [10,11], and their reverse rates.

The model for L-shell Mo has 5505 levels and includes singly-excited states up to $n = 7$ for O- through Ne-like ions and both singly and doubly excited states up to $n = 4$ for Na- and Mg-like ions. Energy level structures and complete radiative and collisional coupling data were calculated with [9,10]. Selected energy levels and radiative decay rates in the Ne-like ion were calculated to high accuracy by [12].

The model for K-shell Ti has 308 levels and includes singly-excited states up to $n = 4$ for the H- and He-like ion, singly-excited states up to $n = 3$ in the Li- through O-like ions, and doubly-excited states up to $n = 3$ in the He- through O-like ions. Energy level structures for all ions were calculated with [11]. Levels within ionization stages are coupled by radiative decay [11] and collisional excitation and de-excitation. For selected transitions from the ground states, collision strengths have been calculated. For all other collisional transitions, the Van Regemorter approximation is used.

The electron distribution function used in this paper consists of a Maxwellian distribution at an electron temperature T_e and a fraction f of hot electrons. The hot electron distribution is given by $C\varepsilon^{-3}$ for $\varepsilon > 10$keV, where ε is the electron energy. Hot electrons represent high-energy, current-carrying electron beams in the pinch plasma and can significantly affect collisional rates.

Mo L-shell Results

Results from more than fifteen Mo x-pinch shots with wire diameters from 24 to 62 μm have been modeled. A typical comparison of experimental and modeled spectra is given in Fig. 1 along with identification of diagnostically important features. The improvement of the Maxwellian fit of Fig. 1(a) with inclusion of hot electrons is illustrated in Fig. 1(b), in which a spectrum with hot electrons, a lower bulk temperature, and a larger electron density fits the experiment quite well. The modeled Mo spectra have distinguishable dependence on each of the three parameters T_e, f, and n_e. The ionization balance of Mo increases with all three parameters, but the electron density has distinct effects on the relative intensity of Ne-like lines. This is shown in Fig. 2(a), in which the ratio of selected Ne-like lines is given for various T_e as a function of n_e. This ratio is fairly insensitive to f and is thus a robust density diagnostic. In contrast, Fig. 2(b) illustrates that ratios of the features Mg/Na and F/Na are quite sensitive not only to T_e but also to f. Opacity effects are not required to fit the experimental spectra. Overall, the fits of all experimental Mo spectra are good and indicate n_e between 1.5×10^{21} and 2×10^{22} cm^{-3}, T_e between 675 and 850 eV, and hot electron fractions up to 8.5%. The fraction of hot electrons tends to increase with the current and the electron density tends to decrease with increasing wire diameter.

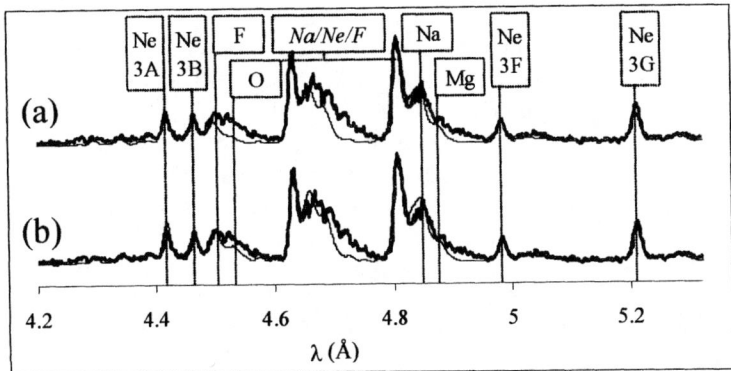

FIGURE 1. L-shell Mo line identification and comparison of modeled spectra (black lines) with experiment (gray lines: 24μm/0.73MA x-pinch). In (a), the modeled spectrum has T_e = 1keV, n_e = 1.5×10^{22}cm^{-3}, and f = 0. In (b), the modeled spectrum has T_e = 0.83keV, n_e = 2×10^{22}cm^{-3}, and f = 3%.

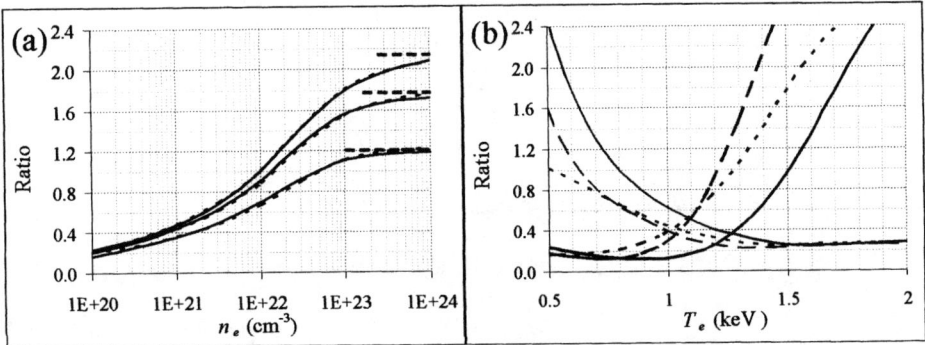

FIGURE 2. Ratios of Ne-like lines (3A + 3B)/(3F + 3G) as a function of n_e are given by the solid lines in (a) for three values of T_e: 0.5, 1.0, and 2.0keV. The ratios at LTE for the three T_e are given by dashed lines in (a) at 1.20, 1.76, and 2.12, respectively. The ratios in (a) are insensitive to f. Ratios of the features Mg/Na (gray lines) and F/Na (black lines) as a function of T_e are given in (b). The solid lines in (b) have f = 0 and n_e = 10^{21}cm^{-3}, the dashed lines have f = 0 and n_e = 10^{22}cm^{-3}, and the dotted lines have f = 5% and n_e = 10^{21}cm^{-3}. Ratios are valid for spectra with Ne-like fwhm from 0.01-0.02Å.

K-Shell Ti Results

Line emission from a wide range of ions is present in the experimental K-shell Ti spectra, from H-like Ti to Ti Kα. It is unlikely that all emission comes from a single region, and the Ti plasmas are taken to have two regions: a hot, dense region with hot electrons that contributes all of the H-like and most of the He-like radiation and a cooler, less dense region that contributes Li- to O-like radiation. Ti Kα and Li- to O-like satellites are most intense in v-pinch and some single wire implosions, while Lyα emission is most intense in x-pinches and conical arrays. Spectra from single wire implosions exhibit wide and uncorrelated variations in Lyα and satellite intensity. Preliminary analysis of the emission from hot plasma regions suggests T_e of 1-2keV for v-pinch and some single wire implosions and 2-3keV for x-pinches and conical arrays. The optically thin ratios of H- and He-like Ti line emission used for these T_e estimates are in good agreement with previous work [2]. However, even modest optical depths can have a profound effect on H- and He-like emission from hot plasma regions [2] and it is found that emission from the hot regions cannot be adequately described without opacity. The cool plasma region, in contrast, can be well-described without opacity by regions with temperature gradients from 0.15 to 0.6keV. Emission from the cool regions can be also be used to diagnose the presence of hot electrons. Fig. 3 shows the distinguishable effects of T_e and f on modeled Ti spectra.

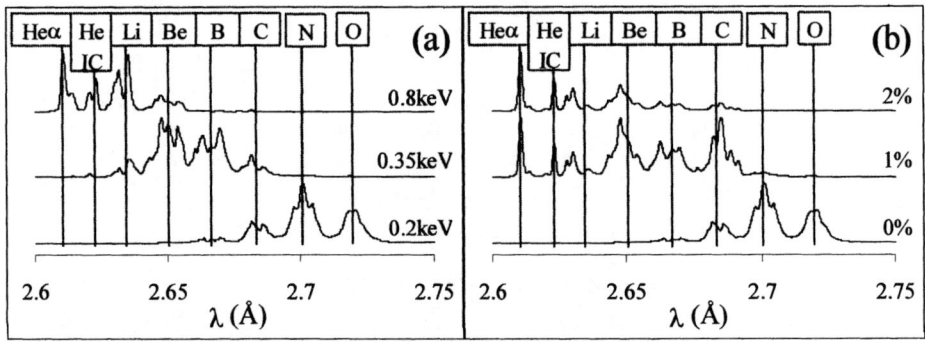

FIGURE 3. K-shell satellite spectra of Ti at $n_e = 10^{21} \text{cm}^{-3}$. The spectra in (a) have $f = 0$ and T_e as indicated. The spectra in (b) have $T_e = 0.2\text{keV}$ and f as indicated.

ACKNOWLEDGEMENTS

We acknowledge program support from Dr. B. Bauer and the NTF team for help with x-pinch experiments. The present work was supported by the DOE-NNSA/NV Cooperative Agreement DE-FC08-01NV14050, SNL, and UNR.

REFERENCES

1. Deeney, C., Coverdale, C.A., Douglas, M.R. et al., *Phys. Plas.* **6**, 2081-2088 (1999).
2. Apruzese, J.P., Pulsifer, P.E., Davis, J. et al., *Phys. Plas.* **5**, 4476-4483 (1998).
3. Lepell, P.D., Coverdale, C.A., Deeney, C. et al., Abstracts of the 43d Annual Meeting of Division of Plasma Physics, Long Beach, CA, Bulletin of the American Physical Society **46**, 235 (2001).
4. Shlyaptseva, A.S., Hansen, S.B., Kantsyrev, V.L., et al., Digest of Technical Papers, 2001 IEEE Pulsed Power Plasma Conference, pp. 753-756 (2002).
5. Shlyaptseva, A.S., Hansen, S.B., Kantsyrev, V.L. et al., *Rev. Sci. Inst.* **72**, 1241-1244 (2001).
6. Kantsyrev, V.L., Bauer, B.S., Shlyaptseva, A.S. et al., *Rev. Sci. Inst.* **72**, 663-667 (2001).
7. Kantsyrev, V.L., Fedin, D.A., Shlyaptseva, A.S. et al.,*Results of x-ray wide-band and time-resolved imaging and spectroscopic studies of 0.9-1.0 MA high-Z x-pinch development* (this conference).
8. Bernshtam, V.A., Ralchenko, Y.V., and Maron, Y., *J. Phys. B* **33**, 5025-5032 (2000).
9. Bar-Shalom, A., Klapisch, M., and Oreg, J., *Phys. Rev. A* **38**, 1773-1784 (1988).
10. Klapisch, M, Schwob, J. L., Freankel, B. S., and Oreg, J., *J. Opt. Soc. Am.* **67**, 148 (1994).
11. Vainstein, V. L. and Safronova, U. I., *ADNDT* **21**, 49 (1979).
12. Safronova, U. I., Namba, C., Marakami, I. et al., *Phys. Rev. A* **64**, 012507/1-13 (2001).

X-ray Wide-Band and Time-Resolved Imaging and Spectroscopic Characterization of Hot Spots and Jets in 0.9-1.0 MA X-pinches

Victor L. Kantsyrev, Dmitry A. Fedin, Alla S. Shlyaptseva,
Stephanie B. Hansen, Nicholas D. Ouart

Physics Department of the University of Nevada, Reno. Reno, NV 89557 USA

Abstract. The new results of time-resolved x-ray imaging and spectroscopic study of the development of jets and hot spots in 0.9-1.0 MA x-pinches during the rise of the current are presented. The results of x-pinch radiation properties studies and possible applications of such an x-ray source in a surface physics and backlighting are reported. Experiments were performed with the NTF "Zebra" machine with a peak current of about 1.2 MA, a rise time of 100 ns, a maximum stored energy of 200 kJ, and 1.9 Ω pulse-forming line impedance. X-ray diagnostics include 12 x-ray/EUV devices[1]: x-ray time- and spatial-resolved or time-integrated spectrometers, time-resolved and time-integrated x-ray imaging devices, x-ray polarimeters/spectrometers, fast x-ray and hard x-ray detectors. X-pinch planar-loop and wire twisted configurations were used.

I. EXPERIMENTAL RESULTS

High resolution x-ray time-gated images, time-integrated pinhole images, transmission grating (TGS) spectra, and spatially resolved x-ray spectra show that the line, continuum and harder radiation comes from the same region where the Ti, Fe K-shell, Fe, Mo L-shell or W and Pt M-shell sources are located. The most compact source was obtained with planar-loop x-pinches. The average size of L-shell ($\lambda \approx$ 30-80 Å) emitting regions for Ti is 3-6 mm, for Fe is 3-10 mm, and of M-shell for Mo is 2-4 mm. These dimensions are close to dimensions of short wavelength sources: Ti K-shell (2.5-4 mm), Fe K-shell (2-3 mm), Mo L-shell and W, Pt M-shell radiation. The size of hot spots in x-pinches depends on a spectral region of registration. For example, for a Mo x-pinch it was reduced from 0.3-3 mm to smaller than 70 µm (pinhole camera resolution) when a region of observation was changed from 4-10 Å (Mo L-shell and continuum) to λ<1.5-2 Å (continuum with planckian spectra). Hot spots are located in a shape of a chain along a central axis of an x-pinch near the cross point of wires. The number of x-ray bursts coordinates with the number of hot spots in images in a spectral region shorter than line radiation of an x-pinch. A size of 0.15 –10 keV emitting regions of high-current x-pinches decreases as the cutoff energy of imaging systems increases.

Energetic electron beams play a significant role in high-current x-pinches development and influence the x-pinch plasma spectra. Due to energetic electron beams, strong jets (Fig.1) in each discharge are observed directed toward the discharge axis perpendicular to the wires in most cases on the anode-sides of the contact. They are luminous due to excitation by fast electrons beams originating from the necked-off contact. In fact, the V-shape x-ray images of spectral lines were observed only for K_α Ti in Ti K-shell x-pinch spectra (Fig. 2a), and for L_α Mo and L_β in L-shell x-pinch spectra (Fig. 2b), but not for spectra of Ti or Mo multicharged ions. These spectral lines were due to excitation of low ionization stages of Ti or Mo ions by energetic electron beams. A stagnation of side-jets on the discharge axis forms also a central jet. The side-jets have a 1-2 mm period. They are presumably spikes of the Rayleigh-Taylor instability of the accelerated wires. Side-jets appear to be formed during all periods of x-ray generation. The time of jets life varied from ten to several tens of ns (Fig.1).

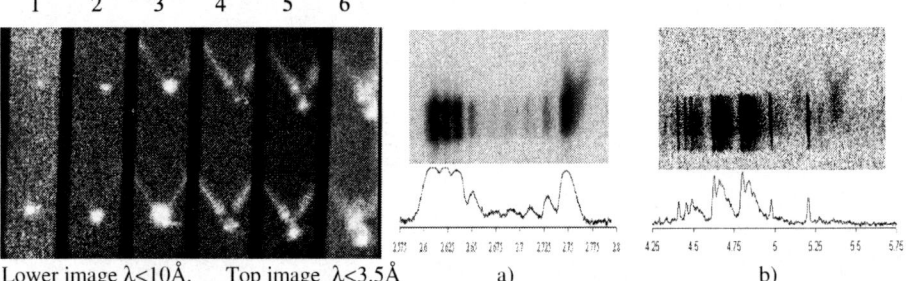

Lower image $\lambda<10$Å.   Top image $\lambda<3.5$Å   a)   b)

FIGURE 1. Gated x-ray images. Anode is at the top of images. Frame is 6 ns. Interval between beginning of frames is 17 ns. Frame #1 correspond 40 ns after current start. X-ray bursts (PCD detector peaks) were in frames 3 and 4. Spatial resolution is 230 µm. Mo planar-loop with a 50-µm wire(Shot # 93).

FIGURE 2. The portion of spatially resolved spectra of Ti (a, shot # 47) and Mo (b, shot # 90) x-pinches. Anode is at the top of images. Spatial resolution in vertical direction is 7 mm.

The estimation of temperature and density of x-pinch plasmas were based on theoretical modeling of K-shell Ti and L-shell Mo spectra[25]. Time-gated x-ray spectroscopy has been applied. In order to fit the experimental spectra, many synthetic spectra were constructed over a range of T_e, N_e, and f (hot electron fraction). In this way, the plasma parameters varied a lot for different x-ray bursts (hot spots) during the same Mo x-pinch discharge (Fig.3): T_e was from 0.7 to 1.1 keV, N_e from 10^{19} to 5×10^{21} cm^{-3}, and f from 0 to 9 %. Therefore, for detailed interpretation of experimental data, x-ray streak-camera spectrometer measurements (time resolution 10-100 ps) should complement the time-gated spectrometer measurements (ns scale resolution). The duration of x-ray pulses decreases as the cutoff energy of imaging system increases. The pulse width decreases as an atomic number of x-pinch material increases. The maximum total energy for x-pinches was more than 10-11 kJ for Mo x-pinches, 7-8 kJ for Fe, Pt and W, and 6 kJ for Ti x-pinches. For Ti, Fe and Mo x-pinches, optimum yields were obtained with wires mass density approximately of 100-200µg/cm.

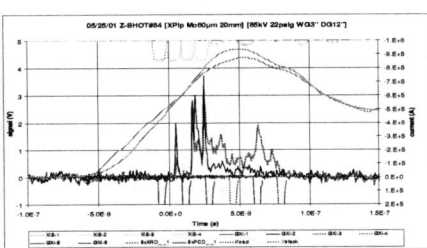

I I 4.5 Å 4.6 Å Gated x-ray spectra (GXS) From the top: Frame 2, 3, and 4. Frame duration 9 ns.	X-pinch current-thick smooth lines, GXI frames from 1 to 6-at the bottom, GXS frames from 1 to 4 -at the top, XRD (wide peaks) with 8 μm Be, and PCD (narrow peaks) with 25 μm kapton filters response versus the time

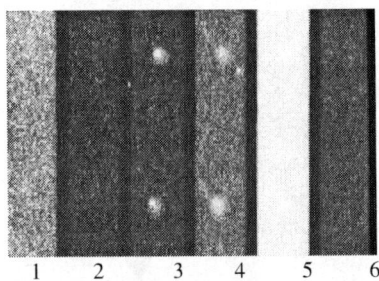
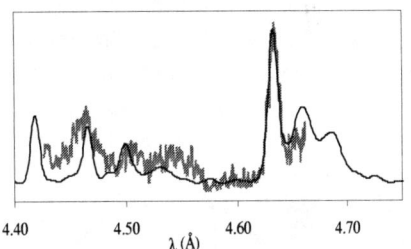

Gated x-ray images (GXI). Frame 6 ns. Spatial resolution 230 μm. Anode is at the top of image.

Frame 2. $N_e= 2 \times 10^{21}$ cm^{-3}, $T_e=0.85$ keV, f =8%

Comparison of experimental time-resolved spectra (gray lines) with modeling.

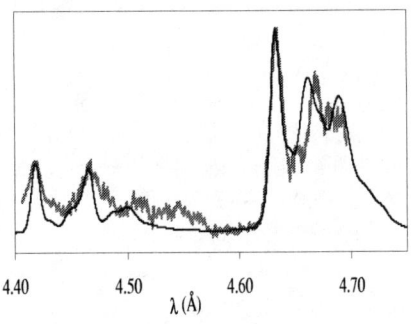
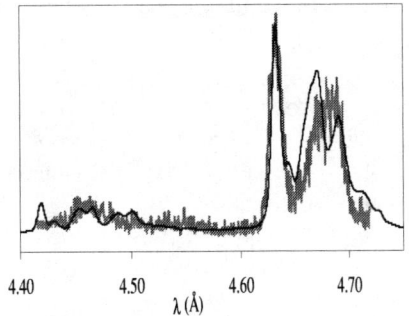

Frame 3. $N_e= 1 \times 10^{21}$ cm^{-3}, $T_e=0.90$ keV, f =0%

Frame 4. $N_e= 1 \times 10^{19}$ cm^{-3}, $T_e=0.95$ keV, f =0%

FIGURE 3. Spectroscopy of a Mo 62-μm planar-loop x-pinch (shot #84).

TABLE 1. Minimum pulse duration and maximum x-ray yield

	PCD (filter)	XRD (filter)	XRD (kimfoil5μm)
Ti	2.2±0.5 ns (Be8μm)/20-30 J	4±1ns(kimfoil 5μm)/3-4.5 kJ	
Fe	1.3±0.7 ns (Be8μm)/100-150 J	3.7±0.7ns(kimfoil 5μm)/1 kJ	
Mo	1.1±0.5 ns (kapton25 μm)/20-30 J	1.2±0.7ns(Be 8μm)250-350 J	15-20 ns
W	1.1±0.7 ns (kapton25 μm)/5-15 J	1.1±0.7 ns(Be 8μm)/15-25 J	15-20 ns
Pt	1.2±0.7 ns (kapton25 μm)	1.1±0.7 ns(Be 8μm)/15 J	

A comparison of presented Mo L-shell 0.9 –1.0 MA x-pinch data with Cornell University 0.2-0.3 MA results [2] shows that the Mo L-shell yield scales with the peak current as I^n to the power about n ≈ 3.

Applications of 1 MA x-pinches for 1-10 keV x-ray backlighting can be limited by a multipulsed scheme with a low (from tens to hundred mkm) spatial resolution.

In a surface physics, a powerful microfocus x-ray 1 MA x-pinch plasma source can be used with focusing glass-capillary converters (GCC) [3]. A relatively small size of the x-pinch emitting region in the wide-band spectral range 0.25- 5 keV will provide the application of the same GCC for different goals. Estimations have shown, with the 1 MA x-pinch high-z source coupled with GCC for collecting and focusing x-ray on a flat surface in a 2-3 mm spot, it is possible to reach a high level of energy concentration. An energy density in a region of 8-10 keV will be 50-100 mJ/cm^2, in 1-1.5 keV region up to 1-2 J/cm^2, and even 5-10 J/cm^2 in a 0.25-0.5 keV region. The threshold of melting of a surface of metals is 1-2 J/cm^2. The flux density can very from 5x 10^6-10^7 W/cm^2 (8-10 keV region) to (1-5) 10^9 W/cm^2 (0.5-1.5 keV region). Such facility can be used for material science and microelectronics research.

ACKNOWLEDGEMENTS

Special thanks to many collaborators whose work and help are very appreciated, particularly K. Struve, C. Deeney, D. McDaniel, T. Nash, J. McGurn, G. Chandler, D. Jobe, and D. Nielson from Sandia National Laboratories. We acknowledge the program support by B. Bauer and the NTF team: S. Batie, S. Fuelling, B. Le Galloudec, G. Newman, R. Presura, H. Faretto, A. Oxner.

This work has been supported by DOE, SNL, DOD, and UNR.

REFERENCES

1. Kantsyrev, V., Bauer, B., Shlyaptseva, A. *et al., Rev. Sci. Instr.* **72** (1), 663-667 (2001).
2. Shelkovenko, T.A., Sinars, D.B., Pikuz, S.A., Hammer, D.A., *Phys. Pl.* **8** (4), 1305-1318 (2001).
3. Kantsyrev, V., Bruch, R., Phaneuf, R., Publicover, N., *J. X-ray Sci. Technol.* **7**, 139-158 (1997).

Analysis of Anisotropy of Spectra and Spatial Distribution of Hard X-Ray Emission from 0.9-1.0 MA High-Z X-Pinches

Victor L. Kantsyrev, Dmitry A. Fedin, Alla S. Shlyaptseva,
Stephanie B. Hansen, David Chamberlain, Nicholas D. Ouart

Physics Department of the University of Nevada, Reno. Reno, NV 89557 USA

Abstract. The presence of collimated 1.5-2.0 MeV electron beams in 0.9-1.0 MA high-z x-pinch plasmas leads to the generation of hard x-ray radiation in the crossing point of the wires. The results of 0.9-1.0 MA Fe, Mo, and W x-pinch studies have shown that an x-pinch is an effective source of hard x-ray radiation (more than 50-100 keV). An anisotropic hard x-ray radiation with an energy of several hundred keV was observed moving upwards along the axis of x-pinches. The size of the source was estimated as small as 1 mm. The time duration of a hard x-ray beam from an x-pinch is 100-150 ns.

The side-on measured electron temperature is higher than the end-on temperature. Introducing of a hard x-ray synchrotron radiation in a side-on direction is used for the explanation of the distortion of side-on spectra. The value of the local magnetic field in plasma hot spots required for the effective hard x-ray synchrotron radiation appearance has been estimated.

EXPERIMENTAL SETUP and RESULTS

The study of 0.9-1.0 MA Fe, Mo and W x-pinch anisotropy of spatial and spectral distribution of hard x-ray radiation and estimation of the possibility of applying such an x-pinch as a hard x-ray backlighter were performed with the NTF x-ray/EUV diagnostics that covered a spectral region from 0.16 keV up to 1 MeV. Hard x-ray diagnostics include time-resolved detectors and filtered x-ray films detectors (Fig.1). To study the hard x-ray emission (50-200 keV), end-on and side-on partially collimated fast (time resolution 0.7 ns) filtered Si x-ray diodes (AXUV-HS5) were used. A fast hard x-ray detector (PM) with a scintillator and a photomultiplier was located outside the vacuum chamber on the x-pinch axis at 0.5 m from plasma. The detector sensitivity was limited to energies above 0.6 MeV. A planar-loop type of x-pinches was used.

The structure of x-pinches driven by a high current (0.9-1.0 MA) includes an energetic electron beam [1]. Its total energy was up to 10-20 kJ (15-20% of the x-pinch discharge energy for Mo and W). These results were based on the experimental estimation of the energy needed to produce the holes of the observed size at the intersection points of the wires with the anode. A diameter of an energetic electron beam was estimated as small as 1mm. The presence of a collimated electron beam in 0.9-1.0 MA high-z x-pinch plasmas leads to generation of hard x-ray radiation in the crossing point of the wires. For estimation of the source characteristics, radiographic tests of Fe and Mo x-pinches with an Al, Cu and steel static objects were carried out. A high-current x-pinch could be a potential point-type backlighter source of hard x-ray

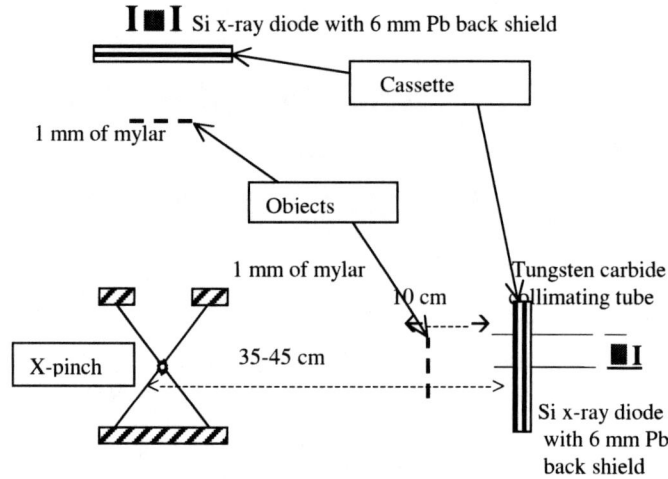
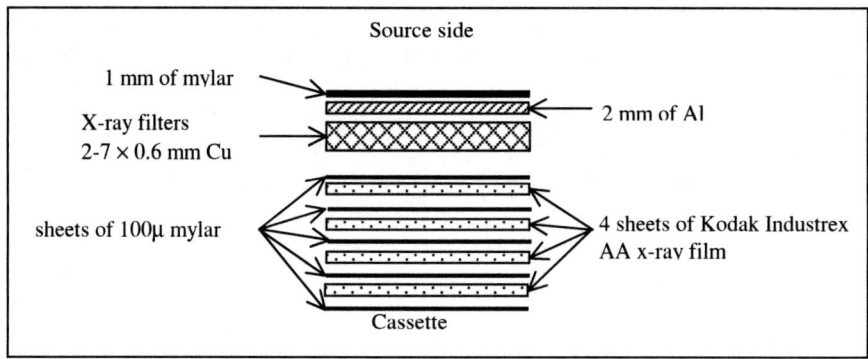

FIGURE 1. The scheme of hard x-ray experiments.

radiation [2]. The best x-ray resolution was 0.2-0.4 mm. A minimum effective size of a 40-100 keV source was 1±0.3 mm. The total time duration of 50-200 keV x-ray radiation was up to 100-150 ns (Fig. 2) and included several pulses with a duration of 10-50 ns. The end-on measured hard x-ray intensity was higher than the side-on. The Fe or Mo x-pinch hard x-ray spectra in a spectral region from 30-40 keV to 80-100 keV correspond to the spectra of hot dense plasma. But end-on measured hard x-ray emission in a region 100-500 keV was mostly due to the interaction of 1.5-2 MeV electron beams with Fe or Mo targets [3], and side-on measured spectra were due to the interaction with 2 MeV and higher electrons. The electron beam energy is higher than the applied anode-cathode voltage (about 1 MV). This result is consistent with the other study [4]. The strong hard x-ray emission and the small size of the hard x-ray source indicate an existence of a compact plasma with a near-solid density in the x-pinch cross-wires region. An observed exceeding of the end-on hard x-ray intensity over the side-on can be explained by the bremsstrahlung mechanism. But the fact that

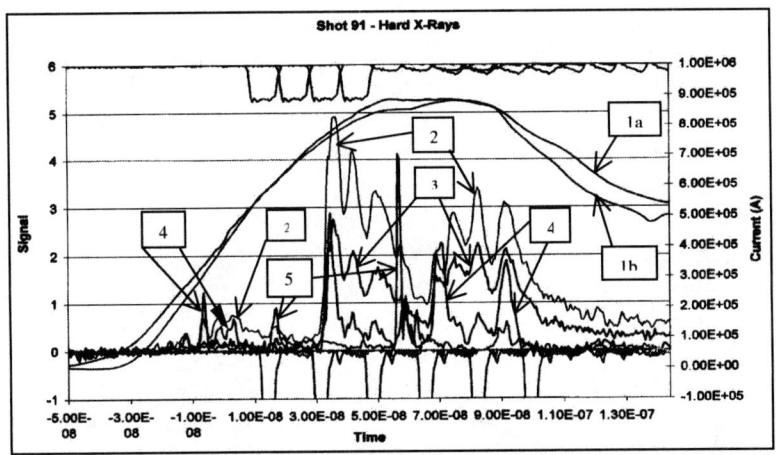

FIGURE 2. Signals from filtered Si x-ray diodes, PM and PCD (photoconductive detector filtered for 1-10 keV measurements). 1. Current (a- load, b- stack); 2. Side-on Si diode; 3. End-on Si diode; 4. Photomultiplier; 5. PCD.

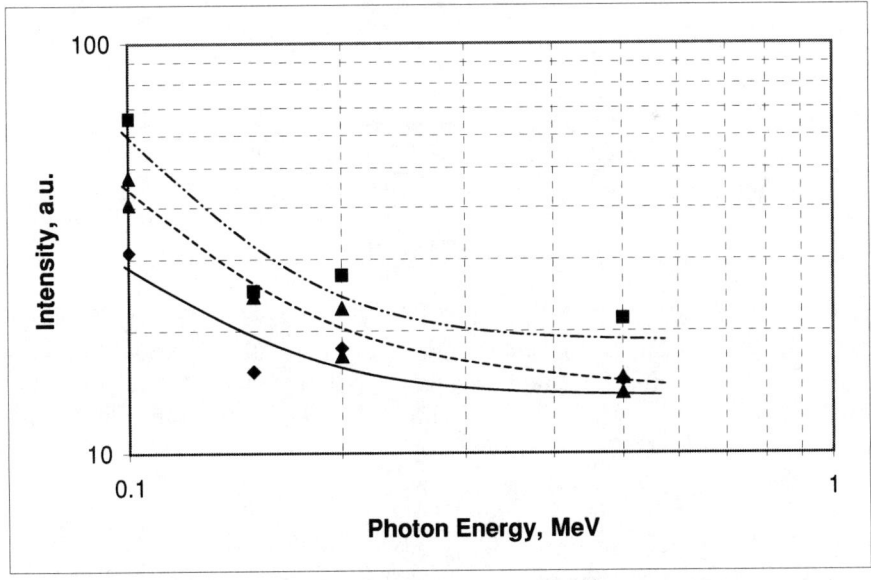

FIGURE 3. Relative spectra of hard x-ray emitted from Mo x-pinches. ▲ ---- Mo, end-on, shot #83; ♦ – Mo, side-on, shot #83; ■ – ⋯ – ⋯ Mo, side-on, shot # 84.

the side-on electron plasma temperature estimated from hard x-ray emission was higher than the end-on (Fig.3) can not be explained in the same way. The following ideas are implemented. The significant part of side-on hard x-rays is a synchrotron radiation formed as a result of interaction of high energy electrons orbiting near a hot spot with a strong local magnetic field (Fig.4). A well-known magnetic field equation was used to estimate a local magnetic field needed for effective hard x-ray synchrotron radiation appearance: $B = 1.864/E^2 \times \lambda_c$, where $\lambda_c = 0.559 \times R \times E^{-3}$ is such that a half of the power is irradiated at longer wavelengths than this, and a half at shorter wavelengths; B - magnetic field (in T); E – electron beam energy (in GeV); R- electron orbit radius (in m). For 1.5-2 MeV electrons (mostly responsible for distortion of spectra) with an orbit radius $R \approx 1\text{-}2$ μm, a magnetic filed will be $B \approx 3 E/R \approx 2300\text{-}6000$ T. This result is comparable in an order of a magnitude with estimation of B [5] based on measurements of an x-pinch current and dynamics of Mo plasma implosion at 1.1 ns before the moment of the intense x-ray burst ($B \approx 1400$ T) and in the moment of the maximum x-pinch implosion ($B \approx 9900$ T).

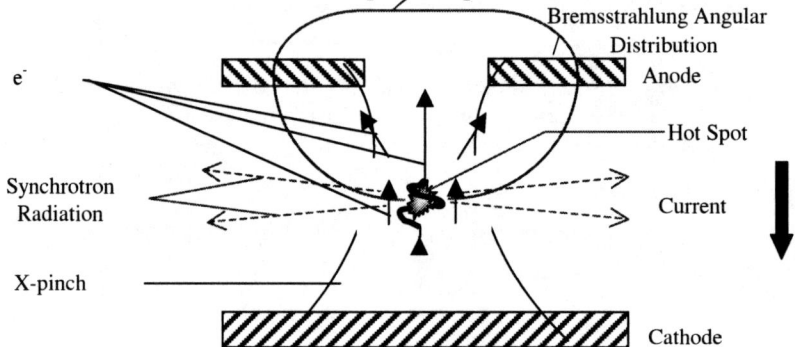

FIGURE.4. The scheme of generation of synchrotron radiation in a hot spot.

ACKNOWLEDGEMENTS

The authors acknowledge the many collaborators: K. Struve, C. Deeney, T. Nash, J. McGurn, G. Chandler, and D. Jobe from SNL for useful discussions and help in experiments, B. Bauer for programming support, S. Bate, S. Fuelling, B. Le Galloudec G. Newman, R. Presura, H. Faretto, A. Oxner for help in experiments. This work was supported by DOE, DOD, SNL, and UNR.

REFERENCES

1. Kantsyrev, V., Bauer, B., Shlyaptseva, A. et. al., *Rev. Sci. Instrum.* **72**, 1(II), 663-667 (2001).
2. Kantsyrev, V.L., Bauer, B.S., Shlyaptseva, A.S. et. al., *SPIE Proc.* **4502**, 62 (2001).
3. Edelsack, E.A., Kreger, W.E., Mallet, W., Scofield, N.E., *Health Physics* **4**(1), 1 (1960).
4. Robledo, I., Mitchell, I.M., Aliaga-Rossel, R., Chittenden, J.P., Dangor, A.E., Haines, M.G., *Phys. Pl.* **4,** 490 (1997).
5. Shelkovenko, T.A., Sinars, D.B., Pikuz, S.A., Hammer, D.A., *Phys. Pl.* **8**, 1305 (2001).

Analysis of Time Evolution of Z-pinch Lines Using an Advanced 5-channel Spectrometer Polychromator

D.A.Fedin, V.L.Kantsyrev, A.S.Shlyaptseva, S.B.Hansen, S.Fuelling.

Physics Department, University of Nevada Reno MS 220, Reno, NV, 89557, USA

Abstract. The measurements with temporal resolution are very important in studies of properties of radiation from hot dense plasmas ($0.8 < T_e < 1.5$ keV, N_e up to $2\text{-}3 \cdot 10^{22}$ cm^{-3}) generated by a pulsed-power z-pinch machine with a current $I_{max} \sim 0.8 \div 1.0$ MA and a current rise time of 100 ns. To perform measurements of radiation from hot dense plasma an advanced 5-channel time resolved spectrometer "Polychromator" was used. All channels are independent from each other and axially symmetric with respect to the main optical axis. An additional sixth channel sited on the main optical axis is used for device alignments and measurements using a time-integrated, spatially-resolved transmission diffraction grating spectrometer. A flexible design allows us to use different types of dispersing elements and detectors. A procedure of optical alignments of the device is described. Experimental data from time resolved channels as well as from a TGS channel are presented and briefly discussed.

FIVE-CHANNEL SPECTROMETER POLYCHROMATOR

Five-channel spectrometer Polychromator is designed for registration of x-ray emission from hot dense plasmas with spectral and spatial resolution [1]. A spectral range of observation is defined by a set of dispersing elements used in each channel. To study K- and L-shell radiation, four different types of crystals were employed: KAP ($2d=26.63$ Å), PET 002 ($2d=8.742$ Å), LiF 200 ($2d=4.027$ Å), and α-quartz ($2d=6.687$ Å). An angular position of each dispersing element varies from 7° to 63° with respect to an optical axis of a particular channel. This set of crystals allows us to study properties of plasma radiation in a wide spectral region (from 1 to 25 Å). 25 μ Be filters are used in each channel for protection from the visible light and debris.

A registration system is based on application of fast Si diodes [2,3]. An additional protection from a visible light was provided by 1 μm mylar coated on both sides by 0.1 μm Al. A flexible design of the registration system allows us to substitute currently used detectors by any type of a time resolving detector, which fits a standard SMA connector.

Because of the axial symmetry of the described device, only one channel is shown in Fig.1. For optical alignments a set of standard tools such as an optical bench, a *He-Ne* laser, a set of neutral optical filters, 100% *Al*- mirrors, a set of diaphragms, a copy of a crystal holder with 100% reflective surface, inserts for "zero point" verifying and

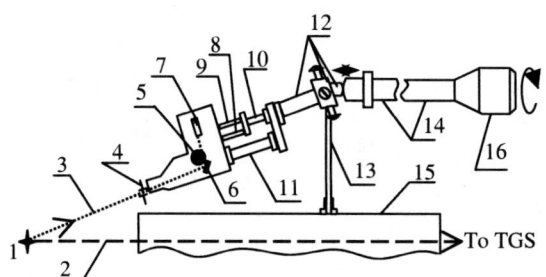

	Poly	TGS
Spectral resolution	≤460	40÷50
Spatial resolution	3÷5 mm	150÷200 μm
Time resolution	≤0.7 ns	N/A; ≤1 ns
Spectral range	1÷25 Å; 120÷240 Å	20÷80 Å

FIGURE 1. Simplified drawing of one of the Polychromator channels.
1. Light source (plasma or test source); 2. Optical axis (axis of symmetry) of Polychromator connected with a TGS assembly; 3. Optical axis of a working channel; 4. Entrance diaphragm; 5. Crystal-detector movement controlling block; 6. Crystal (MLM); 7. Detector (Si-diode or PCD); 8. Crystal (MLM) motion translation stage; 9. Detector motion translation stage; 10. "Zero point" control gap; 11. Assembling rod; 12. Angular translation stage with the optical axis 3 2-D adjustment system; 13. Supporting unit for the system 12; 14. Translation stage; 15. Axial tube for optical alignments and connecting of TGS assembly; 16. Precision micrometer.

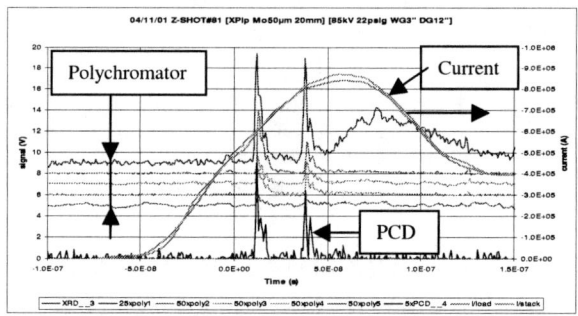

FIGURE 2. X-ray signals from a 50 μm *Mo* x-pinch
Polychromator channels from top to bottom: 1: $\lambda=5.21$ Å, *Ne*-like *Mo*; 2: $\lambda=4.80$ Å, *Ne*-like *Mo*; 3: $\lambda=1.94$ Å, "cold" $K_\alpha Fe$; 4: $\lambda=4.41$ Å, *Ne*-like *Mo*; 5: $\lambda=5.5$ Å, continuum, control line. PCD (with a 25 μm kapton filter) response versus time. Note, that maximum of Polychromator signals was correlated in time with PCD signals except the first channel (*Mo L*-shell radiation).

set of wrenches and screwdrivers are required.

Each channel of the device is adjusted individually by the standard procedure. A protecting filter (not shown in the Fig.1) in front of the entrance diaphragm 4 is substituted by a pinhole and the optical axis 3 is being established. The crystal 6 is substituted by the exact copy of the crystal holder with the 100% reflective surface. Then, using adjustment screws of the assembly 12, the parts 4 through 12 are preliminary matched with the axis 3. After this procedure, the micrometer 16 is set to the reading recommended by a manufacturer in order to establish a "zero point" position. Then, a screw (not shown) connected with the crystal translation stage 8 in assembly 5 is slightly loosen and the proper length of the control gap 10 is being set using the first standard insert. Using the second 45° standard insert, the angle between crystal 6 and axis 3 and tighten crystal translation stage 8 screw the "zero point" is being established. To finalize optical alignments, it is necessary to establish a proper θ -2θ detector 7 movement. For this purpose, a screw (not shown) connected with detector movement translation stage is loosen and detector being set in a position when a reflected from dummy crystal laser beam comes to the sensitive area of the detector. After tighten detector movement translation stage screw the whole crystal-detector system is checked in all available angles to insure that at any angle radiation stays at the same point of the crystal and all the time within the detector sensitive area. Exactly the same procedure is repeated for all remaining channels. An alignment on an experimental site consists only from finding the central optical axis of the experimental setup. For this purpose a point source imitator was placed instead of a real plasma source and a simple telescoping system is used.

An analysis of all factors influencing accuracy of the device gives a total value of the resolution $\lambda/\Delta\lambda \leq 460$. These factors include manufacturing limits, accuracy of alignments and geometry of the experimental setup.

The different behavior of studied spectral lines is clearly seen in Fig.2. Studying of time evolution of "cold" $K_\alpha Fe$ line gives information about hot electrons in x-pinch plasmas. The amount of hot electrons in plasma is very important parameter for correct determination of electron densities and temperatures of hot dense plasmas. The waveforms from Polychromator are also time marks for time resolving spectrometers and imaging systems.

TRANSMISSION DIFFRACTION GRATING SPECTROMETER.

Sixth (axial) channel of the device was equipped with a TGS to make spectroscopic measurements with spatial and spectral resolution in a wide wavelength range. The TGS is very well described and is a reliable diagnostic tool for studying soft x-ray radiation from different plasma sources [4,5]. Spectrometers with transmission gratings used as dispersing elements have number of advantages in comparison with spectrometers based on crystals or reflecting gratings. These advantages are: TGS is easy to align, it usually has a higher aperture, covers a wide spectral range and is relatively non expensive. With a known absolute response of a detector used [6], an absolutely calibrated spectrometer is an excellent device to use with other soft x-ray radiation detectors to prove and calibrate data. With the diffraction grating (period 1 μm, gap to period ratio ½, substrate thickness 0.5 μm) and the detector (Kodak DEF-5 film [6]) used in our experiment the spectral range was limited to $\lambda = 20 \div 80$ Å. A total yield of soft x-ray radiation in this range for Fe shot 50 (Fig.3) was 1.3 kJ. The total yield of x-ray/EUV radiation obtained with help of an unfiltered Ni bolometer was 6.3 kJ, which is consistent with TGS data.

Spatially resolved spectra can provide more information about source structure (Fig.3) and characteristics. An addition of a time resolving gated MCP imager gives an opportunity to obtain time evolution of plasma radiation. Together with waveforms obtained from other five channels one can collect a complete set of information necessary to calculate very important plasma parameters such as electron temperatures and densities. A comparison of calculated spectral shapes with a Plank

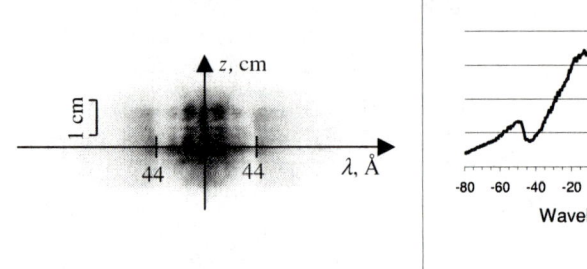

FIGURE 3. Spatially resolved spectrum and its densitometric trace for a central part of *Fe* x-pinch.

radiation spectrum allows evaluating of black body temperature of plasma. Integrated spectral distributions give a yield of x-ray emission in a wavelength range specified. The value obtained and signals from absolutely calibrated XRD and the bolometer are in good agreement, which confirms that methods of data processing are correct.

CONCLUSION

A multichannel crystal spectrometer/transmission diffraction grating spectrometer Polychromator was designed to study a time history of x-ray/EUV spectral line radition from powerful plasma sources such as z-pinch. A flexible design of device allows to use it also with other x-ray sources. High time and spectral resolution proivides an opportunity to study an evolution of very important plasma parameters such as electron densities and temperatures. Five time-resolved channels are also used as relative time markers for the time-resolved gated pinhole camera and spectrometer. An additional time integrated channel with a TGS collects data in a wide wavelength region with reasonable spectral and spatial resolution. The TGS can be easily converted in time-resolved device by substituting an x-ray film by a gated MCP detector.

ACKNOWLEDGEMENTS

The authors acknowledge many collaborators whose work and help is excerpted in this paper, in particular, B.Bauer S.Batie, A.Oxner, R.Presura, H.Faretto, B.LeGalloudec. This study was supported by DOE, DOD, SNL and UNR.

REFERENCES

1. Fedin, D., Kantsyrev, V., Bauer, B., Shlyaptseva, A., Brytov, I., *SPIE* **3764** (III), 80-84 (1999).
2. Korde, R., Cable, J.C., Canfield, L.R., *IEEE Transactions on Nuclear Science* **40** (6), 1655-59 (1993).
3. Canfield, L.R., West, R., Woods, T.N., Korde, R., *SPIE* **2282** (V), 31-38 (1994).
4. Alexandrov, Yu.M., Koshevoi, M.O., Murasheva, V.A. *et al., Laser and Particle Beams*, **6** (3), 561-567 (1988).
5. Baksht, R.B., Oreshkin, V.I., Fediunin, A.V. *et al., Soviet Journal Fizika Plazmy (Plasma Physics)*, **20** (11), 962-967 (1994).
6. Alexandrov, Yu.M., Eidmann, K., Fedin, D.A. *et al., Nucl. Instrum. and Methods in Phys. Research* **A308**, 343-346 (1991).

Emission Produced at Compression of Deuterium Current-Sheath with Wire in Plasma Focus Discharge

Pavel Kubeš,[1] Jozef Kravárik,[1] Daniel Klír,[1] Marek Scholz,[2] Marian Paduch,[2] Krzysztof Tomaszewski,[2] Irena Ivanova-Stanik,[2] Barbara Bienkowska,[2] Leslaw Karpinski,[2] Leszek Ryć,[2] Libor Juha,[3] Josef Krása,[3] Marek J. Sadowski,[4] Lech Jakubowski,[4] Adam Szydlowski,[4] Aneta Banaszak,[4] Hellmut Schmidt,[5] Vera M. Romanova [6]

[1] *Czech Technical University, Prague, Czech Republic*
[2] *Institute of Plasma Physics and Laser Microfusion, Warsaw, Poland*
[3] *Institute of Physics, Prague, Czech Republic*
[4] *The Andrzej Soltan Institute of Nuclear Studies, Otwock-Swierk, Poland*
[5] *International Center on Dense Magnetized Plasma, Warsaw, Poland*
[6] *Physical Lebedev Institute, Moscow, Russia*

Abstract: The implosion of a deuterium plasma toward an Al wire was performed at the current amplitudes of 1.5-1.8 MA. The Al wire of 120 µm diameter and 4-5 cm length was placed on the top of the inner electrode. Pulses of XUV radiation and soft X-rays were detected at the maximum compression of the plasma focus column. They contained lines of Al VI – Al XIII ions. The presence of the Al wire on the axis of plasma focus column caused a decrease in the hard X-ray production and neutron yield to about 30-50%. The neutron pulse was characterized by the FWHM $\approx$ 200 ns and the yield up to 5×10^{10}. The maximum of the neutron production occurred up to 200 ns later than the maximum of the soft X-rays.

The plasma focus discharges are studied due to high efficiency of the X-rays and the neutron yield production (when the deuterium is used as a filling gas). The results of the research on the interaction of the hydrogen current sheath with the Al wire of 120 µm diameter were presented in [1]. The results of the studies of the XUV, X-ray and neutron production at the implosion of the deuteron current sheath in the electrode outlet, were summarized in paper [2]. This contribution describes an influence of the Al wire located on the top of the inner electrode on the X-ray and neutron emission at the implosion of the deuterium current sheath.

The PF 1000 facility was operated at the energy of 600-650 kJ and the maximum currents of 1.5-1.8 MA. The Al wire (3 cm in length and 120 μm in diameter) was located on the axis of the electrode without any galvanic connection.

The radiation, ranging from visible to hard X-rays, was measured side-on means of two gated soft X-ray multichannel plates (MCP - with 3 ns exposure, 10 ns delay between exposures, filtered with 5.2 μm polyster - C_8H_8, 1.11 gcm^{-3}). There were used the PIN silicon detectors (filtered with 10 μm Be), 2 VR framing cameras with 3 ns exposure, two scintillators (Ne102a filtered with 10 μm Al or 20 μm Cu foils for X-rays in the keV range), the XUV grazing incidence spectrograph for the 2-10 nm range (to scan K-shell lines of C, N and O and L-shell lines of Al) and the 2D imaging mica crystal spectrograph for study of Al K-shell lines in the keV range. To perform time-resolved measurements of the hard X-ray (20-50 keV) and neutron emission we used three scintillation probes, located end-on at distances of 6.5 m, 40.2 m, and 85 m from the electrode outlet.

Fig.1 presents a temporal dependence of the soft and hard X-ray and neutron production. The soft X-rays (detected with PIN diodes) were emitted during 300-500 ns with the FWHM equal to 60 – 200 ns. The X-rays within the 8 – 25 keV range were emitted in a temporal correlation with soft X-rays, but usually on 2 narrower peaks of FWHM equal to 30 – 60 ns. The hard X-rays above 50 keV were emitted in a temporal correlation with neutrons. The neutron signals were attributed to plasma focus region placed at a distance of 6 cm from the electrode outlet assuming 2.45 MeV neutron energy. This energy (with very small differences) was evaluated from the temporal positions of neutron pulses recorded at different distances and the same FWHM. The neutron pulses (FWHM of ~ 200 ns) at the trace 1 and 5 in Fig. 1 corresponded to the shifted scintillation signals detected of distances of 6.5 and 40.2 m, respectively. These pulses started usually after the soft X-rays and their maximum occurred up to 200 ns later than the maximum of the soft X-rays. The total neutron yield reached values up to ~ 4×10^{10}, approximately 30-50% of those recorded without wires. High neutron yield was observed in shots with an extremely stabile corona recorded with the streak camera. Thus, the presence of the higher Z-elements of the axis area of the electrode outlet of 1 mm in diameter and 4-5 mm in length did not change the neutron generation considerably.

The radiation of the corona around the wire started ~ 70 ns before the PIN diode recording the maximum in visible, XUV and soft X-ranges. The diameter of the imploding plasma sheath was a few cm and the corona of 2-3 mm in diameter radiated uniformly without perturbations along the total length of the wire. The maximum of the soft X-rays (taken as a reference time zero, t=0) appeared in a temporal correlation with a minimum of the current derivative and with the minimum of the corona diameter imaged in visible, XUV and soft X-ray ranges. The radiation was emitted uniformly along the total length of the wire (Fig. 2) at this time. Later on the soft X-ray intensity was decreasing with an increase in the corona diameter. The increase of the diameter in visible range was imaged by means of the streak camera (the velocity of onset was ~ 6×10^3 ms^{-1}). Small perturbations were observed upon the surface, and the local spots of more intense radiation had larger diameters (Fig.3, 4).

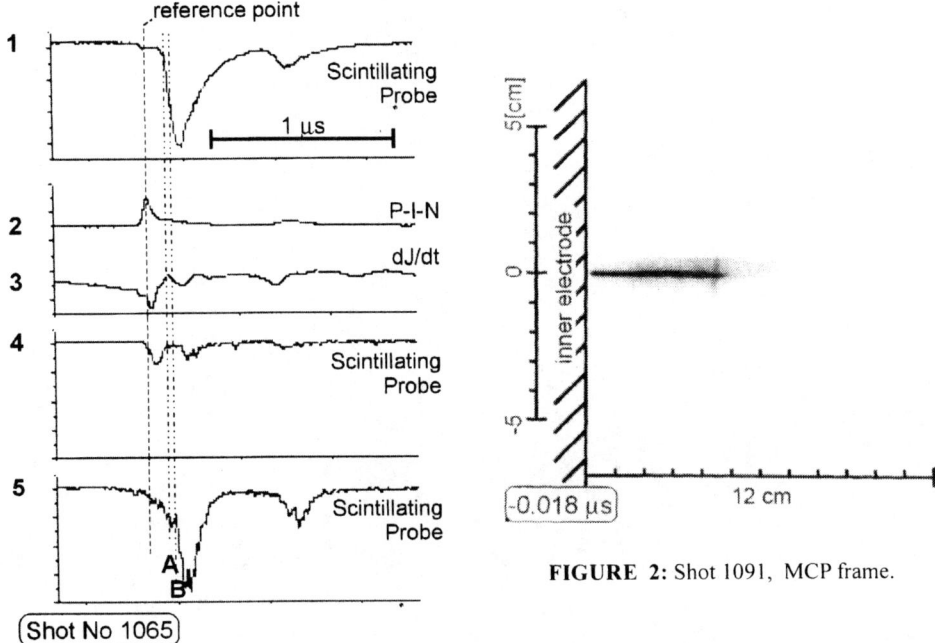

FIGURE 1: Oscillogram: 1&5 - neutron signals, 2 - PIN diode, 3 - current derivative, 4 - hard X-rays.

FIGURE 2: Shot 1091, MCP frame.

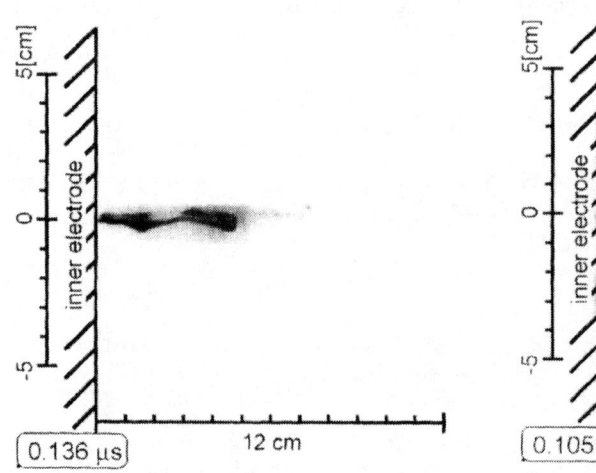

FIGURE 3: Shot 1065, MCP frame corresponding to B in Fig. 1.

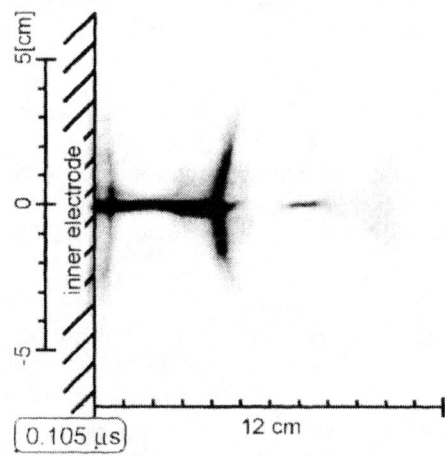

FIGURE 4: Shot 1065, VR frame corresponding to A in Fig. 1.

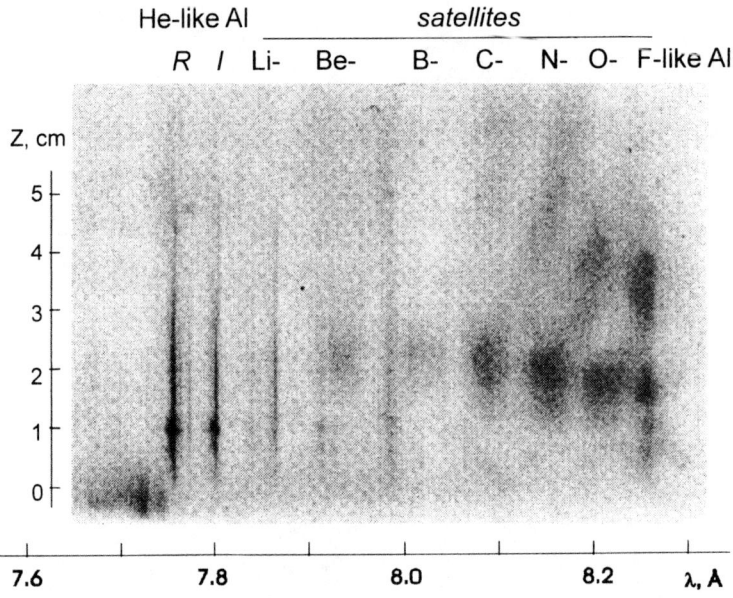

Figure 5: Shot 1113, aluminium K-shell spectrum.

The H- and He- like Al lines (Fig. 5) were emitted from the whole length of the wire corona of ~ 1 mm diameter, but the He-like Al satellites were observed from much brighter and longer forms. The temperature of plasma, calculated with the FLY code, reached values up to 500 eV.

Acknowledgement

This research has been supported by the research program No J04/98:212300017 and programs INGO No LA 055 and LN00A100 of the Ministry of Education of the Czech Republic.

References

[1] Scholz M., Paduch M., Tomaszewski K., Karpinski L., Kubeš P., Kravárik J., Szydlowski A., Romanova V., Pikuz S.: Experimental Studies of Al Corona Plasma Created within the PF-1000 Plasma Focus Facility, Czechoslovak Journal of Physics, Vol. 50 Suppl. 3 (2000), pp. 150-154.

[2] Scholz M., Pisarczyk T., Szydlowski A., Sadowski M.: Recent Studies of Fusion Neutrons withinPF1000 Facility, Czechoslovak Journal of Physics, Vol. 52 Suppl.4 (2002), pp. D93 - D 99.

Energy Transformation in Z-Pinch and Plasma Focus Discharges with Wire and Wire-in-Liner Loads

Pavel Kubeš,[1] Jozef Kravárik,[1] Daniel Klír,[1] Marek Scholz,[2]
Marian Paduch,[2] Krzysztof Tomaszewski,[2] Leslaw Karpinski,[2]
Yury L. Bakshaev,[3] Peter I. Blinov,[3] Andrey S. Chernenko,[3]
Sergey A. Dan'ko,[3] Valery D. Korolev,[3] Andrey Y. Shashkov,[3]
Victor I. Tumanov[3]

[1]*Czech Technical University, Prague, Czech Republic*
[2]*Institute of Plasma Physics and Laser Microfusion, Warsaw, Poland*
[3]*Russian Research Center "Kurchatov Institute", Moscow, Russia*

Abstract: The results of the study of the Z-pinch and plasma-focus plasmas at presence of the axial C, Al, or Cu wires of sufficient high diameter are discused in this paper. The wire was positioned on the top of the inner electrode of the PF 1000 plasma focus (1.8 MA, IPPLM Warsaw), or at the axis with or without the tungsten or alumine wire array load at the S-300 facility (3 MA, RRC Kurchatov Institute, Moscow), and at the axis of the small Z-pinch Z-150 (50 kA, CTU Prague). The plasma corona around the wire was generated both by the current going through the wires and by the implosion of the wire array or of the current sheath. The experiments showed interesting results often observed in some shots of Z-pinch type discharges - existence of helical structures, two relatively long and stable pinch phases, oscillation of pinch diameter, and back return of the plasma exploding from the pinch. All these observed phenomena can be evolved by spontaneous self-generation and transformation of the axial magnetic field in the pinch during the plasma implosion and explosion. A configuration of axial and azimuthal magnetic field confines the plasma and later transforms or dissipates during a few tens or hundreds ns. A fast transformation of internal magnetic fields can induce a sufficiently high electric field for generation of keV particles and radiation.

Study and usage of Z-pinch discharges is connected with solving of two principal problems, limitation of instability development and a way of generation of high energy particles and radiation. The first problem is partially solved by the faster increase of the current, by better cylindrical symmetry of the load and plasma, by higher density of the plasma or by the presence of a stronger magnetized plasma.

Generation of high energy particles and radiation is usually explained by an increase of the plasma resistance, when the main part of the energy at the pinch phase is released in radiation. The energy transformations are usually presented as a chain: compressing magnetic energy → kinetic energy → heating of the pinched plasma → radiation and the stationary conditions are described by Bennett equation. A few percentage of particles and photons is accelerated to the high energy of a keV. In [1] three possible mechanisms for ion acceleration are mentioned - direct acceleration due to generation of a high inductive voltage during current breakdown after formation of the neck (qualitative picture Trubnikov 1986 on the base of diameter decrease), mechanism related to compressional heating in the neck releasing by ejection of hot plasma from its ends (described by Vikhrev 1986 on the base of Bennett equilibrium) and the stochastic acceleration of the tails of the ion velocity distribution function due to microturbulence in a plasma with relatively low density. Acceleration of electrons is usually explained by runaway process when the electrons have high enough initial energy. High energy X-rays are produced by high-energy electrons.

In this presentation some remarks to the active role of magnetic field at energy transformations are discussed.

For study of the wire corona the experiments with the C, Al, Cu fibers of diameter of 30-300 μm have been performed at the small Z-pinch (50 kA, CTU Prague, wire load), at the plasma focus PF 1000 (1.8 MA, IPPLM Warsaw, wire on the top of the inner electrode) and at the fast Z-pinch S-300 (3 MA, Kurchatov Institute in Moscow, wire and wire-liner load). Around the wire with a core in solid state a corona was formed with the dense and magnetized plasma which firstly depress an instability development and secondly slow down the velocity of transformations of the plasma configurations as discussed later.

We would like to show a few observed phenomena, which help to improve the existence and transformation of axial magnetic field: the first is the helical structures in some phases of pinching discharges (Fig.1, Fig. 2), the second is two relatively long and stable pinch phases in some shots (Fig.3), the third is the repeating of the pinch after short explosion (Fig. 4), the fourth is the "second pinching" of the pinch phase (Fig. 5) and the fifth is back return of the plasma exploding from the pinch (Fig. 6).

All these effects could be caused by the existence of an axial magnetic field (B_z) and its transformation inside the plasma. This axial field with a random orientation can be spontaneously self-generated at the Z-pinch implosion in the consequence of fluctuations of plasma density, implosion velocity and cylindrical symmetry of magnetic field (analogy to the magnetic field generation in stars and planets). At the final phase of the plasma implosion the kinetic energy does not transform directly into heat, but partially to the compressed axial component of magnetic field. In some shots with a better developed B_z, the pinch with higher diameter and usually lower X-ray emission is formed. The final helicity of composed azimuthal and axial magnetic field is dependent on the intensity of the spontaneous self-generation of the axial component, on the radius of the pinch layer and on B_z periodicity along the Z-axis. The lower helicity belongs to internal layers of the pinch and the periodicity along the z-axis is related to the periodicity of m=0 instabilities. The final phase of the implosion

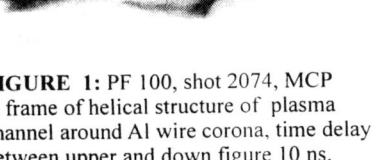

FIGURE 1: PF 100, shot 2074, MCP X frame of helical structure of plasma channel around Al wire corona, time delay between upper and down figure 10 ns.

FIGURE 2: S 300, shot 9805191, helical structure of time integrated pinhole camera figure of Cu wire at current 1 MA, 12 mm mylar.

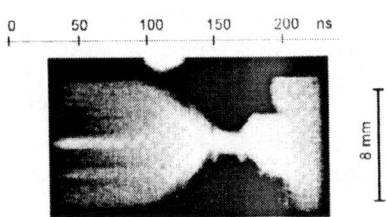

FIGURE 3: S 300, shot 0011011, visible streak camera record of two phases of pinch of imploded W-wire array.

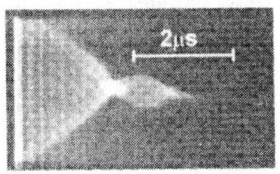

FIGURE 4: PF 1000, oscillation of current sheath pinch.

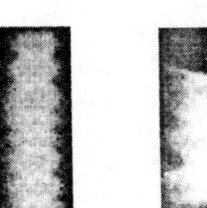

FIGURE 5: S 300, shot 9905193, X-frame pinch of Al wire corona at 1MA current.

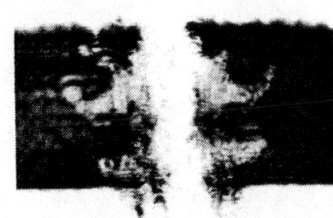

FIGURE 6: shot 0205282, shadowgramm of Cu wire corona at 1MA current.

is connected with development of these instabilities and radial explosion of the plasma from the pinch localities with higher diameter. This expanding plasma is crossing through the external magnetic field with opposite helicity (due to opposite orientation of closed B_z lines in external layers of the pinch) and it is caught into azimuthal current loops with opposite current orientation than that connected with internal helicity of magnetic lines. Inside of these loops the total axial magnetic field is much smaller than inside of the helical structures and the loops can pinch. (The neck of the m=0 instabilities in Fig. 3 could be examples of these loops). The dynamic behavior of the observed helical formations was roughly estimated in paper [2] by a ratio of imploding pressure of azimuthal magnetic field and expanding pressure of axial magnetic field. Then, at the plasma expansion, the influence of B_z is limited due to generation of the opposite azimutal current, while the compression influence of the azimuthal magnetic field is increased and the expansion can be stopped (Fig. 3). In some cases the compression can convert into implosion as it is shown in Fig. 4. During the implosion discussed before the damping of the generated and compressed axial B_z lines is increasing. The expansion pressure of the axial magnetic field is vanishing during the time of diffusion and penetration of magnetic fields of both opposite orientations. This process enables the pinching of the necks (generation of hot spots) and induction of the high electric field (acceleration of high energy particles and radiation). Then the time of steady state and transformations is given by the time of magnetic fields dissipation.

Vanishing of the opposite axial magnetic field is probably accompanied by high electric field induction, and ion and electron acceleration.

The main results can be summarized as follows. In the corona of thick fibers at Z-pinch implosion the azimuthal magnetic field transforms partially into axial component and at the explosion vice-versa. High energy particles can be generated by a high electric field, which is induced by the fast penetration and dissipation of magnetic field inside the plasma.

Acknowledgement

We are grateful to Prof. Kingsep for significant comments. This research has been supported by the research program No J04/98:212300017 "Research of Energy Consumption Effectiveness and Quality" of the Czech Technical University in Prague, by the research program INGO No LA 055 "Research in Frame of the International Center on Dense Magnetized Plasma" and " Research Center of Laser Plasma" LN00A100 of the Ministry of Education, Youth and Sports of the Czech Republic.

References:

[1]. Ryutov D. D, Derzon M. S, Matzen M.K, The physics of fast Z pinches, Review of Modern Physics, vol 72 No. 1 (2000), pp. 167-224.

[2] Kubeš P., Renner O., Kravárik J., Krouský E., Bakshaev Y.L., Blinov P.I., Chernenko A.S., Gordeev E.M., Danko S.A., Korolev V.D., Shashkov A.Y.: Diagnostics of an Al Wire Corona of a MA Z-pinch, Plasma Physics Reports, 28 No.4 (2002), pp. 296-302

Multi-wire Z Pinch Experiments

Min Hu, Bruce R. Kusse

Lab of Plasma Studies, 369 Upson, Cornell University, Ithaca, NY, 14853

Abstract. We report recent results on multi-wire Z-pinch experiments using Cornell's XP pulsed power generator (400kA peak current, 100ns). Planar Al wire arrays were studied. A global magnetic field was provided by the return current path, which was located ~2.5mm above the array. Two channels of shearing interferometer separated by 20ns and an x-ray backlighter were used for diagnostics. The velocity of the plasma driven by the global field, the merging velocity toward the array center and wire expansion velocity were measured.

I. INTRODUCTION

Fast, cylindrical wire array Z-pinch implosions are an efficient way of generating dense hot plasma and are promising x-ray sources for inertial confinement fusion [1-2]. Important questions associated with these Z-pinch implosions concern plasma formation, inter-wire plasma merging and the dynamics of the implosion. Detailed experimental information is required to provide a better understanding of the physics of Z-pinches and to validate simulation models.

II. EXPERIMENTAL SETUP AND DIAGNOSTICS

The experiments were performed using Cornell's XP pulsed power generator (400kA peak current, 100ns pulse duration). Our geometry consisted of planar wire arrays plus a return current ~2.5 mm above the plane of the array (Fig. 1). The return current was used to generate a global B-field to simulate large cylindrical wire arrays. Results reported here were produced using arrays of 4 Al wires, 13μm and 25μm, in diameter with either 1mm or 0.5mm gaps. X-ray images were made using Al x-pinch backlighters with a magnification of about 2.8:1. The viewing angle of the x-ray source relative to the plane of the wire array was ~40°. The return current was in series with a parallel combination of the x-pinch and the

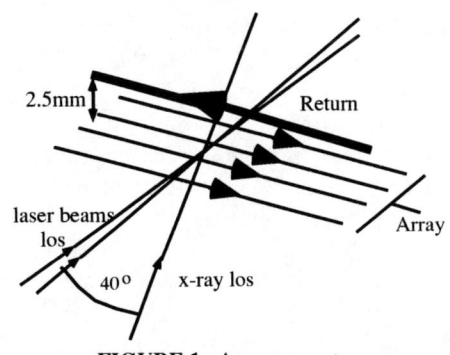

FIGURE 1. Array geometry

wire array. As a result about 3/4 of the total current flowed through the array. The x-pinch radiation was filtered with an 8μm Polyamide foil and a 2μm aluminized Mylar foil (1000 Å Al coated). An 8-step Al wedge with thicknesses ranging from 0.5μm to 2.5μm was deposited on the filter for calibration [3]. Two shearing interferometry images [4] separated by 20ns were also recorded using the second harmonic of a Nd:YAG laser (λ=532ns, 5ns FWHM). The two laser beams were almost in the plane of the wire array and observed the plasma movement in the direction of the $\mathbf{J}\times\mathbf{B}$ force.

III. RESULTS AND DISCUSSION

A. Pushing of Plasma by the Global Field

Figure 2 shows a typical interferogram of 13μm Al wires with 1mm inter-wire gaps

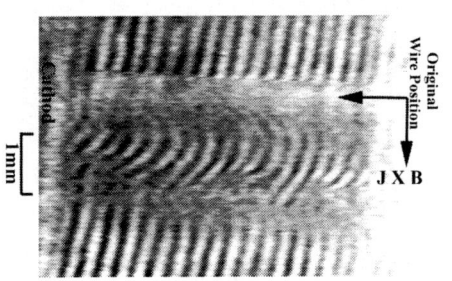

FIGURE 2. Interferogram of 4 Al wire array, 13μm wire diameter and 1mm inter-wire gaps at 23ns.

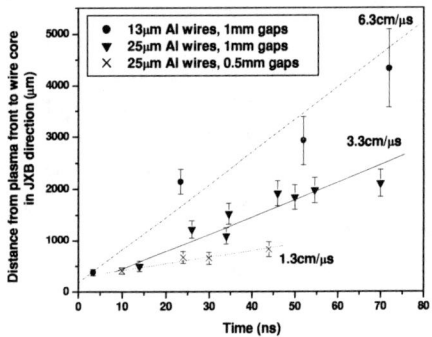

FIGURE 3. Comparison of plasma pushing velocities for three wire configurations

taken 23ns after the start of the current. At this time the plasma front had been pushed about 2mm from the plane of the array. Other measurements, reported below, show that transverse to this motion, at this time, the plasma was only ~0.2mm from the original wire core positions. Experiments were also conducted using 1mm and 0.5mm spaced 25μm Al wires. The plasma front position as a function of time is shown in Figure 3. The pushing effect is more prominent for smaller diameter wires and for larger inter-wire gaps. The pushing of plasma began as the current started. For 13μm Al wires with 1mm inter-wire gaps, plasma areal densities at 23ns and 52ns were also measured (Figure 4). The numbers were obtained for the midpoint of the wires. A high density plasma front was observed, which rapidly moved away from the plane of the array. Unfortunately some plasma densities were too high for our interferometer. The dash lines in Figure 4 are just estimates for those regions. Between the plane of the array and the dense plasma front, the plasma areal density decreased with the distance until hitting the edge of the front. At 52ns, the plasma density near the core was estimated to be 6×10^{18}cm^{-3}.

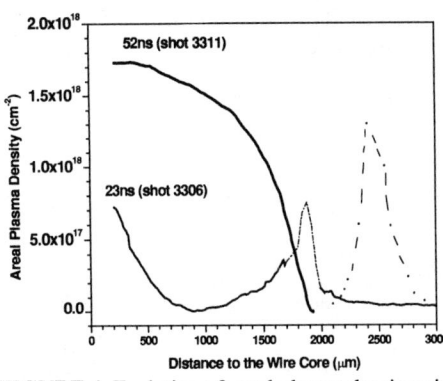

FIGURE 4. Evolution of areal plasma density with time for 13μm Al wires with 1mm inter-wire gaps.

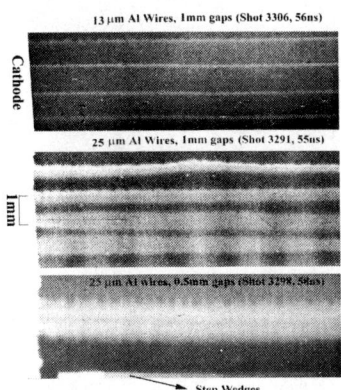

FIGURE 5. Typical radiograms

B. Wire Core Expansion and Plasma Merging

Backlighting images provided information about expansion and the inter-wire plasma merging. In the radiograms of Figure 5, the expansion is measured by the width of the wire images and merging by the central positions of the images. By using Al x-pinches (17μm and 25μm wires), we were able to get image times ranging from 40-70ns. Instabilities were observed for all the configurations with wavelengths, measured at 55 ns, which were about 400μm and seemed to be independent of wire diameter and spacing. The average expansion rate for 13μm Al wire arrays with 1mm gaps was about 3×10^5cm/s. For 25μm Al wire arrays with 1mm and 0.5mm gaps, the expansion rates were 7×10^5cm/s and 9×10^5cm/s respectively.

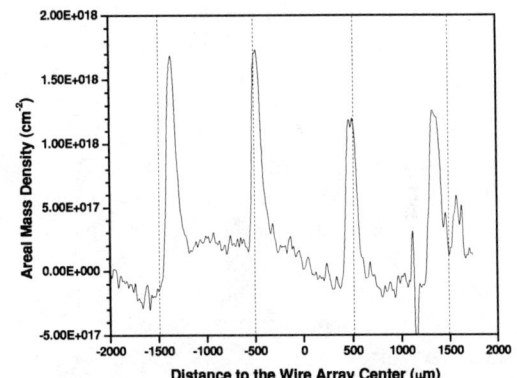

FIGURE 6. Areal mass density distribution of 13μm Al wires with 1mm gaps (shot 3306, 56ns).

With the Al step wedges, the areal mass density distributions were estimated. Figure 6 shows the areal mass density distribution of 13μm Al wires with 1mm gaps, measured from the midpoint of the wires as projected to the plane of original wire array. The original wire positions are indicated with dash lines.

By integrating over the wire core region, the percentage of mass in the cores was estimated. It turned out that, at 56ns, only about 15% of the total mass was in cores for the 13μm Al wire with 1mm gaps. At a similar time, much more mass was in the wire cores for 25μm Al wires (about 56% for 1mm gaps and about 67% for 0.5mm gaps).

For the larger diameter wires more mass was merging at a higher velocity than for the smaller wires. Figure 6 also shows that the wire cores were attracted toward the center. Merging velocities for three configurations are shown in Figure 7. Merging velocities of the outer two wires were about twice as much as for the inner two wires. Generally the merging velocities were smaller than corresponding plasma pushing velocities. But for 25μm Al wire arrays with 0.5mm inter-wire gaps, these two velocities were about the same. Merging motion appeared to be delayed in time relative to the current onset by 30-50ns.

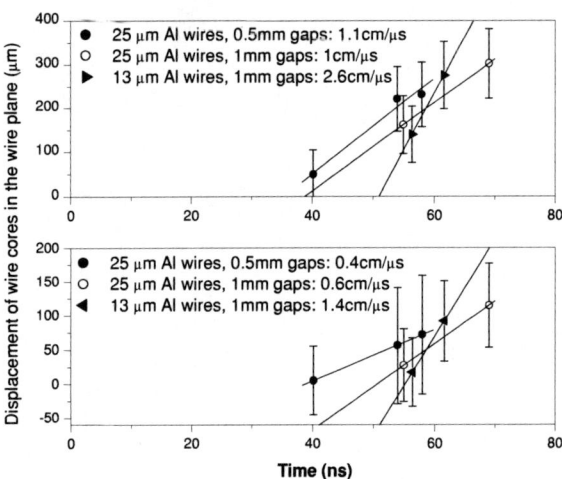

FIGURE 7. Merging velocities for three configurations. (a) average over the outer two wires; (b) over the inner two wires.

IV. CONCLUSIONS

Experimental observations of 4-wire Al array z-pinches were presented. These results provide a 3D picture of the wire dynamics, pushing of plasma by the global magnetic pressure, the inter-wire plasma merging and the expansion of cores. Velocities were measured for two wire sizes, (13μm and 25μm) and two inter-wire gaps (1mm and 0.5mm). The pushing of plasma started almost immediately, while plasma merging seemed to be delayed. The pushing effect was more significant for smaller size wires (13μm Al vs. 25μm Al), or with larger gaps (1mm vs. 0.5mm). Pushing was faster than merging. This difference became smaller with bigger diameter wires. Merging velocities for the outer two wires were twice as large as for the inner two wires. Development of m = 0-like instabilities was observed with the wavelength ($\lambda \sim 400\mu m$) and appeared to be unrelated to wire diameter or inter-wire gaps.

ACKNOWLEDGMENTS

Work supported by DOE grant DE-FG03-98DP00217 & Sandia contract BD-9356.

REFERENCES

1. Deeney, C., Douglas, M. R., Spielman, R. B., et al., *Phys. Rev. Letters* 81, 4883 (1998).
2. Matzen, M. K., *Phys. Plasmas* 4, 1519-1527 (1997).
3. Pikuz, S. A., Shelkovenko, T. A., Mingaleev, A. R., et al., *Phys. Plasmas* 6, 4272 (1999).
4. Pikuz, S. A., Romanova, V. M., Baryshnikov, N. V., et al., *Rev. Sci. Instrum.* 72, 1098 (2001).

Plasma Formation Around Single Wires

Peter U. Duselis and Bruce R. Kusse

Laboratory of Plasma Studies, 369 Upson Hall, Cornell University, Ithaca NY

Abstract. At Cornell's Laboratory of Plasma Studies, single wires of various metals were exploded using a ~250 ns pulser with a rise time of ~20 A/ns. It was found that the wires first experience a resistive heating phase that lasts 50-80 ns before a rapid collapse of voltage. From that point on, the voltage across the wire was negligible while the current through the wire continued to increase. We attribute this voltage collapse to the formation of plasma about the wire. Further confirmation of this explanation will be presented along with new experimental data describing preliminary spectroscopy results, the expansion rate of the plasma, and current flow along the wire as a function of radius. The resistance of the wire-electrode connection will be shown to significantly affect the energy deposition. Various diagnostics were used to obtain these experiments. Ultraviolet sensitive vacuum photodiodes and a framing camera with an 8 ns shutter were used to detect and measure the width of the visible light emitted by the plasma. A special wire holder was constructed that allowed the transfer of current from the wire to the surrounding plasma to be observed.

INTRODUCTION

Single 3 cm long, 25 µm diameter copper wires were driven by a 75 nF capacitor charged to approximately 15 kV. The circuit parameters are such to give a 20 A/ns rate of current rise, similar to the pre-pulse on the large Z-machine at Sandia. The current, voltage and resistance of the wire are shown in Figure 1. Initially the wire had a cold resistance of 2 Ω. As the current and voltage increased, the wire was resistively heated, and at 75 ns the resistance had increased to approximately 16 Ω. At this point in time the voltage collapse occurred and the resistance fell to less than 0.3 Ω. The collapse took only a few nanoseconds. After this collapse the voltage and resistance remained small while the current continued to increase. We present below a set of experimental results that look in detail at the voltage collapse and confirm the formation of plasma and transfer of current to this plasma.

EXPERIMENTS AND RESULTS

The interest in single wire experiments has led Cornell University's Laboratory of Plasma Studies to perform several new measurements involving only one wire and relatively low currents. Most of these experiments were concerned with explaining the voltage collapse as a result of the formation of plasma around the wires. Other experiments were concerned with the composition of the plasma and the amount of energy deposited in the wire before plasma formation.

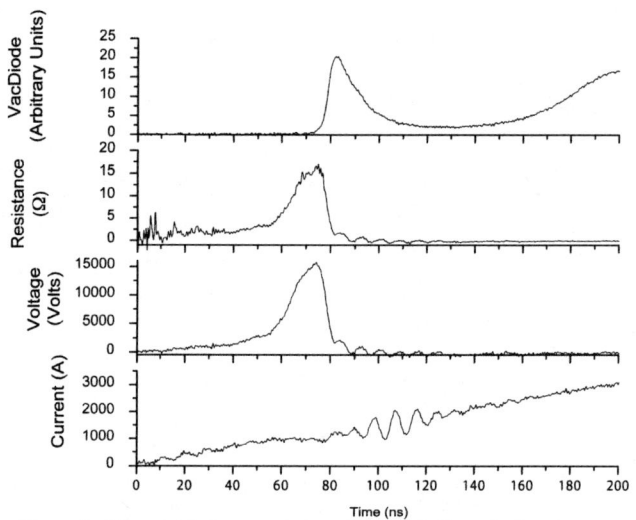
Figure 1: Electrical characteristics of an exploding single wire.

In order to detect plasma at the time of the voltage collapse, a vacuum photodiode and a framing camera were used to record the light given off by the plasma. The simple vacuum photodiode detected UV and had a response time of less then 1 ns. The photodiode was composed of a circular mesh in front and a plate in back that were separated by 2 mm. The plate in back acted as the cathode when it was biased to − 2000 volts with respect the mesh that became the anode. Its response to the exploding wire can be seen in Figure 1. The exploding wire gave off no UV until the time of the voltage collapse. At that point the there was a burst of UV, and there continued to be UV while current was being driven.

A framing camera was also employed to observe the plasma. It had an exposure time that could be as quick as 8 ns and responded to visible light. Its resolution was better than 50 µm. There were several results from this framing camera. It recorded no light emitted from the wire before the voltage collapsed. When the voltage was collapsing, it was able to record plasma as wide as 310 µm surrounding the wire (See Figure 2). However, it usually only recorded light about 250 µm from the wire core. Figure 3 shows the width of the visible light detected as a function of time. Each data point is from a different shot. By decreasing the shutter speed, the framing camera became more sensitive to light but gave up temporal resolution. While using a longer shutter, the camera was able to detect a plasma column that was 700 µm wide. The shutter was opened at 70 ns and closed 35 ns later. This was far wider than the camera recorded with an 8 ns shutter at any point during the wire explosion.

A 0.3 m spectrometer integrated the spectrum over both the time and space for the exploding wire. The spectrometer had three different gratings, 150 g/mm, 1200 g/mm, and 2400 g/mm. For the copper and molybdenum wires, it recorded strong neutral copper and molybdenum lines, respectively. However, both materials emitted a pair of

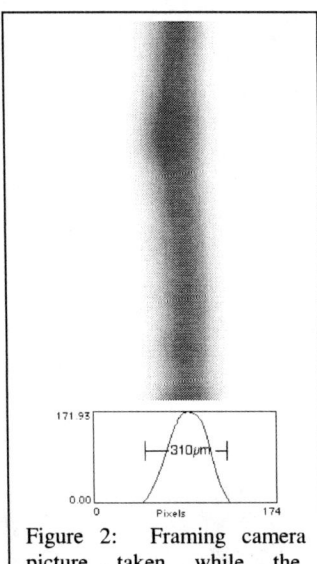

Figure 2: Framing camera picture taken while the voltage was collapsing.

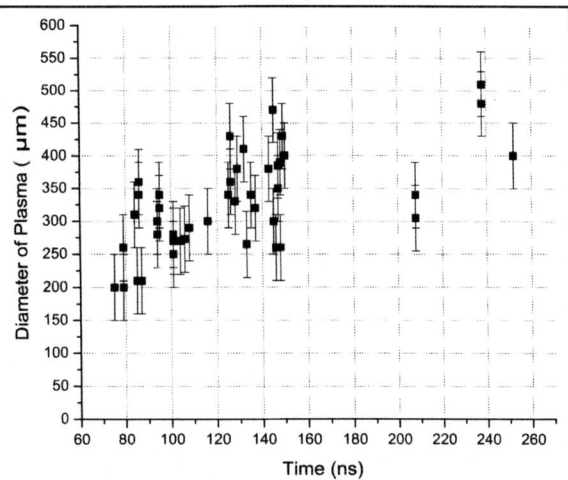

Figure 3: Plasma width as a function of time as measured by the framing camera with an 8 ns shutter.

lines around 465 nm, suggesting that there was a surface contaminant such as oxygen or nitrogen present in the plasma. Background gas in the chamber was ruled out because the chamber was filled with argon, but no argon lines were detected. Further study is required to determine the exact nature of the lines.

It was proposed earlier in this paper that at voltage collapse, the current transfers from the wire core to the surrounding plasma. A new cathode was constructed that consisted of two plates with insulation sandwiched between them. The upper plate of this cathode had a hole that was 740 µm in diameter and lined with 130 µm of insulation. The wire passed through the upper plate first, and then through a concentric hole in the lower plate. A resistive shunt monitored the current flow through the bottom plate while a Rogowski coil measured the total current through

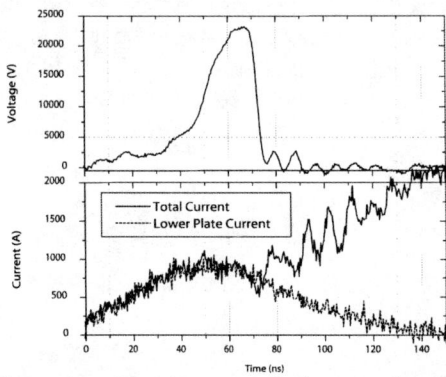

Figure 4: Observation of the plasma transferring away from the wire core.

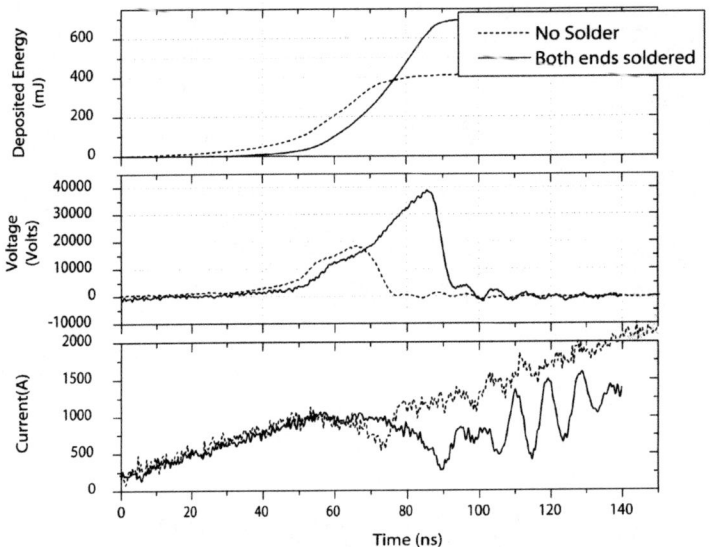

Figure 5: How soldering the wire ends to the electrodes alters the electrical characteristics of the exploding wire.

both the wire and plasma. The idea behind this new cathode was that if current flowed through the upper plate, then plasma must have extended from the wire core to the upper plate. Figure 4 shows the result of this experiment. When the voltage collapsed, not all the current flowed through the bottom plate. As time progressed, the current through the bottom plate eventually went to zero.

Previously experiments at Cornell University relied on dangling weights from the wires to make good electrical connections. This insured that at both ends of the wire, the wire would contact the electrode at several points and the tension of the wire would insure good electrical connections. Instead of this method, a new experiment at Cornell University soldered 25 µm diameter copper wires directly to the electrodes. It was found that the wires developed greater voltages before the voltage collapsed, and heated longer. As a result of this, the deposited energy in the wires also increased significantly. Figure 5 summarizes these results. Early in the current pulse the poor connections between the wires and electrodes could have heated up and ejected electrons. These electrons could strike a plasma, especially if they came from the cathode.

ACKNOWLEDGMENTS

Sandia contract BD-9356 and DOE contract DE-FG03-98DP00217 supported this research.

Investigation of the Initial Stage of Electrical Explosion of Fine Metal Wires

G.S. Sarkisov[a], S.E. Rosenthal[b], K.W. Struve[b], D.H. McDaniel[b], E.M. Waisman[c], P.V. Sasorov[d]

[a] Ktech Corporation, Albuquerque, 2201 Buena Vista SE, Suit 400, N M 87106-4265, USA
[b] Sandia National Laboratories*, P.O. Box 5800, Albuquerque, N M 87185-1194, USA
[c] Alameda Applied Sciences Corporation, 169 Saxony Rd., Ste. 210, Encinitas, CA 92024, USA
[d] Institute of Theoretical and Experimental Physics, Moscow 117259, Russia

Abstract. The results of an experimental investigation of the initial stage of the electrical explosion of thin metal wires are discussed. It is shown that energy deposition into the metal core depends on current rate and voltage polarity. The dependence of the expansion velocity and light emission on the atomization enthalpy of the metal is discussed. The light emission profiles demonstrate the micro conditions of the expanded wire core. Results of experiments for Al wire have been compared with 1D MHD simulations.

The investigation of the initial stage of exploding fine metal wires is important to modern Z-pinch physics. The exciting results that were achieved on the Z-facility at Sandia National Laboratories [1], of 2 MJ of x-rays at 200TW power have stimulated investigations for better understanding of the wire's initiation. The energy deposition into the individual wires of the array can strongly affect the homogeneity of the wire-array plasma shell and on-axis plasma compression. Direct influence of the current prepulse on the x-ray yield for an Al wire array on a 1 MA installation was demonstrated in [2]. Wire initiation research based on a unique radiography technique has been done at Cornell University and summarized in [3]. Extending work [3], we will discuss our results based on optical diagnostics.

Experiments were performed in vacuum using a 100kV Maxwell 40151-B pulser with energy of 12J and a current of 3kA. Two regimes of explosion were studied: fast explosion with a 150A/ns current rise and slow explosion with a 20A/ns current rise. The pulse was applied to a 2 cm long wire (Ag, Al, Cu, Au, Ni, Ti, Pt, Mo, and W) placed in the center of coaxial target unit. Current, voltage, light emission, streak shadowgraphy of wire diameter with a 200ns diode laser backlighter, and 150ps laser shadowgraphy at 523 nm were employed as diagnostics of the exploding wire.

Fig.1 shows the synchronized waveforms for current, voltage, light emission and wire diameter obtained from an exploding 20μm Ti wire. Voltage reaches its maximum value by ~30ns and subsequently collapses to zero during ~5ns. The collapse occurs due to "surface breakdown" related to ionization of the ambient low-density vapor and its fast expansion. During voltage collapse we observe the onset of sharp light emission corresponding to the ionization of the vapor. After voltage collapse the shape of the current becomes the same as in a short circuit. The wire core

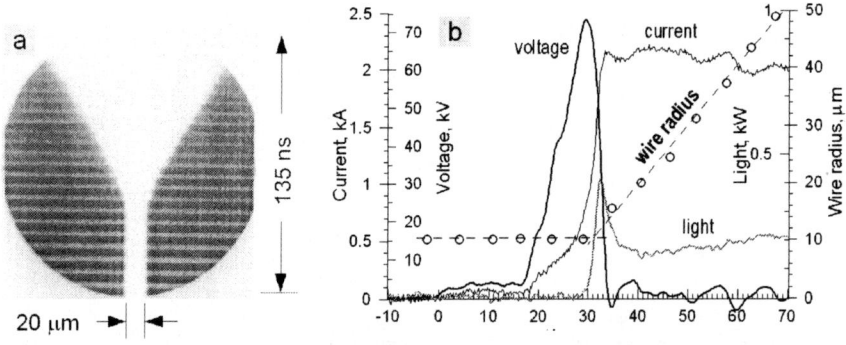

Fig.1 Streak image of the radial current, voltage, light emission

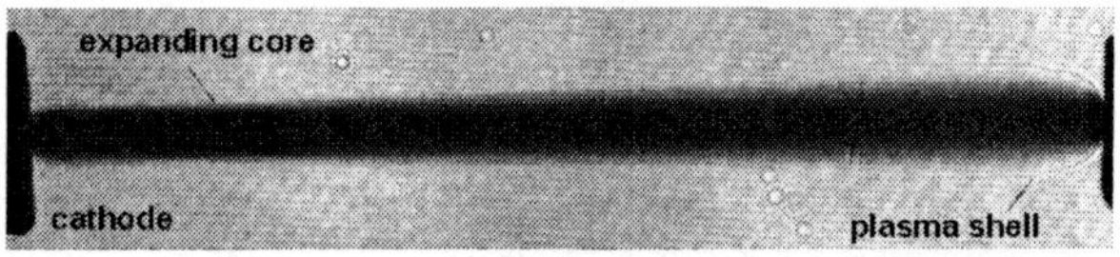

Fig.2 Laser shadowgram for fast exploding 20μm Cu wire at 200ns.

starts expanding with the voltage collapse. This expansion is related to the drop of magnetic field pressure at the core when the current rapidly switches from the wire to the ambient higher-conductance and low-density plasma shell. MHD simulations of exploding aluminum wire, mentioned below, also show this relationship. Fast-dropping magnetic field pressure can cause explosive boiling [4] in the overheated metal core and start wire expansion.

For some metals we can see a shock-wave structure at the anode related to collision of the low-density plasma shell from radial expanding wire and the expanding anode plasma (Fig.2). Heating of the anode by energetic electrons generated along the wire surface during the vapor breakdown process creates the under-anode plasma.

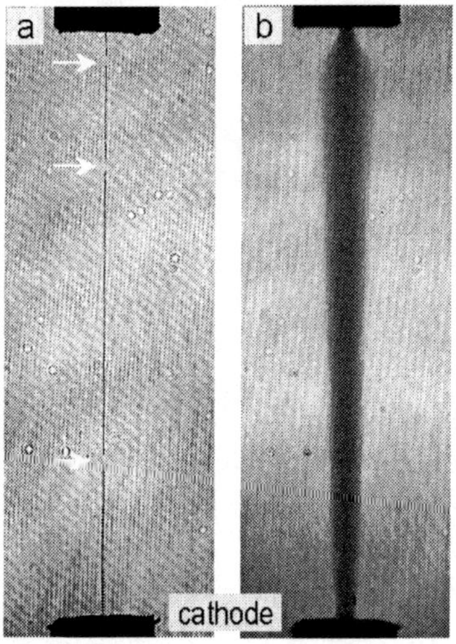

Figure 3 illustrates significant differences between fast and slow explosion modes for 20μm tungsten wire. For the slow explosion mode (Fig.2-a), the current

Fig.3 Laser shadowgram of (a) slow explosion 20μm W wires at 100μs and (b) fast explosion at 1μs. White arrows on (a) show places of wire disintegration.

transferred from the wire to the surface breakdown before melt due to an anomalously high electronic emission [5] and the wire disintegrated into parts. The maximum resistivity in this case does not exceed the melt value [6]. For the fast exploding mode (Fig.2-b), the tungsten wire absorbed significantly more energy before surface breakdown and this energy feeds the core expansion after voltage collapse [6]. The maximum resistivity in this case exceeds the melting level [6], and the expanding core

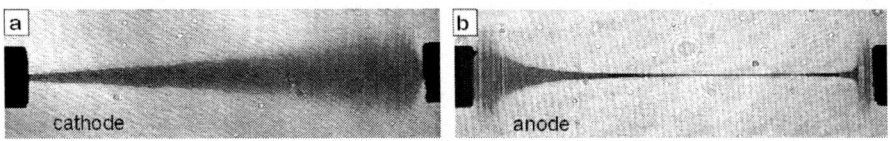

Fig.4 Laser shadowgrams at 1µs of 20µm Ti wire for (a) positive and (b) negative explosions.

can be in a sol-vapor phase [5].

Fig.4 illustrates the "polarity" effect [7] for exploding 20µm Ti wires. Changing the polarity of the electrical pulse reverses the sign of the radial electric field ($E_z \sim (1-5) \cdot 10^4$ V/cm and $E_r \sim 10^6 - 10^7$ V/cm) and greatly influences the structure and value of deposited energy. For positive-polarity explosion E_r increases the potential barrier for electronic emission from the wire surface, hence the surface breakdown starts from the cathode where the radial field is least. The shunting breakdown then propagates toward the anode in the form of an ionization wave and produces a conical envelop of overheating. For

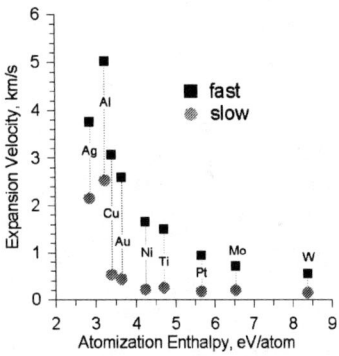

Fig.5 Dependence of expansion velocity of 20µm wires on atomization enthalpy.

negative-polarity explosion the radial field decreases the potential barrier for emitted electrons. In this case breakdown starts from the central part of the wire (with minimum disturbance from electrodes) and propagates to the periphery.

The averaged velocity of wire expansion (Fig.5) increases with decreasing atomization enthalpy E_A. For fast explosion mode this relationship can be approximated by $v \sim (E_A)^{-2}$. This behavior corresponding to the wire "overheating" discussed in [8].

Figure 6 shows the temporal profile of light emission for different atomization enthalpies. All metals exhibit an initial sharp peak of light emission at the moment of voltage collapse, related to the ionization of the surrounding vapor, and the secondary radiation varies strongly with the atomization enthalpy. For metals with small atomization enthalpy the secondary radiation decays after the initial peak. For refractory metals the secondary radiation grows for 300-500ns after

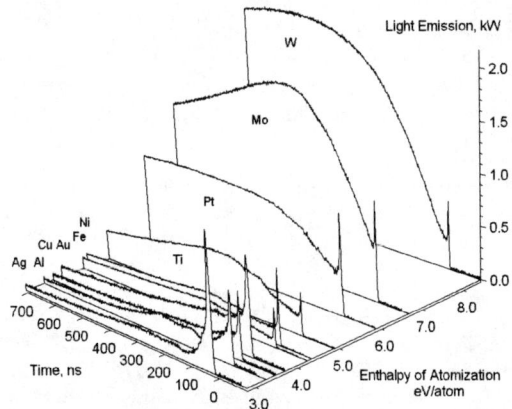

Fig.6 Evolution of the light emission vs. atomization enthalpy for fast exploding wires.

the initial peak. The higher atomization-enthalpy metals produce stronger secondary radiation. Secondary radiation drops fast only for non-refractory metals (Ag-Au) that experience "overheating" [8] (deposited energy ~2.2-3.4 times enthalpy of atomization). In this case the wire core mostly vaporizes and then adiabatically cools by expansion. This process is amenable to MHD simulation. For refractory metals (Ni-W) on Fig.6 the energy deposited before voltage collapse is insufficient to totally vaporize the wire and the expanding core consists of vapor and sol [5]. In this case the radiation dynamics involves two phenomena: increasing of transparency for radiating particles as they fill a larger volume and eventual decreasing radiation due to their radiative cooling. This phenomenon is similar to the conventional firework display and leads to the high-atomization-enthalpy radiation profiles of Fig. 6. For a given metal, increasing of wire overheating leads to transformation of the radiation profile from "particle" to "gas" types.

We have performed a 1D simulation of fast exploding 25μm Al wire using the ALEGRA [9] MHD code. The most accurate EOS and transport tables are presently available for aluminum [10]. The experimental current shape was used in the simulation. The voltage calculated from this simulation is presented in Fig.7. The simulation voltage exhibits good agreement for the time of voltage collapse and correlated with 20-30% error after melting. However, before melt (time < 10 ns) significant discrepancy exists. We will discuss these results in a future publication.

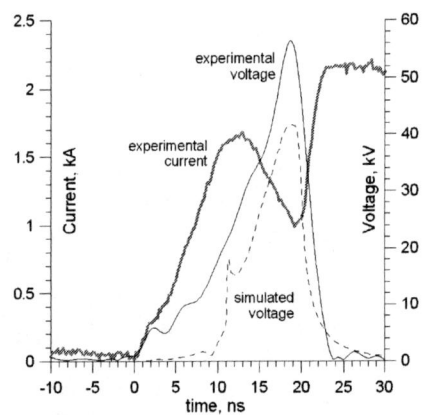

Fig.7 ALEGRA simulation of voltage for explosion of 25μm Al wire.

The authors gratefully acknowledge M. P. Desjarlais for his accurate Al electron transport coefficients; and H. Faretto, A. Oxner and W. Brinsmead for technical help.

*Sandia is a multi-program laboratory operated by Sandia Corporation, a Lockheed Martin Company, for the United States Department of Energy under Contract DE-AC04-94AL8500.

References.
1. Spielman R.B., Deeney C., Chandler G.A., et al. Phys. Plasmas, **5**, pp.2105-2111 (1998).
2. Beg F.N., Lebedev S.V., Bland S.N., et al., Physics of Plasma, **9**, 1, pp.375-377 (2002).
3. Hammer D.A., Sinars D.B., Laser and Particle Beams, **19**, pp.377-391 (2001).
4. Martynyuk M.M., Sov. Phys. - Technical Physics, **19**, pp.793-7, (1974).
5. Lebedev S.V., Savvatimskii A.I., Sov. Phys. Usp., **27** (10), pp.749-771 (1984).
6. Sarkisov G.S., Bauer B.S., DeGroot J.S., JETP Letters, **73**, pp.69-74 (2001).
7. Sarkisov G.S., Sasorov P.V., Struve K.W., McDaniel D.H., submitted to PRE on April 2002.
8. Sarkisov G.S., Rosenthal S.E., Struve K.W., McDaniel D., Waisman E., Sasorov P.V., "Joule energy deposition in exploding wire experiments", in this book.
9. "ALEGRA: User Input and Physics Descriptions – October 1999 Release", Edward A. Boucheron, et. al., SAND99-3012, Sandia National Laboratories, Albuquerque, NM
10. Rosenthal S. E., Desjarlais M. P., and Cochrane K. R., Pulsed Power Plasma Science 2001, IEEE Catalog No. 01CH37251, pp. 781-784

Joule Energy Deposition in Exploding Wire Experiments

G.S. Sarkisov[a], S.E. Rosenthal[b], K.W. Struve[b], D.H. McDaniel[b], E.M. Waisman[c], P.V. Sasorov[d]

[a]*Ktech Corporation, Albuquerque, 2201 Buena Vista SE, Suit 400, N M 87106-4265, USA*
[b]*Sandia National Laboratories, P.O. Box 5800, Albuquerque, N M 87185-1194, USA*
[c]*Alameda Applied Sciences Corporation, 169 Saxony Rd., Ste. 210, Encinitas, CA 92024, USA*
[d]*Institute of Theoretical and Experimental Physics, B. Cheremushkinskaya 25, Moscow 117259, Russia*

Abstract. Measurements of the Joule energy deposition into an exploding wire are presented. Energy deposition into the metal core is strongly dependent on current rate, voltage polarity, wire diameter and coatings, and the environment. The condition of the wire core after breakdown depends on atomization enthalpy of the metal. Results of energy deposition experiments for Al wire are compared with ALEGRA MHD simulations.

The investigation of the Joule energy deposition during the resistive part of the electrical explosion of fine metal wires is very important for Z-pinch physics. The homogeneity and value of deposited energy may influence the stagnation phase in multi-wire array experiments [1]. Value and structure of deposited energy can determine the plasma-ablative processes and amount of wire mass that does not participate in the final compression.

The experimental setup has been described in [2]. Figure 1 presents a laser shadowgram for fast (a) and slow (b) exploding 20μm Al wires.

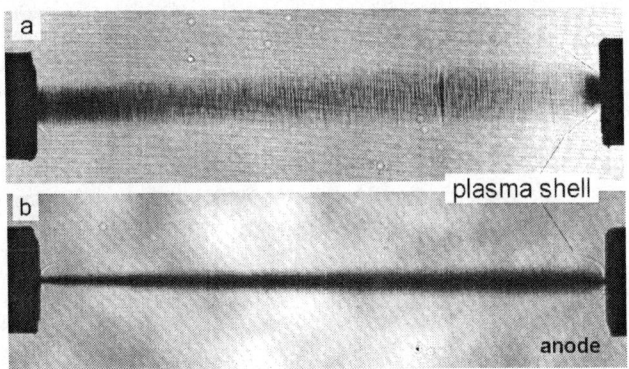

Fig.1 Fast (a) and slow (b) exploding in vacuum of 20μm/ 2cm Al wires at times 190ns (a) and 235ns (b).

We can see that fast exploding wire expands faster then slow one owing to higher energy deposition before breakdown. The expan-ding core (a) exhibits radial stratifications, pos-sibly due to the develop-ment of thermal instabilities [3,4].

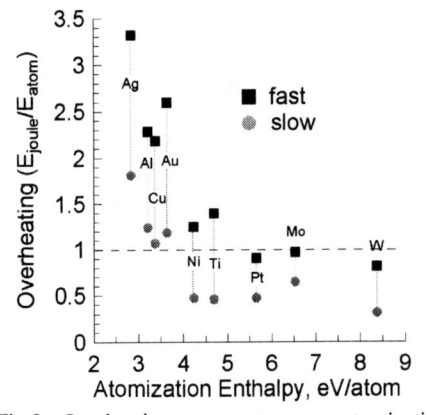 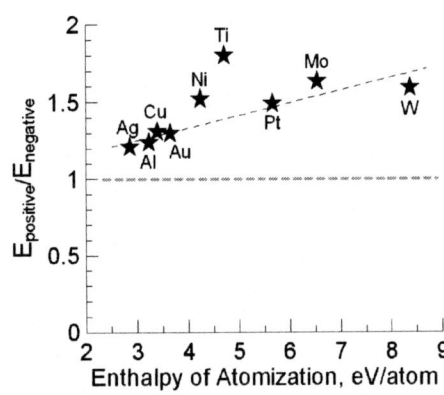

Fig.2 Overheating parameter vs. atomization enthalpy for explosion in vacuum.

Fig.3 Ratio of deposited energy for positive and negative polarity explosion vs. E_{atom}.

Figure 2 presents the averaged value of "overheating" for 9 different 20μm metallic wires vs. each wire's published enthalpy of atomization. The "overheating" parameter is the ratio of deposited energy to the atomization enthalpy of the metal. We can see the systematical increasing of wire "overheating" with decreasing of the atomization enthalpy. The fast exploding mode (150A/ns) gives 1.7-3.1 times higher deposition energy for all metals then the slow mode (20A/ns). Increasing the wire "overheating" results in greater wire expansion velocity [2].

We present evidence for a "polarity" effect in [2]. Figure 3 shows the ratio of deposited energy for positive to negative polarity explosions vs. atomization enthalpy for different metals. We can see that positive polarity explosion, when radial electric field "pushes" electrons back to the wire, results in 20-80% more deposited energy then for negative polarity, when radial field "expels" electrons from the wire.

The next interesting effect is related to the difference between explosion of clean and coated wires. Figure 4 shows the radial laser streak image of a fast exploding 20μm W, and W coated by 0.7μm Au. The onset of wire expansion in all cases coincides with voltage collapse [2]. Velocity of expansion was: 0.66km/s for W (a) and 1.56km/s for Au coating and 0.45km/s for W core (b). The white arrows on Fig.4-b show expansion of dense Au (1) and W (2) boundaries. Structure (3) appears to be a shock wave from low-density W corona propagating inside Au plasma cylinder with velocity 0.45km/s.

The gold coating results in almost 40% overheating of the W wire core before surface breakdown (Fig.5). This experiment demonstrates that the breakdown property of the exploding wires depends on the metal-vacuum interface. The breakdown processes for W+Au wire corresponds to clean Au wire. Less level of electronic emissions for Au than for W delays the formation of a high-conductance plasma shell thus providing more time for energy deposition to the W core than for clean W wire.

Figure 6 demonstrates fast exploding 12μm bare W wire (a) at 1000 ns and slow exploding 12μm W wire (b) at 525 ns with dielectric coating. Slow exploding coated wire demonstrates significantly faster expansion than fast exploding uncoated one (1.3km/s and 0.5km/s). In both cases we have overheating effect directed to the anode.

It implies that shunting breakdown starts from the cathode side and propagates toward the anode [2].

Evolution of the deposited energy before the surface breakdown for coated and bare 12μm W wires is presented in Fig.7. Fast exploding bare wire exhibits 2-4 times larger energy deposition then slow exploding wires. The slow explosion demonstrates an unstable nature. It varies from wire disintegration (curve 2 of Fig. 7) to wire explosion (curve 1 of Fig. 7). In wire disintegration case [5] the deposited energy ~1eV/atom and resistivity do not exceed melting level. For the explosion case the deposited energy reaches ~4eV/atom and resistivity was in a melting level. Fast explosion of bare wire allows a deposition of 9.5 eV/atom.

The slow exploding W wire with dielectric coating experiences 3-8 times more energy deposition then slow exploding bare wires and 1.4 times more than fast exploding bare wires. Some

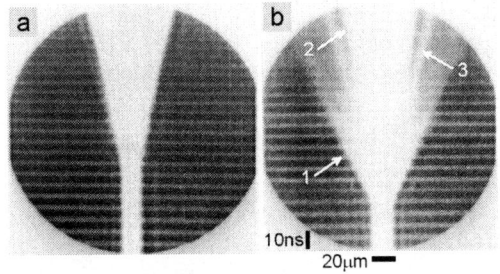

Fig.4 Streak images with laser backlighter of fast explosion of 20μm W (a) and W coated by Au (b) wires.

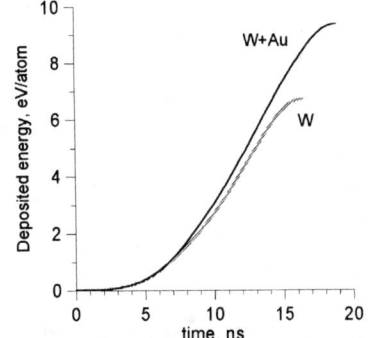

Fig.5 Evolution of the deposited specific energy for fast-exploding W and W+Au wires.

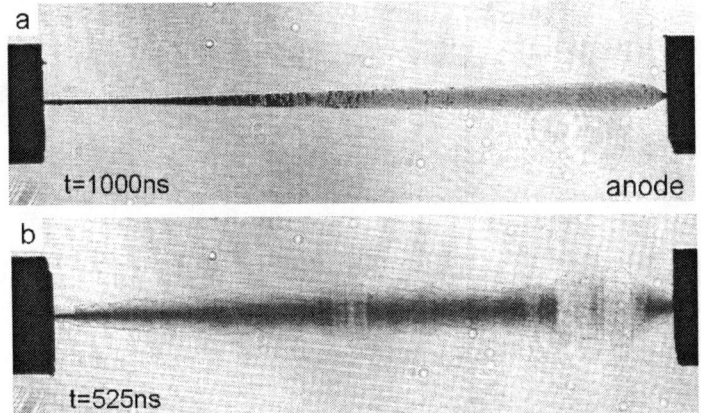

Fig.6 Shadowgrams of fast exploding 12μm bare W wires (a) and slow exploding 12μm coated W wire (b).

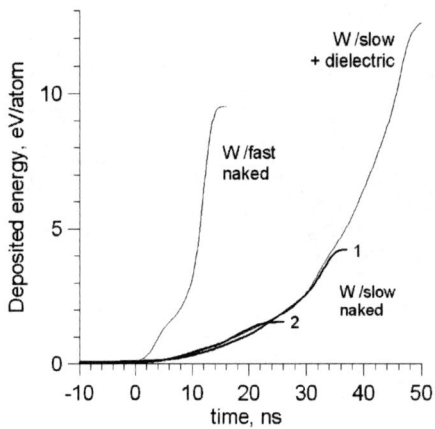

Fig.7 Evolution of deposited energy for fast and slow exploding 12μm W wires. Curve (1) is the best and (2) is the worst slow explosions of bare W wires.

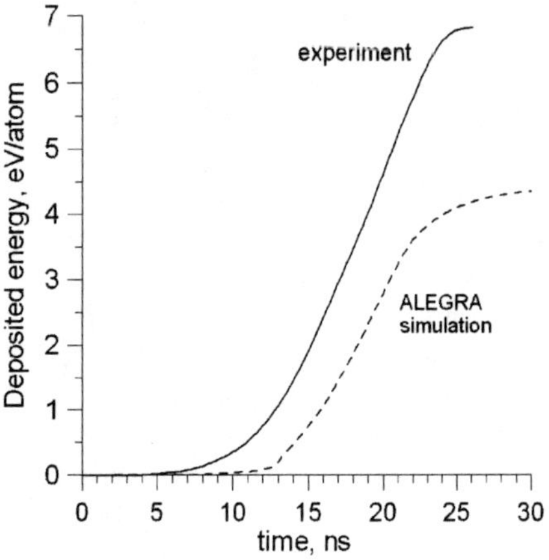

Fig.8 Comparison of the evolution of energy deposition for fast exploding 25μm Al wire with ALEGRA simulation.

aspects of the coating effect have been discussed in [6].

ALEGRA 1D MHD simulation has been done to reproduce fast exploding 25μm Al wire [2]. The most accurate EOS and transport tables are presently available for aluminum. The experimental current shape was used in the simulation. The simulation voltage exhibits good agreement for the time of voltage collapse and correlates within 20-30% error after melting. However, the deposited energy calculated from the simulation (presented in Fig.8) is only 65% of the deposited energy of the experiment. This discrepancy stems from a resistive pre-melt phase (~12 Ω, t < 10 ns) in the experiment that is absent in the simulation (~1 Ω, before melt). Consequently, the simulation misses significant pre-melt energy deposition compared to the experiment. We will discuss these results in detail in a future publication.

The authors gratefully acknowledge M. P. Desjarlais for his accurate Al electron transport coefficients; and H. Faretto, A. Oxner and W. Brinsmead for technical help.

Sandia is a multi-program laboratory operated by Sandia Corporation, a Lockheed Martin Company, for the United States Department of Energy under Contract DE-AC04-94AL8500.

References.
1. Sanford T.W.L., Allshouse G.O., Marder B.M., et all., Phys.Rev.Lett., 77, 25, pp.5064-5066 (1996).
2. Sarkisov G.S., Rosenthal S.E., Struve K.W., McDaniel D., Waisman E., Sasorov P.V., *"Investigation of the initial stage of electrical explosion of fine metal wires"*, in this book.
3. Bennett F.D., in *Progress in High Temperature Physics and Chemistry*, **V.2**, pp. 1-65 (1968).
4. Lebedev S.V., Savvatimskii A.I., Sov. Phys. Usp., **27** (10), pp.749-771 (1984).
5. Sarkisov G.S., Bauer B.S., DeGroot J.S., JETP Letters, 73, pp.69-74 (2001).
6. Sinars D.B., Shelkovenko T.A., Pikus S.A., at all., Phys. Plasma, 7, pp.429-432 (2000).

Results of Experiment on Explosion of W and Al Wires in Water and Vacuum

Alexander G. Rousskikh, Rina B. Baksht, Vladimir I. Oreshkin, Alexander V. Shishlov

Institute of High Current Electronics, 4 Academichesky Ave., 634055 Tomsk, Russia

Abstract. The objective of the present work is to provide an experimental base for constructing the conductivity matrix of tungsten under conditions similar to exploding wire of multi-wire arrays on terawatt systems. This paper reports the results of experiments on nano- and microsecond electric explosions of aluminum and tungsten wires. The generator makes it possible to vary the impedance, the charge voltage, the current rise time, and to explode wires both in water and vacuum. The capacitance of the current generator is 5 µF or 67 nF. The inductance of the current generator is 28 µH, 2.25 µH, or 0.73 µH. This provides the possibility of varying the rise rate of the wire current in the range from $6 \cdot 10^{11}$ to $5 \cdot 10^{16}$ A/(s·cm^2).

INTRODUCTION

Recent progress of the program on the implosion of cylindrical multi-wire arrays with the use of the Z-machine at Sandia National Laboratories, Albuquerque [1], has spurred a number of works on the problem of wire explosions starting from the cold wire state. A wire is exploded when drawing a current of 0.01 ÷ 1 kA for a time of 30 ÷ 100 ns, the most common conductor material being therewith tungsten wires of diameter 7 ÷ 25 µm. This is precisely the range of the parameters with which most of the related studies are dealt [2-4]. Despite great importance and significance of experimental investigations in this field, works on simulation of the processes are no less important. In obtaining a good agreement between the results of simulation and experiment, the main difficulty is a variety of physical phenomena occurring in the explosion of wires. In the case of wires explosion in vacuum, the heating, melting, and evaporation of the wire material is accompanied by desorption of the gas. This leads to the gas breakdown. The high voltage lead to thermoelectricity that, in turn, affect the conductivity of the A-K gap. The foregoing suggests that it would be appropriate first to simplify the conditions of the problem by exploding wires in a liquid dielectric medium and to gain adequate experience in simulating the explosion of conductors itself. In what follows, attempts should be made to simulate all physical processes developing in the explosion of wires in vacuum with the same current generator and the same geometry. The main objective of our experiments was to obtain experimental data on the explosion of tungsten wires on which to base the conductivity matrix of tungsten.

EXPERIMENT DESCRIPTION

The equivalent circuit of the generator is shown in Fig.1.

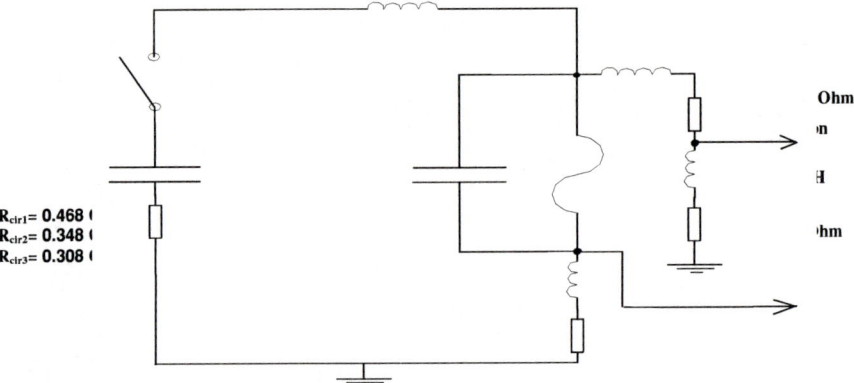

FIGURE 1. Equivalent circuit of the generator

The current generator parameters, such as the capacitance, the circuit inductance, and the charge voltage, are variable. In our experiments, we actually used capacitances of 67 µF or 5 µF, inductances 730, 2250, or 28000 nH, and capacitor charge voltages of 3, 5, 10, 20, 30 kV. The same figure shows schematically the measuring circuit of a high-voltage divider and shunt with account of the stray capacitance of the A-K gap. The chamber of the current generator (Fig.2) is suited for the explosion of wires both in vacuum and in water.

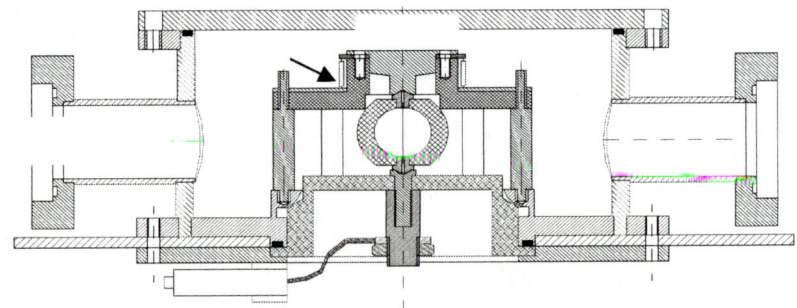

FIGURE 2. Vacuum/Water Chamber for Experiments on Wire Explosion

The length of the wires can also be varied. The wire is stretched in a removable device and the design of the electrodes remains unchanged. This design has turned out to be very convenient for fixing the wires in the A-K gap to ensure a good contact.

Actually, use was made of wires 26 and 9 mm long. The diameter of the W wires was 7.3 μm and that of the Al wires was 15, 30 μm. The insulation properties of the water were no worse than 150 kΩ/cm that was attained by deionizing of the water upon each explosion. In the case where the wires were exploded in vacuum, the pressure in the chamber was no greater than $4 \cdot 10^{-5}$ Torr.

To make sure that the obtained results could be relied on, it was decided to realize a series of test explosions of Al wires. In the course of test shots, the stray capacitance of the electrode separation was found to contribute significantly to the readings taken from electric probes at the moments dU/dt was high (Fig. 3).

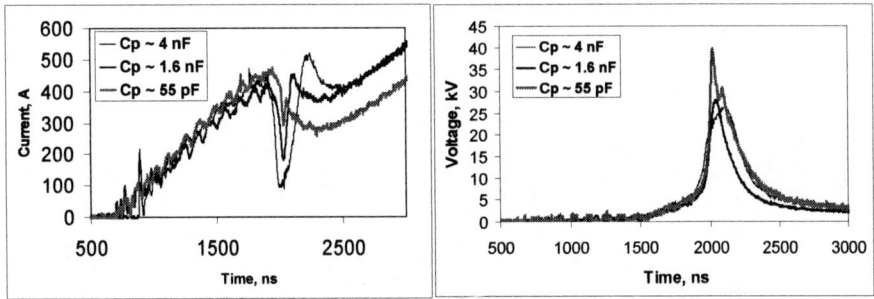

FIGURE 3. Effect of the stray capacitance of the electrode separation on the wire current and voltage. Explosion of 30 μm Al Wire in Water, $U_0 = 10$ kV, $C_{gen} = 5$ μF, $L_{gen} = 28$ μH

This fact should be taken into account in designing similar-type generators, the condition of the minimum interelectrode capacitance should be provided and its actual value should be allowed for in calculations. Once the flaws found during the test explosions of wires were remedied, comparison of the experimental data with the results of simulation [5] and with the results reported elsewhere.

It is known that in vacuum the A-K gap is broken down along the wire within 20 ÷ 50 ns after start of a current. We have made some attempts to ascertain the cause for the given effect and to prolong the resistive phase of the explosion of wire. Two facts which influence on the duration of the resistive phase have been found: 1) an external transverse magnetic field and 2) heating of the conductor. The application of a transverse magnetic field makes it possible to prolong the resistive phase by 10 ÷ 20 ns (Fig. 4), whereas the heating of the wire makes it possible to attain the current pause phase and the breakdown occurs even after of the secondary current rise (Fig.5). The W wires were heated within 1.5 h with a direct current of 25 mA. To attain a positive effect the wire should be exploded without switching off the heating.

CONCLUSION

The results of experiments performed on the explosion of Al and W wires agree well with our model concepts of the character of developing physical processes. This fact shows promise of interpreting adequately the data on the explosion of W

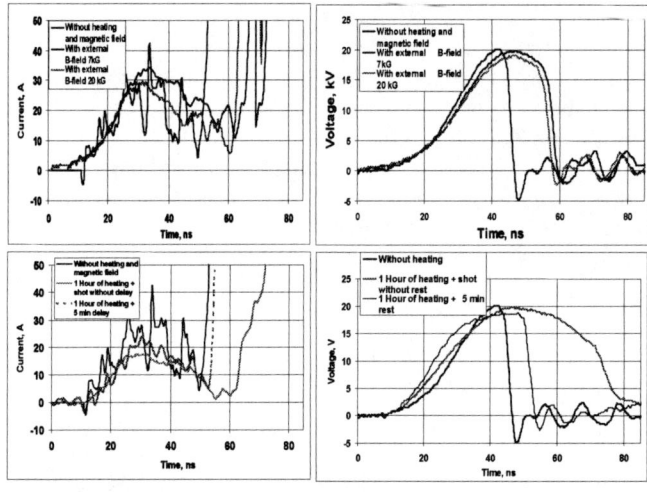

FIGURE 4. Effect of the external transverse magnetic field on the time the electrode separation is breaken down. Explosion of 7 μm W Wire in Vacuum, $V_0 = 20$ kV, $C_{gen} = 67$ nF, $L_{gen} = 2.25$ μH

FIGURE 5. Effect of the preliminary heating of the wire on the time the electrode separation is broken down. Explosion of 7 μm W Wire in Vacuum, $V_0 = 20$ kV, $C_{gen} = 67$ nF, $L_{gen} = 2.25$ μH

conductors in planned experiments and of creating the conductivity matrix corresponding to the actual state of affairs. The fact that the heating of a wire substantially affects the breakdown along it speaks in favor of considerable importance of the absorbed gas in this process. The influence of the transverse magnetic field and the low potential in breakdown is also indicative of the development of electric-arc processes. The further study of the effect of external factors (such as the transverse magnetic field) on the character of the explosion of wires and on the attendant physical processes is of doubtless interest for understanding the physics of cold start in the implosion of multi-wire arrays.

ACKNOWLEDGMENTS

The work is supported by International Science and Technology Center, Project #1826.

REFERENCES

1. R. B. Spielman, S. F. Breeze, G. A. Chandler *et al., Proceedings of the 11th International Conference on High-Power Particle Beams,* Prague, Czech Rep., 1996, edited by K. Jungwirth and J. Ullschmied Czech Academy of Sciences, Prague, 1996, Vol. 1, pp. 150–153;
2. S. A. Pikuz, T. A. Shelkovenko, D. B. Sinars, *et al*, Phys. Rev. Lett. **83**, 4313, 1999;
3. D. B. Sinars, T. A. Shelkovenko, S. A. Pikuz *et al,,* Phys.Plasmas 7, p. 429, 2000;
4. T. A. Shelkovenko, S. A. Pikuz, A. R. Mingaleev, *et al,*Rev. Sci. Instrum. **70**, 667, 1999;
5. V.I. Oreshkin, R.B. Baksht, A.G. Rousskikh *et al,,* Modeling of wire explosion, *Proceedings of the International Conference on Physics of extremal states of matter* -2002", Chernogolovka 2002, p.14-17.
6. D. B. Sinars, Min Hu, K. M. Chandler, T. A. Shelkovenko, *et al,,* Phys. of Plasmas **8**, N 1, p. 216, 2001;

Wire Explosion with 0-1kA per Wire

Bruce R. Kusse, Min Hu, Katherine M. Chandler, David A. Hammer

Lab of Plasma Studies, 369 Upson, Cornell University, Ithaca, NY, 14853

Abstract. Explosions of fine wires driven by a 4.5kA peak current peak were studied. The formation of a plasma shell around the wire core was observed. Mass density distributions as a function of time were measured, which provided more quantitative information about wire expansion. Combining these measurements with refractive index measurements from interferometry the phase state of the wire core was determined.

INTRODUCTION

Wire array z-pinch implosions on the Z-machine at Sandia National Laboratories have produced extremely high energy density plasmas [1]. These experiments were initiated by a ~100ns, 0-1kA per wire ramped current prepulse. Our experiments simulate this current prepulse and study the initial wire explosion that results.

EXPERIMENTAL DESCRIPTION

The experiments described here involved 1 cm long, 25μm Ag, Cu and Au wires. The wires were driven by a 4.5kA peak current pulser [2]. A shunt monitor was used to measured the wire current, and a resistive voltage monitor was used to measure the voltage across the wire. Figure 1 shows typical current and voltage traces. Radiographs of exploded wires were also made with 17μm to 30μm Mo wires. Filters composed of 12.5μm Ti foils with deposited step wedges for density calibration were used [3]. Three channels of Schlieren images and two shearing interferometry [4] images separated by 50ns were recorded. The optical setup is shown in Figure 2. The

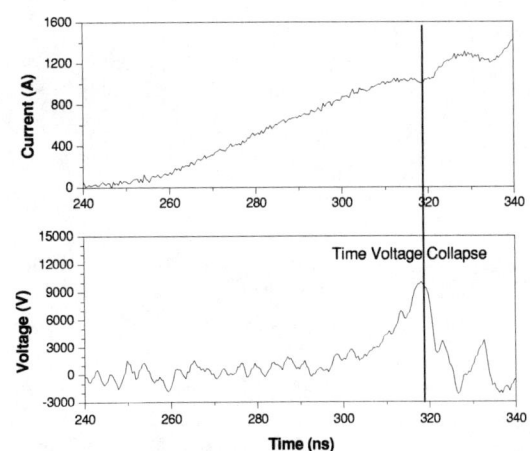

FIGURE 1. Typical Current and voltage traces, shot 1693, 25μm Ag wire

plasma detection limit of interferometry was about $4.2 \times 10^{16} \mathrm{cm}^{-2}$. The bright field Schlieren channel was composed of a big diameter hole, which essentially recorded shadow images. The sensitivity to plasma was estimated to be $3 \times 10^{17} \mathrm{cm}^{-3}$ for the dark field strip Schlieren, and $6 \times 10^{17} \mathrm{cm}^{-3}$ for the knife edge Schlieren.

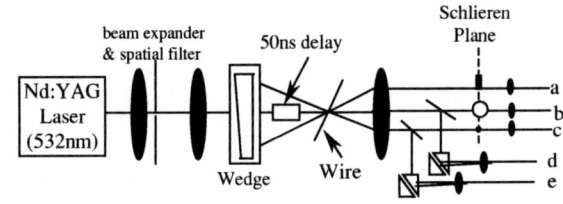

FIGURE 2. Optical Setup. a. dark field knife edge Schlieren; b. bright field Schlieren; c. dark field strip Schlieren; c. first interferometry; e. second interferometry, delayed 50ns.

EXPERIMENTAL RESULTS

As shown in Figure 1, the resistive heating phase was terminated about 70ns after the current started by a fast voltage collapse. This voltage collapse was due to the formation of a plasma shell around the wire core, which had much lower resistance than the wire core, as indicated by the Schlieren and interferometry data. Figure 3 shows example Schlieren and interferometry images for 25μm Au at 284ns. The peak plasma density measured from interferometry was $9 \times 10^{17} \mathrm{cm}^{-3}$ at a radius of 350μm, which was consistence with estimates from Schlieren images. For most of the experiments, the plasma densities were between the detection limits of strip Schlieren and knife edge Schlieren. An interesting fact to be pointed out is that wires usually appeared more expanded near to the anode end, while higher density plasmas were observed near the cathode end, as shown in Figure 3(b) and (c).

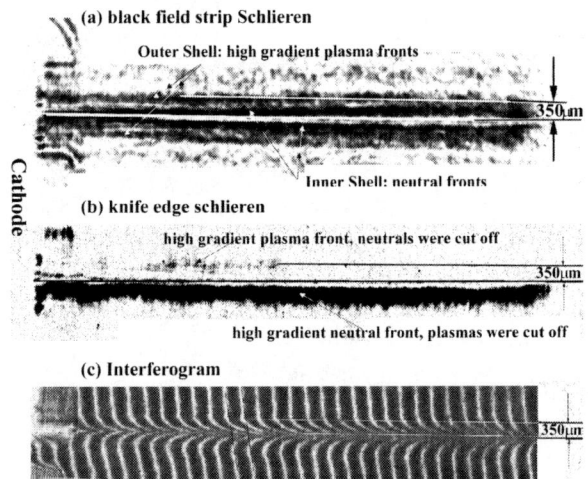

FIGURE 3. Schlieren and interferometry images show plasma shell around the wire core. Shot 9068, 25μm Au, 284ns

The experiments described above were conducted without x-pinch backlighting. This is necessary to study the properties of the plasma shell. It appeared that the UV radiation from an x-pinch was able to enhance plasma formation on the low current wire. An increase of plasma by 4-6 times in the experiments with x-pinch was observed. However, since we usually had the x-pinch fire after the time of voltage collapse, this additional plasma should not affect the wire core behavior a lot, as far as the plasma part was not concerned. The experiments we will discuss will generally be experiments with x-pinch backlighting images as well as optical diagnostics.

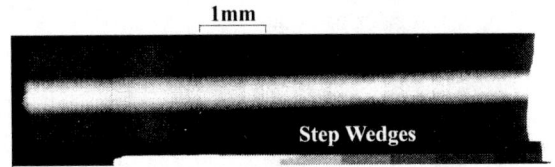

FIGURE 4.. Typical radiogram, Shot 1074, 25μm Ag coated with 1μm Polyester, 144ns

Even with the x-pinch UV we believe that the wire diameter measured at the neutral-plasma edge was still a good parameter for indicating the wire core expansion rate. This was confirmed by comparing the data with data without the x-pinch. The expansion rates of 25μm Ag, Au and Cu wires, bare and insulated, were measured with interferometry. The results are very similar to expansion rates measured from x-pinch radiographs that reported [2].

With step wedges deposited on the filters for x-ray imaging, the mass density of the expanding wire cores were measured quantitatively. As an example, we discuss the behavior of 25μm Ag wires coated with 1μm Polyester. A typical radiograph is shown in Figure 4. Figure 5(a) shows the mass density profiles at three different times. They were all measured at the midpoint of the wires. By integrating over the mass density, the mass enclosed in a certain wire core diameter can be plotted versus the radius of wire cores (Figure 5(b)), and it can be seen that the outer shell (bigger mass ratio inside) expanded faster than the inner shell. The axial flux $\Gamma = n\ v$ was estimated and was found to be constant at a number $\sim 4\times 10^{25}/cm^{2}\cdot s$.

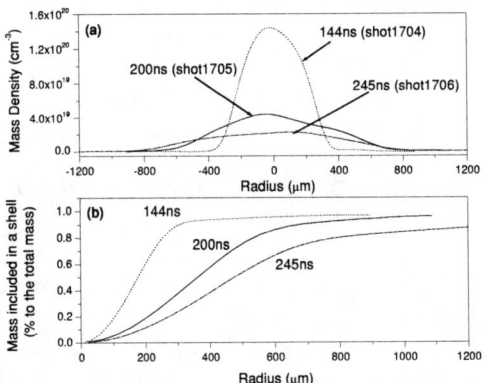

FIGURE 5. 25μm Ag coated with 1μm Polyester. a) Mass density distribution developing with time. (b) Shell expansion.

Consequently, the kinetic energy can be calculated with MHD equations, which turn out to be inversely proportional to n^2, where n was the number density of the atoms. Kinetic energy at ~10eV in the innermost core and increase to 1keV at wire edges were estimated. The kinetic energy of all the atoms was estimated to be about 225mJ, which is in very good agreement with the energy deposition for this kind of wire (366mJ, [2]), considering 138mJ in addition was needed to vaporize the wire material. It also led us to believe that those wires were fully vaporized.

Mass density can also be obtained with interferometry if the assumption that no plasma exists in the neutral region can be made. Figure 6 shows the density distributions calculated both with radiograph and interferogram at similar times. The polarizability used for converting the refractive index to density was the theoretical number for free Ag atoms, found in the CRC handbook [5]. The two curves fit pretty well, which again proved that the wire was fully vaporized. Deviations between the two curves probably indicate the mixture of plasma and neutrals. In principal, the actual plasma and neutral densities could be resolved with two photon interferometry.

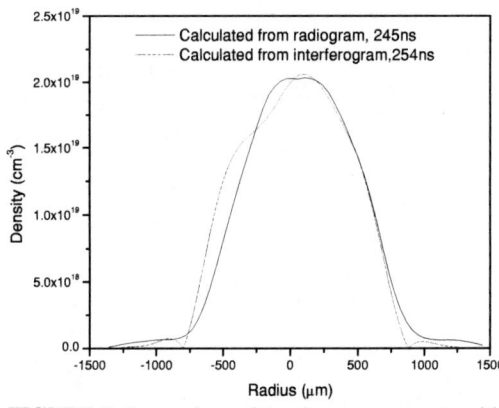

FIGURE 6. Comparison of density measurements with radiograms and interferograms. 25μm Ag coated with 1μm Polyester, shot 1706.

The last thing to emphasize is that, for most kinds of wires, the energy deposition was quite low compared to the vaporization energy. Even with more than 100% of the vaporization energy deposited, full vaporization is not guaranteed since part of energy was converted to the kinetic energy for axial expansion. Under those conditions, the measured polarizabilities were smaller as compared to the polarizabilities of a pure neutral vapor. We therefore believe that the wire was in a phase of a mixed super heated liquid and vapor, and the liquid droplets contributed much less polarizability than free atoms.

CONCLUSION

Single wire explosions with ~1kA/wire were studied. Plasma shells around the wire core with density ~6×10^{16}cm^{-3} were observed. This is consistent with the idea that current is transferred to this plasma shell at the time of voltage collapse. The behavior of the wire expansion was also investigated. The expansion was observed to be faster at larger radii while retaining a constant axial flux. The temperature of wire particles increases with radius from ~10eV to 1keV. We also claim that information on the phase state of expanded wire can be determined by measuring the polarizabilities.

ACKNOWLEDGMENTS

This research was supported by Sandia contract BD-9356 and DOE contract DE-FG03-98DP00217.

REFERENCES

1. Deeney, C., Douglas, M. R., Spielman, R. B., et al., *Phys. Rev. Letters* 81, 4883-4886 (1998).
2. Sinars, D. B., Hu, M., Chandler, K. M., *Phys. Plasmas* 8, 216-230. (2001).
3. Pikuz, S. A., Shelkovenko, T. A., Mingaleev, A. R., et al., *Phys. Plasmas* 6, 4272-4283 (1999).
4. Pikuz, S. A., Romanova, V. M., et al. *Rev. Sci. Instruments* 72, 1098-1100 (2001).
5. *CRC handbook of Chemistry and Physics*, edited by Lide, D. R., 10-201 (1997).

Layering of an Annular Z-pinch Sheath in the Presence of an Axial Magnetic Field

Stanislav A. Chaikovsky and Aleksei Yu. Labetsky

Institute of High Current Electronics, 4 Akademichesky Ave., Tomsk 634055, Russia

Abstract. Optimization of the neon K-shell radiation yield and power was performed in double-shell implosion experiments with and without an axial magnetic field. The experiments were carried out on the IMRI-5 generator, which provides a short-circuit peak current of 470 kA with a risetime of 430 ns. This paper presents experimental results which concern the z-pinch implosion time shortening in the presence of an axial magnetic field, as observed in experiments.

INTRODUCTION

Double shell neon gas puff experiments with and without axial magnetic field on the IMRI-5 generator at the current level of 400 kA and the current rise time of 430 ns were performed [1]. In course of the experiments it was found that z-pinch implosion times in shots with an axial magnetic field are 10-15% lower than that in the shots without the magnetic field (Fig. 1).

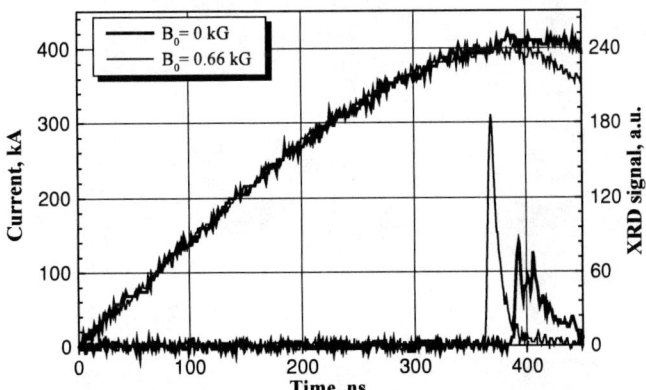

FIGURE 1. Load current and XRD traces of the double gas puff implosions with and without an axial magnetic field.

This paper is aimed to present additional experimental results, which concern the effect observed.

EXPERIMENTAL

The experiments were carried out on the IMRI-5 generator, which provides a short-circuit peak current of 470 kA with a risetime of 430 ns. The load region was the same

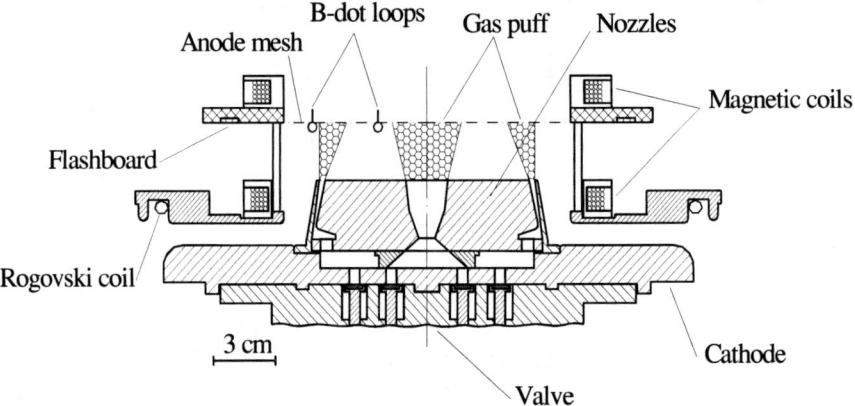

FIGURE 2. Schematic drawing of the IMRI-5 generator load unit with 100/20 nozzle configuration.

as in the experiments [1]. Neon gas puff z-pinches were formed using a fast valve and interchangeable nozzles. Four sets of nozzles with outer shell mean diameters of 4.4, 6.0, 8.0, and 10 cm were used. The inner shell was a solid fill with a nozzle exit diameter 1/5 of the outer shell mean diameter. The distance between the anode and the cathode was either 1.7 or 2.2 cm (for the nozzle with an outer shell initial diameter of 10 cm). A schematic drawing of the IMRI-5 generator load unit with a 10-cm diameter outer nozzle is shown in Fig. 2.

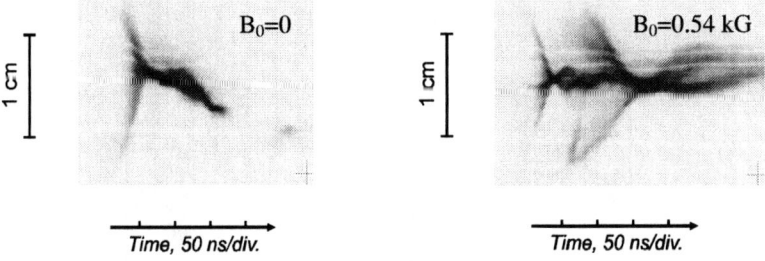

FIGURE 3. Streak camera pictures obtained for double shell gas puff with outer shell initial diameter of 4.4 cm in shots with and without an axial magnetic field.

The z-pinch current was measured by a Rogowski coil. B-dot probes were used to observe current sheath dynamics during implosions. One of the probes was located at a diameter a little greater than the initial outer shell diameter and another was placed well inside the outer gas puff (Fig. 2). The initial axial magnetic field was varied in the range 0.25–2.5 kG.

The onset of the z-pinch implosion was fixed by the dip in the current and B-dot traces and by the maximum of XRD's signals. The implosion dynamics was observed with a visible light streak camera. The streak camera triggering was synchronized with the current and XRD signals.

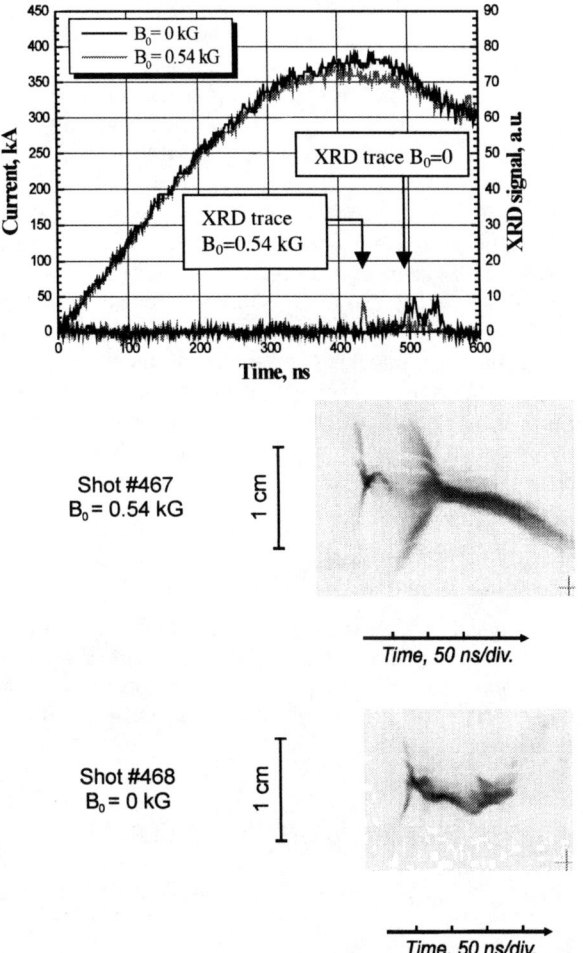

FIGURE 4. Streak camera pictures of single gas puff implosions with and without an axial magnetic field synchronized with current and XRD traces. The initial shell diameter was 4.4 cm

RESULTS AND DISCUSSION

Typical current and XRD traces in shots without and with a magnetic field are shown in Fig. 1. For all nozzle sets, the onset of the z-pinch stagnation for the shots with B_z field was 10-15% earlier than for the shots without a magnetic field.

In the streak photos, it can be seen that in the course of an implosion with B_z one more imploding shell is formed (Fig.3). This second shell stagnates a few tens of nanoseconds later than the first one. Such a situation was observed only for shots with B_z and initial outer shell diameters of 4.4 and 6 cm.

It is notable that this effect is observed even if the inner shell is absent (Fig. 4). Thus, the layering of the outer shell occurs. Streak camera images synchronized with current and XRD traces are shown in Fig. 4 for the implosion of only the outer shell with an initial diameter of 4.4 cm.

For an initial magnetic field of 0.54 kG, two implosions were observed in the streak images (Fig. 4): the first implosion occurred at the 440th nanosecond and the second one took place 100 - 125 ns later, that is, in the process of implosion, the imploding gas puff splitted into two shells. The K-shell radiation pulse corresponds to the first implosion. In the absence of an axial magnetic field, the streak camera records only one implosion with a corresponding burst of K-shell radiation at the 500 ns.

At present, we do not have an explanation of the outer shell layering effect. It should be noted that this effect is perhaps responsible for the efficient K-shell x-ray production in the presence of an axial magnetic field. Actually, application of a magnetic field allows one to increase the neon K-shell radiation power and reduce the shot-to-shot K-shell yield variations [1].

SUMMARY

Layering of an annular z-pinch shell was observed in the presence of an axial magnetic field for shell initial diameters of 4.4 and 6 cm and implosion times of 450-500 ns. When imploding, the initial gas puff shell separates into two sheaths. The first "fast" sheath stagnates and emits a K-shell x-ray pulse 50-70 ns earlier than a z-pinch does without an axial magnetic field.

ACKNOWLEDGMENTS

We are very grateful to team of Pulse Technique Department of our institute for technical support and useful discussion at construction of IMRI-5 generator. Work is supported by Contract DTRA01-01-P-0102.

REFERENCES

1. Chaikovsky S. et al., *see a companion paper in this volume*.

Dense Transient Plasmas Driven by a Mega-Ampere Device in the Chilean Nuclear Energy Commission

Leopoldo Soto[1], Walter Kies[2], Gustavo Sylvester[1], Günter Ziethen[2], Marcelo Zambra[1], José Moreno[1], Patricio Silva[1], Lipo Birstein[1], Richard M. Muñoz[1], and Renato Saavedra[3]

1 Comisión Chilena de Energía Nuclear, Casilla 188-D, Santiago, Chile
2 Heinrich-Heine-Universität, Düsseldorf, Germany
3 Universidad de Concepción, Chile

Abstract. Recently, the pulse power generator SPEED 2, a medium energy and large current device (187kJ, 4MA, 300kV, 400ns, dI/dt ~ 10^{13} A/s), has been transferred from the Düsseldorf University to the Plasma Physics Group of the *Comisión Chilena de Energía Nuclear (CCHEN)*. The SPEED 2 arrived at CCHEN in May 2001 and was in operation in January 2002, being the most powerful device for dense transient plasma in the Southern Hemisphere. Experiments in different Z-pinch configurations using the SPEED 2 generator will be carried out at CCHEN in the future. Possible objectives using the SPEED 2 devices are discussed in this work: a) Neutron flux characteristics from plasma focus discharges operating in D_2 (with temporal and spatial resolution) correlated to discharge parameters, plasma dynamics and instabilities. Particular investigation of the effect of insulator surface preparation and conditioning on pinch behavior and neutron yield, b) High brightness and soft X- ray radiation from transient electrical discharges, especially in wire arrays, and c) Magnetic confinement in a quasistatic z-pinch at mega ampere peak current. Plasma dynamics and stability in an original quasi-static z-pinch configuration (a gas embedded compressional Z-pinch) and in a combination scheme that use gas puff and plasma focus will be studied at high current using the SPEED 2 generator. Results about SPEED 2 performance, obtained in Chile, are presented.

INTRODUCTION

At present the Plasma Physics and Plasma Technology Group of the *Comisión Chilena de Energía Nuclear (CCHEN)* has the experimental facilities in order to study dense transient discharges in the extremes of: I) energy, tens of joules and hundred of kilojoules, II) current, tens of kiloamperes and some mega-amperes. Both I) and II) in the same time scale, hundred of nanoseconds. On one hand we have in operation a very small plasma focus device (160nF, 25 kV, 50kA, 50J, 200ns rise time, dI/dt~2-3x10^{11}A/s) [1, 2]. On the other hand, recently, the pulse power generator SPEED 2, a medium energy and large current device (4.1µF, 300kV, 4MA, 187 kJ, 400ns rise time, dI/dt~10^{13} A/s) [3], is in operation at CCHEN. The SPEED 2 arrived at the CCHEN in May 2001 from Düsseldorf University, Germany, and it is in operation since January 2002, being the most powerful and energetic device for dense transient plasma in the Southern Hemisphere. Moreover, SPEED 2 is the unique dense plasma transient experiment operating at currents of Mega-amperes in Chile. Simultaneously an intermediate device is at present being constructed (1.25µF, 100kV, 700kA, 250ns rise time, dI/dt~10^{12} A/s). Thus, experimental studies in dense transient plasmas in a wide range of parameters (and in the same range of time) can be developed in our laboratory [4]. In particular experiments in different Z-pinch configurations, at current of hundred of kiloamperes to mega-amperes, using the SPEED 2 generator will be carried out. A device designed 15 years ago will be used for the research of scientific topics relevant today and for the research and development of new ideas.

RESEARCH PROGRAM

Most of the previous works developed in SPEED 2 at Düsseldorf were done in a plasma focus configuration. The Chilean operation has begun implementing and developing diagnostics in a conventional plasma focus configuration. Then, after getting the experimental expertise with SPEED 2, new experiments in a quasi-static Z-pinch, gas puffed plasma focus, and wires array will be developed to extend the device capabilities.

Some specific topics that will be studied using the SPEED 2 at CCHEN are the following:

Instabilities and neutron emission. The origin of the neutron emission from a plasma focus operating in Deuterium is attributed mainly to two mechanism: a) thermonuclear reactions and b) beam-target reactions. The second mechanism is associated to ions acceleration processes in a m=0 instability and the subsequent gap developed in the plasma column, similar to early Z-pinch experiments. Many authors indicate that this would explain the anisotropy observed in spatial distributions of the emitted neutrons. Similar instabilities occur in cryogenic Deuterium fiber discharges, where diagnosed experiments, with current rising to 1MA in 200ns, show that disruptions can occur at late stages, leading to relativistic electron beams, which are associated to hard X-rays burst and neutrons yield. However, this is a controversial area for theory.

SPEED 2 experiments for neutron flux characterization (with temporal and spatial resolution), operating in a plasma focus configuration filled with Deuterium gas, will be carried out. The aim is to correlate discharge parameters, plasma dynamics and instability. It is proposed to attempt to establish an experimental correlation between the spatial anisotropy of the neutron emission and total yield with hard X-ray emission and hot spot formation. Particularly, the interest will also involve a research on the effect of the preparation and conditioning of the insulator surface on pinch behavior and neutron yield.

We will intend to find net results related with neutron pulse emission processes, and plasma properties loci where they are being generated (linked to the hot spot), to establish validity range with theoretical models. The results will be outstanding from the transient dense plasma physics perspective, and they will contribute to establish design criterion of plasma focus optimized devices in order to generate pulsed neutron beams with potential applications in detection and analyze systems.

In addition, to separate the two principal neutron production mechanisms in a Deuterium pinch plasma (thermonuclear and the so-called beam-target), it is proposed to establish a comparison between measurements of neutron spatial distributions in a plasma focus (in which a m=0 instability occurs), and measurements of neutron spatial distribution from a stable z-pinch to a m=0 mode, under similar density and temperature parameters. A quasi-static Z-pinch as the gas embedded compressional z-pinch [5, 6, 7], as described operating in Deuterium with current rising to ~1MA in ~200ns is a candidate to this experimental proposal driven by SPEED2. Also the gas puffed plasma focus [8].

Quasi-static Z-pinch. A quasi-static and stable Z-pinch in order to obtain a dense and hot plasma for thermonuclear fusion applications is a foundational goal for the Z-pinch research.

The dynamics and stability of the double column pinch [14, 15], will be studied at high current in the SPEED 2. The initial impedance of the double column pinch is compatible with the SPEED 2 impedance in order to obtain a maximum current about 1MA or higher. That is not the case for a single fiber, where a maximum current of 700 kA was obtained in previous experiments in Düsseldorf. So, once again, a device designed 15 years ago, can be used for the research and development of new ideas.

The neutron spatial distribution from a gas embedded compressional Z-pinch, operating in Deuterium, will also be measured and compared to the neutron spatial distribution from a plasma focus with a comparable density and temperature.

Z-pinch as intense X-ray sources. X-ray radiation is emitted from high energy density plasma generated by pinching phenomenon. The Z accelerator (11.4 MJ, 20MA, 100ns) at Sandia Natinal Laboratory, USA, is the most powerful X-ray source in the world. A current of 20 MA rising in 100ns over a tungsten wire arrays produce a pinching plasma that emits a X-ray pulse (10ns FWMH) with a peak power of 200 TW. The final goal of that mega-project is to produce inertial confinement fusion irradiating D-T targets using the VUV radiation emmited from the Z-pinch plasma (hohlraum device). Even though at Sandia huge powerful X-ray pulse have been obtained, the related physical mechanisms are yet an open field.

For high-temperature plasmas the bulk of the emission spectrum is distributed in VUV and soft X-ray regions and its application in various fields is being expected. Also, hard X-ray emission is observed during the discharge. The physical mechanism related with these emissions are an open field yet, both experimentally and theoretically. Radiative collapse, and population inversion are interesting phenomena. Radiative collapse occurs if the power radiation losses are greater than the input power on the plasma. A hot (hundred of eV), small size (microns to submicrons sizes) and high density (solid density) plasma is predicted in theory.

PRELIMINARY RESULTS

The Chilean operation of the SPEED 2 device has begun implementing and developing diagnostics in a conventional plasma focus configuration. Discharges in plasma focus mode have been performed at +/-30kV charging voltage in a six step

Marx generator (i.e. 180kV, 100kJ). A peak current greater than 2MA is achieved. Figure 1 shows the corresponding electrical traces. Fig. 1 a) corresponds to a discharge 5.9 mbar deuterium, at this pressure the load is practically the inductance of the central collector and electrodes (~5nH). Figure 1 b) 2.5mbar, at this pressure the pinch phenomena is clearly observed in the peak voltage and in the dip of the dI/dt.

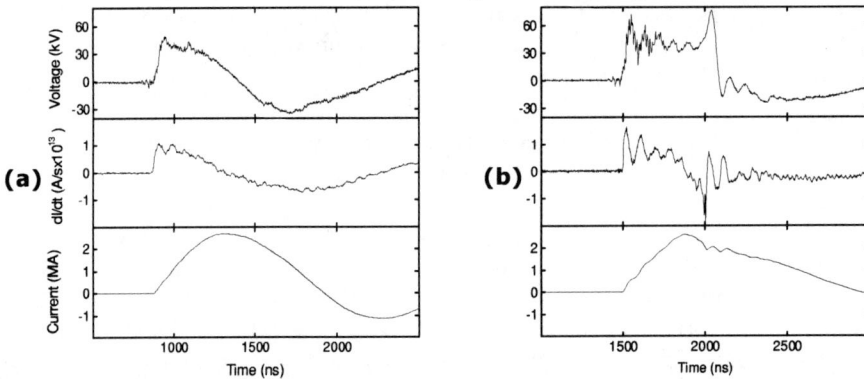

Figure 1. Electrical signals in SPEED 2 generator operating in Chile at CCHEN. Discharges in plasma focus mode have been performed (+/-30kV charging voltage in a six step Marx generator, i.e. 180kV, 100kJ). A peak current greater than 2MA is achieved. a) 5.9 mbar deuterium, b) 2.5mbar, at this pressure the pinch phenomena is clearly observed in the peak voltage and in the dip of the dI/dt.

Time integrated and time resolution neutron detection are now being implementing. The future diagnostics program considers, time integrated and time resolution X-ray images, and PIN diode and particles detectors, X-ray spectroscopy, as well as pulsed interferometry [4, 7].

ACKNOWLEDGEMENTS

This work has been funded by the grant *Cátedra Presidencial en Ciencias* awarded to L. Soto by the Chilean Government, by IAEA and by the grant CCHEN 573.

REFERENCES

1. P. Silva, L. Soto, J. Moreno, G. Sylvester, M. Zambra, L. Altamirano, H. Bruzzone, A. Clausse, and C. Moreno, Rev. Sci. Instrum. **73**, 2583 (2002).
2. L. Soto, P. Silva, J. Moreno, A. Clausse, and W. Kies, "A Very Small Plasma Focus Operating at Tens of Joules", 14th BEAMS and V Dense Z-pinches Conferences (Albuquerque, USA, 2002)
3. G. Decker, W. Kies, M. Mälzig, C. Van Calker, and G. Ziethen, Nucl. Instrum. and Methods, A249, 477 (1986).
4. L. Soto, P. Silva, J. Moreno, M. Zambra, G. Sylvester, A. Esaulov, and L. Altamirano, Brazilian Journal of Physics **32**, 139-154 (2002).
5. L. Soto, H. Chuaqui, M. Favre, and E. Wyndham, Phys. Rev. Lett. **72**, 2891 (1994).
6. L. Soto, H. Chuaqui, M. Favre, R. Saavedra, E. Wyndham, M. Skowronek, P. Romeas, R. Aliaga-Rossel and I. Mitchell, IEEE. Trans. Plasma Science **26**, 1179 (1998).
7. L. Soto, A. Esaulov, J. Moreno, P. Silva, G. Sylvester, M. Zambra, A. Nazarenko, and A. Clausse, Physics of Plasma **8**, 2572 (2001).
8. W. Kies, G. Decker, U. Berntien, Yu. V. Sidelnikov, D. A. Glushkov, K. N. Koshelev, D. M. Simanovskii, and S. V. Bobashev, Plasma Sources Sci. Technol. **9**, 279 (2000).

Laser Initiated Hollow Gas-Embedded Z-Pinch

Cristián Pavez, Hernán Chuaqui, Raúl Aliaga-Rossel, Mario Favre, Ian Mitchell and Edmundo Wyndham

Departamento de Física, Pontificia Universidad Católlica de Chile, Casilla 306, Santiago 22, Chile

Abstract. A preliminary study of a shell gas-embedded Z-pinch initiated by a 8 ns Nd-YAG laser pulse (1.06 μm) is presented. The experiment was carried out on a pulsed power generator (GEPOPU) capable of delivering a peak current of 200 kA with a rise time of 35 ns, 120 ns pulse length. The experiments reported here were carried out in Hydrogen gas at 1/3 atm. Flat electrodes were used with a 10 mm separation, with a 3 mm central hole at the anode to allow the passage of the preionizing laser. The laser was focussed onto the cathode, creating a circular metallic plasma from which an annular preionization is established after the discharge voltage is applied. Different optical configurations were used to obtain an annular focal spot were tested, the results shown here were obtained using a diverging lens with its central part blocked off in combination with a converging lens. The second harmonic of the Nd-YAG laser used to preionize the plasma was used to do the optical diagnostics. A compression from 1.5 to .5 mm of the resulting plasma is observed, no instabilities are observed during the compression phase. Further work is planned to study the stability properties of the discharge after peak compression.

INTRODUCTION

The Z-pinch is the simplest geometry for confinement of plasmas, in which the plasma current generates its own magnetic field for confinement, but it has been observed to be unstable under MHD instabilities [1]. Early experiments of compressional Z-pinches also showed instability to mode $m = 0$ (sausage).

The idea of a gas embedded Z-pinch [2],[3] and the availability of a new class of pulse power generators led to renewed interest in Z-pinch discharges. Experiments carried out at Imperial College in London, demonstrated that the gas embedded Z-pinch is $m = 1$ unstable, and the instability grows from the axis of the discharge. Even in a situation in which there is evidence of compression, in a modified gas embedded Z-pinch, there is evidence that at late times a $m = 1$ is beginning to develop at the axis [4]. Experimental evidence on other Z-pinch geometries, such as plasma on wire [5] indicates that it is possible to obtain a stable discharge. In the latter case an Al plasma is generated around a solid fibre and no evidence of $m = 1$ is observed. In a gas liner pinch there has been extensive work, showing that a double shell configuration has improved stability [6].

Theoretical work on Z-pinch discharges was reviewed in [7]. The conclusion on this paper is that traditional pinches will not be stable, as they fail to satisfy the Kadomtsev criteria. The present work puts forward a possible configuration to avoid both $m=0$ and $m=1$ instabilities. This is achieved by having a gas embedded Z-pinch, where the background gas suppresses the $m=0$ mode, and generating a central hollow discharge which has a high density inner core, which would prevent the growth of the $m=1$ instability. In this paper the formation of the hollow pinch is considered.

EXPERIMENTAL CONFIGURATION

The overall layout of the experiment is shown in Fig. 1. The transmission line and discharge chamber are on the right side of the figure and the laser is on the left. Preionization is achieved using the Nd-YAG fundamental at 1.06 µm, with an 8 ns laser pulse, 200 mJ. The second harmonic output at 532 nm is delayed and used for optical diagnostics. The two polarizations of the laser beam (vertical and horizontal) are separated using polarizing beamsplitters, one of the beams is delayed relative to the other one, in order to obtain two frames of the plasma state at different times. Both beams are sent through the discharge chamber in a direction perpendicular to the axis of the plasma column. The two images are separated by the used of polarizers in front of the CCD cameras. The delay between frames is 20 ns. Interferometry, shadowgraphy and schlieren photography were performed on the experiment. Current is obtained from an integrated single turn Rogowski loop.

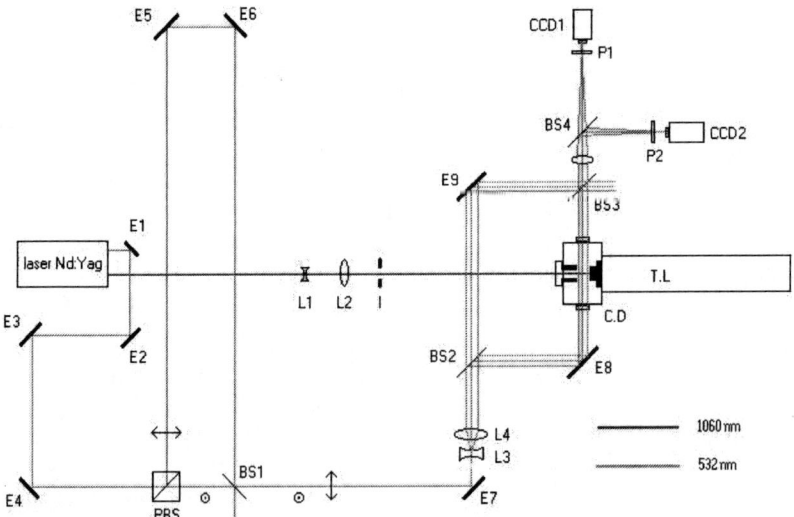

Figure 1. Optical configuration of the experiment showing the trajectory of the diagnostics beam (solid) and the preionizing laser beam (dashed). L1 is the diverging lens with central blockage, M are mirrors, PBS polarizing beamsplitters, BS beamsplitters, DC discharge chamber and TL transmission line.

A set of experiments to determine whether a preionizing circular and uniform plasma could be established were carried out using diffractive and refractive optical components. The diffractive elements proved unsuitable for this purpose due to the low diffraction efficiency, at best 70%. Therefore, all the rest of the experiments were carried out with refractive components, over 90% of the laser energy is available for preionization. The optical setup is shown in Figure 2

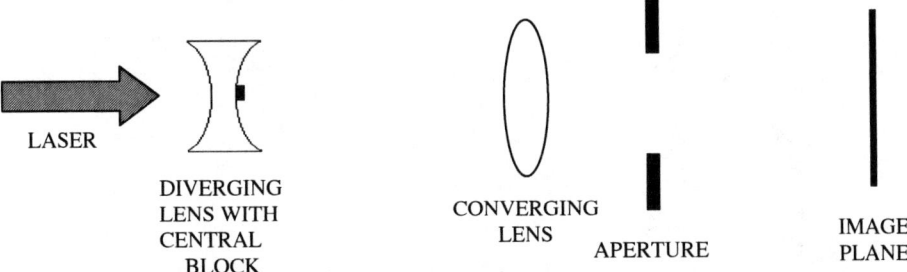

Figure 2. Circular focal spot imaging system used to generate the preionizing metallic plasma at the cathode. A diverging lens with its central area blocked off was used to generate the circular line focusing spot.

In order to asses the uniformity and symmetry of the initial preionizing plasma a set of burn marks on smooth stainless steel plates were taken, showing that a circular and uniform burn mark is indicative of a uniform and symmetrical beam. Additionally, shadowgrams of the plasma formed by the laser were obtained with no electric field applied, to study the laser generated plasma profile. Figure 3 shows one of these shadowgrams which indicate that a uniform ring plasma is formed. The initial ring diameter is 2 mm, as can be seen on the shadowgram. Shadowgrams obtained at earlier times show the same degree of symmetry, but as the plasma size is significantly smaller we are showing an example of an image obtained at 40 ns. From these images it is apparent that the initial preionizing plasma is symmetrical.

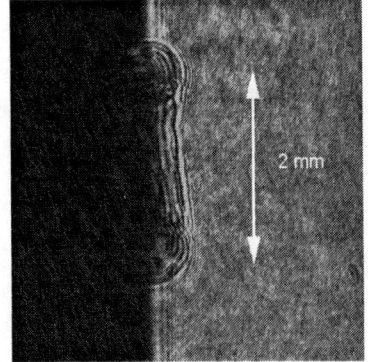

Fig 3. Shadowgram of the preionizing plasma 40 ns after the laser pulse.

The plasma evolution was studied with a double frame interferometry scheme, employing the second harmonic of the laser, while the fundamental was used to generate the preionizing plasma. The optical configuration is shown on Figure 1. The laser pulse was 8 ns long and the time between frames was 20 ns. Figure 4 shows an interferogram sequence of the discharge. At times earlier than 20 ns after the beginning of the discharge the integrated density along the line of sight is too small to allow detection of a fringe shift. The interferogram sequence shown provides a clear indication that a hollow plasma profile is established and compression is observed. At times of maximum compression and

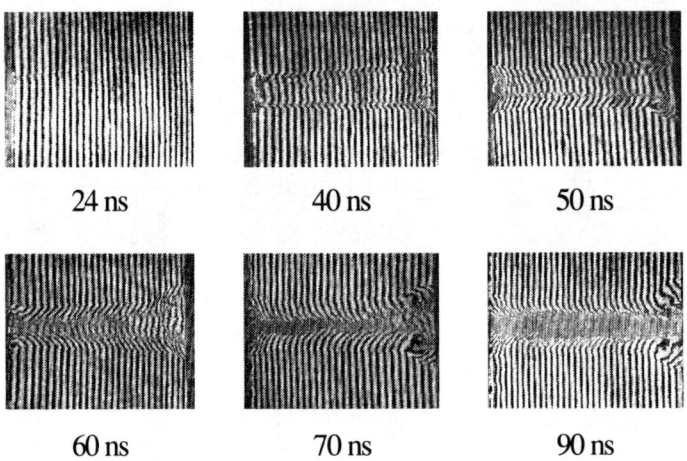

Figure 4. Interferogram sequence obtained from different discharges of the pinch column. At times earlier than shown the phase shift generated by the electron density is not sufficient to cause fringe deflection. From maximum compression onwards it is not possible to follow the evolution of the plasma at small radii.

later it is not possible to observe fringes, probably due to a changing density which results in fringe motion during the laser pulse (~8 ns).

The maximum density observed is $\sim 10^{-19}$ cm^{-3}, indicating that there is compression of the initial gas. In the central region during and after maximum compression it would appear that an $m=1$ instability does develop, which agrees with the fact that at small radii the Kadomtsev criterion is not met. Diagnostics using a shorter laser pulse is planned. Future work requires the modification of the optical configuration in order to be able to introduce a plastic solid core, thus avoiding $m=1$ instabilities.

ACKNOWLEDGEMENTS

This work was carried out under FONDECYT grants 7980023 and 8980011.

REFERENCES

1. Cousins, S. W and. Ware, A. A, *Proc. Phys. Soc.* **B 64**, 161, (1951).
2. Fälthammar, C-G., *Phys. Fluids* **4**, 1145 (1961).
3. Smårs, E. A., *Ark. Fys.* **29**, 97 (1964).
4. Soto, L. Chuaqui, H. Favre, M. and Wyndham, E., *Phys. Rev. Lett.* **72**, 2891 (1994).
5. Wessel, F. J. Etlicher, B. and Choi, P., *Phys. Rev. Lett.* **69**, 3183 (1992).
6. Wrubel, Th. Ahmad, I. Büscher, S. and Kunze, H.-J., *Fourth International Conference on Z-pinches*, Vancouver (1997).
7. Coppins, M., *Fourth International Conference on Z-pinches*, Vancouver (1997).

Discharge Formation in Fast Pulsed Capillary Discharges

Mario Favre[a], Peter Choi[b], Ana María Leñero[a], Francisco Castillo[a], Francisco Susuki[a], Hernán Chuaqui[a], Ian Mitchell[a], and Edmund Wyndham[a]

[a]*Facultad de Física, Pontificia Universidad Católica de Chile, Casilla 306, Santiago 22, Chile*
[b]*EPPRA sas, 16 Avenue du Québec, 706 SILIC, Courtaboeuf 91961, France*

Abstract. We present a study of discharge formation in a fast pulsed capillary discharge operated in Argon or Methane, in a 50 mm long, 1.6 mm internal diameter capillary, at −15 kV applied voltage. A pressure gradient is used along the capillary, with pressures between 0.2 and 1.0 Torr in the cathode region and a pressure one tenth lower in the anode side. The diagnostics include a capacitive probe array, Faraday cup and beam-target scintillator-photomultiplier detectors. It is found that following the emission of electron beams from the hollow cathode region, a fast ionization wave propagates from the cathode towards the anode, with characteristic velocities of the order of 10^6 to 10^7 m/s. The propagation of the ionization front is assisted by the electron beams, which reach a peak current of around 200 mA.

INTRODUCTION.

Fast pulsed capillary discharges are currently been investigated as efficient radiation sources emitting in the vacuum ultraviolet to soft X-ray region [1-3]. We have reported on a new type of pulsed hollow cathode capillary discharge with on-axis initiation [4], which combines the physics of the traditional capillary discharge with that of the transient hollow cathode discharge (THCD)[5], to achieve a high rate of current rise with very short current pulse duration. The e cathode aperture in THCD createsconditions for local ionization in the hollow cathode region (HCR) before any significant ionization takes place in anode-cathode gap. The appearance of the hollow cathode plasma leads to the formation of on-axis electron beams, which play an important role in breakdown formation inside the capillary. For fast pulsed discharges, the capillary is surrounded by a coaxial return conductor, in close contact with the outer surface of the capillary, in order to minimize circuit inductance. This leads to a geometry in which a large portion of the capillary length is shielded from the applied electric field. Breakdown formation in such shielded geometry requires the propagation of an ionization structure, which penetrates the shielded and practically field free region before complete voltage breakdown can occur. Our experimental results indicate that pre-breakdown processes in shielded capillary discharges are characterized by the formation of a fast ionization wave. We present time and space resolved measurements of ionization growth, in correlation with measurements high energy electron beams. A qualitative model based on the hollow cathode effect (HCE) is proposed to explain the initial formation and later time evolution of the observed ionization wave.

EXPERIMENTAL SET-UP

The pulsed capillary discharge device has a low inductance high voltage discharge geometry, obtained by integrating the energy storage medium directly onto the discharge electrodes and surrounding most of the capillary length by a tight conducting shield, which is kept at ground potential and provides the current return path [4]. The capillary is an alumina tube, 50 mm long, 1.6 mm internal diameter, located on axis between the two electrodes. Argon and Methane were used, at pressure 0.2 to 1.0 Torr in the cathode side and one tenth lower in the anode side. The applied voltage was - 15 kV. Plasma formation inside the capillary is measured with a capacitive probe array[6]. A scintillator-photomultiplier (sc-pm) arrangement is used to monitor high energy electron beam activity of characteristic energy above a few keV. A non biased Faraday cup is used to measure the electron beam current. In the operational conditions the capillary shield is at anode potential, and the HCR is located outside the capillary volume, similar to that in conventional THCD [5].

EXPERIMENTAL RESULTS

Figure 1 shows characteristic signals for discharges in Argon and Methane at 600 mTorr. In both cases the e-beams are measured with the sc-pm detector. Signals P1 to P3 correspond to the capacitive probe array, with P1 being closer to the grounded electrode plate. Probe separation is 10 mm. In both cases the e-beam signal precedes the capacitive probe signals. The capacitive probe signal is seen to rise in sequence in the three capacitive probes, starting at P1. A single peak characterizes the signal in all capacitive

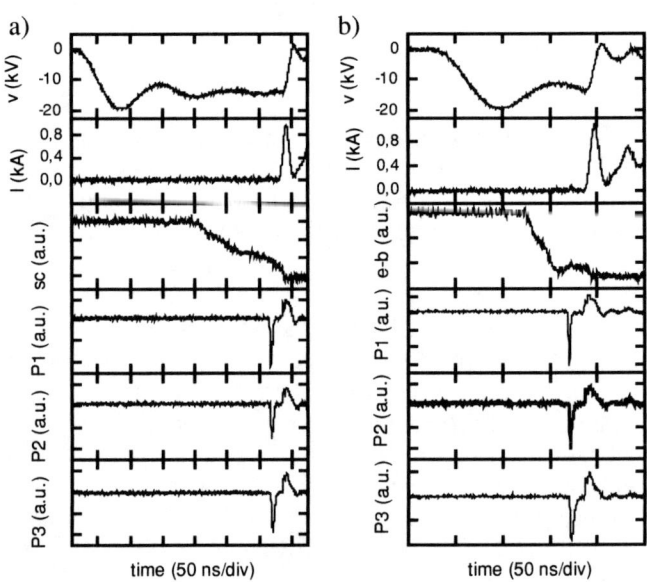

Figure 1: characteristic signals at 600 mTorr. a) Argon, b) Methane.

probes, with the same polarity of the applied voltage pulse.

To assist with the physical interpretation of the probe signals, a numerical integration has been performed. This is shown in figure 2, for both charging polarities. The upper trace, V, is the voltage across the electrodes, and Q1 to Q3, are the numerical integration of signal P1 to P3 in figure 1.

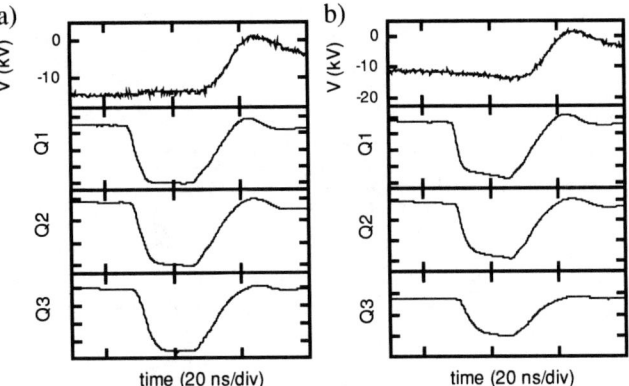

Figure 2: voltage and time integrated capacitive probe signals. a) argon, b) Methane.

Q is proportional to the product between the plasma potential V and the coupling capacitance C of the ring probe relative to a capillary plasma. A rise in Q indicates that a plasma has formed at the position of the corresponding probe and that the external potential has moved up to the position of the probe. In both cases the integrated probe signals shows the displacement of the external potential along the capillary, with the signal collapsing at voltage breakdown. Characteristic speed for the potential to move along the capillary is of the order 10^7 m/s.

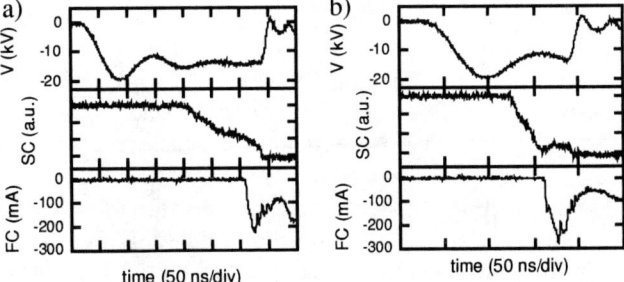

Figure 3: voltage, scintillator and Faraday cup signals, at 600 mTorr. a) Argon, b) Methane.

Time resolved measurements using a non-biased Faraday cup indicate that the time evolution of the e-beams is complex. This is shown in figure 3. The upper trace is the voltage, SC is the e-beam signal measured with the sc-pm detector, and FC is the Faraday cup signal. As the scintillator detector is sensitive only to high energy electrons, the signals in figure

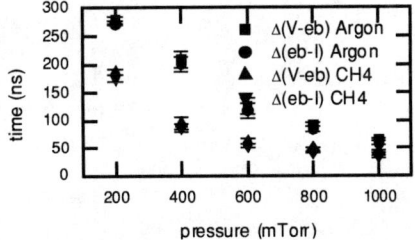

Figure 4: relative time delay between the start of the voltage pulse (V), the onset of the high energy electron beam Δ(eb), and the start of the discharge current (I).

3 indicate that initially the e-beams are of high energy, but low current, but at later times there is a significant increase in the e-beam current. The fact that the sc-pm

signals does not increase significantly indicates that the increase in electron flux is of low characteristic energy.

Measurements of the relative time delay between the start of the voltage pulse and the onset of the high energy e-beam (Δ(V-eb)), and the time delay between the onset of the e-beam and the start of the discharge current (Δ(eb-I)) were performed. The results are shown in figure 4, for Argon and Methane, over the pressure range investigated. As expected, characteristic delays decrease as the pressure increases, due to a reduction in the mean free path for electron impact ionization. Differences in characteristic time delays between voltage rise and electron beam generation in both gases are explained in terms of ionization processes inside the HCR, which are determined by the E/N condition (ratio of local electric field and gas density), and ionization cross section for the particular filling gas.

DISCUSSION AND CONCLUSIONS

The tight aspect ratio of the discharge geometry and the fact that the external capillary shield is kept at ground potential define a geometry in which the capillary region, surrounded by the external shield, is essentially a field free region (FFR). The HCR is located just inside the opening in the cathode electrode. Under these conditions, a normal HCR process, with e-beam assisted ionization, will govern the ionization growth inside the capillary. When the e-beam is formed, it propagates into the long capillary, penetrating the FFR. A nearly free field, weakly conducting plasma is then formed inside the capillary. This weakly ionized plasma can not be detected by the capacitive probe array, due to the near ground local potential condition. The plasma region extends until the weakly conducting plasma bridges the cathode and the outer conducting capillary shield. At this time a fast penetration of the potential into the FFR takes place, which is then detected by the capacitive probe array as a fast e-beam assisted potential wave. The substantial increase in e-beam current, with low characteristic electron energy, plays an important role in increasing the ionization rate at this stage. Electric breakdown occurs after the high voltage potential moves close to the end of the capillary.

ACKNOWLEDGMENTS

This work has been funded by FONDECYT grants 8980011 and 7980024. We thank our late colleague and friend Igor Rutkevich for his fundamental contribution to the understanding of our experimental results.

REFERENCES
[1] M.A. Klosner and W.T. Silfvast, Optics Letters **23** 1609 (1998)
[2] L Juschkin, A Hildebrand and H-J Kunze, Plasma Sources Sci. Technol. **8** 370 (1999)
[3] I. Krisch, P. Choi, J. Larour, M. Favre, J. Rous, and C. Leblanc, Contrib. Plasma Phys. **40**, 135 (2000)
[4] P. Choi and M. Favre, Rev. Sci. Instrum. **69** 3118 (1998)
[5] P. Choi, H. Chuaqui, M. Favre, and E. Wyndham, IEEE Trans. Plasma Sci. **15** 428 (1987)
[6] M. Favre, H. Chuaqui, A.M. Leñero, E. Wyndham, and P. Choi, Rev. Sci. Instrum **72** 2186 (2001)

Experimental Observation in a High Current Capillary Discharge

Edmundo Wyndham[a], Hernán Chuaqui[a], Mario Favre[a], Ian Mitchell[a], Raúl Aliaga-Rossel[a], Peter Choi[b].

[a]*Facultad de Física, Pontificia Universidad Católica de Chile, Casilla 306, Santiago 22, Chile.*
[b]*EPPRA sas, Silic 706, 91961, Cortaboeuf, France.*

Abstract. Experimental observations of the soft X-ray emission from a 5 mm diameter, 60 mm length capillary discharge in an initial vacuum show the effectiveness of transient hollow cathode mechanisms in establishing the initial plasma conditions for a 120 kA Z-pinch. Hollow cathode geometry is used. A laser spark behind the cathode orifice initiates an axial electron beam when an external bias provides a current-limited preionizing current in the capillary. These beams intensify during the first few ns of the main discharge. The Z-pinch dynamics are measured using time-resolved soft X-ray spectroscopy and filtered pinhole photography. These are found to vary according to the rate of rise of the current, 1.5 and 3.0 x 10^{12} A/s, the polarity and level of the prionizing current. A stable pinch is formed with a preionizing current of ~50A as an e-beam from the hollow cathode and with the higher value of di/dt. With higher values of preionizing current the e-beams show diode rather than hollow cathode origin and the stability of the following Z-pinch deteriorates notably. The pinched plasma is predominantly of wall material at ~100 eV.

INTRODUCTION

An Argon Z-pinch formed in a capillary discharge when operated at high current has been shown to be a successful and repeatable soft X-ray lasing medium [1,2]. Even at much lower driver energies of order 1J, a small diameter capillary discharge can produce a ns timescale soft X-ray source [3]. Transient hollow cathode (THCD) generated electron beams have been shown to be a key effect in these discharges [4]. These experiments operate and pinch in the filling gas, generally Argon. THCD effects can be important in a good initial vacuum, but with a laser generating a transient plasma in the hollow cathode (HC) volume [5]. In this work we present work for a physically longer discharge than ref. [5] and for a range of preionizing currents. The rate of rise of the main discharge current may be fixed at 1.5 or 3 x 10^{12} A/s and its effect on the minimum pinch diameter will be discussed. The THCD effects are found to extinguish if the preionizing current exceeds a threshold value, but e-beams are obtained with a different temporal behavior, characteristic of a diode, and the pinch column is markedly less homogenous at all times. Without the preionizing current no soft X-ray emission is obtained. The species observed by means of soft X-ray spectroscopy show that the wall material is the source of the pinch plasma and that the injected Ti plasma from the HC volume is, as expected, not significant.

EXPERIMENTAL DETAILS AND RESULTS

The experiment is performed in a 5mm ID alumina ceramic tube with a length of 6 cm. A laser with an incident energy of 0.2J is focussed onto a Ti metal bar just behind the cathode orifice, which connects the HC volume to the capillary. A small peaking gap isolates the cathode from the coaxial line of the generator so that a preionizing current may be established once the laser plasma is formed until the main current pulse is applied. During this ~10 µs period, in which the preionizing voltage is applied via a resistor used to limit the current, e-beams are extracted from the HC, maintaining a constant potential of ~1 kV across the tube. Pinching is observed for both positive and negative applied voltage, however clear differences of behavior are seen as will be discussed below. The main current pulse may be applied at different values of di/dt, while maintaining the same maximum current of ~120 kA: 1.5 and 3 x 10^{12} A/s. During the initial 5 –10 ns of the applied voltage a greatly intensified e-beam current is observed with a Faraday cup until the current exceeds ~5 kA. From ~20 ns into the current pulse soft X-ray emission from plasma within the capillary is observed.

The plasma is observed by a number of diagnostics. A combined Faraday cup and XRD observes in a qualitative way the e-beams and soft X-rays at a distance of ~15 cm from the anode. Filtered Si planar X-ray diodes resolve the emission temporally. A four-frame 5 ns gated MCP on axis provide spatial and temporal resolution of the soft X-ray emission. Either a series of pinholes, filtered into five different energy bands, or a Rowland circle grazing incidence spectrometer was used to register emission from ~30-300 Å.

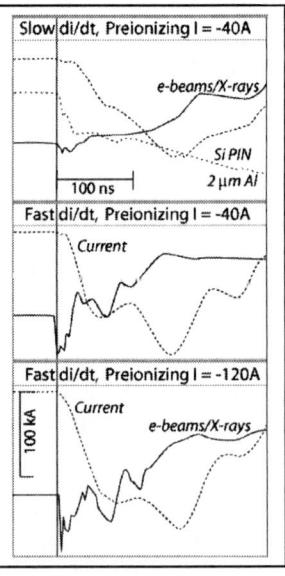

In figure 1 the e-beams/X-ray behavior is shown together with the main current. The two different regimes of the generator di/dt are also compared with two values of the prionizing current, I_P. For the lower value, the e-beams occur while the discharge holds off the applied voltage during the first 10 ns of the applied voltage. In contrast at I_P = -120 A, the e-beams occur immediately on the application of the voltage. Also repeated and intense e-beams are observed during the following 100 ns of the current pulse. With a positive applied prionizing voltage the traces at I_P = 40 A are similar, but no pinching is observed at I_P = 120A.

In figure 2 the pinch dynamics are compared at I_P = -40 A, but at the two values of di/dt. The upper two sets of images in both cases are from the same shot. The filter sets are different because the emission intensity and the temperature are both lower for the lower value of di/dt. This is easily seen on comparing the Mylar and

FIGURE 1. The main discharge current is shown under the two values of the di/dt used. The combined Faraday cup/XRD indicate the e-beams and soft X-rays. The behavior for two values of the prionizing current are shown.

polycarbonate images. The pinch at 37 and 57 ns is seen to be much more compact at higher di/dt. The pinch is better centered under these conditions. Taking into account the pinhole diameters, the projected pinch diameter in the Ti filter, which detects the highest ionization states of Al present (Al X and XI), is approximately 1.0 mm. While the polycarbonate/Ag filter, which passes Al VII to X ionization stage emission, projects a diameter of about 2 mm, This filter indicates a less regular structure characteristic of an instability present in the column. At 70 ns the plasma is expanding at the higher di/dt, but is emitting strongly close to 50 Å in the slow di/dt case.

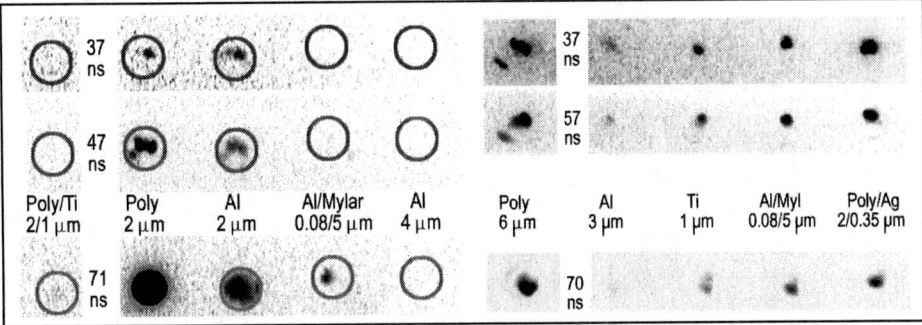

FIGURE 2. Pinhole images using the indicated filters. Left at low di/dt and right high di/dt. Times are measured with respect to the fiducial line in Fig. 1. The preionizing current in both cases is –40 A.

The behavior of the pinch at high di/dt is shown for two cases of I_P. In the left hand group the I_P = -120 A. The pinch is very unstable, occupying most of section of the capillary, and appears to have hot spots. There is appreciable emission in the Ti filter from the central part of the tube projection, and there is appreciable diffuse emision in the Ni filter pass-band, probably from the dominant O VII line at 21 Å. The right hand group of images show a very unstable pinch when a positive preionizing current of 40 A is applied. There appears to be the formation of a series of hot spots. The temperature of these is similar to the left hand group.

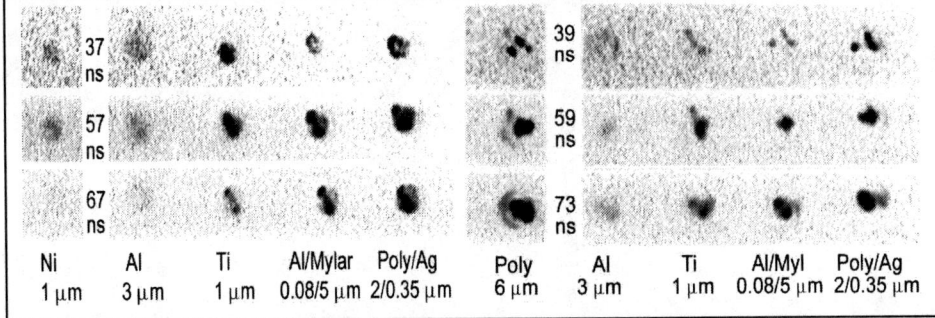

FIGURE 3. Pinhole images using the indicated filters. Both sets are at high di/dt. Times are measured with respect to the fiducial line in Fig. 1. The preionizing current is (left is) –120 A and right –40 A.

Information of the species present in the whole section of the discharge at different times is shown in figure 4. These spectra are taken at I_P = –40 A and at the higher value of di/dt. The sensitivity of the grating and the spill-over from the zero order become noticeable below 40 Å, so some of the lines contributing to the Ni and Al filter images are not seen with this diagnostic. All the species present are those of alumina, indicating that the greater part of the plasma is from ablated wall material. The coexistence of a wide range of species at the same time, for example at 94 ns Al VII to XI ionization stages are seen suggests a rather inhomogeneous plasma, possibly with hot spots in the necks of the instabilities. From the abundance of the different species the plasma attains a temperature of approximately 80 eV between 40 and 60 ns.

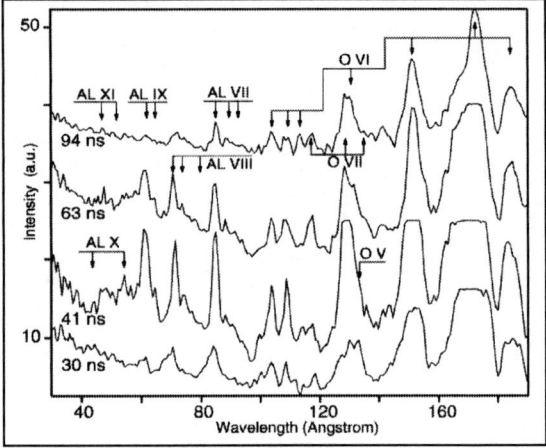

FIGURE 4. The soft X-ray spectra at different times in the discharge.

DISCUSSION

The usual lower operating pressure operating range for a THCD discharge is about 100 mT. However essential characteristics of the discharge have been reproduced here by establishing a transient plasma in the presence of an electric field which penetrates the HC through the cathode orifice. The e-beam generated in the THCD discharge has been shown to be well collimated and is clearly effective at ablating plasma from the ceramic walls. Other work in well-formed Argon capillary discharges [6] at pressures of around 500 mT, while not explicitly referring to THCD effects, has found that a prionizing current of ~50 A to be optimum when applied for similar times to the present work. The higher level saturates the HC processes and the diode-like e-beams may well be far less collimated. The uniformity of the pinch with negative preionizing and high di/dt is quite good. If an effective means of injecting a different species than the wall material, it may be possible to establish longer discharges in a variety of species. The consideration of the THCD processes discussed here may be relevant in defining initial conditions in capillary discharges optimized for soft X-ray lasing.

REFERENCES[1]

1. Rocca J.J., Shylapsev V., Tomasel F.G., Cortazar, O.D., *Phys. Rev. Letters* **74**, 2192-2195, (1994).
2. Rocca J.J., *Rev. Sci. Instrum.* **70**, 3799-3827, (1999).
3. Choi P., Favre M., *Rev. Sci. Instrum.* **69**, 3118-3122, (1998).
4. Favre M., Chuaqui H., Wyndham E., Choi P., *IEEE Trans. Plasma Sci.*, **20**, 53-56, (1992).
5. Wyndham E., Aliaga-Rossel R., Chuaqui H., Favre M., Mitchell I., Choi P., *IEEE Trans. Plasma Sci.*, **30**(2), (2002).
6. Ben-Kish A., Shuker M., Nemirovsky R., Fisher A., Ron A., *Phys Rev. Lett.* **87**(1), (2001)

---

[1] The work has been funded by the chilean government research fund, FONDECYT #8980011.

Properties of Hot-Spots in Plasma Focus Discharges Operating in Hydrogen-Gas Mixtures

Patricio Silva[a] and Mario Favre[b]

[a]*Comisión Chilena de Energía Nuclear, Casilla 188 D, Santiago, Chile*
[b]*Facultad de Física, Pontificia Universidad Católica de Chile, Casilla 306, Santiago 22, Chile*

Abstract. We have investigated the properties of hotspots formed in low energy Plasma Focus discharges operating in Hydrogen-Argon mixtures, at 140 kA current level. The results show that best conditions for reproducible and localized hotspot formation are obtained by adjusting the base pressure in such a way that the mass load allows the time of first radial collapse to coincide with peak current. When the Plasma Focus is operated with 20% Argon content, rather uniform hotspots, of 115 µm characteristic size and 300 eV characteristic temperature, are produced with a better than 80% reproducibility in their axial localization. A significant fraction of the radiation is emitted in the 3.2 to 3.88 keV region, corresponding to K_α emission from highly ionized Argon.

INTRODUCTION

The Plasma Focus (PF) device is a known source of dense transient high temperature plasmas. The X-ray emission is associated with the formation of an approximately cylindrical plasma column, with some brighter embedded structures: the so called hotspots, plasma points or micropinches [1-3]. Typical size of the hotspots varies from tens to hundreds of micrometers, with chacteristic temperatures in the 100 to 1000 eV range, higher than that of the surrounding plasma column [4]. These features have stimulated experimental research in order to characterize the PF as a suitable radiation source emitting in extreme ultraviolet (EUV) to soft X-ray region. The main difficulties in attaining a reliable PF based radiation source have been found in the lack of spatial reproducibility of the hot-spot formation and in the shot to shot variation of the hot-spots characteristic temperature. In this work we have investigated the parameter conditions that result in a more reproducible generation of hotspots, in terms of size, temperature and spatial distribution.

EXPERIMENTAL APPARATUS

The experiments have been performed in PFP-I, a Mather type PF. Typical current pulse at 20 kV charge is 140 kA peak, with 2 µs quarter period. Further details about the PF device have been published elsewhere [4]. The experiments were performed in mixtures of Hydrogen and Argon, in different ratios, ranging from 5 to 20% Argon. A filtered multi-pinhole and Slit-Wire X-ray camera [5] is used for time integrated, space resolved observations of the plasma emission. The experimental set-up included up to six pin-holes of 100 and 200 µm diameter which are used in pairs with six 100 µm slits. Thin wires of 25 to 250 µm diameter were used to measure the hotspot sizes. Time integrated X-ray images were recorded in Ilford HP5 film. Matched pairs of filters of equal mass density are used to look at different spectral windows within the 0.7-5.0 keV region. This spectral region includes Argon lines from highly ionized states, and characteristic K_α emission from the anode electrode. Up to six different filters were used in the pin-hole Slit-Wire camera combination: 12 and 16 µm Beryllium, 15 µm Aluminium, 17 µm Mylar, 12.5 µm Titanium, and 3 µm Silver plus 24 µm Mylar. An X-ray code, XRAYFIL [6], is used to analyze the X-ray emission from the focus plasma. The plasma temperature is inferred from the experimental data by comparing the measured ratio of signal, optical density or photon flux, through the different filters with the calculated values. Filtered Silicon PIN diodes (BPX65) of 1 mm^2 sensible area per diode, which are sensitive to characteristic radiation above 1.91 keV, with a lower energy window around 1 keV for the case of an Aluminium filter, are used to monitor the time evolution of X-ray emission from the focus plasma.

EXPERIMENTAL RESULTS

The focus design is optimized to obtain strong focusing at peak current with a base pressure of 176 Pa in pure Hydrogen, which corresponds to a mass density of $0.15 \cdot 10^{-4}$ µg/cm^3. To keep a constant mass load during the axial phase of the focus discharge, the base pressure for optimal focus with constant mass load is 91, 61, and 37 Pa, for 5, 10 and 20% Argon mix, respectively.

A representative time integrated single shot slit-wire and pinhole images for 20% Argon mixing is shown in figure 1. Six wires of different diameters are used across each slit to produce a shadow in the X-ray image. The different wire shadows are labeled in the figures as follows: 1) 250 µm Silver, 2) 160 µm Tin-Copper alloy, 3) 80 µm Tungsten, 4) 50 µm Tin-Copper alloy, 5) 25 µm Tungsten, and 6) 100 µm Copper. A well defined plasma column with an embedded hotspot is observed. Anode emission is seen at the bottom of each image. A qualitative assessment of the pinhole and slit-wire images indicates that hot-spots are much brighter at the higher Argon content. In fact, at higher Argon mixing clear images are obtained with harder filters, as Aluminium or Titanium, whereas at lower Argon

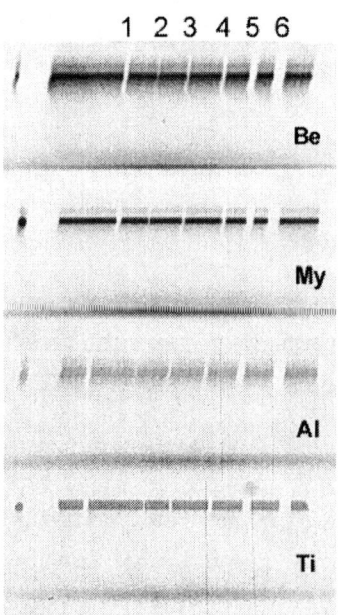

Figure 1: pinhole and slit-wire images obtained with 20% Argon mixing

content, only softer filters, such as Beryllium or Mylar produce clear enough images. Images through the harder Silver/Mylar filter were obtained only at the higher Argon content, but the quality was not enough to be used quantitatively. If the operating pressure is increased with respect to the optimal value for a given Argon content, even though bright spots still are formed, a less uniform plasma develops and the hotspots are not aligned in straight column.

The characteristic size and position of the hotspots is determined from the slit-wire images. Results for 20% Argon contents are shown in figure 2. The axial position is measured from the top of the anode electrode. The characteristic hotspot size at the near optimum pressure condition, taken as an average over the different filters used, is around 115 µm, without significant variations over the Argon content investigated. As the characteristic size depends on the particular spectral window associated with each filter, differences in individual hotspot size for different filters at a given condition is a rough indication of temperature non-uniformity across the hotspots. At the near optimal pressure and lower Argon content hotspots form in region extending about 20 mm from the anode, with a tendency for hotspots formed near the anode to be bigger than those formed at higher axial positions. However, at 20% Argon mix and near the optimal pressure condition, figure 2-a shows that the hotspots become highly localized, with an average axial position of 9.7±0.8 mm above the anode. As the pressure is increased keeping the Argon content fixed at 20%, the hotspots become much less localized and the dispersion in their characteristic size increases, as seen in figures 2-b and 2-c. The results show clearly that an optimal condition can be achieved in terms of size variation and axial distribution of hotspots, which in this particular case corresponds to a mixture with 20% Argon at a pressure which corresponds to the optimum mass load for focusing at peak current.

Pinhole images of the hotspots obtained with three filter combinations, 16 µm Beryllium/17 µm

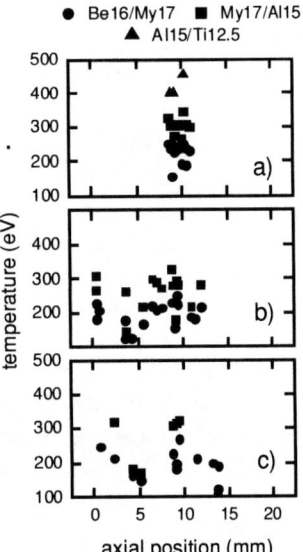

Figure 2: size and position of hotspots over several shots at 20% Argon content. a) 40-44 Pa, b) 56-68, and c) 80-88 Pa.

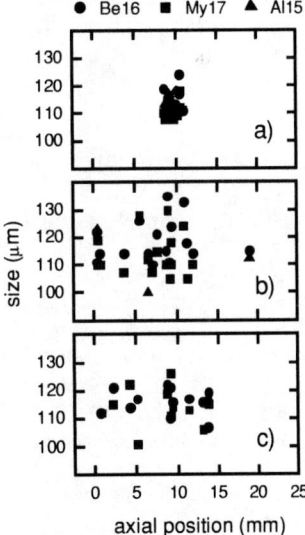

Figure 3: temperature versus position of the hotspots at 20% Argon content. a) 40-44, b) 56-68 Pa, and c) 80-88 Pa.

Mylar, 17 μm Mylar/15 μm Aluminium and 15 μm Aluminum/12.5 μm Titanium, where used in conjunction with the code XRAYFIL to measure the plasma temperature. Results for 20% Argon content are presented in figure 3. At the lower Argon content only the softer filters Beryllium and Mylar produced good enough images, which are suitable for temperature estimations. The average temperature obtained with the Beryllium/Mylar filter ratio increases from 197±47 eV at 5% Argon content to 204±55 eV at 10%, and 221±31 eV at 20%, at the near optimum pressure condition in all cases. The Mylar/Aluminium ratio gives temperatures of 280±94 eV and 302±23 eV at 10% and 20 % Argon respectively, whereas the Aluminium/Titanium ratio gives a characteristic temperature of 418±31 eV at 20% Argon mix. An increase in operational pressure at 20% Argon content reduces the characteristic temperature of the resulting hotspots to around 196 eV and 250 eV, when measured with Beryllium/Mylar and the Mylar/Aluminium filter combinations respectively. The highest characteristic temperature and lower temperature dispersion is obtained at the higher Argon content investigated, and in a pressure range which is close to the optimal mass load condition.

CONCLUSIONS

Our results show that by adjusting the base pressure in such way that the mass load allows the time of first radial collapse to coincide with peak current, best conditions for reproducible and localized hotspots are produced. When the PF is operated in mixtures of Hydrogen and Argon at 140 kA level, the higher Argon content produces rather uniform hotspots, of 115 μm characteristic size and 300 eV temperature, with a better than 80% axial localization. A significant fraction of the radiation is emitted in the 3.2 to 3.88 keV region, corresponding to K_α emission from highly ionized Argon.

ACKNOWLEDGMENTS

P. Silva acknowledges support from a Presidential Chair in Science awarded to L. Soto (CCHEN) by the Government of Chile. Part of this work was funded by FONDECYT grant Nº 8980011.

REFERENCES
[1]   Choi P.,.Wong C.S, and Herold H., *Laser Part. Beams* **7**, 763 (1989).
[2]   Bayley J., Decker G., Kies W., Mälzing M., Müller F., Röwekamp P., Westheide J., and Sidelnikov Y.V., *J. Appl. Phys.* **69** 613 (1991).
[3]   Koshelev K.N., Krauz V.I., Reshetnyak N.G., Salukvadze R.G., Sidelnikov Y.V., and Khautiev E.Y., *J. Phys. D: Appl. Phys.* **21** 1827 (1988).
[4]   Favre M., Silva P., Choi P., Chuaqui H., Dumitrescu C., and Wyndham E, *IEEE Trans. Plasma Sci.* **26** 1154 (1998).
[5]   P. Choi, C. Dumitrescu, E. Wyndham, M. Favre, And H. Chuaqui, *Rev. Sci. Instrum.* **73** 2276 (2002).
[6]   Dumitrescu C. and Choi P., *Proc. 4th Int. Conf. On Dense Z-pinches* (*AIP CP 409*) 491 (1997).

Energy Dissipation in the Run-down Phase of Plasma Focus Discharge

Mehrdad A.M.Kashani and Tetsu Miyamoto

Nihon University, College of Science and Technology, Kanda-Surugadai, Chiyoda-Ku,Tokyo 101, Japan

Abstract. The energy dissipation processes in the Mather-type plasma focus discharge, especially in the run-down phase of discharge, were studied. In the present paper the mechanism of energy dissipation and influence of the dissipation on production of neutron yield in a 7 kJ Mather-type plasma focus device have been investigated by using two types of cathode electrode, (1) Sixteen copper-bars arranged along the envelope of cylinder (bar cathode) and (2) Tubular cathode. The energy dissipation in the run-down phase was much lower in the bar cathode than in the tubular one and as a result, the neutron yield was higher in the bar cathode. The present results clearly show that under the same conditions, except for the cathode structure, the energy dissipation correlates with the contact area and the plasma density near cathode.

INTRODUCTION

One of the most important characteristics of the plasma focus is intensive neutron emission that is produced when deuterium is used as filling gas. There are a lot of experimental parameters required for optimizing neutron yield, such as length of center electrode, radii of electrodes, gas pressure, impurities, insulator material, geometry and so on [1]. The optimum condition for each device is roughly speaking, obtained by pinching the plasma in front of the anode at the phase near the current maximum. When a strong focus occurs, the impedance of the pinch region increases dramatically, and the circuit energy is concentrated there. In order to increase the input energy on the pinch phase, the energy dissipation and energy loss in run-down phase of discharge should be small. The relation between cathode structure and energy dissipation in the run-down phase has not been well studied. However, it is important for complete optimization of the device. In this paper, the energy dissipation mechanism is investigated in a 7 kJ Mather-type plasma focus device mainly the energy dissipation near the cathode by using two types of cathode structure, 1) bar and 2) tubular cathodes.

EXPERIMENTAL SETUP

The present experiments were carried out using a 7 kJ Mather-type plasma focus device. The coaxial anode and cathode electrodes are 35 mm in internal diameter and 80 mm in external diameter, respectively. The anode and cathode electrodes were separated by a Pyrex tube insulator, which is 2 mm in thickness and 26mm in breakdown length. The anode from the edge of the Pyrex tube to the top of the anode and the cathode were 13 cm and 12 cm in length, respectively. The device is operated with two types of cathode electrodes, the bar cathode that consists of 16 bars, each 10 mm in diameter arranged cylindrically and the tubular cathode, which is a cylindrical copper tube with 10 mm in thickness. The current I and dI/dt were measured using electrostatically shielded Rogowski and magnetic probe coil. Using the silver activation counter, we measured in each shot the time integrated neutron flux emitted from focus. The detector was calibrated using an Am neutron source to obtain the absolute value.

EXPERIMENTAL RESULTS

The short-circuited current at the bottom of the cathode and anode I_0 (we call this the zero inductance current) was 500 kA with a quarter cycle time of 1.4 μsec. In Fig.1, the average neutron yield versus deuterium filling pressure is shown for both types of cathodes. The neutron yields were much lower in the tubular cathode than in the bar one. These curves were obtained from more than 200 discharges. System specification and experimental parameters at 25 kV for the bar and tubular cathodes are summarized in Table1.

TABLE 1. System specification and discharge parameters at optimum condition (Charging voltage = 25 kV

Parameter	Bar cathode	Tubular cathode
Bank capacitance C (μ F)	22.5	22.5
Bank energy W (kJ)	7	7
Bank inductance L_0 (nH)	34	34
Bank impedance Z_0 (mΩ)	39	39
Zero inductance current I_0 (kA)	500	500
Short-circuit inductance L_{Sc} (nH)	51	50
Short-circuit current I_{Sc} (kA)	413	439
Initial short-circuit current rise $\dot{I}_{Sc}$ (kA / ns)	0.50	0.57
Short-circuit quarter-period $\tau/4$ (μs)	1.69	1.66
Optimum D_2 gas pressure P_o (torr)	4.5	2.5
Current just before pinch I_p (kA)	390	286
Time to pinch t_p (μs)	1.74	1.60
Neutron yield Y_N (per shot)	5.80×10^8	2.35×10^8

INTERPRETATION OF EXPERIMENTAL RESULTS

The experimental results show that the short-circuited current I_{Sc} (short-circuited at top of the anode and cathode) was lower in the bar cathode and the current just before pinch (I_p) in both types of cathodes decreased naturally in comparison with those in the short-circuited case, because of dissipation. However, the decrease in

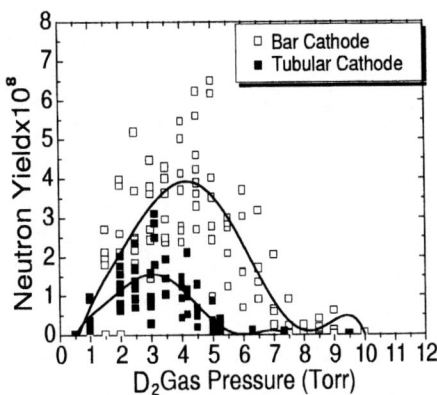

FIGURE 1. Neutron yield as function of deuterium filling pressure for the bar and tubular cathode

the tubular cathode was larger. The current just before pinch I_p and the optimum gas pressure are lower in the tubular cathode and the latter means that the plasma escaped through the separation between adjacent bars in the case of the bar cathode. The ratios of the energy loss in run-down phase to the bank energy (the square I_p/I_{Sc}) are about 0.42 (that corresponds to 2.97 kJ) for tubular cathode and 0.94 (that corresponds to 6.61 kJ) for bar cathode. Surprisingly, considerable fraction of the bank energy was lost before pinch in the tubular cathode case. It will be important to diminish this loss in order to improve the capability of the plasma focus device. The dissipation near the anode does not differ much in both types of cathode, because of the same anode configuration in both cases. Therefore, we consider that the difference resulted from the cathode structure. In the plasma sheet compressed near the cathode, the joule heating and the thermal condition are estimated as:

$$\eta i^2 = \overline{\eta} T^{-3/2} i^2 \sim \overline{\eta} T_{out}^{-3/2} i^2 \qquad (1)$$

$$\left| \frac{1}{r} \frac{\partial}{\partial r} r \kappa \frac{\partial T}{\partial r} \right| \sim \overline{\kappa} n \frac{T_{in}^{7/2} - T_{out}^{7/2}}{\delta^2} \sim \overline{\kappa} n \frac{T_{in}^{7/2}}{\delta^2} \qquad (2)$$

where η $(=\overline{\eta}\, T^{-3/2})$, κ $(=\overline{\kappa}\, n\, T^{5/2})$, T_{in}, T_{out} $(\ll T_{in})$, n and δ are the resistivity, the thermal conductivity, the temperature of plasma sheet at both anode and cathode side, the characteristic density and thickness of the sheet. In the quasi-steady state, where the joule heating is nearly equal to the thermal conduction loss, we obtain the relation:

$$\left(\frac{I\delta}{S}\right)^2 \sim (i\delta)^2 \sim \frac{\overline{\kappa}}{\overline{\eta}} n T_{in}^{7/2} T_{out}^{3/2} \qquad (3)$$

Therefore, the total dissipated energy is estimated as:

$$RI^2 \approx \int \eta i^2 dV \sim \overline{\eta}\frac{\delta}{S}\frac{I^2}{T_{out}^{3/2}} \sim \overline{\kappa}\frac{T_{in}^{7/2}}{\delta}nS \tag{4}$$

This means that the total dissipated energy increases with the contact area and the density, if $T_{in}^{7/2}/\delta$ does not change so much. In the tubular cathode, most swept plasma was compressed near the inner surface of the tubular cathode. This cold, dense plasma contacted to the cathode in wide region, and was heated and cooled constantly. It seems that most of the energy would be lost due to thermal conduction or due to ablation of the surface material of the electrode. On the other hand, in the bar cathode a considerable fraction of this plasma escapes through the separation between adjacent bars, and the plasma moves to the tip of the anode, forming a thin sheet plasma. The contact area and the plasma density near the cathode are smaller, and as a result, the energy dissipation during axial rundown phase decreases. It was considered to be the reason why the plasma current just before pinch was higher in spite of lower short-circuited current in the bar cathode.

CONCLUSION

The experimental results show that the energy dissipation in run-down phase is lower in the bar cathode than in the tubular cathode. This mechanism is interpreted as in the following:

(1) The dissipated energy in run-down phase correlates with the contact area, the plasma density and the temperature gradient near the cathode.

(2) The energy dissipation was lower in the case of the bar cathode. Because both contact area and density are considerably smaller in the bar cathode than in the tubular cathode, the neutron yield was higher in the bar cathode

(3) The important parameter for the neutron yield is the plasma current just before pinch rather than the bank energy.

The results of cathode effect on production of neutron yield show that the bar cathode is much better than the tubular one if experiments are carried out with these devices under the same conditions except for the cathode, but the previous experimental results obtained by many authors and devices do not necessarily seem to support our experimental results. The difference between the neutron yield in the bar and tubular cathodes is masked by the deviation (about an order) of previous experimental results. The reason of this deviation results from the fact that the previous experimental results were obtained in different devices under different optimum conditions with the bar or tubular cathode. It should be noted here, that for sufficient optimization, cathode structure should be investigated more carefully.

REFERENCES

1. Deker, G., Flemming, L., Kaeppeler, H. J., et al, Plasma Phys. **22**, 245 (1980).

FUSION, NEUTRON PRODUCTION AND OTHER APPLICATIONS

X-ray and Neutron Emission from PF-1000 Facility

Marek Scholz[a], Barbara Bienkowska[a], Irena Ivanova-Stanik[a],
Leslaw Karpinski[a], Ryszard Miklaszewski[a], Marian Paduch[a],
Krzysztof Tomaszewski[a], Ewa Zielinska[a], Marek J. Sadowski[b],
Lech Jakubowski[b], Adam Szydlowski[b], Aneta Banaszak[b],
Hellmut Schmidt[c], Pavel Kubes[d], Josef Kravarik[d], Vera Romanova[e]
Silvia Vitulli[f]

[a] *Institute of Plasma Physics and Laser Microfusion; Warsaw, Poland*
[b] *Soltan Institute for Nuclear Studies (IPJ) 05-400 Otwock-Swierk, Poland*
[c] *International Center of Dense Magnetised Plasmas; Warsaw, Poland*
[d] *Czech Technical University, Prague, Czech Republic*
[e] *Physical Lebedev Institute, Russian Academy of Sciences, Moscow, Russia*
[f] *ENEA Centro Ricerche Brasimone, Bologna, Italy*

Abstract: The large plasma focus facility (PF-1000) which has been operated at the Institute of Plasma Physics and laser Microfusion for about five years, was recently modernised and optimised. At present the PF-1000 facility is equipped with electrodes of considerably larger dimensions then those used previously for starting experiments. The both electrodes are now about 600 mm in length. The outer one (cathode), formed of 24 stainless steel rods is 400 mm in diameter. The inner electrode (anode), made of copper in a tubular form and embraced at the base with an alumina insulator, is 230 mm in diameter. The PF-1000 device can operate up to 1 MJ stored energy in the condenser bank.

The main goal of the experiments performed with new electrode assembly was to study the neutron emission characteristics such as: total neutron yield in each single shot, neutron flux angular anisotropy, energy spectra of neutrons emitted up-stream etc. It was also important to find how those characteristics depend on discharge parameters (D_2 – gas filling pressure, voltage, current, and electrical energy of the discharge).
In a dense plasma focus (DPF) devices a pulsed, high current electrical discharges in gases are produced. Development of the plasma sheath leading to formation of the short living, rather dense plasma can be divided into following phases: in the first one, a plasma sheath is generated on the insulator which separates both coaxial electrodes of the gun. In the second phase the sheath is pushed by the Lorentz force along the coaxial cavity. At the end of the gun a radial implosion begins, the current sheath

moves towards the gun axis and forms a plasma column (pinch). The propagation time from the gas breakdown (creation of a current sheath) to the pinch formation is a few microsecond. The pinch phase is much shorter, of the order of few tens of nanoseconds. In general sense, a plasma focus can be considered a power transformer: the energy stored as magnetic energy is abruptly converted into pinch plasma. The properties of the pinch plasma are dominated by the occurrence of macroscopic and microscopic instabilities. In consequence, it leads to generation of powerful beams of electrons, ions, large emission of X-rays and of fusion neutrons when the filling gas is deuterium. There are two mechanisms for fusion reactions in plasma focus devices. The first is a thermal mechanism for which the D-D reactions are produced by the thermal collisions between deuterons in the pinch plasma. The second is the reaction produced by accelerated deuterons colliding with the deuterons in the plasma and/or the neutral gas atoms outside the pinch (beam – target reaction).

Two different kinds of x-ray emissions are observed during pinch evolution. The first is a soft x-ray emitted by electron bremsstrahlung in the thermal energy range from pinch column. The second, emission of a hard x-ray, results from the electron collisions of the inner electrode by electron beams emitted from the pinch.

In this presentation we study, through four frame x-ray camera the presence of regions of intense soft (few keV) x-ray radiation correlated with time resolved measurements of current waveform, neutron, soft and hard x-ray radiation. We also investigate the possible existence of both mechanisms of productions of neutrons (thermal and non-thermal), through measurements neutron pulse at different distances from the electrode outlets, in comparison with time resolved measurements of soft and hard x-ray radiation.

The PF-1000 facility was equipped with the large electrodes [1] and it was operated at an energy level 0.75 MeV in the capacitor bank when it is worked at a voltage of 33 kV. At this voltage the maximum discharge current I reaches 1.8 MA and, using D_2 at 4 hPa as the filling gas, the highest fusion neutron yield achieved so far was about $2.7 \cdot 10^{11}$ neutron/shot. A like in the previous experiment [2,3] in order to measure the total neutron yield and to determine the neutron angular distribution, the use was made of four silver counters. Those counters were placed around the main experimental chamber of the PF-1000 facility, and they were calibrated using an Am-Be neutron source of a precisely defined activity equal $1.5 \cdot 10^7$ neutrons/(s·4π). Time resolved measurements of hard x-ray and neutron pulses were performed by means of four scintillation probes which were located on the symmetry axis of the PF-1000 device, at different distances from the electrode outlets (6.5 m downstream, 40.22 m and two probes at 84,59 m upstream). To improve accuracy of the time resolved measurements, internal time delays (photomultipliers transmission times) of each scintillation probe were determined by means of a known laser pulse, a fast photo diode, and appropriate optical cables. Simultaneously with the neutron measurements, there were also used other diagnostic tools:

- Soft x-ray emission was measured with PIN diode, which was covered with a 10 μm Be foil filter;

- A current derivative was used in order to measure discharge current waveforms;
- Pulsed electron beams were registered by means Cerenkov-type detectors.

In Fig.1, two typical oscilloscope registrations of the signals are shown, displaying from the top to bottom, Fig. 1a dI/dt signal, soft x-rays and neutrons. In the group of traces in Fig. 1b it can be observed neutron signals from different PMT for the same shot. One can easily observe that the soft x-ray pulse recorded with the PIN diode coincides with the signal from dI/dt probe. The first neutron pulse, which were recorded with the scintillation probes, shows a characteristic pre pulse which precedes the main neutron emission. The pre pulse coincides with the signal from soft x-ray emission and minimum of dI/dt signal.

The neutron energy values were roughly estimated from the measured time-of-flight signals (Fig. 1b). Taking as a reference for the assumed t=0 neutron emission time the x-ray pulse it was shown that the pre pulse of neutron signal could have energies about 2.45 MeV (thermal neutrons). The bulk of the recorded neutrons (see Fig. 1,2) could have energies below 2.45 MeV (non thermal neutrons). We observed the same situation for another shot (Fig. 2). It seems probably that this double structure shows its temporal correlation to the instant of the first compression and occurrence of the m=0 instability. This double structure of neutron pulses is in agreement with results also found on the POSEIDON device in Stuttgart [4].

The surprising effect observed for PF-1000 device, is the appearance of two neutron pulses separated by a long time of about 2.5 μs (Fig. 1) in many shots. Two groups of x-ray signals (and dI/dt) about 2.5 μs one after another and accompanying neutrons pulses were registered, too.

Our experimental studies have shown that, simultaneously with the emission of x-ray pulses, intense electron beams also appear (Fig. 3). The pulsed REBs were registered by means Cerenkov-type detectors, equipped with radiators made of rutil crystals, which were placed at a distance of 100 cm from the electrode output, at angles of 68°, and 90° to z-axis. Cerenkov – light signals were transmitted through appropriate optical cables and recorded with fast photomultipliers. The correlation of two signals can be observed which corresponds to two neutron pulses (Fig. 3.)

These results from the experiment performed and discussed in the paper can be summarized as follows:
- PF-1000 operates regularly with comparable neutron yield on the level of 10^{11} neutron per shot;
- The first neutron pulse mostly shows a double structure. In this double structure pulse, the second one is more intense than the first one. The results confirm the coexistence of both mechanism (thermal and beam-target), and show the possible competition between them for the plasma focus neutron yield;
- Two neutron pulses, shifted by about 2.5 μs were recorded It seems that second pulse, is caused by re-breakdown of the discharge current at the end of the electrode region, near plasma column disruption.

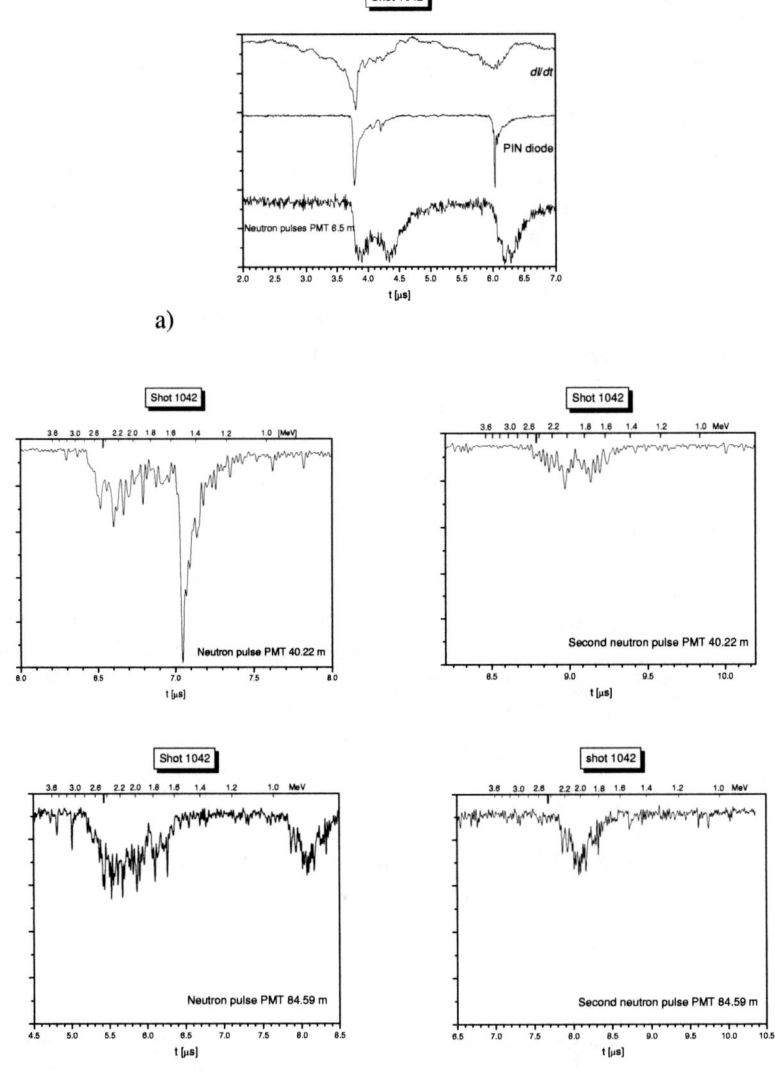

FIGURE 1. a) Soft X-ray, neutron signal (6.5 m PMT covered by 7.5 cm lead) and dI/dt signal. Condenser bank energy: 0.65 MJ, pressure of deuterium: 3.36 mb, maximum current: 1.7 MA, neutron yield: $4.3 \cdot 10^{10}$; b) Neutron signals (40.22 m PMT, 84.59 m PMT)

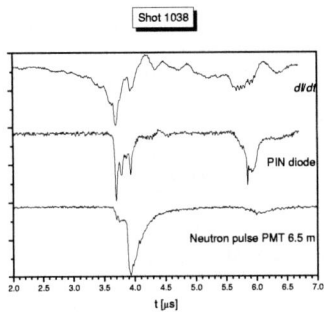

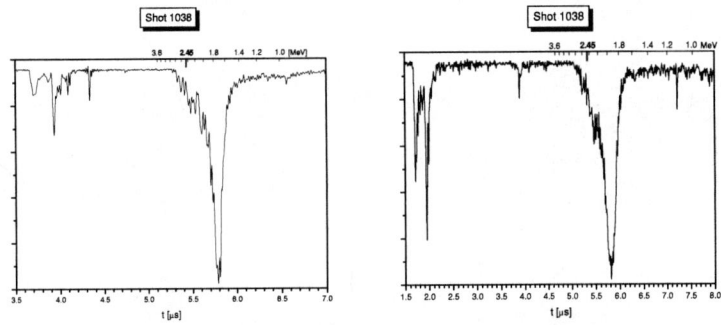

FIGURE 2. Soft X-ray, neutron signal (6.5 m PMT covered by 7.5 cm lead) and dI/dt signal. Condenser bank energy: 0.65 MJ, pressure of deuterium: 3.36 mb, maximum current: 1.6 MA, neutron yield: 10^{11}. Hard X-ray and neutron signal from PMT (40.22 m and 84.59 m downstream)

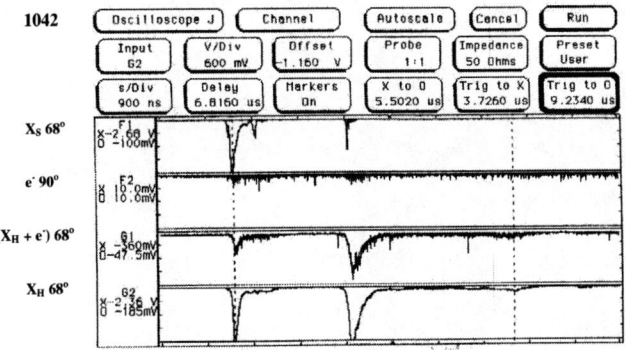

FIGURE 3. Time resolved signals from X-ray and electron detectors (recording electrons of $E_c > 130$ keV, soft X_s of 3-8 keV and hard X_n of 6-30 keV)

ACKNOWLEDGMENTS

This paper has been supported by IAEA Research Contract No. 11941/R0/Regular Budget Fund (RBF) and by the research program INGO No LA 055 "Research in Frame of the International Center on Dense Magnetised Plasmas"

REFERENCES

1. Mather J.W. *Phys. Fluids* **7** (Supplement) (1964) 28 – 34
2. Scholz M., Karpinski L., Paduch M. Et al. Proc. 27th IEEE Int. Conf. Plasma Science ICOPS 2000, New Orlean, USA 1C 09 – 94.
3. Szydlowski A., Scholz M., Paduch M., et al, *Nukleonika* **46** (supplement) (2001) 61.
4. Jager U. And Herold H., *Nucl. Fusion* **27** (2, special issue) (2000) 93.

Neutron Emission Characteristics of a High-Current Plasma Focus: Initial Studies

Bruce L. Freeman[a], John C. Boydston[a], Jim M. Ferguson[a],
Brent Lindeburg[a], Alvin D. Luginbill[a], James C. Rock[a], Teresa E. Tutt[a],
E. Chris Hagen[b] and Lee Ziegler[b]

[a]*Texas A&M University, TAMUS-3133, College Station, Texas 77843-3133 USA*
[b]*Bechtel Nevada, P.O. Box 98521, Las Vegas, Nevada 89193-8521 USA*

Abstract. The Texas A&M University plasma focus machine is operational and is beginning to provide good experimental data. It has its origins in several earlier machines and is located in a former service station building with a shield wall that provides a good geometry for neutron measurements. We are operating in the high pressure mode for a plasma focus, similar to previous efforts in the US. Early neutron measurements are providing some insight for the machine's operation.

INTRODUCTION

The new plasma focus machine (TAMU DPF) at Texas A&M University is operational and has a good geometry configuration for performing neutron time-of-flight diagnostics of the DPF itself and for diagnostic development. Thus, without any requirement for scaling, the TAMU DPF is an appropriate system for providing support for pulsed neutron needs in the $10^{11} - 10^{12}$ n/pulse range from the D(d,n)T reaction.

The capacitor bank to power this plasma focus stores 480 kJ at a maximum charge voltage of 60 kV. The capacitance of the bank is ~268 µF, and its inductance is about 25 nH through the header insulator. The rise time of this system is ~4.3 µs. Machine currents for these data are 1.4 to 1.8 MA.

The capacitors are the 1.85 µF, 60 kV cans from the SHIVA I capacitor bank at Kirtland AFB. The bank is configured into six modules with 24 capacitors each, with a single ATLAS rail-gap for each module. The trigger subsystem is a Maxwell, shorted, charged cable design. The control system for this machine is Labview®-driven and fiberoptically coupled.

Physically, the TAMU DPF is located in a WW-II era base service station building with a characteristic room dimension of ~9 m square. The plasma focus device is located in the center of the room and aligned vertically downward. All time dependent data are measured at 90° to the gun axis. The walls of the building are unfilled concrete block. Located 1.2 m from the outer wall, a 60-cm thick concrete shielding wall extends around three sides of the experimental end of the building. In three locations, steel line-of-sight pipes extend from the inside of the building through the concrete shield wall.

PLASMA FOCUS OPERATION

There are two distinct modes for operating a plasma focus. These are the low and high pressure modes, and they differ in several distinctive ways. Of course, there are mixtures of these modes, depending on the exact conditions for the machine dimensions, the maximum current, and the initial gas pressure before the discharge. The discussion is intended to magnify the extremes of these modes for clarity.

In the low-pressure mode of operation, the rz-pinch becomes current carrier starved much earlier in its history, probably due to the early onset of instabilities. The measured neutron source from this mode of operation includes a region that is roughly 2 cm in diameter and about 20 cm long in a 2-3 MA machine [1]. The neutron distribution tends to be forward directed with as much as a 50% anisotropy. Unfortunately, one frequently observes somewhat erratic and unreproducible behavior. The neutron yield scales as $I^{3.3}$, with saturation effects observed in the Posidon machine [2]. This is the operational mode that many of the European machines have used since the Fracatti [3] and Lemiel [4] systems were functional in the late 1970's.

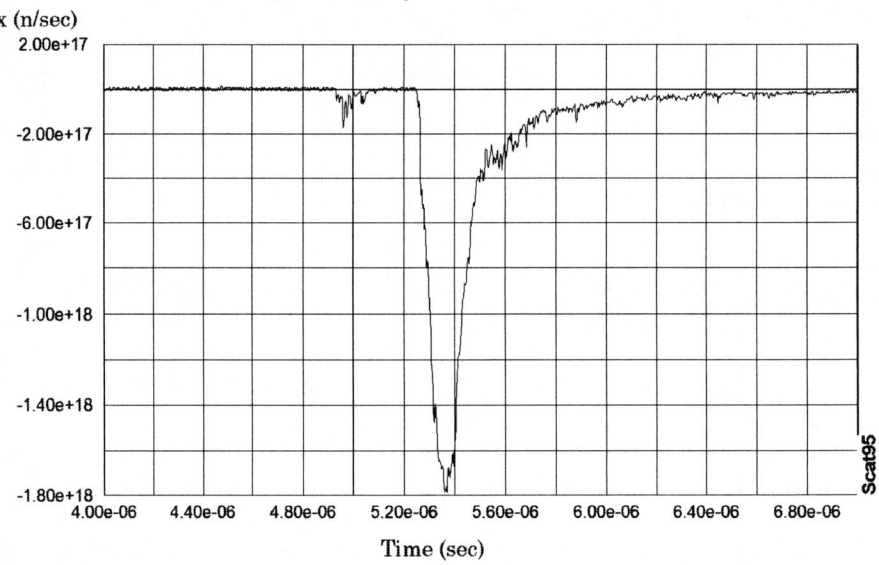

FIGURE 1. Time of flight trace for the 8.85-m line-of-sight for shot 5 on 5/30/2001.

In the high-pressure mode of operation, the rz-pinch still experiences instabilities, but the growth rate appears to be slower due to mass flow through the pinch. The neutron emitting region from a pinch operating in this regime is about 1 to 2 mm diameter and ~1 cm long [5]. The neutron distribution is within about 10% isotropic, and the pinch behavior may be very reproducible. In this mode, the neutron yield scales as I^{4-5}. This is the mode that essentially all of the early machines at Los Alamos used in the 1960's and 1970's[6].

NEUTRON MEASUREMENTS – INITIAL STUDIES

The neutron measurements have included total yield detectors and time resolved measurements. The total yield detection has been performed using the standard LASL silver detectors [7]. The active elements of these units are Geiger tubes. The detection system uses silver activation that enables a quiet counting environment after the machine pulse. The two detectors are located 3.05 m and 4.37 m from the focus center line and at about a 25° back angle from the end of the anode.

The prompt neutron detectors use photomultipliers (PMT) coupled with plastic scintillators. The PMTs are Amperex XP2020 tubes and are 12 stage tubes with gains of about 10^7:1. Linear currents are about 1 ampere. The scintillation material is Pilot U plastic. The combination of the Amperex XP2020 and the Pilot U scintillator is a good match in performance for rise and fall times, ~1.2-1.4 ns. The two detection stations used so far are 8.85 m and 17.4 m from the plasma focus and are at 90° to the machine axis. With few exceptions, both prompt neutron detectors are housed in 5-cm thick lead shields to reduce the (n,n') interaction signal.

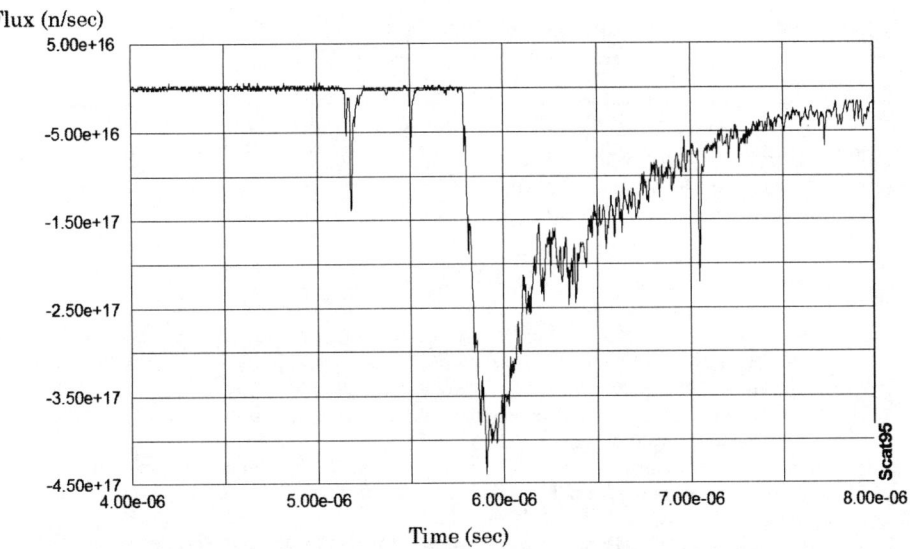

FIGURE 2. Time of flight trace for the 17.4-m line-of-sight for shot 2 on 5/30/2001.

The scintillator thickness for the 8.85-m line-of-sight (LOS) is 0.635 cm, approximately 25% of the mean free path (MFP) for a 2.45 MeV neutron. However, it is only about 0.32% of the MFP of a 14.1 MeV neutron. By contrast, the 17.4-m LOS scintillator is 2.54 cm thick, about a MFP thickness for 2.45 MeV and about 12.7% MFP for 14.1 MeV neutrons. Thus, the shorter line of sight is very insensitive to the higher energy neutrons, while the longer line of sight will have a better sensitivity but a correspondingly lower flux incident on the detector.

Figure 1 shows a trace recorded from the 8.85-m line-of-sight. In this trace, the alignment through the penetration in the shield wall is reasonably good, so the neutron pulse width of about 138 ns is not distorted by stray scattering in the geometry. While some of the hard gamma radiation from the (n,n') interactions is still recorded, the 2.45 MeV neutron pulse is the dominate feature. This trace, with a total neutron yield of about 3.9×10^{11} n/pulse, shows some indication of multiple pulses from the focus but provides no indication for the presence of 14.1 MeV neutrons.

However, another shot with a total neutron pulse of about 2.6×10^{11} n/pulse provides a possible 14.1 MeV neutron pulse from the partial reaction of the tritium produced Fig. 2. Other laboratories have reported the 14.1 MeV daughter neutron pulse for higher yield experiments, so this is not a new result. The timing of the observed pulse would indicate that it originated well after the 2.45 MeV pulse began, which is consistent with the partial reaction of a daughter isotope. Unfortunately, this pulse is entirely absent from the 8.85-m trace. Thus, one cannot provide a convincing analysis from this data to support a firm neutron energy identification.

SUMMARY

The TAMU DPF is operating at low voltages and beginning to mature as a facility that can provide useful neutron pulses for several applications. The presence of a thick concrete shield wall enables good geometry neutron time of flight measurements. We have acquired data to support earlier observations that the full-width-half-maximum neutron pulse from a higher current plasma focus is about 100 ns. As the machine is optimized and a wider array of diagnostic tools are employed, we expect to produce significant new data to support advanced applications.

REFERENCES

1. Bernard, A., et.al., "Study of Neutron Emission and Turbulence in the Focus Experiment With a Time Resolution on the Order of a Nanosecond," *Plasma Physics and Controlled Nuclear Fusion Research*, Vol. 3, 1974, pp. 83-98.
2. Herold, H., et.al., "Comparitive Analysis of Large Plasma Focus Experiments Performed at IPF, Stuttgart and at IPJ, Swierk," Nuclear Fusion, Vol. 29, **4**, 1255-1269, (1989).
3. Rager, J.P., "Observations of Soft X-Ray Emitting Plasma Structures During the Main Neutron Emission of Plasma Foci," Third Topical Conference on *Pulsed High Beta Plasmas*, Edited by D. Evans, Pergamon Press, 1975, pp. 391-394.
4. Bernard, A., et.al., "The Dense Plasma Focus – A High Intensity Neutron Source," Nuclear Instruments and Methods, **145**, 191-218 (1977).
5. Trusillo, S. V., Guzhovskii, B. Ya, Makeev, N. G., and Tsukerman, V. A., "Determination of the Region of Neutron Generation in Bubble Chambers with Plasma Focus," JETP Letters, Vol. 33, No. 3, February 1981, pp. 140-143.
6. Mather, J., "Dense Plasma Focus," in *Methods in Experimental Physics, Plasma Physics*, Vol. 9B, Edited by R.H. Lovberg and H.R. Griem, (Academic Press, New York) 1971, pp. 187-249.
7. Lanter, R. J. and Bannerman, D. E., "The Silver Counter A Detector for Bursts of Neutrons," Los Alamos Scientific Laboratory report LA-3498-MS, July 14, 1966.

A Very Small Plasma Focus Operating at Tens of Joules

Leopoldo Soto[1], Patricio Silva[1], José Moreno[1], Alejandro Clausse[2], and Walter Kies[3]

1 Comisión Chilena de Energía Nuclear, Casilla 188 D, Santiago, Chile
2 PLADEMA- CNEA-CONICET and Universidad Nacional del Centro, 7000 Tandil, Argentina.
3 Heinrich-Heine-Universität, Düsseldorf, Germany

Abstract. A very small plasma focus device has been designed and set-up. The plasma focus operates in the limit of low energy (160nF capacitor bank, 65nH, 25-40kV, ~ 32-100J). The design of the electrode was assisted by a simple model of a Mather plasma focus. A neutron yield of 10^4 - 10^5 is expected from discharges in deuterium. Experiments in H_2 have been performed at pressures from 0.1 to 2 mbar. The diagnostics used in the experiments include current derivative, voltage monitor, and plasma image using a ICCD camera gated at 5 ns. The umbrella-like current sheath running over the end of the coaxial electrodes and the pinch after the radial collapse can be clearly observed in the photographs. The velocity of the radial collapse is of the order of 10^5 m/s. The observations are similar to the results obtained with devices of much higher energies.

INTRODUCTION

Small plasma pinches can reproduce scenarios of high energy densities, radiation emission, and instability phenomena, which trigger fundamental and applied research with relatively low cost [1, 2]. An important part of the recent experimental research on dense plasma pinches is oriented to novel application apart from fusion [3].
Plasma focus attracted the attention of the plasma research community for pulsed radiation applications. The emitted neutrons can be applied to perform radiographs and substance analysis, taking advantage of the penetration and activation properties of neutral radiation. The intense x-ray pulses produced by bremstrahlung radiation from localized electron beams and from hot spots are excellent candidates for radiography of moving or wet objects and for microelectronic lithography. On the other hand, ions and electrons beams from plasma focus are used in surface treatment.
On one hand, had small portable PF devices been available, the added value of the emissions would substantially increase, for a number of nuclear techniques can be produced in wider domains of applications. On the other hand, in spite of all the accumulated research, there are several questions still waiting for an answer, particularly those concerning the sheath formation, insulator conditions and impurities, and those concerning the radiation emission mechanisms involved in the transient plasma processes occurring during the pinch. Experimental research with a plasma focus driven by a capacitor bank of tens of joules is useful for both, basic studies and applications. As first stage of a program to design a repetitive pulsed radiation

generator for industrial applications we constructed a very small plasma focus operating at an energy level of the order of ~ 100 joule or less (160nF capacitor bank, 65nH, 20-35 kV, 32-100J) [1, 4]. The design calculations indicate that neutrons yields of $10^4 - 10^5$ neutrons per shot are expected with discharges in deuterium.

In this article we present optical observations of the plasma motion (radial and axial) in our very small plasma focus operating in hydrogen at 25kV (50J). Single frame image converter camera (5 ns exposure time) was used to obtain plasma images in the visible range.

EXPERIMENTAL DETAILS

The very small plasma focus device constructed at *Comisión Chilena de Energía Nuclear (CCHEN)* is a simple device, a capacitor bank of 160nF is discharged over the coaxial electrode through a spark gap. The device is operates with charging voltages of 20 to 35 kV (160nF, 20-40kV, ~32-128J). A total inductance of 65 nH was measured. The total impedance of the generator was of the order of 0.6Ω. The optimum size of the electrodes was then determined using a theoretical model of plasma focus whose results compare well with experimental data from several different devices [5]. The model predicts $10^4 - 10^5$ neutrons per pulse, using electrodes lengths of 0.5 to 2 cm, internal and external radii of $r_1 = 0.8$ cm and $r_2 = 1$ cm, and Deuterium pressure in the range 0.1 – 1.0mb. Figure 1(a) shows a diagram of the electrodes of the device, which resulted in an hybrid Filipov-Mather configuration, with aspect ratio equal to 0.625. Details of the criteria design and the assembling parts of the device appear in a previously reported work [4].

Several discharges have been performed in hydrogen at different pressures with a charging voltage of 25 kV, ie. 50 J storage in the capacitor bank. Peak current of 39±3 kA is obtained at those conditions. The voltage between the electrodes was measured with a resistive divider and the current derivative was measured with a Rogowskii coil. Figure 1 (b) shows the electrical signals of a shot in hydrogen at 0.47 mbar.

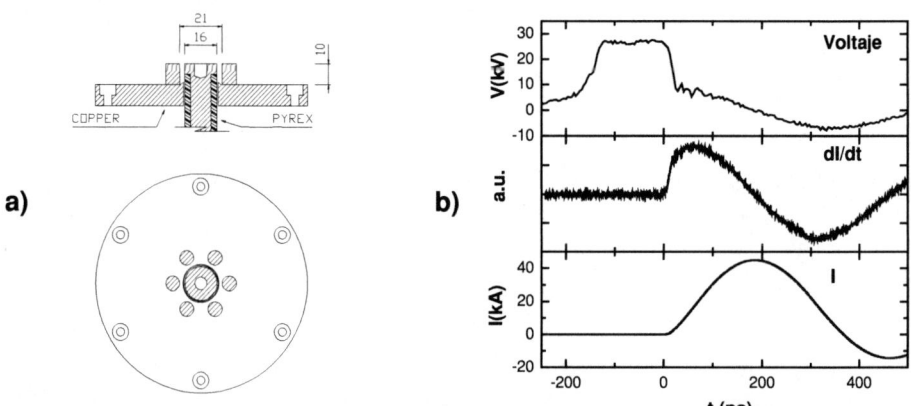

Figure 1. a) Diagram of the electrodes configuration (in mm). Top: side view, Down: top view (the electrodes are shadowed). The anode is the central electrode and the cathode are the six bars around the anode. b) Electrical signals for a shot in Hydrogen. Pressure = 0.47 mbar, charging voltage = 25 kV (50 J energy storage in the capacitor bank). The voltage and current were measured with a 5% of error.

Plasma images in the visible region have been obtained with a 5 ns gated ICCD camera for different time of the discharges and for different Hydrogen pressures. The plasma images have been correlated with the derivative current signals, and the effect of the Hydrogen pressure on the implosion time (pinch timing) was obtained [4]. In the practice discharges operating in Hydrogen at 0.4±0.1 mbar produce compression close to the peak current in the present device. A sequence of the plasma images obtained at 0.47 mbar in hydrogen is presented in Figure 2. The umbrella-like current sheath running over the end of the coaxial electrodes and the pinch after the radial collapse can be clearly observed in the photographs.

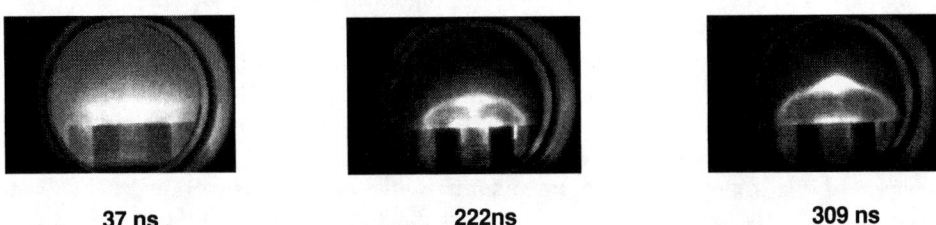

37 ns   **222ns**   **309 ns**

Figure 2. Sequence of the plasma images obtained with a ICCD camera gated at 5ns of exposure time from different discharges in hydrogen at 0.47 mbar.

RESULTS AND DISCUSSION

From the photographs the radius of the plasma column vs. time and the axial plasma motion were obtained. The plasma radius, a, was measured at position z_o=4mm from the anode top. The frontal axial edge of the plasma was measured at distance z_1 from the anode top. Once the plasma column is disrupted, a rear axial edge appears, which was measured at a distance z_2 from the anode top. Figure 3 shows the plasma radius temporal evolution, $a(t)$, at 0.47 mbar in Hydrogen. The radial motion shows three stages: a) a first stage between 40 and 160 ns with a velocity of the order of 2×10^4 m/s, b) the radial collapse (from 160 to 220 ns) with a velocity of the order of 10^5 m/s, and c) plasma column remain apparently stable with a constant radius of the order of 0.8mm by some 40ns, after which the plasma column disrupts. Figure 3 also shows the motion of the frontal and rear axial plasma edge, $z_1(t)$ and $z_2(t)$. The velocity of the frontal axial plasma edge $z_1(t)$ is 4.5×10^3 m/s between 50 to 240 ns. Correlated with the plasma disruptions this velocity increases to 6×10^4 m/s. The velocity of the rear axial plasma edge $z_2(t)$ is of the same order.

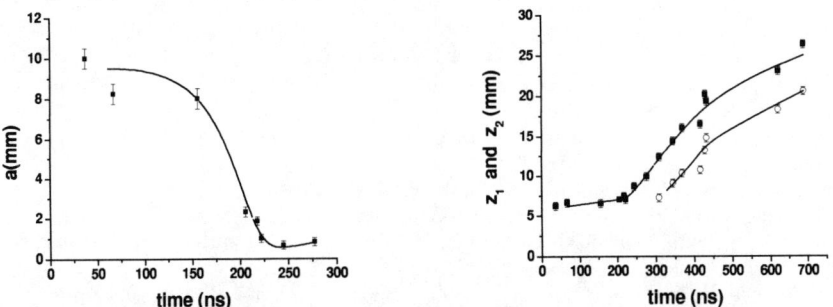

Figure 3. Radial plasma motion a(t) and motion of the frontal and rear axial plasma

The pinch ion density can be estimated from the final radius observations. In a plasma focus in which the radial phase dominates over the sheath formation stage and over the axial run-down, the number of ions per unit length (line density, N) of the plasma pinch is roughly the filling density multiplied by the cross section of the anode. For 0.47 mbar in H_2, considering a fully ionized diatomic gas the filling pressure gives an ion density of $2.54 \times 10^{22} m^{-3}$, so the line density should be about $N \sim 5 \times 10^{18} m^{-1}$. If the plasma is compressed to a 0.6 mm radius cylinder, ion densities of about $4 \times 10^{24} m^{-3}$ are expected. From the measured current and from the estimated line density N, the effective Bennett temperature in the pinch can be estimated, resulting about 48 eV using I=40kA and $N \sim 5 \times 10^{18} m^{-1}$. It is interesting to note that a device of only tens of joules can produce a dense plasma with parameters of density and temperature similar to those obtained in gas embedded Z-pinches operated by pulsed power generators of hundred to kilojoules energies [6]. Thus, very small plasma focus as the reported herein could be a good candidate for basic studies in dense plasmas.

The typical dip in the signal of the current derivative observed in most plasma-focus devices at the moment of the maximum compression does not appear in our experiments. This feature could indicate that the final radius of the pinch does not produce an important change in the inductance load, the pinch impedance being rather small in comparison with the external circuit. A new device in which the inductance has been reduced to 38nH has been constructed. Also the anode radius has been reduced.

The device developed by us is useful both for basic research and applications. The observations presented show the typical plasma dynamics obtained in plasma focus facilities at various energy levels, but at very low energy, thus opening the possibilities of understanding the physics related to the plasma focus working at this unexplored limit of energy. Future work will include plasma diagnostic experiments using time integrated and time resolution neutron and X-ray detection, PIN diode and particles detectors, and pulsed interferometry. Particularly important will be the characterization of the X- ray radiation spectrum.

ACKNOWLEDGEMENTS

This work has been funded by the grant *Cátedra Presidencial en Ciencias* awarded to L. Soto by the Chilean government and Bilateral Agreement: Comisión Nacional de Energía Atómica (Argentina) – Comisión Chilena de Energía Nuclear, (Chile).

REFERENCES

1. L. Soto, A. Esaulov, J. Moreno, P. Silva, G. Sylvester, M. Zambra, A. Nazarenko, and A. Clausse, Physics of Plasma **8**, 2572 (2001).
2. L. Soto, P. Silva, J. Moreno, M. Zambra, G. Sylvester, A. Esaulov, and L. Altamirano, Brazilian Journal of Physics **32**, 139-154 (2002).
3. C. Moreno, M. Vénere, R. Barbuzza, M. Del Fresno, R. Ramos, H. Bruzzone, F. P. J. Gonzales, and A. Clausse, Brazilian Journal of Physics, **32**, 20 (2002).
4. P. Silva, L. Soto, J. Moreno, G. Sylvester, M. Zambra, L. Altamirano, H. Bruzzone, A. Clausse, and C. Moreno, Rev. Sci. Instrum. **73**, 2583 (2002).
5. C. Moreno, H. Bruzzone, J. Martinez and A. Clausse, IEEE Trans. Plasma Sci. **28**, 1735 (2000).
6. L. Soto, H. Chuaqui, M. Favre, and E. Wyndham, Phys. Rev. Lett. **72**, 2891 (1994).

Characteristics and dynamics of a 215-eV dynamic-hohlraum x-ray source on Z

T. W. L. Sanford,[a] D. L. Peterson,[b] R. W. Lemke,[a] R. C. Mock,[a]
G. A. Chandler,[a] J. P. Chittenden,[c] R. E. Chrien,[b] G. C. Idzorek,[b]
R. J. Leeper,[a] C. L. Ruiz,[a] and R. G. Watt[b]

[a]*Sandia National Laboratories*, P. O. Box 5800, Albuquerque, NM 871850 USA*
[b]*Los Alamos National Laboratory, Los Alamos, NM 87545 USA*
[c]*Blackett Laboratory, Imperial College, London SW7 2BZ United Kingdom*

Abstract. A radiation source has been developed on the 20-MA Z facility that produces a high power x-ray pulse (9.7±1.8 TW, 52±10 kJ), generated primarily from the interior of an imploding dynamic-hohlraum target centered along the z-axis. The radiation pulse, characterized here together with its underlying dynamics, is used for performing radiation-flow and ICF experiments with drive temperatures in excess of 200 eV.

INTRODUCTION

The dynamic hohlraum (DH) developed and discussed here [1, 2] is currently the most powerful x-ray source being used for radiation flow and ICF studies in the laboratory (Fig. 1). It utilizes two arrays of tungsten wires in a cylindrical z-pinch configuration [3] to form an imploding plasma shell [4], which upon impacting a low-opacity cylindrical target centered on the z-pinch axis, generates x-rays. In this arrangement, the high atomic-number plasma shell acts as a radiation case, trapping radiation generated within the target. As the atomic levels of the target begin to burn through, the trapped radiation flows from its interior through an REH [radiation exit hole (Fig. 1)] into a region of interest, such as into a secondary hohlraum for ICF studies [1, 5]. Alternatively, an ICF capsule placed internal to the DH offers the possibility of studying capsule implosions directly [6].

For these applications the wire arrays have lengths of ~10 mm, are mounted at radii of 20 and 10 mm, are composed of 240 and 120 wires, and have total masses of ~2 mg and 1 mg, respectively. The high numbers of wires help ensure that the resulting imploding plasma shell acts as a relatively uniform hohlraum wall that develops high power as it impacts the target [4, 7, 8]. The target is a solid cylinder of 14 mg/cc CH_2 foam with a radius of 2.5 mm. Its mass approximately equals the 3-mg

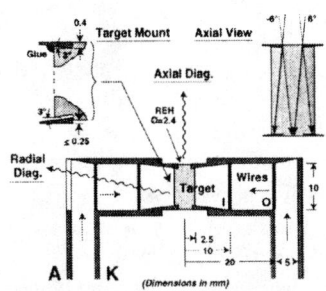

FIGURE 1. Experimental arrangement.

combined mass of the two tungsten arrays. Centered above the target, within the anode, is the REH, which has a radius of 1.2-mm. The target foam extends into and tamps the REH to prevent closure resulting from the low level of radiation generated during the run-in of the arrays. The slope of the array glide plane between the inner array and the REH is set to 3 degrees to prevent premature sliding of the tungsten plasma across the REH [9].

The characteristics of the radiation exiting the REH and emitted radially from the pinch are measured using diagnostic suites that include bolometers, XRDs, PCDs, and fast-framing pin-hole cameras. The dynamics of the implosion, x-ray generation, and measurement interpretation are explored primarily using two radiation-magnetohydrodynamic-code simulations in the r-z plane that include the development of the magnetic RT (Rayleigh-Taylor) instability. In one code [10], a 3T (three temperature) approximation along with radiation diffusion is used. In the other [11], a MG (multi-group) flux-limited diffusion approximation is used. In general, both models provide similar descriptions of the underlying physics [2]. Details of the load, wire array, target, diagnostics, and simulations are given in Refs. 1 and 2.

The next two sections discuss first the characteristics of the radiation exiting the REH and then the current interpretation of the dynamics responsible for its generation. The paper is concluded by a brief summary.

CHARACTERISTICS

Figure 2 illustrates the remarkable reproducibility of the axial radiation pulse. The pulse has a risetime of 2.7±0.3 ns, a pulsewidth [FWHM (full-width half-maximum)] of 4.5±0.7 ns, and a peak value of 9.7±1.8 TW (assuming Lambertian emission—see Ref. 2). These uncertainties, as elsewhere in the paper, refer to the RMS shot-to-shot variation. The associated temperature of the radiation field internal to the hohlraum at this peak is inferred to be 215±10 eV. For the most part, the field exiting the REH is consistent with that of a Planckian distribution (Fig. 3). Above 1.5 keV, however, the distribution deviates from the Planckian. At peak axial power this high-energy portion of the spectrum contains ~6% of the power. It can be spectrally characterized as either that of a Planckian distribution generated from a small area at a higher temperature, or that of a non-thermal electron distribution with a temperature in excess of 600 eV (Fig. 3).

Peak axial power occurs at -1.4±0.4 ns, where zero time is defined as the peak of the off-axis (radial) power (Fig. 2). The time −1.4 ns correlates with the start of the rapid rise in the radial power, which is typically associated with the onset of stagnation. The subsequent decrease in axial power, despite the higher temperatures being developed by the continued convergence of the target, likely results from

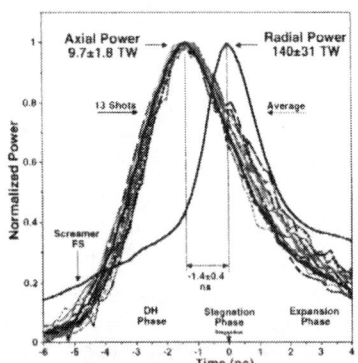

FIGURE 2. Thirteen normalized and shifted axial power pulses relative to the average radial power.

the colder and opaque tungsten beginning to restrict the axial radiation flow. As is shown in the next section (Fig. 6), at this time the pinch also exhibits a rapid radial deceleration, supporting the association with stagnation onset. The radial extent of the pinch reaches its smallest size at stagnation (t=0 ns), which is followed by pinch expansion. The initial impact of the combined arrays on to the foam target occurs near −5 ns, which accounts for the rapid rise in the radiation observed in the axial power pulse. The intervals between about −5 ns and −1.4 ns, -1.4 ns and 1.4 ns, and greater than 1.4 ns are referred to as the DH, stagnation, and expansion phases, respectively (Fig. 2).

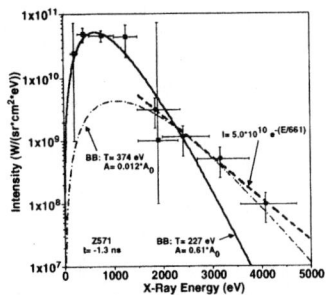

FIGURE 3. Measured axial spectrum at peak axial power. A_0 is the area of the REH. Black-body (BB) fit to spectra below 2000 eV (solid line), BB fit to spectrum above 1500 eV (dash-dotted line).

DYNAMICS

The radiation measured and simulated by the MG-model near the collision of the outer array with the inner array is illustrated in Fig.4, which occurs near −21 ns. At the collision, the 17 kJ modeled (shaded area in Fig. 4B) (or the 38 kJ simulated by the 3T-model) is consistent with the 25 kJ measured (shaded area in Fig. 4A) within the factor of two estimated uncertainty of the measurement. Both models indicate that the radiation generated between about −15 and −6 ns is due to array collision with a pre-expanded target plasma. (Compare the modeled radiation in this region generated with and without the target in place Fig. 4B). In the models the expansion arises from the absorption by the target of radiation generated from Joule heating in the plasma arrays. The measurements with and without a target, in contrast, exhibit little difference in this region (Fig. 4A), suggesting minimal target expansion. The measurements show also a level of run-in radiation over the interval −35 to −5 ns not present in either model, suggesting perhaps array interaction with precursor plasma. This precursor plasma may arise from the explosion of the wires in the array prior to their merger to form a quasi plasma shell [12-14]. Preliminary simulations of a precursor using a resistive MHD code in xy geometry [12] suggest the momentum imparted to the target is sufficient to prevent the target from expanding despite the internal energy generated by its absorption [15].

Stereoscopic images of the REH together with drawings (and associated lineouts) beside the images in Fig. 5 illustrate the evolution of

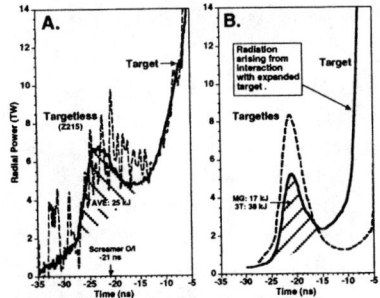

FIGURE 4. (A) Measured and (B) MG-model of radial power near outer-inner array collision.

the DH once the arrays have collided and begin to impact the target. In the models, the axial radiation (prior to stagnation) is generated through shock heating of the target. Figure 5A shows the inferred position of the shock-front and photo-surface (when viewed from outside the pinch) at about –5 ns. At this time the opacity of the target is low enough to permit the front to be observed. By –3 ns the images suggest that the hottest part of the target is on-axis with the region nearest the REH being cooler than that deeper inside the target. The models indicate the cooling is a result of radiation flow through the REH. Again, the images suggest the opacity is low enough to discern these temperature gradients. By stagnation, however, the centering of the images within the REH suggests the target is opaque with the target center being hottest. The radius of the shock front extracted from such images agrees well with that modeled prior to –2 ns (Fig. 6). Similarly, Fig. 6

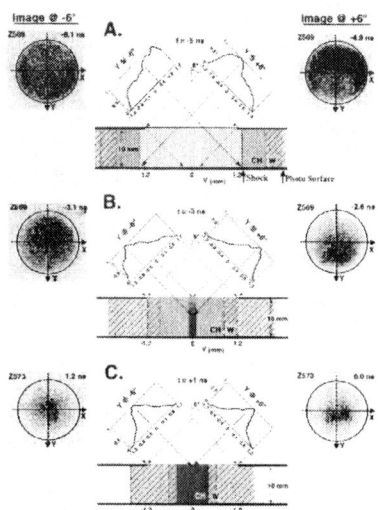

FIGURE 5. Measured stereo emission images of the REH and associated y-axis lineouts at (A) –5 ns, (B) –3 ns, and (C) +1 ns, illustrating emitting region within the DH.

illustrates that radii extracted from a radially resolved streak-camera image of the exterior of the DH surface agree with that simulated prior to –2 ns. The measurements indicate, however, that the pinch converges to only about half the value calculated, with significant expansion following stagnation. The temperature of the inside and outside of the DH based on averages of such measured radii (Fig. 6) and associated radial and axial powers (Fig. 2) is plotted in Fig. 7. The plot shows that the temperature of the DH interior exceeds that of its exterior from about –3 ns to stagnation, and shows definitively that the configuration is behaving as a hohlraum, trapping radiation within its interior. At peak axial power, the DH interior is about 230 eV in agreement with the Planckian spectra measured (Fig. 3).

In the modeling the peak axial power occurs as the tungsten/CH$_2$ interface just slides past the outside edge of the REH at –3 ns (Figs. 6 and 8B). Despite the continued increase in the internal DH temperature with time, the reduction in the radiation flow through the REH due to the presence of the high-opacity tungsten limits and reduces the modeled radiation through the REH. In contrast, the

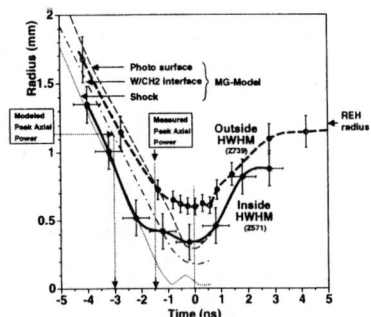

FIGURE 6. Measured and MG-model of DH radii.

measurements indicate that the peak in the axial power occurs at the onset of stagnation (-1.4 ns in Figs. 6 and 8B).

The models discussed here [10, 11] have evolved from simulations of bare pinches on Z. For these early models, a random variation in the mass density of the outer plasma shell was applied to seed the magnetic RT instability. The perturbation levels in the codes were adjusted so that the resulting instability development produced radiation pulses similar to experimental results. For the 3T-model this level was 6% [9, 16]. Applied to the current geometry, this model (as well as the MG-model) significantly under-predicted the axial

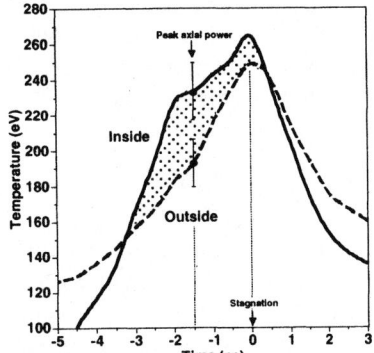

FIGURE 7. Average measured outer and inner DH surface temperature as a function of time.

radiation output (Fig. 8B). Reducing the perturbation level by an order-of-magnitude (to 1% in the case of the 3T-model) brought the simulations into somewhat better agreement with experiment (Figs. 8A and 8B). The results from these reduced RT perturbation simulations is that shown in Figs. 4B and 6. The reduction in perturbation level, however, did not bring the peak axial time into agreement with the measurements (Fig. 8B). Moreover, the enhancement generated in the axial emission very near stagnation remained, in contrast to experiment. In the simulations this enhancement results from the partial transparency of the very hot tungsten, which permits emission from the exterior of the DH to also pass through the REH in contrast to the experiment. Empirically, it was found that by artificially increasing the "roughness" of the W/CH$_2$ interface, the pulse shape could be brought into better agreement with the experiment (Fig. 8C). Investigation showed the detailed pulse shape is very sensitive to the delicate interplay between increased internal heating and restricted radiation flow. In the rough-interface model (Fig. 8C), peak power occurs just as a bubble near the REH begins to stagnate. Even though the flow from the lower

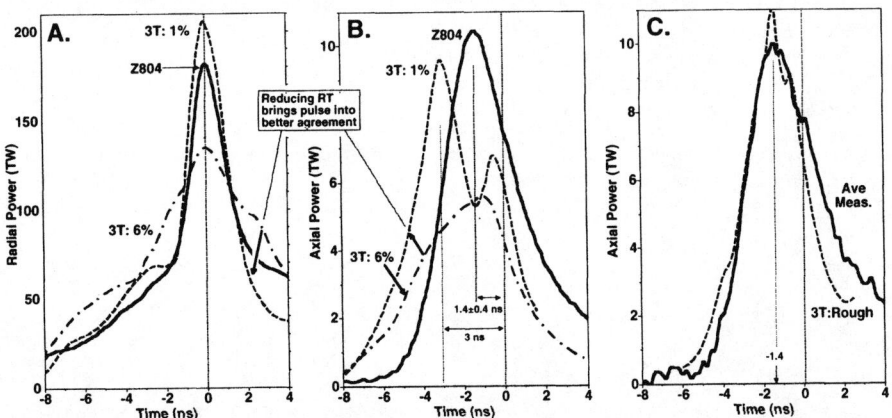

FIGURE 8. Measured and modeled (A) radial and (B) axial power. (C) Average measured axial power and 3T-model with roughened W/CH$_2$ interface.

portion of the hohlraum is restricted from escaping, the increased temperature in this region dominates the radiation output. In contrast, at this time the nominal 3T-model shows the emission through the REH originates as deep as 3 mm* below the REH. The temperature at this time is somewhat lower and the power is reduced relative to the rough model (see Figs. 21 and 22 in Ref. 2).

SUMMARY

We have developed and characterized an x-ray source that is useful for radiation transport and ICF studies at drive temperatures in excess of 200 eV. The 2D plasma-shell modeling when using a reduced level of the random-density to seed the RT instability (relative to that required for targetless pinches) is approximately consistent with the general timing and power levels measured. Inconsistencies with the 2D modeling, however, remain. Modeling comparisons with the measurements suggest the interface between the tungsten and CH_2 foam may be rougher than calculated using the standard RT development. Moreover, the target is inferred to remain close to its original position prior to impact with the tungsten plasma, and modest run-in radiation is generated prior to impact, all in contrast to that modeled. These differences may indicate the presence of a precursor plasma, arising from wire material driven ahead of the main implosion. Such a precursor has been measured in other tungsten wire-array systems [13-15]. It has the potential to tamp the target, reducing its expansion from run-in radiation, to generate radiation via collisions with the array and target, and perhaps to roughen the tungsten-target interface due to axial variations in the precursor itself [12, 13, 15].

REFERENCES

1. Sanford, T. W. L., et al., *Phys. Plasmas* **7**, 4669 (2000).
2. Sanford, T. W. L., et al., in press, *Phys. Plasmas* **9**, (August 2002).
3. Deeney, C. et al., *Phys. Rev. Lett.* **81**, 4883 (1998).
4. Sanford, T. W. L., et al., *IEEE Trans.Plasma Sci.* **26**, 1086 (1998).
5. Sanford, T. W. L., et al., *Phys. Rev. Lett.* **83**, 5511 (1999).
6. Bailey, J. E., et al., *Bull. Am. Phys. Soc.* **45**, 164 (2000).
7. Sanford, T. W. L., et al., *Phys. Rev. Lett.* **77**, 5063 (1996).
8. Deeney, C. et al., *Phys. Rev. E.* **56**, 5945 (1997).
9. Peterson, D. L., et al., *Phys. Plasmas* **6**, 2178 (1999).
10. Peterson, D. L., et al., *Phys. Plasmas* **3**, 368 (1996).
11. Lemke, R. W., et al., 28th Inter. Conf. Plasma Sci., 17-22 June 2001, Las Vegas, NV, IEEE Cat. No. 01CH37255, p. 183.
12. Chittenden, J. P., et al., *Phys. Plasmas* **8**, 2305 (2001).
13. Lebedev, S. V., et al., *Phys. Plasmas* **8**, 3734 (2001).
14. Cuneo, M. E., et al., *Bull. Am. Phys. Soc.* **46**, 234 (2001).
15. Palmer, J. B. A. et al, this conference proceedings.
16. Peterson, D. L., et al., *Phys. Plasmas* **5**, 3302 (1998).
* Sandia is a multiprogram laboratory operated by the Sandia Corporation for the US DOE under Contract No. DE-AC04-94AL85000.

Direct Drive Inertial Confinement Fusion in a Z-Pinch Plasma

Robert W. Clark, Jack Davis, Alexander Velikovich,
Leonid Rudakov[a], and John L. Giuliani, Jr.

Radiation Hydrodynamics Branch, Plasma Physics Division
Naval Research Laboratory, Washington, DC 20375
[a]*Berkeley Research Associates, Inc., P. O. Box 852, Springfield, VA 22150 USA*

Abstract. The recent successes with the Saturn and "Z" facilities at Sandia National Laboratory have renewed interest in Z-pinch fusion as a means of producing an abundance of high-energy photons. We have estimated that, in a nuclear fusion pulsed Z-pinch, peak currents in excess of 20-30 MA may produce magnetic fields sufficient to confine α-particles. We performed a series of numerical simulations with Au/CH/DT loads for devices with peak currents ranging from 20 to 60 MA. A detailed ionization model for Au was employed, and includes a forest of transported emission lines or line groups. For each case, we will give the calculated D-T yield and the yield of the α-particles deposited in the plasma.

INTRODUCTION

The "Z" facility at Sandia National Laboratory is capable of putting 18-20 MA into a load, and in a few years, the "ZR" machine will be able to produce load currents in excess of 25 MA. The recent successes with large Z-pinch devices have renewed interest in Z-pinch fusion as a process for creating a hot dense plasma. However, the prospects for designing an operating fusion device depend on the results from some very basic research. Can substantial thermonuclear burn be produced in a Z-pinch? How efficiently will the α-particles, which carry much of the fusion energy, be confined in the plasma? At what values of the magnetic field will magnetic confinement of the α-particles occur? If breakeven cannot be achieved, will such devices be useful for enhanced high-energy photon PRS output? In previous studies [1,2] we explored at the prospects for ignition in a high-current Z-pinch and found that (1) The magnetic fields produced in a 30 MA pinch may be sufficient to confine α-particles, (2) Low-Z Bremsstrahlung (characteristic of fusion plasmas) may be an attractive alternative for enhancing the photon spectra for PRS applications, and (3) Stabilization in the early phase and high-compression (exceeding Bennett equilibrium) in the final stage is a promising path to Z-pinch fusion. Several DT/CH configurations were considered to quantify these hypotheses, including a D-T shell with a CH pusher and a B_z interlayer. A summary of the Au/CH/DT configurations and D-T yields is given in Figure 1.

MODEL

The dynamics of the radially imploding Z-pinch plasma was simulated using a one-dimensional multi-zone non-LTE radiation-magnetohydrodynamics code, DZAPP [3], which uses a transmission line circuit model to represent the driving generator. The thermonuclear burn was calculated using rates from Glasstone and Lovberg [4], and a diffusion model was used for the transport of the α-particles. The diffusion equation for the α-particle energy density E_α takes the form

$$dE_\alpha / dt = (1-\gamma) E_\alpha \, div \, \mathbf{v} + div \, (D \, grad \, E_\alpha) + S - 2 \nu_e E_\alpha$$

where $\gamma = c_p/c_v$ is the gas constant and $\mathbf{v}$ is the velocity. The α-particle diffusion coefficient and the electron-ion collision frequency are given by

$$D = Q / [(\nu_e m_i) (9 + \Omega_i^2 / \nu_e^2)]$$
$$\nu_e = n_e \Lambda Z^2 (m_e/m_i) / (3.44 \times 10^5 \, T_e^{3/2})$$

where T_e and n_e are the electron temperature and density, Λ is the Coulomb logarithm, and Q is the initial α-particle energy (3.52 MeV). The α-particle source term $S = -Q \, d[T]/dt$, where [T] is the local tritium concentration (we assume that [D] = [T]). This treatment takes the local transverse magnetic field into account via the term involving the ion cyclotron frequency $\Omega_i = ZeB/m_i c$ in the diffusion coefficient D. The circuit model used in the simulations is based on an equivalent circuit for the Z generator. An extensive atomic database was used for each of the elements in the target, and detailed radiation transport was performed [5-7].

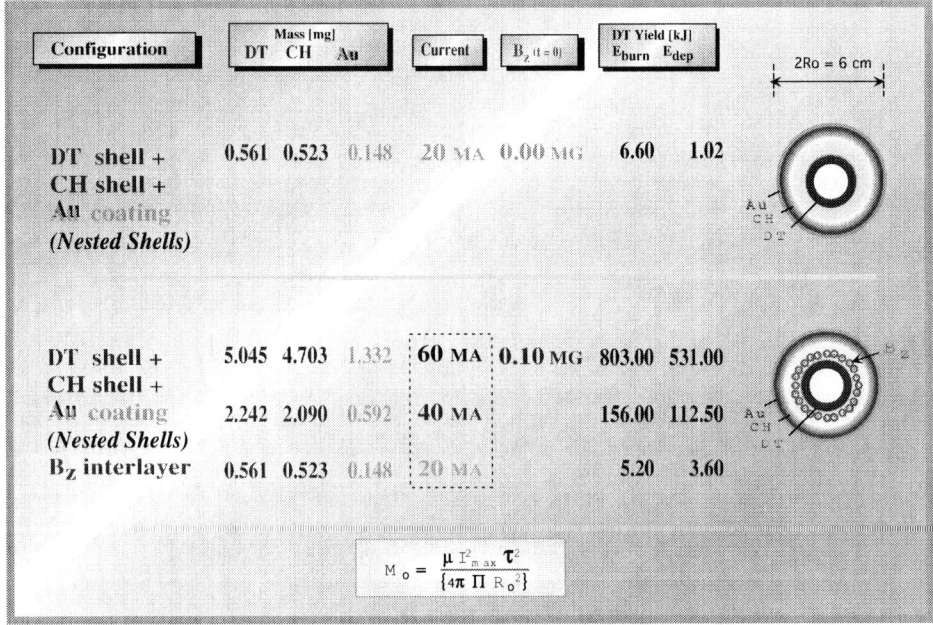

FIGURE 1. Initial configurations and yields for Au/CH/DT Z-pinch implosions

RESULTS

Although a number of different load designs were investigated, we will describe only the results with a D-T shell, an Au-coated CH Pusher and B_z Interlayer. In simple D-T implosions, the thermonuclear burn is inefficient, even at the highest currents. The ion temperature, which approaches 100 keV for the 60 MA case, is certainly sufficient to provide nearly optimum D-T reaction cross-sections. Unfortunately, the plasma density is not very high in the hottest regions. Also, the plasma bounces, quickly reducing the temperature and density; the D-T becomes too cool for fusion after only a few nanoseconds. A mechanism which could delay the bounce would likely improve the burn efficiency. The most complex (and promising) series of simulations involved concentric D-T and Au/CH shells separated by a B_z interlayer. The axial magnetic field is embedded in the plasma at time t=0, such that it is uniform between the D-T and Au/CH shells and zero elsewhere. Such a field configuration could be established between two wire arrays if they were twisted slightly in opposite directions [8]. The initial magnitude of the B_z interlayer was varied over a fairly large range, and it was found that values of B_z near about 100 kilogauss gave the best yields. We considered three peak currents: 20, 40 and 60 MA. In all cases, as the implosion progresses, the axial field diffuses into (and eventually through) the D-T and Au/CH shells. In addition, when the Au/CH shell impinges on the underlying D-T plasma, the B_z interlayer is compressed until it is comparable in magnitude to the azimuthal field B_θ driving the implosion. Thus, the B_z interlayer ultimately becomes quite thin. The magnetic interlayer also has the property of confining the α- particles in the D-T region. The yields vary strongly with current, $E_{D-T} \sim I_{max}^{4.5}$.

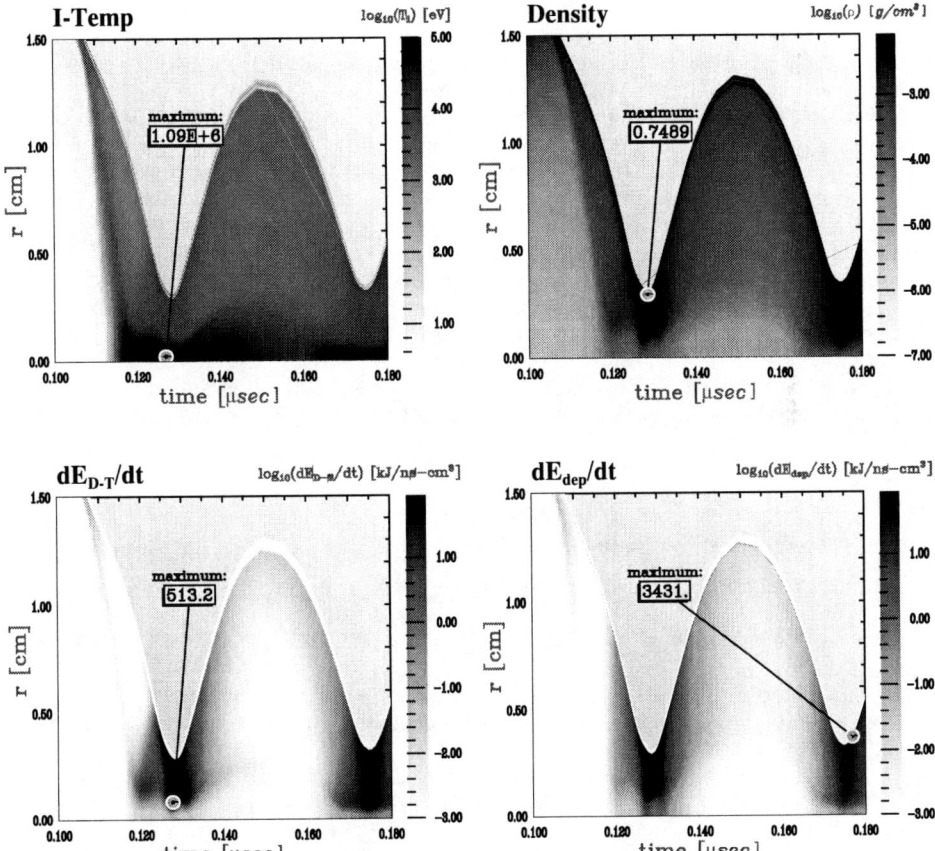

FIGURE 2. History of ρ, T_e, fusion heating and α-deposition for the 60MA case.

In the three simulations, although the *peak ion temperature* does not vary by very much, the temperature in the vicinity of maximum burn increases with current. In addition, the ion density increases with current. The masses were scaled so that the bounce occurs at about 125 ns in each case. At 20 MA, a second bounce occurs, but it is very weak, and little D-T burn occurs. At 40 MA, the subsequent bounce is stronger, and the burn is more robust (but the second pulse of α-particles is much less than the first). For the 60 MA case, the burn in the second pulse of α-particles is nearly as large as the first. The histories of several important quantities from the 60 MA case are shown in Figure 2: ion temperature T_i, density ρ, fusion heating dE_{D-T}/dt, and α-particle deposition dE_{dep}/dt. As is characteristic of z-pinch implosions, the electrons remain substantially cooler than the ions. Most of the burn occurs away from the axis, where temperature and density are *both* favorable. The α-particles produced in the reaction are stopped most efficiently in the cool dense outer region of the target, particularly the Au layer. The yields are given in Fig 1.

CONCLUSIONS

We have investigated load designs that might be employed to produce D-T fusion in proposed Z-pinch devices and found that high currents (I > 40 MA) reduce α-particle losses, but that they are not necessarily deposited where they will do the most good (in terms of increasing the burn fraction). The α-particle deposition occurs mostly in cool, dense plasma near the outer edge of the pinch. D-T burn in these devices seems to be limited more by density (and the rapidity of the bounce) than by the ion temperature. At the highest currents, it is easy to obtain 100 keV ions near the axis. The yield is improved with a massive Au/CH pusher, and improved still further when an axial magnetic interlayer was introduced. The load parameters (initial radii, array masses, etc.) were selected to give good energy coupling to the load, but were not optimized. It is likely that substantially improved fusion-like plasma conditions capable of radiating high-energy photons could be obtained after some adjustment of these parameters. In particular, the nested array simulations assign approximately equal masses to the inner (fuel) and outer (pusher) plasmas. However, the fusion yields do not reach breakeven (even at 60 MA). Substantially higher ion temperatures in the fuel and increased yields could be obtained by decreasing the fuel mass M_{D-T} and increasing the pusher mass M_{pusher} such that the total mass $M_o = M_{D-T} + M_{pusher}$ is unchanged, but with $M_{pusher} / M_{D-T} \sim 10$.

ACKNOWLEDGMENTS

Work supported by the Defense Threat Reduction Agency.

REFERENCES

1. *"Advanced Radiation Theory Support Annual Report 1998, Final Report"*, Radiation Hydrodynamics Branch, Naval Research Laboratory.
2. S. M. Gol'berg et al., *"Compression, heating and fusion in Megagauss Z-Q pinch systems"*, in Dense Z-Pinches, ed. N.R. Periera, J. Davis and N. Rostoker (AIP, NY, 1989). Phys. Rev. Lett. **81**, 4883 (1998).
3. J. Davis, J. Giuliani, Jr., M. Mulbrandon, Phys. Plasmas **2**, 1766 (1995).
4. S. Glasstone and R. H. Lovberg, *"Controlled Thermonuclear Reactions"*, (Van Nostrand, NY, 1960), Chapt. 2.
5. D. Duston, R. W. Clark, J. Davis and J. P. Apruzese, Phys. Rev. A **27**, 1441 (1983).
6. J. P. Apruzese, J. Davis, D. Duston and R. W. Clark, Phys. Rev. A **29**, 246 (1984).
7. J. P. Apruzese, J. Quant. Spect. Rad. Transf. **34**, 447 (1985).
8. R. E. Terry and R. W. Clark *"Analysis of Magnetic Interlayer Staged PRS Loads"*, in Dense Z-Pinches, ed. N. R. Periera, J. Davis and P. E. Pulsifer (AIP, NY, 1987).

The Inverse Z-Pinch as a Physics Test Bed, and a Possible Target Plasma for Magnetized Target Fusion (MTF)

Irvin Lindemuth [a], Bruno Bauer [b], Stephen Fuelling [b], Ronald Kirkpatrick [a], Volodymyr Makhin [b], Radu Presura [b], Peter Sheehey [a], Richard Siemon [a]

[a] *Los Alamos National Laboratory, Los Alamos, New Mexico, 87545 USA,*
[b] *University of Nevada, Reno, Reno, Nevada, 89557-0058 USA.*

Abstract. From an overall fusion system perspective, there remains an untested and interesting possibility of compressing a magnetized target plasma with beta greater than unity by a magnetically driven imploding liner, or other target pusher driver. This approach, known as Magnetized Target Fusion (MTF), operates in an intermediate density regime and time scale between magnetic and inertial fusion, which are separated by twelve orders of magnitude. Even if magnetized plasma transport is Bohm-like, fusion gain in the MTF parameter space appears accessible with existing drivers, which means MTF does not require a major financial investment in driver technology.

The physics of plasma confinement by material walls, and the thermal transport of magnetized high-beta plasma in the MTF regime, has received relatively little study theoretically, computationally, or experimentally. This paper describes a proposed experiment to test wall confinement in a regime of plasma parameters relevant to MTF. The geometry being considered is an inverse pinch designed to heat plasma to 100-eV temperatures. By using a current crowbar, the plasma formed in the pinch can be held against an outer wall (the return conductor) and the rate of cooling can be measured and compared with predictions from theory and numerical models.

The well-benchmarked two-dimensional radiation-MHD code MHRDR, is being used to guide the design activity. The existing 2-terawatt Zebra generator at the Nevada Terawatt Facility is the power supply under consideration. Results from the code show adequate heating, formation of a quasi-static magnetic equilibrium, and a near-classical cooling rate to a room temperature boundary, even in the presence of substantial plasma convection.

INTRODUCTION

In the context of seeking a low-cost development path for fusion energy, the possibility of compressing a magnetized target plasma with beta greater than unity by a magnetically driven imploding liner, or other target pusher driver, appears very exciting [1,2]. Even Bohm-like turbulent transport in a magnetized plasma appears to be acceptable.

Beta (plasma pressure relative to magnetic pressure) can in principle exceed unity if direct plasma confinement is provided by material walls. However, plasma behavior with high-beta wall confinement has been studied only a little, theoretically [3], computationally [4,5], and experimentally [6]. This paper describes a proposed experiment to test wall confinement in a regime of plasma parameters relevant to MTF. The geometry being considered is an inverse pinch designed to heat plasma to 100-eV temperatures. By using a current crowbar, the plasma formed in the pinch can be held against an outer wall (the return conductor) and the rate of cooling to the wall can be measured.

The well-benchmarked two-dimensional radiation-MHD code, MHRDR, is being used to guide the design activity. The existing 2-terawatt Zebra generator at the Nevada Terawatt Facility is the power supply under consideration. Results from the code show adequate heating and a near-classical cooling rate to a room temperature boundary, even in the presence of substantial plasma convection.

PLASMA FORMATION AND HEATING

The geometry of the inverse pinch is shown in Fig. 1. The conductor dimensions for a proposed experiment would be length, outer radius and inner radius of 40 cm, 10 cm, and 5 cm. When voltage is applied at the bottom between the inner and outer cylinders, a current sheet forms at the insulating boundary and accelerates outward.

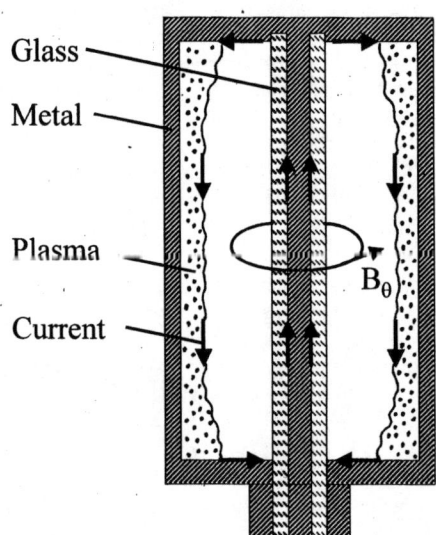

FIGURE 1. Geometry of the inverse pinch. Plasma is shown after it has been accelerated outward and settles down against the outer boundary. In the force balance achieved, pressure on the outer walls is comparable to the magnetic pressure, and a thin layer of plasma is confined with density increasing towards the wall as temperature decreases.

As a general rule in pinches, ion heating results from thermalization of the directed kinetic energy. Temperature scales as $kT/e \sim (c/\omega_{pi}) E$ where ω_{pi} is calculated for the initial gas density, and E is the inductive electric field (V/L for the inverse pinch). Typical parameters are fill pressure of 50 mtorr ($c/\omega_{pi} \sim 0.5$ cm) and $E \sim 2$ kV/cm, so that $T \sim 1$ keV. According to MHRDR with the actual Zebra current pulse and plasma dynamics taken into account, a few hundred eV temperatures are usually obtained.

COOLING RATE TO ROOM-TEMPERATURE BOUNDARY

After the plasma collides with the outer wall it oscillates radially for a few Alfven transit times while thermalization takes place. In the radial direction pressure balance is expected to become established between jxB forces pressing outwards, and pressure gradient and wall forces pressing inward, and a sheath is expected to form with steep temperature and density gradients near the wall [3,4].

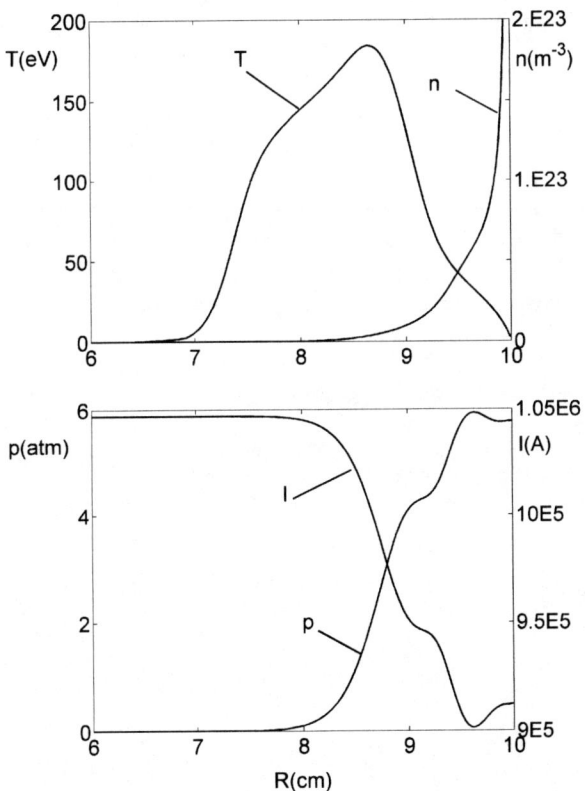

FIGURE 2. Typical plasma parameters near the outer wall (R=10 cm) according to 1D MHRDR calculations.

In one-dimensional MHRDR results, the thin sheath that forms at the outer wall has typical parameters as shown in Fig. 2. The pressure supported mechanically at the outer boundary is a calculated value in MHRDR simulations, and is generally found to be comparable to the peak plasma pressure and magnetic pressure as shown.

A rough estimate of the classical cooling time for a magnetized plasma is $\tau \sim \delta^2/\chi$ where δ is the sheath thickness indicated in Fig. 2, $\chi = \rho_i^2/\tau_{ii}$ is the thermal diffusivity perpendicular to B, ρ_i is the ion gyro radius, and τ_{ii} is the ion Coulomb collision time. For the parameters of this proposed experiment, τ is about 5 us, which is similar to results from detailed calculations using MHRDR. This cooling time is orders of magnitude longer than would result if free-streaming losses of ions or unmagnetized-electron conduction losses were considered. The detailed MHRDR calculations give faster cooling by a factor of a few when convection is considered in two dimensions. In two-dimensional runs, the central region of the 40-cm pinch is reasonably similar to the one-dimensional runs. However, the room-temperature electrodes at each end spoil the 1D equilibrium and convection cells are generated in which plasma flows axially and radially in the convective cells. Additional work with MHRDR is needed to better quantify the enhanced cooling by convection.

CONCLUSIONS

The usefulness of an inverse pinch geometry for wall-plasma studies has been demonstrated. Fusion-relevant parameters appear achievable using the Zebra high-voltage power supply at UNR's Nevada Terawatt Facility. Simple estimates for expected heating processes and thermal cooling rates agree qualitatively with detailed calculations using the MHRDR code. An experiment in the indicated parameter regime should yield new and definitive results on the intriguing physics of wall-plasma confinement of relevance to MTF and other applications.

REFERENCES

1. R. Siemon, I. Lindemuth, K. Schoenberg Comm. Plas. Phys. Cont. Fusion **18**, 363 (1999).
2. I. Lindemuth, R. Kirkpatrick, Nuc. Fus. **23**, 263 (1983).
3. G. Vekshtein, Rev. Plas. Phys. **15**, Consultants Bureau (1990).
4. I. Lindemuth et al., Phys. Fluids **21**, 1723 (1978).
5. I. Lindemuth et al., Phys. Rev. Lett. **75**, 1953 (1995).
6. B. Feinberg, Plas. Phys. and Contr. Fusion **18**, 265 (1976).

Ignition of Fusion Burn Wave by Nonthermal Plasma of Z-pinch

Victor V. Vikhrev

Russian Research Cente , Kurchatov Institute, 123182, Moscow, Russia

Abstract. Neutron emission from a Z-pinch is anisotropic and the energy distribution function of the ions is nonmaxwellian. The analysis in this paper emphasizes that a non-maxwellian distribution of ion energy corresponds to an enhanced nuclear fusion production compared to the maxwellian case. The criterion for burn wave ignition is discussed under the assumption of nuclear energy release from a non-thermal plasma in a Z-pinch.

Introduction

Irradiation of spherical targets by X-rays from powerful lasers or Z-pinches is presently considered the main approach for inertial confinement fusion. In these approaches the initial energy is first stored in electric or magnetic fields, then converted into the energy of visible light or of X-rays. For example, the Z-pinch transforms the electric energy of capacitor banks into magnetic field energy, then the magnetic energy is transformed into the energy of X-rays, and lastly the X-rays are focused on a spherical target [1].

In our opinion, a more attractive way to get thermonuclear reactor is the direct compression of fuel by magnetic field and further heating of this fuel up to thermonuclear parameters [2,3]. The Z-pinch is the most appropriate system to realize this idea, because the Z-pinch plasma is already compressed, contained and heated by the magnetic field.

In the RRC Kurchatov Institute, Z-pinch experiments are carried out on direct transformation of magnetic energy into the energy of high temperature plasma with further initiation of thermonuclear burn wave [4]. A high temperature plasma is created in a z-pinch neck. In turn this plasma initiates the thermonuclear burning wave which propagates along the z-pinch axes.

The initiation and propagation of burn wave in Z-pinches with D-T load can be described in the following way. Assume, that the thermonuclear reaction releases a certain minimum amount of power in some plasma volume. Alpha particles that leave this plasma volume heat the D-T fuel in the neighboring region. These are also heated up to thermonuclear temperatures and become the source of still more alpha particles. The thermonuclear burn wave propagates if the plasma produces more energy in alpha particles than is needed to heat that plasma volume up to thermonuclear parameters.

Spectra of ions in a Z-pinch

The ion energy spectrum of ions measured in Z-pinch experiments has never been

a Maxwellian. In a Maxwellian distribution the amount of high energy particles decreases exponentially with increasing particle energy. However, experimentally measured ion spectra show larger numbers of high energy particles than corresponds to a Maxwellian distribution. More detailed measurements have show that ion spectra do not contain ion beams with some particular energy [5,6]. Instead, the number of high energy ions decreases rapidly according to the power law,
$$f \sim 1/E_d^k \qquad (1)$$
The exponent k can be from 2 up to 4, depending on the spectral energy region.

References 7 and 8 present one possible mechanism for the generation of high energy ions in Z pinches, namely, high temperature plasma outflow from Z-pinch neck. The outflowing ions interact with the main plasma column, and the final ion energy spectrum is the result of this interaction. It is assumed that the energy distribution of ions in the low energy region is very close to Maxvellian, and that in the high energy region the distribution is dominated by high energy particles coming from Z-pinch necks. Without anomalous resistance the spectrum of the high energy particles is given by the power law dependence (1) with exponent k=2. With anomalous Joule heating the exponent is 2 ÷ 3. The total number of high energy particles measured in experiments agrees with the theory presented in [7].

Yield of thermonuclear fusion reaction for the case of power dependence of high energy ions

The intensity of thermonuclear d-d reaction was analyzed for Maxwellian ion distribution in the low energy region ($E_d < (k+1) T_{eff}$) and in the high energy region ($E_d > (k+1) T_{eff}$) the power dependence for the number of particles was used. In other words, we assumed a Maxwellian distribution that goes over smoothly into a power law (Fig1).

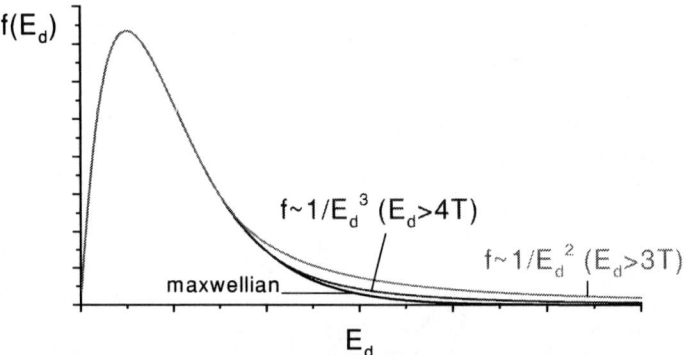

FIGURE.1 Energy distribution function of ions with various tails: 1) Maxwellian tail, 2) power tail $f \sim 1/E_d^2$ 3) power tail $f \sim 1/E_d^3$

In the connection point of these two distributions the continuity of both distribution function and its derivative is maintained.

Fig.2 shows the rate of thermonuclear reactions at constant pressure, nT_{eff} = const. For comparison the reaction rate is also shown for a Maxwellian energy distribution.. At an effective temperatures for ions $T_{eff} > 3$ keV the thermonuclear reactions in a plasma with a power-law tail are just as strong as in a plasma with a pure Maxwellian distribution. However, in the lower temperature region $T_{eff} < 3$ keV the thermonuclear reaction reactions are much more intense in a plasma with ions that have a power law distribution function than in a Maxwellian plasmas with the same pressure. In the low temperature region $T_{eff} < 1$ keV the thermonuclear reaction rate can be expressed as:

$$R = A(nT_{eff})^2 \, T_{eff}^{k-2.5}. \qquad (2)$$

Usually the life time of the pinch plasma τ in the pinch column is determined by $\tau = r/v_s$, where v_s is the sound velocity and r is radius of plasma column. Therefore the typical neutron generation time increases as the effective ion temperature decreases. The resulting neutron yield is $Y = RV\tau \sim T_{eff}^{k-3}$, where V is the plasma volume. This means that for a given mechanism of neutron generation, and for an exponent k=3, the neutron yield does not depend on effective temperature of the plasma column.

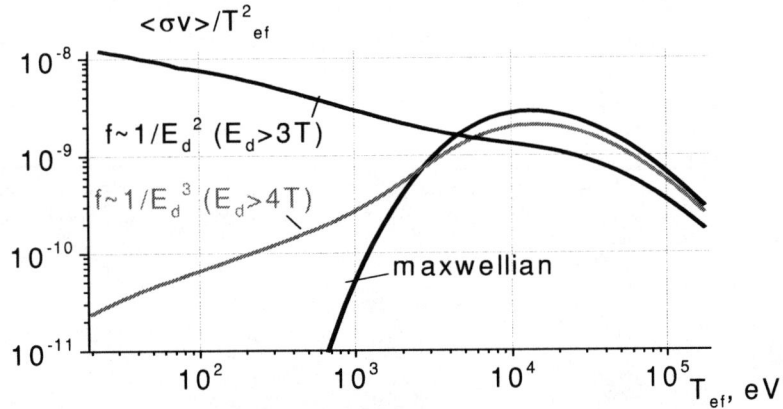

FIGURE 2. Intensity of the fusion reactions in a Z-pinch plasma with various distribution at the same pressure

Ignition of fusion burn wave by nonthermal plasma of Z-pinch

Paper [9] describes the thermonuclear burning wave initiation in Z-pinch neck for the case of Maxvellian energy distribution of ions. This paper gives the parameters needed to initiate a burn wave with currents in the range of 10 MA to 100 MA, and $\rho r = 0.23$ g/cm^2, assuming an ion temperature of 10 keV.

The present paper asserts that the current may be lower than the value suggested in Reference 9 if the plasma ions were to have a power-law high energy tail. The actual value of the current can be obtained only after detailed kinetic modelling of the Z pinch plasma, in the range of 100-1000 eV ion temperature and including the thermonuclear yield from the nonmaxwellian tail. The model should take into account the important thermonuclear yield from localised plasma regions, even though the bulk of the plasma column has not yet been heated to thermonuclear temperatures.

Paper [10] analyses the criterion for burning wave initiation. It was shown that the principal total number of emitted neutrons with DD load is given by formula:
$$Y_{DD} > Y^* = 1.67 \; 10^{13} hI^2 \qquad (3)$$
where h is the length of the zone radiating neutrons in cm, and I is the current in MA. Once the neutron yield measured in the experiment exceeds this value it indicates that a thermonuclear wave would occur if the DD load were to be replaced by DT.

The burn wave criterion is unaffected by the presence of a nonthermal ion component, except that for a nonthermal plasma the same neutron yield occurs in the range of 100 to 1000 eV, a much lower temperature than for a pure Maxwellian energy distribution: this demands 10 keV.

Conclusions

1. The rate of thermonuclear reactions in a Z pinch with average ion energy 100-1000 eV but including the thermonuclear yield of a non-maxwellian tail is a few orders of magnitude higher than that of a Maxwellian plasma with the same average ion energy, see Fig.2.
2. Initiation of a thermonuclear burn wave in a DT Z-pinch may already occur when a small local region in the plasma with enough ρr reaches an effective ion temperature ~ 1000 eV.
3. We would like to see whether a deuterium Z-pinch can produce more neutrons than given by formula (3). When criterion (3) has been exceeded, replacing the D-D fuel with D-T should result in a thermonuclear burn wave that propagates along the Z-pinch. This conclusion does not depend on the mechanism of neutron generation in Z-pinch.

Acknowledgement

This work is supported by Russian Fund of Fundamental studies under No. 02-02-16840.

References

1. Ryutov D.D, Derzon M.S., Matzen M.K., Reviews of Modern Physics **72**, 167-223 (2000).
2. Vikhrev V.V., and Ivanov V.V., Sov. Phys. Dokl. **30**, 492-495 (1985).
3. Linhart J.G., Bilbao L, Nuclear Fusion **40**, 941-954 (2000).
4. Bakshaev Yu.L., Blinov P.I., Chernenko A.S. et. al., Czehoslovak Journal of Physics **52**, Suppl. D, 212-220 (2002).
5. Filippov N.V., Filippova T.I , Pis'ma v ZHETF, 262 (1977).
6. Rapique M.S., Patran A., Springham S.V. et al., Czehoslovak Journal of Physics **52**, Suppl. D, 156-160 (2002).
7. Ananin C.I., Vikhrev V.V., Fiz. Plasmy **7**, 494 (1981).
8. Vikhrev V.V., and Baronova E.O., Proc.of BEAM's 98, Haifa, Israel, **2**, 662-665 (1998).
9. Vikhrev V.V., Rozanova G.A., Plasma Phys. Rep. **19**, 40-43 (1985).
10. Vikhrev V.V., Nucleonica **46**, Suppl. 1, 9-12 (2001).

On Perspectives of Creation of High-Power Neutron Source at Deuterium Plasma Compression in Z-Θ Pinch Geometry

V.D.Selemir, A.V.Ivanovsky, A.P.Orlov, V.F.Yermolovich,
G.V.Dolgoleva, V.A.Demidov, V.I.Karelin, P.B.Repin

RFNC-VNIIEF, Sarov, 607190, Russia

Abstract. Calculation results of thermonuclear neutrons yield at compression of cylindrical column of deuterium plasma with longitudinal magnetic field, compressed by external tungsten plasma liner (Z-Θ pinch geometry) are presented. Numerical simulations are presented for conditions of PBFA-Z facility in complete one-dimensional magneto-hydro-dynamical setup with radiation transfer in diffuse approximation. Scaling dependencies of neutron yield from the value of initial longitudinal magnetic field and mass density of deuterium gas are obtained.

INTRODUCTION

It is known that existence of longitudinal magnetic field stabilizes development of bulk convective and Rayleigh-Taylor instabilities in cylindrical plasma liner, powered in Z-pinch configuration [1]. This effect allows us perform stable radial compression (~20-times) of captured magnetic flux by imploding plasma liner, formed both from hollow annular gas jet [2] and from multi-wire cylindrical array [3].

At location of solid cylindrical column (target) from D_2 (Fig. 1) on the system axis, its compression by amplifying longitudinal magnetic field (θ pinch geometry) takes place. As it is shown in [4], the value of stable radial compression of the target is determined mainly by maximum obtainable compression of the external liner. It is supposed that the mentioned target is created either by pulsed gas puff, or it is deuterium-containing porous cylinder (made of $C_{17}D_{16}O_2$, for example).

In this work we numerically study perspectives of thermonuclear neutrons generation at compression of pure deuterium plasma column in Z-θ pinch geometry for conditions of PBFA-Z facility. Initially, as an external liner, we considered tungsten foil, simulating in one-dimensional case cylindrical multi-wire array with similar linear mass.

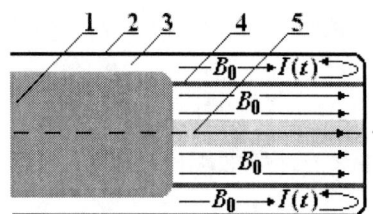

FIGURE 1. Scheme of compression of magnetized deuterium plasma column in Z-θ pinch geometry. 1 – current guide; 2 – reverse current guide; 3 – vacuum gap; 4 – plasma liner with initial radius R_0, length L; 5 – cylindrical target made of D_2 with initial radius r_0, length L.

PROBLEM STATEMENT

One-dimensional magneto-hydro-dynamical model, taking into account radiation transfer in one-group diffusive approximation [5] was used for numerical analysis of perspectives of obtaining of neutron yields in Z-θ pinch scheme. In order to perform through calculations at initial time moment $t=0$ in radial direction, three areas (fig. 1) with similar temperature $T_0=2.5 \cdot 10^{-5}$ keV were selected: I-column D_2 ($0<r<r_0$) with constant density ρ_1 and linear mass $m = \pi r_0^2 \rho_1$; II-"vacuum" interlayer made of D_2 ($r_0<r<R_0$) with constant density $\rho_2=0.01 \cdot \rho_1$; III-cylindrical tungsten liner with density $\rho_3 =19.3$ g/cm^3 and linear mass M.

Calculations were performed in two different setups: with additional external power source and without it. Power source was switched on at time moment $t_s>0$, count off from the beginning of current growth through the external liner and during 5ns it uniformly heated deuterium plasma in the range I up to the temperature $T_s=50$эВ. Neodymium glass laser, for example, could be additional energy source.

According to [1] initial longitudinal magnetic field with strength B_0(kGs)=(10÷30)·I_{av}(MA)/R_0(cm), where I_{av} average current value through the liner in imploding process, R_0 – initial radius of cylindrical liner, is necessary for stable compression of hollow plasma liner. In numerical investigations the value of initial magnetic field was selected within the limits of the specified range.

Equivalent electrical circuit of PBFA-Z facility is described in [6]; initial configuration in all calculations was similar: R_0=2 cm; r_0=0.5 cm; L=2 cm; M=2 mg/cm; m – variable. With these conditions amplitude of current pulse through tungsten liner was I_m=16÷18 MA.

CALCULATION RESULTS

Fig. 2(a) presents total neutron yields from 1 cm of deuterium plasma column length depending on the value of initial density ρ_1 at different values B_0 and absence of additional plasma heating source. Fig. 2(b) presents similar dependencies for the case of additional energy source, "switching on" in time moment t_s=100 ns. One can see that maximum neutrons yield ~10^{14}-10^{15} n/cm is observed at initial densities of deuterium ρ_1~10^{-4}g/cm^3 in area I and it monotonously decreases at increase of initial magnetic field B_0. Preliminary plasma heating leads to increase to neutrons generation. Considerably this effect becomes apparent within the limits of low ($\rho_1<10^{-5}$g/cm^3)

deuterium densities. This is associated with different regimes of plasma compression. Quasi-adiabatic compression regime of preliminary heated plasma is realized in the limit of low densities. Ideally, it is characterized by plasma temperature growth dependence on degree of its radial compression as $T(t)=T_0 \cdot (r_0/r(t))^{4/3}$. Within the limits of high initial densities ($\rho_1 > 10^{-4}$ g/cm^3) of deuterium gas the regime of quasi-adiabatic plasma compression passes to shock compression regime.

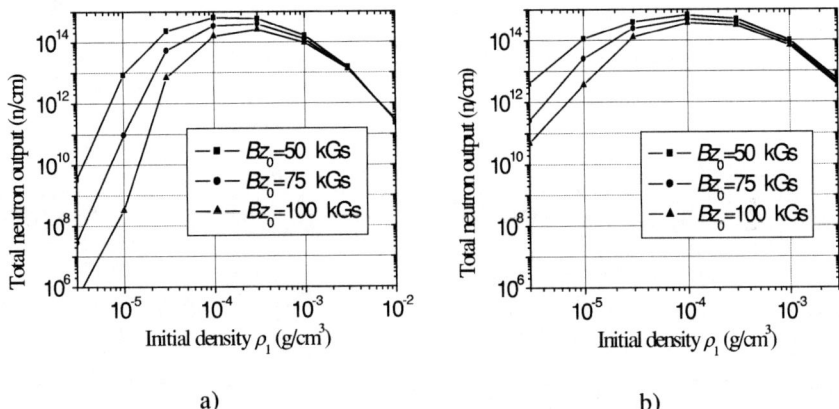

a)   b)

FIGURE 2. Neutrons yield from 1 cm of the length depending on plasma initial density: a) – without preliminary heating; b) – with preliminary heating.

Peculiarities of compression in the described regimes are presented on Fig. 3-4, where evolution of radii of areas I-III boundaries and neutrons generation intensity in area I for the case of preliminary plasma heating and $B_0=100$ kGs are presented. Comparing Fig. 3-4 one can see that relatively small plasma column boundary oscillations takes place at small initial deuterium density in the regime of quasi-adiabatic compression on the background of uniform compression of plasma column. Boundary oscillations are stipulated by appearance of magneto-sonic waves [4], whereas at high initial density plasma is compressed almost synchronously with tungsten liner. Specified peculiarities are shown on time dependencies of neutrons yield. Smooth time dependence (see Fig. 4b) is characteristic for stage compression (shock heating with further adiabatic compression). In case of quasi-adiabatic compression, time dependence of neutrons yield intensity has spiking character (see Fig. 3b with spikes, correlating with magneto-sonic waves passage.

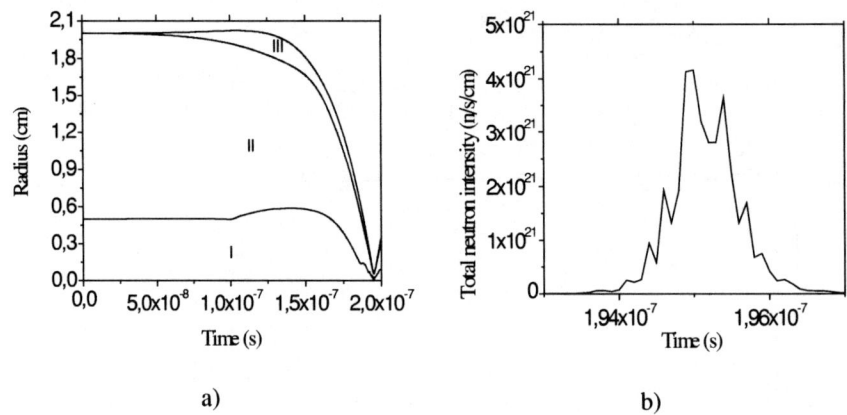

FIGURE 3. Time dependence of boundary radii – a) and neutrons yield intensity – b) for preliminary heated plasma ((ρ_1=10 µg/cm^3, B_0=100 kGs). Maximum degrees of compression: liner ~33; plasma ~32.

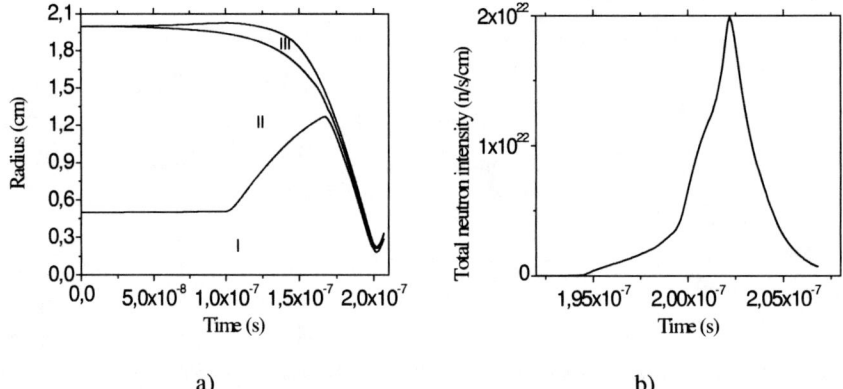

FIGURE 4. Time dependence of boundary radii – a) and neutrons yield intensity – b) for preliminary heated plasma (ρ_1=1000 µg/cm^3, B_0=100 kGs). Maximum degrees of compression: liner ~9; plasma ~3.

1. Bud'ko A.B., Liberman M.A., Velikovich A.L. and Felber F.S. ,*Phys. Fluids B* **2**, 1159-1169 (1990).
2. Felber F.S., Malley M.M., Wessel F.J. et all. , *Phys. Fluids* **31**, 2053-2056 (1988).
3. Leon J.F., Spielman R.B., Asay J.R. et al., *Proc. of 12th IEEE Intern. Pulsed Power Conf.* **1**, 275-278 (1999).
4. Ermolovich V.F., Ivanovsky A.V., Orlov A.P. and Selemir V.D., *Technical Physics* **45**, 1241-1250 (2000).
5. Antonenko E.M., Dolgoleva G.V., Ermolovich V.F., *VANT, Ser.: Mat. Mod. Fiz. Proc.* **4**, 62-66 (1996).
6. Struve K.W., Martin T.H., Spielman R.B. et al., *Proc. of 11th IEEE Intern. Pulsed Power Conf.*, 162-167 (1997).

The Production of Hypersonic, Radiatively Cooled Plasma Projectiles of Extremely High Energy Density in Imploding Z-pinches.

J.P. Chittenden[a], A.M. Dunne[b], M. Zepf[a], S.V. Lebedev[a], A. Ciardi[a] and S.N. Bland[a]

[a] *Imperial College, Blackett Laboratory, Prince Consort Road, London, SW7 2BZ, U.K.*
[b] *AWE Aldermaston, Reading, Berkshire, RG7 4PR, U.K..*

Abstract. Experiments with conical wire array Z-pinches, have shown that a highly supersonic tungsten plasma jet can be formed, with the presence of strong radiative cooling resulting in a very small flow divergence. In this work, we explore the possibilities of exploiting a similar effect in imploding wire arrays to produce a high velocity, dense plasma projectile. 2D(r,z) MHD simulations of imploding Z-pinches on the 'Z' generator, with a shaped mass per unit length, indicate the production of a projectile with velocity, density and diameter comparable to the final implosion velocity and stagnated plasma density and diameter in present wire array Z-pinch experiments. This represents an enormous (~100TW) kinetic power delivered out of the end of the pinch. 2D radiation hydrodynamic simulations of the projectile impacting a solid tungsten foil are used to estimate the potential for generating both compact radiation sources and target plasmas of extremely high energy density. Results indicate that pressures of 2 Gbar and radiation temperatures of ~660eV are produced in the target. 3D view factor simulations of a double ended hohlraum system with twin projectiles and converter foils are used to evaluate these radiation sources for use in indirect drive inertial confinement fusion research. Results suggest that a flux-equivalent peak temperature of ~225eV could be delivered to a NIF scale ICF capsule.

INTRODUCTION

The double ended, Z-pinch driven, vacuum hohlraum concept represents a clearly scalable method of achieving thermonuclear fusion ignition. The intrinsically large size of these hohlraums, however means that the concept naturally scales to high yield fusion experiments and requires larger scale machines than are presently available. An alternative approach is to try and find a method of concentrating the power of an imploding Z-pinch into a smaller volume, thus achieving higher energy densities and allowing fusion ignition experiments at lower yields on smaller facilities.

In this paper we investigate the potential for using high velocity projectiles of plasma to transport power away from the imploding Z-pinch, into a smaller scale target hohlraum, where it's kinetic energy can then be converted into X-rays in close proximity to the capsule. This scheme provides not only the potential for substantial increases in hohlraum temperatures but also a means of generating target plasmas of extremely high energy density.

PROJECTILES IN FUSION AND PLASMA JETS

The use of high velocity projectiles has been proposed in a number of inertial fusion concepts, either for direct bombardment of the target, to provide a fast igniter pulse or as an X-ray radiation source for indirect drive. In this work we consider an approach which makes use of intense radiative cooling in a dense plasma projectile to produce an extremely high Mach number and low divergence flow, thereby keeping the plasma compact after launch. This is analogous to the formation of radiatively cooled plasma jets as observed in scaled experiments [1] investigating proto-stellar jet collimation using the MAGPIE pulsed power facility at Imperial College. In these experiments a ~1MA current is applied to a conical array of sixteen 18μm tungsten wires. The flow of current steadily ablates the wires which therefore act as sources of low density tungsten plasma. The jxB force then accelerates this plasma perpendicular to the wires, producing a supersonic conical flow of material which converges towards the axis. Upon stagnation on the axis, the kinetic energy associated with the radial momentum is thermalised and for high Z elements this energy is almost entirely lost to radiation. The axial momentum, however is conserved, resulting in a high velocity outflow. The tip of the jet has an apparent velocity of $~2 \times 10^5$ ms^{-1} and a Mach number of ~30. With the correct choice of load, the conical plasma stream can be sustained for up to 200ns, producing a supersonic tungsten jet of ~40mm length and <2mm diameter.

FORMING HIGH ENERGY DENSITY PLASMA PROJECTILES

For hohlraum heating applications, we need a significant fraction of the generator energy to be converted into the kinetic energy of a small fraction of the load mass over a very brief acceleration period. This can be achieved using the comparatively simple load geometry of a cylindrical liner with a variable mass per unit length. Figure 1 shows results from a 2D(r,z) Eulerian resistive magneto-hydrodynamic (MHD) of a liner driven by the 'Z' generator at Sandia National Laboratory. In this case, the initial mass per unit length undergoes a smooth doubling (from 1 to 2mg/cm) between z = 2mm and z = 5mm. This essentially produces a single, large scale and controllable Rayleigh-Taylor driven imploding bubble. Collapse of the bottom 5mm of the liner occurs as the current approaches maximum value, effectively producing a miniature coaxial plasma gun. At 93ns, the constricted region at z = 5 mm produces an extremely large magnetic field (several thousand Tesla) which further accelerates the small fraction of the liner material immediately above this region to high velocity. The physical structure and characteristics of this plasma are similar to those obtained in the cross-point of X-pinch plasmas [2]. In the X-pinch case, however, the presence of a large axial precursor jet inhibits the formation of a high velocity projectile. In this calculation, 50μm radial resolution was used close to the axis in order to resolve the highly compressed regions. A ± 5% random density perturbation was applied at t=0, but had little effect on the result. The formation of a large scale magnetic bubble, in which the plasma surface is continuously stretching, appears to inhibit the development of shorter wavelength perturbations in a similar fashion to that observed in 2D simulations of uniform-fill Z-pinch implosions with curved surfaces [3].

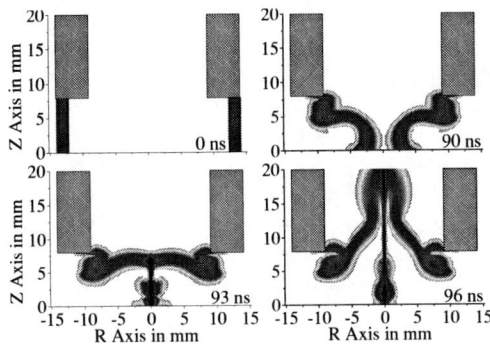

FIGURE 1. Density contours from a 2D(r,z) simulation of a liner with variable mass per unit length (from Ref. 6, © 2002 APS).

Strong radiative cooling means that after launch, the projectile rapidly cools and therefore the internal Mach number is high (~ 200), the divergence is small and the plasma remains compact over a flight of several cm. In order to estimate the efficiency with which the kinetic energy flux of the projectile can be converted to a thermal radiation source, we make use of some conservative estimates of the plasma parameters based on the simulation results just after the projectile is launched. The projectile is 4 mm long, 1mm in diameter, with a velocity of 2×10^6 ms^{-1} and a kinetic energy of 250kJ. The mass density has a roughly Gaussian distribution along it's length with an average density of 43 kg/m^3. The flux of ions is equivalent to a 33MA (4.2×10^{19} A/m^2) beam of singly charged 4MeV tungsten ions, except that the plasma is of course charge neutral. The tungsten ions have a short stopping range in the target and their kinetic energy is converted to radiation within a photon mean free path of the target surface, thus producing an efficient surface emitter.

X-RAY POWER PRODUCTION USING PROJECTILES

The impact of the projectile with a 2mm diameter solid tungsten converter foil was modeled using the 2D Eulerian radiation hydrodynamics code PETRA with 10μm resolution. A strong shock is launched into the converter foil with pressures reaching 2Gbar. Within the dense target, the ion, electron and radiation temperatures remain closely coupled and reach a maximum of 660eV in ~1ns. A supersonic radiation wave propagates from the target surface back into the plasma projectile to create an extended x-ray emission region. Within the stagnating projectile material, the three temperatures become partially decoupled with $T_i > T_e > T_r$. Figure 2 shows the radiation temperature from the interaction region as a function of time. The instantaneous radiation output power is roughly equal to half the incoming kinetic power, ½ ρv^3, throughout the impact. Lateral disassembly times for the impact plasma, due to radial expansion, were of order 2-3ns, as expected from the calculated sound speeds.

The X-ray source generated from this interaction was applied in a 3D viewfactor code model of a double-ended hohlraum (see Figure 2) containing two converter foils and a central plastic pellet of 2mm diameter. Simultaneous emission from each foil was assumed. The time-dependent wall re-emission from each discrete surface was determined from self-consistent 1D radiation-hydrodynamics calculations using SESAME data. The magnitude and distribution of the radiation flux around the capsule were obtained from static-wall viewfactors based on the time-dependent surface absorption and emission terms and losses to the entrance holes and the capsule. These indicated that a flux-equivalent peak temperature of ~225eV could be obtained on the capsule (see Figure 2), whilst maintaining an instantaneous and time-integrated flux uniformity of better than a few percent. This compares with ~130 eV obtained in similar scale end-on hohlraums using single sided radiation flow coupling [4] on 'Z' and ~155 eV predicted for double sided radiation flow.

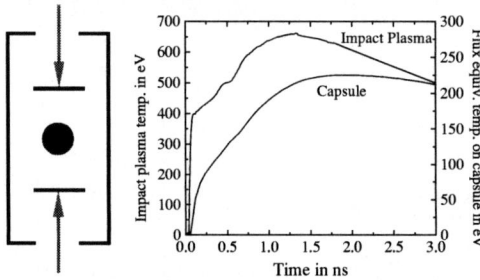

FIGURE 2. Double ended hohlraum design plus radiation temperature versus time from target interaction region and flux equivalent radiation temperature on capsule (from Ref. 6, © 2002 APS).

The simulations presented in this letter merely serve to provide an illustration of the proposed concept and do not represent a quantitative prediction of future experiments. The predicted projectile parameters are however, not unreasonable, the velocity, density and diameter being comparable to the final implosion velocity and stagnated plasma density and diameter in present wire array Z-pinch experiments [5]. In this sense, generation of the projectile simply represents a redirection of ~¼ of the radial implosion kinetic energy into the axial direction. Of this roughly half is converted to X-ray in the high energy density impact plasma over 2-3ns with a peak radiative power of 150 TW per side (more than ten times higher than radiative flow experiments [4]).

The use of hypersonic, radiatively cooled, plasma projectiles, clearly has potential for generating high temperature compact hohlraums and therefore reducing the size of Z-pinch facility required to achieve thermonuclear ignition. In addition, the 2Gbar plasma pressures produced during the impact of the projectile with a solid target could provide an opportunity to study material properties, shock physics and radiation hydrodynamics at far higher energy densities than are presently available. This work was supported by AWE Aldermaston through the William Penney Fellowship scheme.

REFERENCES

1. S.V. Lebedev, *et al.*, ApJ. **564**, 113 (2002).
2. T. A. Shelkovenko, *et al.*, Phys. Plasmas, **8**, 1305 (2001).
3. A. L. Velikovich, *et al.*, Phys. Plasmas **5**, 3377 (1998).
4. T.W.L. Sanford, et. al. Phys. Rev. Lett. **83**, 5511 (1999).
5. C. Deeney, et. al., Phys. Rev. Lett. **81**, 4883 (1998).
6. J. P. Chittenden, et al., Phys. Rev. Lett. **88**, 235001 (2002).

Fusion Criteria for Dense Cylindrical and Sheet Z-Pinches

Tetsu Miyamoto

Atomic Science Research Laboratory, 2-13-19, Higasi-Kokubu, Ichikawa, Chiba, Japan, and College of Science and Technology, Nihon University, Kanda-Surugadai, Chiyoda-ku, Tokyo, Japan

Abstract. Fusion criteria for dense cylindrical and sheet z-pinches [1] are investigated and compared with each other. It is difficult to achieve the inertia fusion limit of cylindrical z-pinches. The fusion criterion is considerably relaxed in the dense sheet z-pinches. Hence, one dimensional liner compression by means of two plane conductors is more effective than the two dimensional compression by imploding cylindrical liner.

1. INTRODUCTION

There are three approaches to thermonuclear fusion -- "external magnetic confinement fusion (EMCF)" that requires solenoid coils, "self-magnetic confinement fusion (SMCF)" that use no solenoid coil and "inertia confinement fusion (ICF)". The SMCF is essentially a pulsed system that is different from the EMCF in both the target density and confinement time, and the reactor concept is also completely different. In the EMCF, the inclusion of solenoid coils makes the reactor large and expensive. Only by scaling-up the device, it seems that better plasma parameters are obtained. On the other hand, the SMCF based on z-pinches has the advantage of being smaller in size and higher in density of plasma. Therefore, the device will be compact and economical, if the fusion plasma is successfully generated. However, conventional z-pinch plasmas are inherently unstable, and it seems that no stabilizing effect is effective enough for creating the fusion plasma. Nevertheless the above-mentioned feature that differs from other approaches inspires us to the SMCF research.

The difficulty of z-pinches results from the cylindrical geometry. The growth rate of mhd instabilities is inversely proportional to the radius. We suppose the poorest limit of SMCF, in which no stabilizing effect exists. This limit corresponds to the ICF, in which the plasma is heated directly by electric power. When the line density is constant, the density is inversely proportional to the square of radius. Hence, the product of the density and sustaining time increases with decreasing radius. The radius that satisfies the Lawson condition certainly exists, if we do not mind whether it is realistic or not.

To suppress the growth rate of instabilities, it was proposed to change from the cylindrical geometry to a sheet one, compressing by adjacent return current conducting walls, otherwise the sheet will revert to a cylindrical column. In this paper we will investigate the fusion criteria for both the cylindrical and sheet z-pinches.

2. FUSION CRITERION

The ratio Q of the released energy to the input one must satisfy $Q = \dfrac{\int P_r dSdt}{w+\int P_B dSdt} \geq \dfrac{1-\eta_{ec}}{\eta_{ec}}$, for the energy multiplication, where the Bremsstrahlung radiation loss and the released fusion power are $P_b = \overline{P}_b n^2 T^{1/2}$ ($\overline{P}_b = 1.57 \times 10^{-40} Z^3$ [J·m^6/(K$^{1/2}$s)]) and $P_r = \overline{P}_r(T) n^2$ ($\overline{P}_r = q_{DT} <\sigma v>_{DT}/4$ [J·m^6/s] for DT reaction). The total input energy w is the sum of the plasma $w_p = 3NkT$ and the magnetic energy w_m that depends on the shape of pinch column, i.e., $w = w_p + w_m = w_p/\eta_p$. The efficiency for recovering energy η_{ec} is 1/3 in the Lawson criterion, which corresponds to $Q > 2$. We introduce the function $F(T,Q) = 3kTQ/\left[\overline{P}_r(T) - Q\overline{P}_b T^{1/2}\right]$. The criterion is expressed as $n_0 t = F(T,Q)/f_n \eta_p \geq F(T,2)/f_n \eta_p$, where $f_n = \int_S n^2 dS / n_0 \int_S ndS$ is the density correction factor and n_0 is the peak density.

3. DENSE CYLINDRICAL Z-PINCHES

The Bennett relation for cylindrical z-pinches are given as $\mu_0 I^2 = 16\pi N^{(c)} kT$, where I, N and kT are the current, the line density and the thermal energy of pinched plasma, and the superscript (c) denotes the quantities about cylindrical z-pinches. The steady state for cylindrical z-pinches is achieved at the Pease-Braginskii current (the PB current) $I_{PB}^{(c)}$ [2], and the line energy density w_{PB} corresponding to $I_{PB}^{(c)}$, $I_{PB}^{(c)} = 4.33 \times 10^5 \sqrt{\ln\Lambda} \approx 1.4 \times 10^6$ (A) and $w_{PB}^{(c)} \equiv (3NkT)_{PB}^{(c)} \approx 1.47 \times 10^5$ (J/m).
In the steady state, the current density is nearly uniform, the density distribution is inversely hyperbolic and the line density is $N = \pi r_0^2 n_0 / 2$, where r_0 is the plasma radius, and the correction factor for density is $f^{(c)} = 1/3$.

The current ratio $\varsigma^{(c)} = I/I_{PB}^{(c)}$ is not necessarily an unity in a quasi-steady state, in which the process changes keeping only the pressure equilibrium. The line plasma and magnetic energy are $w_p^{(c)} = 3N^{(c)} kT = w_{PB}^{(c)} (\varsigma^{(c)})^2$ and $w_m^{(c)} = (w_p^{(c)}/3)[1 + 4\ln(r_w/r_0)]$. Hence, the total line energy is expressed as

$$w^{(c)} \equiv w_p^{(c)} + w_m^{(c)} = \dfrac{4}{3} w_{PB}^{(c)} (\varsigma^{(c)})^2 \left[1 + \ln\dfrac{r_w}{r_0}\right] = \dfrac{w_p^{(c)}}{\eta_p^{(c)}}, \text{ where } \dfrac{1}{\eta_p^{(c)}} = \dfrac{4}{3}\left[1 + \ln\dfrac{r_w}{r_0}\right]$$

The ratio $\eta_p^{(c)}$ is in the range of 0.13 – 0.15 for a conventional ratio of the return current radius r_w to the plasma radius. It will be difficult to recover the magnetic energy at the maximum compression, because a large fraction of the magnetic energy is lost during the plasma-decaying phase. We adopt $\eta_p^{(c)} = 0.15$ in the following comparison although η_p will be smaller due to other losses.

We express the sustaining time of the cylindrical z-pinch plasma column as $t_s^{(c)} = \alpha^{(c)} \frac{r_0}{v_{th}} = \alpha^{(c)} r_0 \sqrt{\frac{m}{2kT}}$. When $\alpha^{(c)} = 1$, t_s corresponds to the theoretical growth time of m=0 instability. Many experiments have shown that $\alpha^{(c)} \geq 1$ due to stabilizing effects, but at the same time, $\alpha^{(c)}$ is not so large for extremely hot, dense z-pinch plasmas. Hence, we should consider as $\alpha^{(c)} \sim 1$. This limit of z-pinches corresponds to a kind of the ICF, where the current plays only the role of heating and gathering the plasma into a tiny volume, i.e., of a driver in the ICF. Considering $F(2 \times 10^8, 2) \sim 0.7 \times 10^{20}$, this limit is achieved at the plasma radius $r_0 = 6.29 \times 10^{-9} \varsigma^2$. The input energy $w = 0.98 \times 10^6 \varsigma^2$ and the line density $N = 1.78 \times 10^{19} \varsigma^2$ are reasonable. It will, however be impossible to compress the plasma column to the density $n_0 = 2.87 \times 10^{35} \varsigma^{-2}$ and this radius.

4. DENSE SHEET Z-PINCHES

The basic features of sheet z-pinches were given in ref.1. When the thickness $2a$ is much smaller than the width $2b$ ($\gg 2a$), we can approximate the plasma sheet by the width $2b$ of the infinite width sheet z-pinch. Then, the pressure equilibrium relation is $\mu_0 (I^{(s)})^2 a/b = 24 N^{(s)} kT$, where $I^{(s)} = 4abi$ and $N^{(s)} = 8 n_0 ab/3$, and the superscript (s) is added in order to distinguish from the quantities used in the cylindrical z-pinches. The current density i is assumed to be uniform, and n_0 is the density at the center plane. In the steady state of sheet z-pinches, the current and the line energy density depend on thickness and width, differing from the fiber z-pinches, and are given as

$$I_{PB}^{(s)} = \frac{b}{a} \overline{I}_{PB}^{(s)} \quad \text{and} \quad w_{PB}^{(s)} \equiv 3(N^{(s)} kT)_{PB} = \frac{\mu_0 (\overline{I}_{PB}^{(s)})^2 b}{8a} = \frac{b}{a} \overline{w}_{PB}^{(s)}, \text{ where we introduce}$$

(note that we use definitions different by factor 2 from ref.1)

$$\overline{I}_{PB}^{(s)} = 2\sqrt{\frac{120 \overline{\eta}}{P_B} \frac{k}{\mu_0}} = 2.18 \times 10^5 \sqrt{\ln \Lambda} \text{ (A)} \quad \text{and} \quad \overline{w}_{PB}^{(s)} = \frac{\mu_0 \overline{I}_{PB}^{(s)^2}}{8} = \frac{60 \overline{\eta} k^2}{P_B \mu_0} = 7.57 \times 10^3 \ln \Lambda.$$

The ratio of the plasma thickness $2a$ to the separation of return current $2d$ is to be nearly 0.75 [1]. The plasma sheet is compressed to this ratio in thickness if the ratio is higher. On the other hand, if the ratio is lower, the plasma expands in thickness to this ratio, and contracts in width. This contraction of the sheet in width will be accompanied with area shock or tearing instabilities of the sheet.

In the quasi-steady state, as the line plasma energy is $w_p^{(s)} = \overline{w}_{PB}^{(s)} (b/a)(\varsigma^{(c)})^2$, the total energy is expressed as

$$w^{(s)} = w_p^{(s)} + w_m^{(s)} = \frac{\overline{w}_{PB}^{(s)}}{\eta_p^{(s)}} \frac{a}{b} (\varsigma^{(c)})^2 = \frac{w_p^{(s)}}{\eta_p^{(s)}}, \text{ where } \frac{1}{\eta_p^{(s)}} = \frac{1}{3} \left(1 + \frac{3d}{a} + \frac{3a}{2\pi b} \ln \frac{d}{a} \right).$$

When $b \gg a$, $\eta_p^{(s)} \approx 3/5$, and $f^{(s)} = 4/5$. The sustaining time of the sheet z-pinch plasma column is appropriate to express as

$$t_s^{(s)} = \alpha^{(s)} \frac{b}{v_{th}} = \alpha^{(s)} b \sqrt{\frac{m}{2kT}},$$

for the sheet z-pinch. Although we do not have any experimental data about $\alpha_s^{(s)}$ at present, we can expect $\alpha_s^{(s)} \gg 1$.

5. COMPARISON OF BOTH Z-PINCHES

There are various comparisons between both z-pinches. We normalize both the z-pinches by equating the ratio of the Larmor radius to the plasma radius (or the thickness). This means that $N^{(s)}/N^{(c)} = 0.852(b/a)$. Using the above mentioned values for η_p and f_n, and assuming the same temperature in both z-pinch, we obtain the following relations

$$\frac{Q^{(s)}}{Q^{(c)}} \approx \frac{F(T,Q^{(s)})}{F(T,Q^{(c)})} = 4.81 \frac{\alpha_s^{(s)}}{\alpha_s^{(c)}} \frac{r_0 b}{a^2}, \quad \frac{n_0^{(s)}}{n_0^{(c)}} = 0.104 \times 4.81 \left(\frac{r_0}{b}\right)^2,$$

$\frac{w^{(s)}}{w^{(c)}} = 0.640 \frac{b}{a}$, $\frac{I^{(s)}}{I^{(c)}} = 0.637 \frac{b}{a}$. The criterion is relaxed in sheet z-pinch even if $\alpha_s^{(s)} \sim 1$ as well as $\alpha_s^{(c)} \sim 1$. As $b/a \gg 1$, the input total energy and the current must be larger in the sheet z-pinches, but the Q value is much higher. For example, we can expect $Q^{(s)} \sim 5$ for $Q^{(c)} \sim a/b$ and $a \sim r_0$ that is too large to satisfy the Lawson criterion in the cylindrical z-pinch.

6. CONCLUSIONS

It will be difficult to expect high value of $\alpha_s^{(c)}$ and also to achieve the ICF limit of cylindrical z-pinches. On the other hand, the fusion criterion is considerably relaxed in the dense sheet z-pinches. There are several methods for producing the dense sheet z-pinches. For example, the sheet z-pinch plasma is produced by compression of a z-pinch plasma column by means of two colliding metal plates. The above consideration shows that the one-dimensional liner compression will be more effective for fusion plasma production than the conventional cylindrical liner compression, if we take into account the stability problem.

REFERENCES

1. Miyamoto, T., *J.Phys.Soc.Japan,* **68**, 1238-1258 (1999).
2. Pease, R.S., *Proc.Phys.Soc.* **B70**, 11 (1957) ; Braginskii, S.I., *Sov.Phys.JETP* **6** 494 (1958).

Considerations For Generating Up To 10 Mbar Magnetic Drive Pressures With The Refurbished Z-Machine (ZR)

Raymond W. Lemke, Marcus D. Knudson, Allen C. Robinson, Thomas A. Haill, Kenneth W. Struve, Thomas A. Mehlhorn and James R. Asay

Sandia National Laboratories
P. O. Box 5800 / MS 1186
Albuquerque, NM 87185-1186

Abstract. The intense magnetic field generated in the 20 MA Z-machine is used to accelerate flyer plates to high velocity for EOS experiments. A peak magnetic drive pressure on the order of 2 Mbar can be generated, which accelerates an approximately 0.2 g aluminum (Al) disc to 21 km/s. In a planned refurbishment of Z, called ZR, it is expected that up to 26 MA will be delivered to certain loads (e.g., dynamic hohlraums). We have used magneto-hydrodynamic (MHD) simulations to predict the peak magnetic drive pressure (and flyer velocity) that can be generated in a shock physics load on ZR. MHD simulations show that motion of the electrodes during the rise to peak current significantly increases the load inductance, which limits the peak current to values less than the expected maximum. This reduces the peak drive pressure to a value below what would be estimated using the expected peak current in a static geometry. However, MHD simulations also show that starting with a load geometry that maximizes magnetic flux on one wall of the anode, it is possible to reduce dynamic geometry effects by tapering the rise of the voltage pulse in combination with using a stiff material for the cathode. Simulations predict that peak magnetic pressures of 6 Mbar in Al, and 10 Mbar in tungsten are possible on ZR. In addition, it is predicted that flyer velocities of 40 km/s and larger can be achieved.

2D MHD SIMULATIONS AND RESULTS

Material science experiments on the Z machine have yielded: (1) equation of state (EOS) measurements of deuterium for pressures up to 700 Kbar in conventional shock experiments [1], (2) measurements of the isentrope of Al up to 1.5 Mbar [2], and of Cu up to 400 Kbar in isentropic compression experiments (ICE) [3] and (3) flyer velocities up to 28 km/s. Indeed, experiments on Z have set benchmarks for future performance. We have performed a computational study using 2D MHD simulation to determine how material science experiments on Z will scale to ZR, the refurbished Z machine. In this paper we discuss physics and design issues at multi megabar magnetic pressures, and present results of 2D MHD simulations which constitute predictions for the performance of material science experiments on ZR.

The main physics issue is a consequence of the compressibility of the conductors that form the load for material science experiments. A typical load is shown in Fig. 1, and is comprised of a stainless steel cathode and Al anode (for example). The deformation that the electrodes undergo due to compression by the magnetic pressure results in significant inductance increase during the current rise time. The time dependent inductance is an effective resistance that creates kinetic energy at the expense of magnetic energy. Thus, significant conductor motion during the rise time of the current pulse reduces peak current and magnetic pressure.

 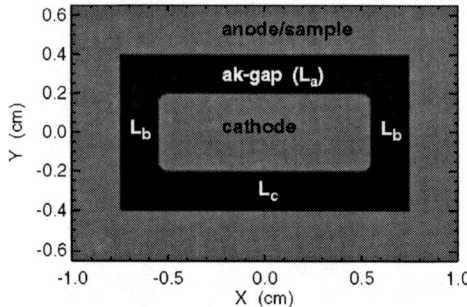

FIGURE 1. (a) Photograph of material science experiment load showing anode housing. Cathode (not shown) is centered on duct formed by anode walls and is connected to top of anode in a short circuit, which is accomplished with a cap. (b) Top down schematic of load showing cathode and anode. The material sample of interest is part of the anode. The L_i represent inductances of various current paths.

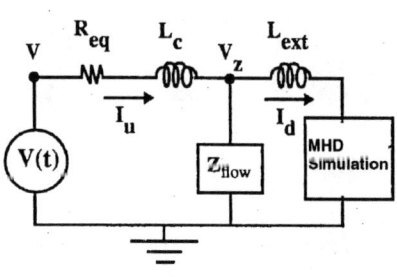 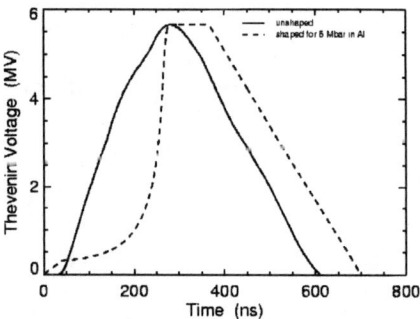

FIGURE 2. (a) Schematic of circuit driven 2D MHD simulation. Values of circuit parameters appropriate for ZR are: R_{eq}=0.18 W, L_c=10.18 nH and L_{ext}=2.3 nH. (b) Time dependent voltages V(t) used to drive the MHD simulation. The solid curve is the anticipated raw (unshaped) ZR voltage, and the dashed curve is an ideal waveform that contains the same amount of energy. The rise time of the ideal pulse is shaped to achieve isentropic compression to 6 Mbar in Al. Power enters the simulation from the 3rd (transverse) dimension in Fig. 1b. The effective length in this direction is 3.6 cm.

We have used the Sandia developed, radiation magneto hydrodynamic, finite element code ALEGRA [4] to investigate time varying inductance (L-dot) effects on ZR. Simulations are 2D MHD and are driven by an external circuit model of ZR. The simulation model is shown schematically in Fig. 2a. The Z_{flow} loss element is ignored, which can result in an over-estimate of the load current of about 5% (depending on voltage). The dynamic load feeds back on the circuit to affect the input power. Voltage waveforms used to drive the simulations are shown in Fig. 2b. Sesame equations of state are used for materials [5], in addition to models for the thermal and electrical conductivities [6,7]. The Ohm's law used includes the resistive electric field and the velocity cross magnetic field term.

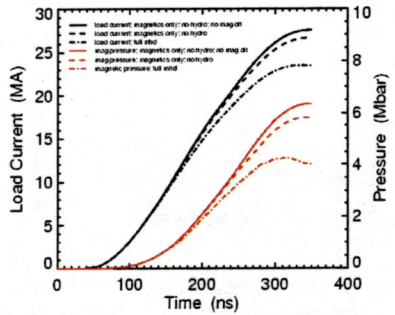 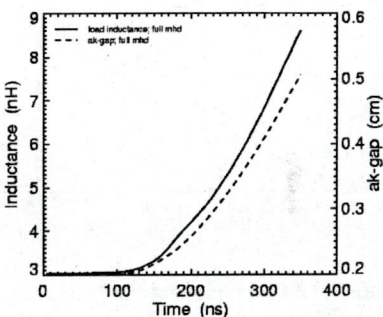

FIGURE 3. (a) Load currents (black curves) and magnetic pressures (red curves) from 2D MHD simulations with: (1) no hydro motion, magnetic diffusion minimized, magnetics only (solid lines), (2) no hydro motion, magnetics with magnetic diffusion (dashed lines) and (3) full MHD (dash-dot lines). (b) Load inductance and ak-gap vs. time in full MHD simulation.

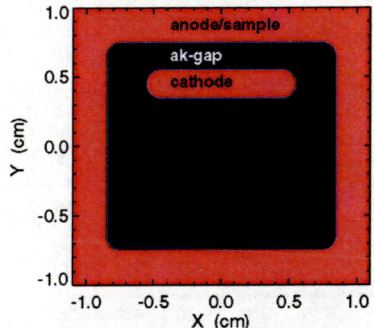 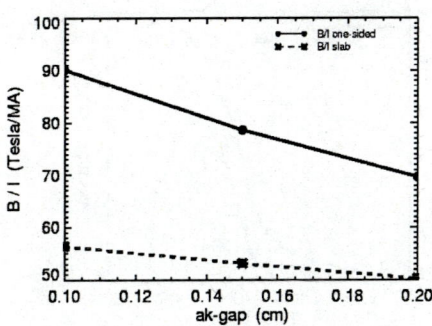

FIGURE 4. (a) Schematic of material science load optimized to reduce alternative current paths thereby concentrating magnetic flux under material sample region (ak-gap) where maximum magnetic pressure is desired. (b) Comparison of magnetic field per unit current for Fig. 4a (solid curve) and the analogous slab geometry (see Fig. 1b).

Figure 3 shows results from a simulation of an unoptimized geometry (similar to Fig. 1b) in which the electrodes are Al; cathode dimensions are 11x4 mm, anode inner dimensions are 15x15 mm and the ak-gap is 2 mm. The plots show that in the case of a purely inductive static geometry 28 MA peak current is delivered to the load.

However, in the full MHD case conductor motion during the current rise time produces an L-dot that reduces the peak current to ~23.5 MA, with a corresponding 33% reduction in peak magnetic pressure. Comparison of Figs. 3a and 3b shows that the inductance increases by factor of 2.3 during the current rise time, which reduces peak current (and magnetic pressure) relative to the ideal case.

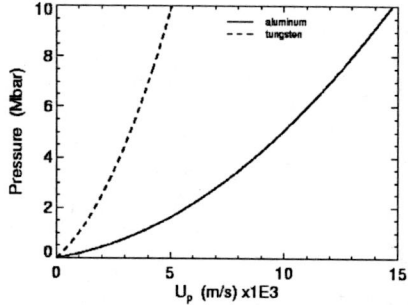
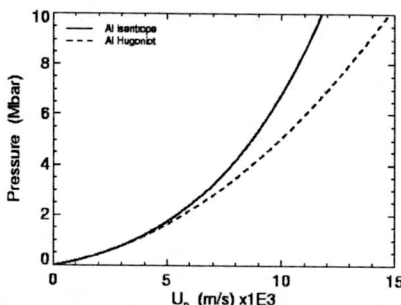

FIGURE 5. (a) Shock hugoniots for W and Al (solid curve) in pressure vs. material velocity space. Plot shows that there should be much less material motion (due to compression) for the same pressure when a stiff material (W) is used for the electrodes. In addition, a much higher pressure can be achieved for the same material velocity when W electrodes are used. (b) Hugoniot and isentrope (solid curve) for Al. Since shock formation is to be avoided, it is necessary to compress the material sample isentropically. The plot shows that this further reduces the material motion for the same pressure.

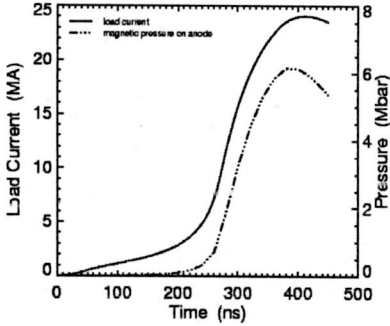
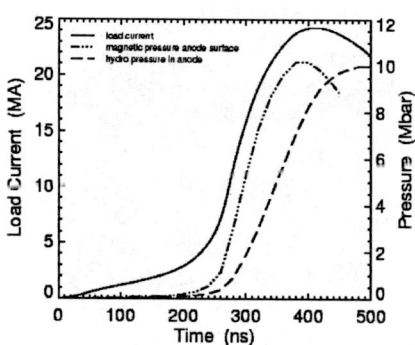

FIGURE 6. Results of full 2D MHD simulations for optimized load (Fig. 4a) with W cathode. Ideal voltage waveform for ZR (Fig. 2b) was used. Load current (solid curve) and magnetic pressure on anode surface vs. time for (a) Al anode and (b) W anode. In the latter case the hydrodynamic pressure in the material (anode) is also shown.

To achieve maximum magnetic pressure in material science experiments requires both electrical and hydrodynamic optimization of the load. An advantage of symmetric loads (as in Fig. 1b) is that data can be taken from multiple samples in one shot. However, the magnetic pressure on the sample can be maximized by minimizing alternative current paths, which produces a load like that shown in Fig. 4a. Figure 4b shows that this one-sided geometry yields a factor of 2.6 higher magnetic pressure for the same current when the ak-gap is 0.1 cm (compared to the symmetric design).

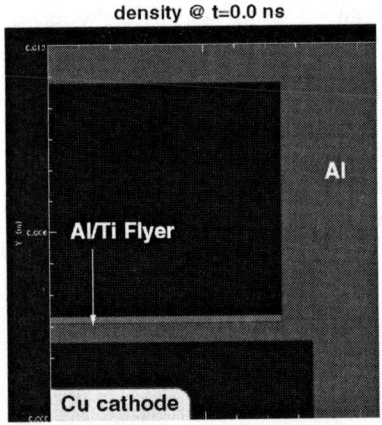
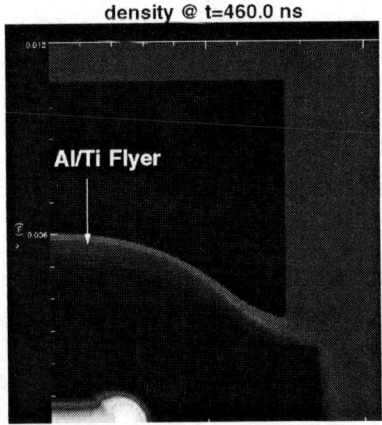

FIGURE 7. Configuration for 2D flyer simulation in quarter symmetry. Flyer is a slab comprised of 600 mm of Al and 200 mm of Ti (in red). Surrounding anode material is Al. (a) Configuration at t=0 ns. (b) Configuration just after peak current. Although the flyer is severely bowed away from the center, the central 2 mm remains uniform until impact with the target (top of plot). Bowing is a result of nonuniformity in the magnetic field in the horizontal direction.

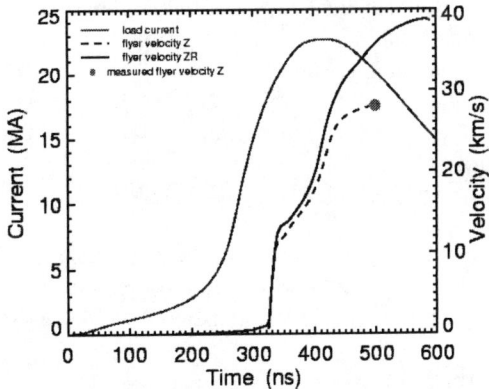

FIGURE 8. Comparison of simulated flyer velocities for Z and ZR (solid curve) voltage pulses with peak velocity measured on Z (the green dot). The ideal voltage waveform (Fig. 2b) was used for the ZR simulation. The simulated current for ZR (red curve) is superimposed. In view of the Z result, the peak velocity for the ZR simulation is probably accurate if the ideal voltage waveform is accurate.

Hydrodynamic optimization is necessary to minimize electrode motion during the current rise time, and to avoid shock formation. Figure 5a shows that using stiff materials for the electrodes results in significantly less conductor motion to achieve the same pressure. To avoid shock formation, which significantly modifies the material sample, isentropic compression is necessary. As shown in Fig. 5b this further reduces material motion.

Isentropic compression requires that the current rise slowly initially, which can be accomplished with a shaped voltage waveform. An example of an ideal shaped voltage waveform for ZR is shown in Fig. 2b.

To determine the maximum isentropic pressures that might be achieved on ZR simulations were performed using the Fig. 4a geometry with a tungsten (W) cathode, and the ideal shaped voltage waveform. Results for Al and W anodes are shown in Figs. 6a and 6b, respectively. The use of an optimized configuration results in peak isentropic pressures of 6.2 and 10.2 Mbar in Al and W, respectively (an increase of 48% relative to the unoptimized case for Al shown in Fig. 3a).

Simulations were also performed to determine the peak flyer velocity that might be achieved on ZR for conventional shock physics experiments. In this case a configuration that was shot on Z was used, which was driven by the ideal shaped voltage waveform for ZR. The simulation geometry is shown in Fig. 7. Figures 7a and 7b respectively show the initial geometry, and the geometry just after peak current when the flyer is traveling at ~35 km/s. As shown in Fig. 8, the peak velocity reaches ~39 km/s. Superimposed on the plot is the predicted velocity and measured peak velocity for the Z shot, in addition to the predicted load current for ZR. The peak velocity increases to ~41 km/s when the optimized geometry (Fig. 4a) is used.

Results of this computational study predict that a major impediment to achieving maximum magnetic pressure in material science experiments on ZR will be current reduction caused by the significant increase in load inductance associated with electrode deformation. This effect can be minimized, and maximum magnetic pressure achieved, through hydrodynamic and electrical optimization of the load. Significant modification of the material sample due to joule heating and/or shock formation is a concern at multi-megabar pressures. Shaping the voltage (current) waveform will be necessary to avoid modifying the material sample before useful data can be obtained.

ACKNOWLEDGMENTS

The authors thank C. A. Hall (SNL) and J. P. Davis (SNL) for contributions. Sandia National Laboratories is a multi-program laboratory operated by Sandia Corporation, a Lockheed Martin Company, for the US Department of Energy under Contract DE-ACO4-94AL85000.

REFERENCES

1. Knudson, M. D., et al., *Phys. Rev. Letters* **87**, 22550-1 (2001).
2. Hall, C. A., et al., *Rev. Sci. Instrum.* **72**, 3587 (2001).
3. Reisman, D. B., et al., *J. Appl. Phys.* **89**, 1625 (2001).
4. Summers, R. M., et al., *Int. J. of Impact Engng.* **20**, 779 (1997).
5. Kerley, G. I., Kerley Publishing Services Report No. KPS98-1, 1998 (unpublished).
6. Desjarlais, M. P., *Contrib. Plasma Phys.* **41**, 267 (2001).
7. Desjarlais, M. P., et al. (to be published).

3-D Modeling of Modifications to the Z Accelerator for Generating Shaped Pulses

Timothy D. Pointon, Mark E. Savage, and Henry C. Harjes III

Sandia National Laboratories, PO Box 5800, Albuquerque, NM 87185 USA

Abstract. One option to temporally shape the power pulse at the load on the Z accelerator at Sandia National Laboratories is timing delays between the 36 pulse-forming lines. However, this can lead to the formation of magnetic nulls in the vacuum section, with the potential for greatly increasing electron losses to—and possibly damaging—the anode. Three-dimensional computer simulations are now being conducted to study this concern. The simulation geometry models a single level of Z, with a radial transmission line driven by nine parallel-plate lines. Every third line is driven early relative to the other six. Results from preliminary runs without particle emission are presented. Voltage and current diagnostics agree quite well with circuit simulations, and spatial field profiles illustrate the evolution of the magnetic nulls in detail.

INTRODUCTION

A requirement for the refurbishment project [1] of the Z accelerator at Sandia National Laboratories [2] is the ability to temporally shape the power pulse delivered to the load. This capability is necessary to optimally drive isentropic compression experiments.[3] Although Z was not designed for pulse shaping, it is being modified to test the feasibility of pulse shaping by independent triggering of the 36 pulse-forming lines that transfer energy from the Marx banks towards the load. At the junction with the azimuthally symmetric center section at r ~ 1.95 m, there are four vertical levels of nine parallel-plate lines, equally spaced in azimuth. In the vacuum section, inside the insulator stack at r ~ 1.65 m, power flows inward on each level along a radial, magnetically insulated transmission line (MITL). The four MITLs are connected in parallel with a double post-hole convolute at r ~ 0.1 m, and the total current flows to the load along a single inner MITL.

To maintain reasonable symmetry, the plan is to drive every third line in azimuth (i.e. 120° apart) on a single level with the same power pulse. Each line can be configured to deliver either a short (~100 ns) or long (~300 ns) pulse. With three sets of three lines per level, there are 12 sets of lines, each in either the short or long-pulse mode, with its own timing delay. Simulations with the Microcap circuit code [4] demonstrate that required pulse shapes can be matched quite well with this scheme.

One concern with this scheme, not addressed by circuit modeling, is the possibility of "magnetic nulls", paths connecting the anode and cathode where $|\mathbf{B}| = 0$, in the vacuum section, where electrons emitted from the cathode are not prevented from reaching the anode. With a high electron flux, expensive hardware could be damaged.

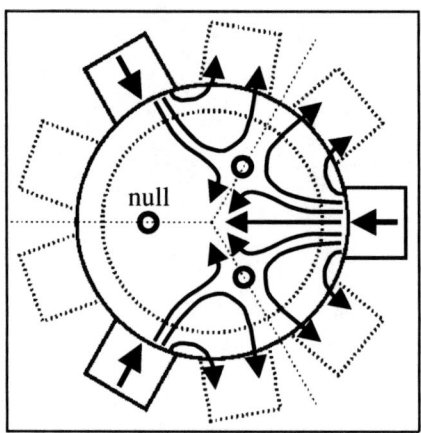

FIGURE 1. The "3-point feed" geometry, drawn to scale for Z, with the dotted circle showing the outer radius of the vacuum region. With every third line driven early (solid lines), the current streamlines qualitatively illustrate the formation of the three magnetic nulls.

There are two types of null formed in this system. First, with timing differences between levels, some of the current in an early level will flow back up the MITL of each late level. At the radius where this current is equal and opposite to the forward-going current in the level, there will be a null, extending over all azimuth. The second type of null occurs with timing differences between modules on a single level. Fig. 1 qualitatively illustrates the "3-point feed" geometry, in which every third line is early relative to the other six. Some of the current in the early lines flows back up the six late lines, and it is clear by symmetry that there must be points on the three radial lines midway between the early lines where the current, and therefore the magnetic field, vanishes. In this paper, we focus on this second type of null.

SIMULATION SETUP

We model a single-level, "3-point feed" system with the 3-D, electromagnetic, particle-in-cell code QUICKSILVER.[5] The simulation geometry uses the D-level of Z, on which the first experiments will soon be conducted. If the six late lines have identical current pulses, the system has six-fold symmetry in azimuth, with half of an early line and a full late line. We model the sector from 0 to 60° in cylindrical coordinates, as shown in Fig 2a. For these simulations, the insulator stack is modeled as a simple annular dielectric, $1.59 \leq r \leq 1.67$ m, with $\varepsilon_r = 2.53$ for rexolite. The stack separates the inner vacuum region from the outer water region, $\varepsilon_r = 80$. The two parallel-plate feed lines connect to the azimuthally symmetric center section at $r = 1.96$ m, and extend out to $r = 2.8$ m, with a gap of 0.14 m. Between $r = 2.8$ and 3.1 m, each line morphs into a "rectangular coaxial" line in the r-ϕ plane, which extends out to $r_{max} = 3.4$ m. Power is fed in at the $r = r_{max}$ plane; with the anode of each line completely enclosing its cathode, the two lines can be driven independently.

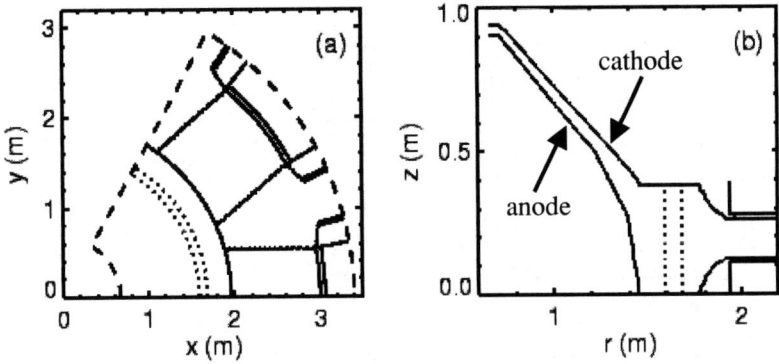

FIGURE 2. The simulation geometry: (a) top view, with the dashed line showing the system x-y boundary, and the dotted line showing the stack, and (b) r-z view through the center of a feed line.

On Z, the D-level MITL extends down to the convolute at $r \sim 0.1$ m. Experience has shown that the null forms at $r \sim 1$ m. The 3-D geometry accurately models the MITL only down to $r = 0.75$ m, as shown in Fig. 2b. Inside this radius, we bend the MITL into a purely radial line that ends at $r_{min} = 0.666$ m. The rest of the MITL is modeled with a 1-D transmission line, terminated with an inductive load. For the runs described here, $L_{load} = 37.9$ nH. This artificial treatment of the inner radial boundary of the geometry has no noticeable effect on the behavior of the null, but allows us to use a larger timestep, $\Delta t = 5$ ps, than if the entire MITL was included in the 3-D geometry.

RESULTS FROM FIELD-ONLY SIMULATIONS

We present results of a simulation in which the early lines are driven in the long pulse mode, and the late lines in the short pulse mode with a 250 ns delay. Fig. 3 compares time histories with a corresponding Microcap run. The agreement is very good for the load current and voltage, and fair for the feed line currents. Between 60 and 300 ns, some current from the early line is clearly flowing back up the late line.

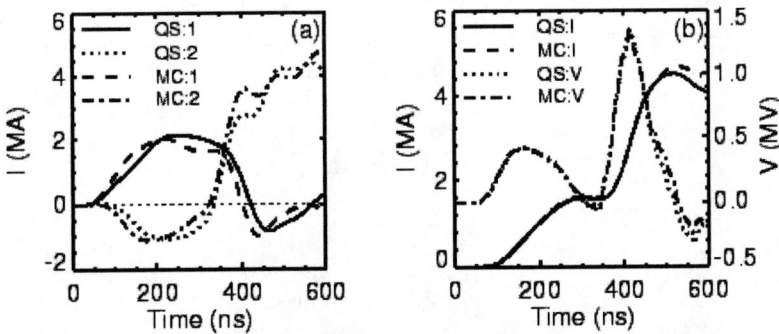

FIGURE 3. Comparison of QUICKSILVER and Microcap time histories: (a) current in each of the feed lines (line 1 is the early line), 0.5 m upstream of the center section, and (b) load current and voltage.

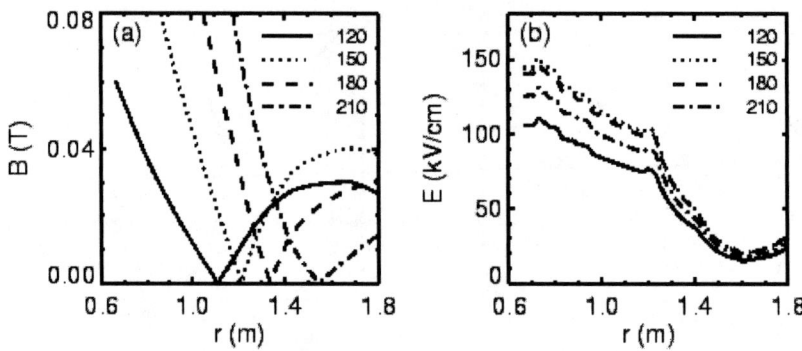

FIGURE 4. Radial lineouts of (a) |**B**| and, (b) |**E**|, at t = 120, 150, 180, and 210 ns. The magnetic field is along the symmetry plane $\phi = 60°$, while the electric field is averaged over all azimuth.

Two-dimensional contour plots of the magnetic field give a detailed picture of the null evolution. A localized null forms on the 60° symmetry plane at r ~ 1.1 m and t ~ 120 ns, well *inside* the vacuum region. It then slowly drifts outward, leaving the vacuum region, r = 1.59 m, at t ~ 210 ns. The four lineouts of |**B**| on the symmetry plane in Fig. 4a illustrate this behavior. During this time period, the MITL voltage passes through its peak value of ~0.5 MV for the early pulse, and starts to decrease. Fig. 4b shows that the electric field does not exceed 100 kV/cm upstream of the null.

CONCLUSION

Although a localized null exists in the vacuum region for ~90 ns, it is not certain that we would have electron emission upstream of the null, since the electric field is relatively low. However, since ZR would be operated at a voltage ~1.4 times higher, it could be more of a concern. We have just started new simulations with particle emission enabled to study electron losses at these nulls.

ACKNOWLEDGMENTS

Sandia is a multiprogram laboratory operated by Sandia Corporation, a Lockheed Martin Company, for the U.S. Department of Energy under contract DE-AC04-94-AL85000.

REFERENCES

1. D. H. McDaniel, *et al.*, "The Z Refurbishment Project", in *these proceedings*.
2. R. B. Spielman, C. Deeney, *et. al.*, *Phys. Plasmas* **5**, 2105-2111 (1998).
3. C. A. Hall, J. R. Asay, M. D. Knudson, *et al., Rev. Sci. Instrum* **72**, 3587-3595 (2001).
4. Microcap is a commercial circuit analysis program, available from Spectrum Software at http://www.spectrum-soft.com.
5. J. P. Quintenz, D. B. Seidel, *et al., Lasers and Particle Beams* **12**, 283-324 (1994).

The RT Instability in Cylindrical Implosion of A Jelly Ring

Yang Libing, Liao Haidong, Sun Chengwei, Ouyang Kai, Li Jun, Huang Xianbin

(Institute of Fluid Physics, CAEP, P.O. Box 919-108, Mianyang, Sichuan, 621900,China)

Abstract: The interfacial RT instability experiments on imploding jelly liners in cylindrically convergent geometry have been performed. The liner's instability growth was observed clearly with a high-speed framing camera. Jelly liners had different initial perturbation forms on their inner and outer interfaces, smooth one or sinusoidal one. The initial perturbations also had different magnitude and spatial frequency (for example, mode n=5, 10, 20). The experimental results show that the growth and coupling of perturbations on inner and outer surface are remarkably different. Meanwhile, the relevant 2D numerical simulation of hydrodynamics combined with Level Set method has been performed.

1 INTRODUCTION

Hydrodynamic instability and turbulent mixing are still challenging problem up to now. There are three mainly kinds of traditional interfacial instability. RT (Rayleigh-Taylor) instability, RM(Richtmyer-Meshkov) instability, KH(Kelvin-Helmholtz) instability. The interfacial instability is widely encountered in ICF (Inertial Confinement Fusion), detonation physics, electromagnetic implosion and other fields. It is one of keys to achieve high energy density in laboratories [1,2].

Three stages are included for interfacial instability in general: exponential increase of small perturbation, nonlinear growth and turbulent mixture. The theoretical and experimental studies on the first linear stage are comparatively sophisticated. However there are many difficulties to be overcome for the study of the latter stages. A primary question is how to distinguish different fluids and to form the initial perturbation in experiments.

The jelly liner(ring) driven by imploding gas detonation is a new experimental technology used in this paper to study the nonlinear growth process of interfacial perturbation and turbulent mixture in cylindrical geometry. In addition, various initial perturbations can be easily settled because of the jelly ring's plasticity. The evolutions of the ring's inner and outer surfaces are clearly observed by high speed framing photography, including perturbation growth, rebound of inner surface and turbulent mixing. The numerical simulation was based on a 2D hydrodynamic code with TVD scheme and in cooperation with level set method to treat the interfaces properly, and well reproduces the experimental observations for different initial perturbations[3,4,5].

2 EXPERIMENTAL SETUP

The scheme of experimental device is shown in Fig. 1. A jelly ring with outer diameter of 110mm, inner diameter of 80 mm, and thickness of 10 mm was sandwiched between two 20mm thick PMMA plates which were transparently polished.

There is a steel ring backing the PMMA plates to ensure that they did not expand in experiments. The experimental implosion chamber is shown in Fig. 2. There was air at 1 atm pressure inside the jelly ring, the outside was filled up with oxygen/acetylene mixture at 1 atm and of volume radio 2.5 : 1.

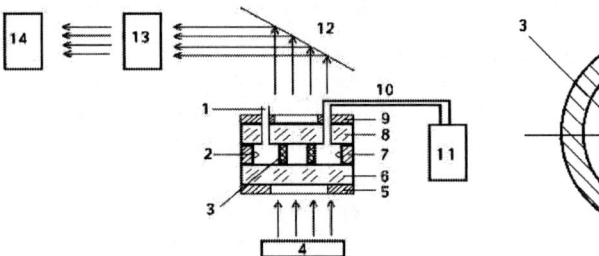

FIGURE 1 Scheme of experimental setup
1,10-gas in and exit; 2-spark gaps; 3-jelly ring; 4-light source; 5,7,9-steel ring; 6,8- PMMA plates; 11- mixture gas can; 12- mirror; 13- optical system; 14- high speed camera

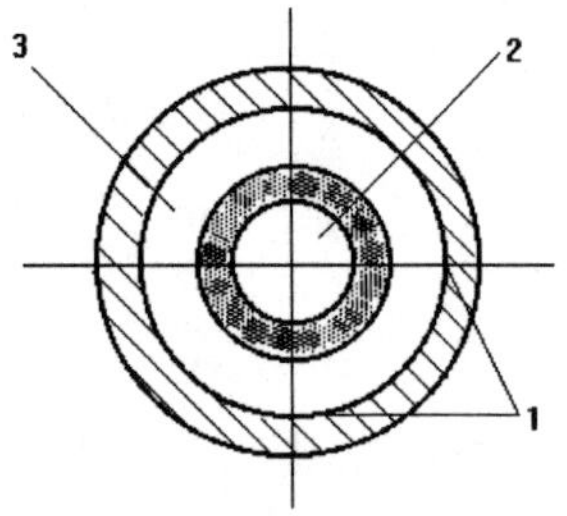

FIGURE 2 Cylindrical imploding chamber. 1-spark gaps array; 2-air; 3- oxygen/acetylene mixture gas

The preparation of jelly is simple, whose initial perturbations on both surface were determined by the shape of model. Sinusoidal patterns were used in the experiments. Because of the transparent PMMMA plates, the high speed framing photography with Schlieren technology has been utilized to record the flow pattern. The framing rate was 6.25×10^4 f/s, the time interval between two frames is 16μs.

3　EXPERIMENTAL AND CALCULATED RESULTS

In our experiments, the jelly rings had different initial perturbation forms with different magnitude and different spatial frequency. For example, azimuthal mode number n=0~20, perturbation magnitude 0~2mm. When the initial perturbations on both outer and inner boundaries had the same mode number, there was still alternation match of them: peak to peak, or peak to bottom.

Fig. 3 shows a typical experimental result. The outer boundary of jelly ring bore a sinusoidal perturbation of n=10 and amplitude 1mm. The inner boundary was smooth. It can be concluded that the outer boundary was accelerated toward the central axis until 520 μs from initiating detonation beginning. In this stage, the outer surface was instable and the inner one was stabile. From 520μs to 720μs, the outer surface was still accelerated toward the central axis, but the inner one is rebounded, both of them were stabile. After 720μs, both the outer and the inner surface were accelerated against the axis. Then the inner surface became instable and the outer one stabile. The calculated results confirm those issues. The perturbation growth obtained from compressible or incompressible model is in accord with the experimental results(see Fig. 4).

The calculated results of n=6, 10, 20 obtained with the incompressible model are shown in Fig.4 and Fig. 5. The initial perturbation amplitude is 1mm. The calculated results are well in agreement with the experimental data in the linear stage.

A feature of the experiments is the investigation of interaction between perturbations on outer and inner boundaries. Two kinds of experiments have been designed for this purpose, where both initial perturbations on outer and inner boundaries had the same mode number n(5,10) and amplitude, but differed in the spatial matching of them, the two initial perturbations matched in the form of "peak to peak"(phase matching), For the first kind, and "peak to bottom"(unmatched phase) for the second kind. The results of these experiments and relative simulations are shown in Fig. 6, Fig.7, Fig. 8 and Fig. 9. It is found significantly that the phase of inner interfacial perturbation became the opposite one during jelly ring implosion. Meanwhile that of outer one was still the same. Consequently, the liner with phase matching will be easier destroyed than the unmatched one. In fact, the matched liner broke down, but the unmatched liner was still of integrity after experiments. The numerical simulation reproduces correctly these interesting phenomena, though their physical reason is not unknown now.

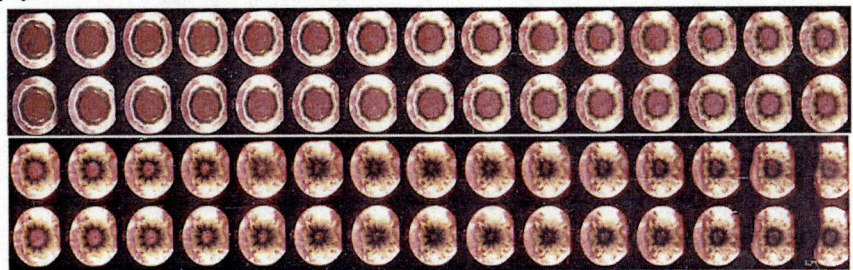

FIGURE 3 Framing photograms of shot No. 19 with initial perturbation n=10 and amplitude 1mm

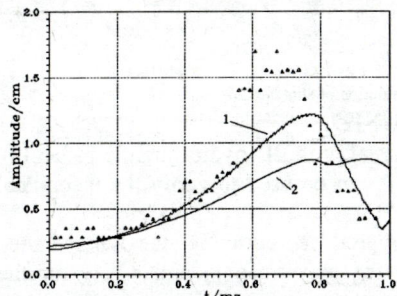

FIGURE 4 Numerical simulation and experimental measurement of perturbation on outer boundary in shot No.19. calculated:1- incompressible model, 2- compressible model; measured: Δ

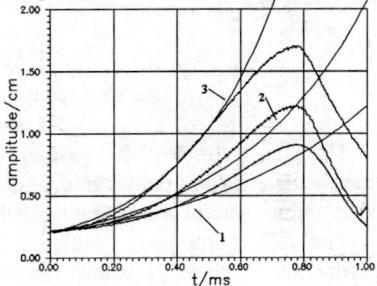

FIGURE 5 Experiments and simulations on different mode number of perturbation on outer boundary. 1. n= 6, 2. n= 10, 3. n= 20

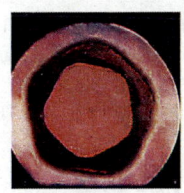 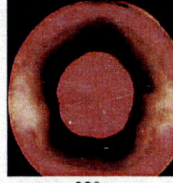 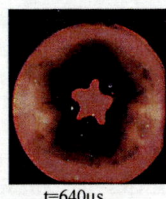 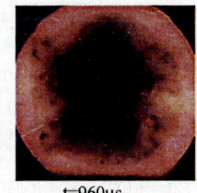

t=0µs   t=320µs   t=640µs   t=960µs

FIGURE 6 Photograms of shot No. 27 with initial unmatched perturbations of n=5

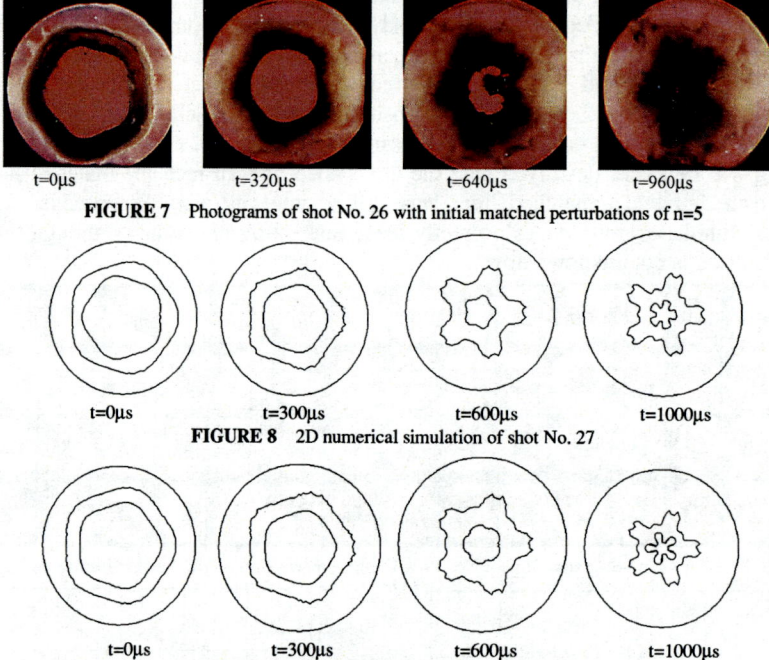

FIGURE 7 Photograms of shot No. 26 with initial matched perturbations of n=5

FIGURE 8 2D numerical simulation of shot No. 27

FIGURE 9 2D numerical simulation of shot No. 26

5 CONCLUSION

The jelly technology has distinctive advantages in 2D hydrodynamic instability studies, especially for converging implosion. It can be used to verify the theoretical models and numerical codes quantitatively.

The 2D numerical simulation can reproduce correctly the complicated experimental process. Its results accord with the experiment quantitatively for the linear stage, and qualitatively for the nonlinear stages. In the experiments on the interaction of perturbations on outer and inner boundaries, an interesting phase reverse of inner perturbation has been observed.

Reference

1. Meshkov, E.E., et al. Turbulent Mixing Development Investigation with Converging Jelly Rings. *The Proceeding of 4th International Workshop on the Physics of Compressible Turbulent Mixing*, 1993
2. Bakhrakh, S.M., et al. Hydrodynamic Instability in Strong Media, *UCRL-CR-126710*, 1997
3. Meshkov, E.E., et al. Computational and Experimental Studies of Hydrodynamic Instabilities and Turbulent Mixing. *DE95006837/HDM*, 1995
4. Baltrusaitis, R.M., et al. Simulation of Shock Generated Instabilities. *Phys. Fluids 8*. 2471, 1996
5. Liao Haidong, The Numerical Simulation of Rayleigh-Taylor Instability of A Finite Thickness Fluid. *Chinese Journal of Explosion and Shock Waves*. Vol 19. No.2. 1999

Phase Transitions in Metal under Fast Selfheating by High-Power Current Pulse

Konstantin V. Khishchenko, Svetlana I. Tkachenko, Vladimir E. Fortov,
Pavel R. Levashov, Igor' V. Lomonosov, Vladimir S. Vorob'ev

*Institute for High Energy Densities, Russian Academy of Sciences,
Izhorskaya Str. 13/19, Moscow, 125412, Russia*

Abstract. A numerical simulation of the initial stage of tungsten wire selfheating by a high-power microsecond current pulse has been carried out. A wide-range semiempirical equation of state to account for the effects of melting and evaporation of tungsten at high temperatures was used. The simulation results demonstrate that the phase transitions in metal, as well as possibility of realization of metastable liquid states have a considerable influence on the dynamics of the initial stage of electrical explosion under the effect of high-power current pulse.

INTRODUCTION

Electrical explosion of metals by a high-power current pulse is an accessible object for experimental studies of the numerous physical phenomena those occur in the matter at high energy densities [1–4]. Therefore, modeling of pulsed-power selfheating of metals is of great interest [5–8]. And it is essential to describe adequately an initial stage of this phenomenon, at which a dense cold core is formed, which exists throughout the process until the developed stage of metal explosion [4].

We carried out a numerical simulation of the initial stage of tungsten wire selfheating by a high-power microsecond current pulse. We implemented a semiempirical equation-of-state model [9], which describes the thermodynamic properties of tungsten, accounts for melting and evaporation, and covers the entire range from normal conditions to very high temperatures and pressures. The metastable (superheated or superexpanded) states of liquid tungsten were accounted for, as in Ref. [8].

MODELING

A one-dimensional cylindrical wire explosion has been modeled. We used the following set of magnetic-hydrodynamic (MHD) equations for the Lagrangian description,

$$dm/dt = 0, \quad \rho\, dv/dt = -\partial P/\partial r - (2\mu r^2)^{-1}\partial(r^2 B_\varphi^2)/\partial r,$$
$$\rho\, d\varepsilon/dt = -P\,\partial(rv)/\partial r + r^{-1}\partial(\kappa r\, \partial T/\partial r)/\partial r + j^2/\sigma_w + k(r)E_w,$$
$$d(\mu B_\varphi)/dt = \partial\left[(\sigma_w r)^{-1}\partial(rB_\varphi)/\partial r\right]/\partial r,$$

where v, m, ρ, and T are the velocity and the specimen mass, density, and temperature, respectively; $\varepsilon(\rho, T)$ and $P(\rho, T)$ are the specific internal energy and pressure given by the tungsten equation of state [8]; B_φ and μ are the magnetic induction and the absolute magnetic permeability; $\sigma_w(\rho, T)$ and $\kappa(\rho, T)$ are the electrical and thermal conductivities defined by the semiempirical formulas [3]; $j = (\mu r)^{-1}\partial(rB_\varphi)/\partial r$ is the current density; E_w is the specific energy loss due to heat radiation from the surface; and $k(a) = 1$ and $k(r) = 0$ at $r \neq a$, where a is the wire radius. The heating current $I = I(t)$ is determined by the equation describing the electrical circuit

$$d^2(LI)/dt^2 + d(R_l I)/dt + I/C = 0,$$

where L is the inductance; C is the capacitance of the capacitor; and R_l is the resistance of the specimen. Initial and boundary conditions are as in Ref. [8].

The numerical simulations of electrical explosion were carried out for the parameters of tungsten wire and electrical circuit from Ref. [3]: $a_0 = 0.175$ mm, $l = 8.7$ cm, $L = 4.5$ µH, $C = 6$ µF, and $U_0 = 20$ kV.

CALCULATION RESULTS

The calculated radial distribution functions of pressure, temperature, particle velocity, and weight concentration of the solid phase $v = (\rho_m^{-1} - \rho^{-1})(\rho_m^{-1} - \rho_s^{-1})^{-1}$ (where ρ_s^{-1} and ρ_m^{-1} are densities of the solid and liquid phases on melting curve) as a function of heating time are shown in Fig. 1 and 2. It can be seen that compression and expansion waves alternate with each other and propagate along the wire radius with velocity of about the sound velocity of liquid tungsten (~ 3.6 km/s). A perturbation originates on the free surface due to the inhomogeneous distribution of pressure in the wire at the beginning of the melting transition.

The phase tracks of some wire layers in the density–temperature and pressure–temperature planes are shown in Fig. 3. The relative temperature and density variations increase (to 6% and 1.5%, respectively), when melting starts, decrease (to 0.6% and 0.1%) when melting is completed, and increase again after the initiation of intense surface boiling. These variations are not significant and the ρ–T phase tracks of the different wire layers almost coincide with the liquid binodal curve. At the same time, the pressure varies by an order of magnitude along the wire radius. Such behavior is typical for condensed matter.

The liquid phase is known to exist for a while in metastable states under certain conditions [10]. The total duration of the metastable states in the simulated process varies along the wire radius, and the maximal total duration $\tau \sim 20$ ns corresponds to

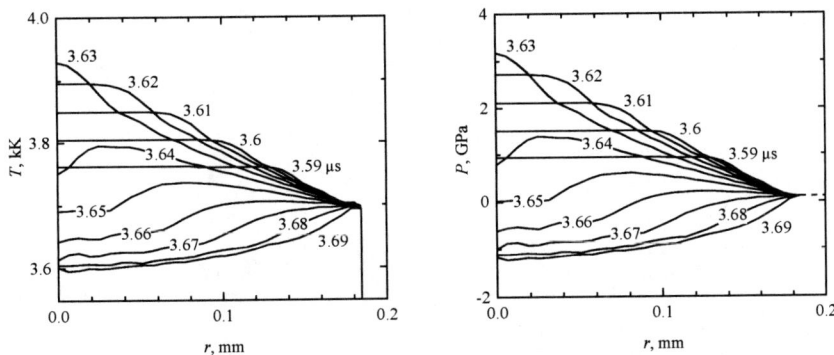

FIGURE 1. Temperature (T) and pressure (P) of the heated wire as a function of radial position and time, from $t = 3.59$ to 3.69 μs. The dashed line denotes the ambient pressure P_a.

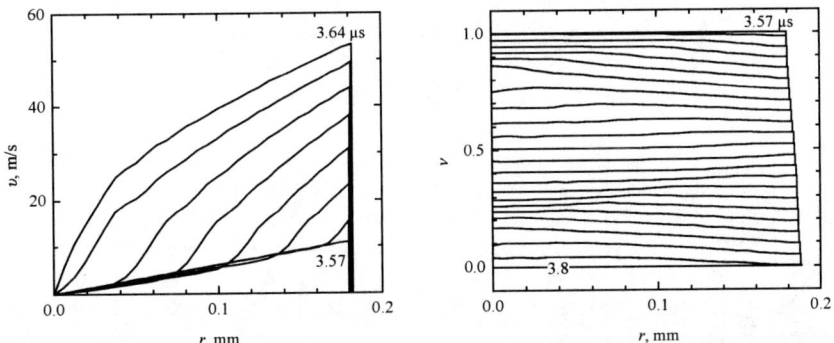

FIGURE 2. Particle velocity (v) and weight concentration of the solid phase (ν) of the heated wire as a function of radial position and time, from $t = 3.57$ to 3.8 μs. The time interval between neighboring curves is 0.01 μs.

the innermost layer of the wire. According to the theory of homogeneous nucleation [10], the time τ_n of nuclei formation in the absence of external fields can be written as $\tau_n = (BnV)^{-1} \exp G$, where $G = A_c/T$ is the Gibbs number, A_c is the work of nuclei formation, $B \approx 10^{10}$ s^{-1} is the kinetic factor, n is the number of nucleation centers per unit of volume, $V = \pi a_0^2 l$ is the wire volume. As shown in Ref. [11], the mean time of the expectation for the appearance of a nucleus becomes comparable with duration of the wire heating at $T > 13.5$ kK. This is the reason why we have to account for the metastable states when modeling the wire heating.

The typical time of MHD instability growth for the modeled regime of wire heating is about $\tau_{MHD} \sim 2$ μs [12]. The melting transition is completed at $t \approx 3.8$ μs. Consequently the growth of the MHD instability and an abrupt increase of the nucleation rate are expected shortly after $t \sim 5.6$–5.8 μs. The current model is not sufficient for the simulation of later stages of these processes. Therefore, the simulation was terminated at $t \approx 5.8$ μs.

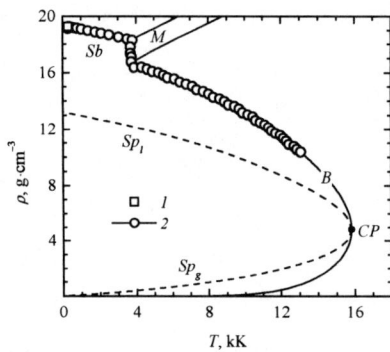 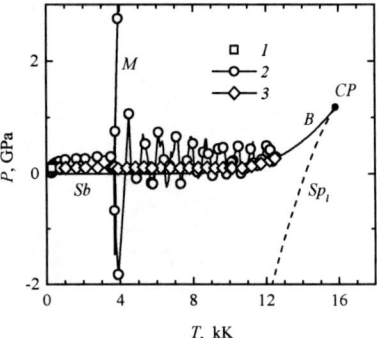

FIGURE 3. Tracks of the wire layers with $r/a = 0.98$ (3) and 0.1 (2) in the ρ–T and P–T planes. M, Sb, and B are the phase boundaries of the solid–liquid, solid–vapor, and liquid–vapor transitions, Sp_l and Sp_g are the spinodals of the liquid and gas phases, and CP is the critical point of tungsten from Ref. [8]; *1* — the initial state.

CONCLUSION

The phase transitions, as well as possibility of realization of metastable states of liquid have a considerable influence on the dynamics of the initial stage of electric explosions of conductors under the effect of high-power current pulse. The melting process discussed here gives rise to alternating compression and expansion waves. These waves propagate from the outer surface to the wire axis and in the reverse direction. The density and temperature are only slightly perturbed by these waves, but the pressure varies by an order of magnitude.

REFERENCES

1. Lebedev, S. V., Savvatimskii, A. I., *Usp. Fiz. Nauk.* **144**, 215 (1984)
2. Gathers, G. R., *Rep. Progr. Phys.* **49**, 341 (1986).
3. Koval', S. V., Kuskova, N. I., Tkachenko, S. I., *High Temp.* **35**, 863 (1997).
4. Pikuz, S. A., Shelkovenko, T. A., Sinars, D. B., Greenly, J. B., Dimant, Y. S., Hammer, D. A., *Phys. Rev. Lett.* **83**, 4313 (1999).
5. Peterson, D. L., Bowers, R. L., Brownell, J. H., Greene, A. E., McLenithan, K. D., Oliphant, T. A., Roderick, N. F., Scannapieco, A. J., *Phys. Plasmas* **3**, 368 (1996).
6. Chittenden, J. P., Aliaga-Rossel, R., Lebedev, S. V., Mitchell, I. H., Tatarakis, M., Bell, A. R., Haines, M. G., *Phys. Plasmas* **4**, 4309 (1997).
7. Ivanenkov, G. V., Stepniewski, W., *Plasma Phys. Rep.* **26**, 21 (2000).
8. Tkachenko, S. I., Khishchenko, K. V., Vorob'ev, V. S., Levashov, P. R., Lomonosov, I. V., Fortov, V. E., *High Temp.* **39**, 674 (2001).
9. Bushman, A. V., Fortov, V. E., *Sov. Tech. Rev. B: Therm. Phys.* **1**, 219 (1987).
10. Skripov, V. P., *Metastable Liquids*, Wiley, New York, 1974.
11. Vorob'ev, V. S., Malyshenko, S. P., Tkachenko, S. I., and Fortov, V. E., *JETP Lett.* **75**, 373 (2002).
12. Gupta, A. S., *Proc. Roy. Soc. A* **278**, 214 (1964).

Experiments With Radiatively Cooled Supersonic Plasma Jets Generated in Conical Wire Array Z-Pinches.

S.V. Lebedev, D.J. Ampleford, S.N. Bland, J.P. Chittenden, A. Ciardi, N. Naz, M.G. Haines, A. Frank[a,b], E. Blackman[a,b], T. Gardiner[a,b]

The Blackett Laboratory, Imperial College, London SW7 2BW, UK
[a]*Department of Physics and Astronomy, University of Rochester, Rochester NY 14627-0171 USA*
[b]*Laboratory for Laser Energetics, University of Rochester, Rochester NY 14627-0171 USA*

Abstract. We present results of astrophysically relevant experiments where highly supersonic plasma jets are generated via conically convergent plasma flows in a conical wire array Z-pinch. Stagnation of plasma flow on the axis of symmetry forms a standing conical shock effectively collimating the flow in the axial direction. This scenario is essentially similar to that discussed by Canto and collaborators [1] as a purely hydrodynamic mechanism for jet formation in astrophysical systems. Experiments using different materials (Al, Fe and W) show that a hypersonic (M ~ 20), well-collimated jet is generated when the radiative cooling rate of the plasma is significant.

INTRODUCTION

Highly collimated supersonic plasma jets are a ubiquitous phenomena occurring in many astrophysical environments. These jets are observed propagating from sources as diverse as Active Galactic Nuclei, Young Stellar Objects and Planetary Nebulae. The progress in high energy density devices used for ICF allows for the possibility of scaled [2] experimental studies of hypersonic plasma flows which are relevant to astrophysical situations [3]. Recent work [4,5] using high power lasers have produced radiative flows whose scalings appear to allow contact with astrophysical parameter regimes including radiative jets. In this paper we present the first results of a series of experiments [6] designed to study high Mach number radiative plasma jets using fast z-pinch devices.

EXPERIMENTAL CONFIGURATION

The schematic of the experimental set-up is shown in Fig 1. A fast rising current, reaching 1MA in ~240ns, is applied to a ~1cm long conical array composed of 16 fine metallic wires. The small radius of the array is 8mm, and wires are inclined at an angle of 30^0 to the array axis. Three different wire materials were used: Al, stainless steel and W of 25µm, 25µm and 18µm diameter, respectively. The resistive heating of the wires by the current rapidly, in a few nanoseconds, converts the wires into a heterogeneous structure with dense, practically neutral cold cores surrounded by a low

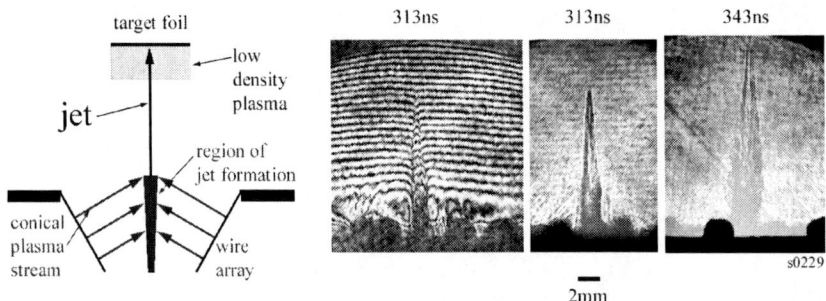

FIGURE 1 Schematic of the experiment and results of laser probing of a jet generated in tungsten conical wire array. Interferogramm (left) and two schlieren images are obtained in the same experiment.

density hot coronal plasma. The global magnetic field generated by the current flowing through the array produces the net **JxB** pinching force, which accelerates the coronal plasma towards the array axis. The characteristic density of the coronal plasma near the wires is $\sim 10^{17}$ cm^{-3} and the measured inward streaming velocity is $\sim 1.5 \times 10^7$ cm/s [7]. The wire cores act as a reservoir of material, allowing the process of formation and sweeping of coronal plasma from the cores to continue for the entire duration of the current pulse in the experiment, thus producing a quasi-stationary converging flow of plasma.

For purely cylindrical wire array z-pinches, stagnation of the coronal plasma flow on the array axis forms a dense, narrow and stable precursor plasma column with characteristic density of $\sim 10^{-3}$-10^{-2} g/cm^3 and electron temperature of T~50eV. The column is confined by the kinetic pressure ρV^2 of the plasma flow [7,8,9]. In a conical array the flow of coronal plasma retains an axial component of momentum after collision on the array axis. The conical standing shock that forms on the axis will effectively redirect the flow in the axial direction and form a plasma jet. This process has been well studied in astrophysical contexts [1].

EFFECT OF RADIATIVE COOLING ON JET FORMATION

Fig 1 shows results from laser probing measurements of a jet formed in the tungsten wire array. The jet has a sharp, well-defined boundary and propagates with a velocity significantly higher than its radial expansion. The measured displacement of the jet tip on the two schlieren images gives velocity $V_z = 2 \times 10^7$ cm/s, while the radial expansion of the jet is much smaller, $V_r < 7 \times 10^5$ cm/s. This allows estimate the internal Mach number of the jet as $M \sim V_z/V_r \sim 30$.

Variation of the wire material (Al, stainless steel (Fe) or W) in the experiments allows us to examine the importance of radiation losses on jet collimation. Previous experiments [7,8] with cylindrical arrays indicate that the parameters of the plasma flow (an inward velocity and the mass flux) are essentially insensitive to the material

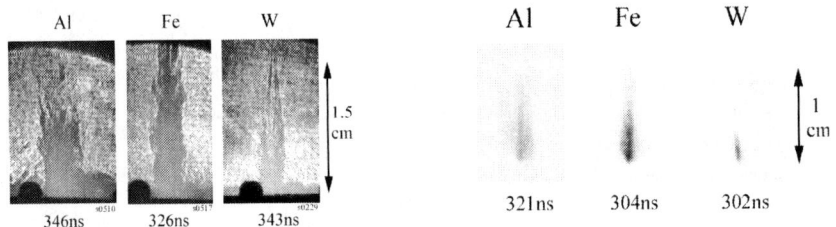

FIGURE 2. Laser probing (left) and soft x-ray (right) images of plasma jets formed in Al, stainless steel and W wire arrays show that degree of collimation increases for elements with higher atomic number, in which rate of radiative cooling is higher.

used. The increase of the atomic number (A), however, leads to a significant increase in the rate of energy loss through radiation [10].

Typical images of jets formed in experiments using conical arrays of Al, Fe and W wires are shown in Fig 2. These images demonstrate that the degree of jet collimation does strongly depend on the atomic number, increasing with A. The diameters of the jets at the end of the formation region, where the converging plasma flow is still present, decrease with increasing atomic number, being ~3.5, 3 and 2.5mm for Al, Fe and W, respectively. For Al the jet is slightly diverging, while for stainless steel the observed opening angle is close to zero, and the jet radius remains constant over the propagation length of >15mm. For tungsten, an apparent convergence of the jet occurs at the jet head. The radius of the jet decreases towards the tip and an opening angle becomes negative there. The shape of the tip may reflect the temporal history of the formation of conical standing shock, however more study of this point is required. The observed dependence of jet collimation on atomic number is consistent with predictions of Canto's model [1] and indicates that changes in jet radius/opening angle are due to a higher levels of radiative losses in high A plasmas.

Experimental evidence of fast radiative cooling of the jet is also seen from gated soft x-ray images (Fig.2), filtered to transmit radiation in the interval ~190-290eV. The data show that emission from the jets is rapidly decreasing along the beam. The length of the emitting part of the jet is significantly smaller than that measured at the same time by laser probing for all materials tested in the experiments (Al, Fe and W). The characteristic length of the emission decay (from the base of the jet) is ~7~mm for Al, ~5mm for Fe and ~2mm for W, while the characteristic lengths of the jets, seen in laser probing images at the same time, are ~1.5cm.

JETS IN TWISTED CONICAL WIRE ARRAYS

The dynamics of jet formation in the twisted conical wire arrays could be modified by the presence of axial magnetic field generated in this configuration. This could add an angular momentum to the converging conical flow, due to non-zero azimuthal component of the JxB force accelerating the coronal plasma from the wires. Fig.3 shows the side-on and end-on XUV images of jets formed in conical tungsten wire arrays. It is seen that even a relatively small twist angle ($\pi/8$) affects the divergence of

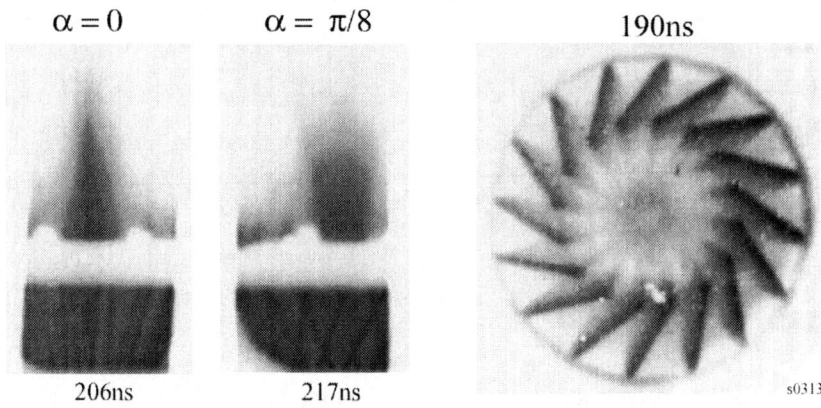

FIGURE 3. Side –on XUV images (left) of plasma jets generated in conical wire arrays twisted by 0 and π/8 angles. End-on XUV image (right) of plasma flow in twisted conical wire array.

the jet. In contrast, experiments with cylindrical twisted wire arrays, in which a comparable axial magnetic field is generated, show that the precursor column is formed in the same way as in non-twisted cylindrical arrays. This indicates that the angular momentum in the plasma flow is the main reason for the observed difference in the jet formation in twisted conical wire arrays.

CONCLUSIONS

We have presented results from a series of experiments designed to create and study high Mach number collimated plasma flows using conical configuration of wire array z-pinch. Translation of this technique of formation of highly supersonic plasma jets onto Z-pinch facilities with a significantly higher drive current than in the present experiments (e.g. 20MA Z facility at SNL) would open additional opportunities for laboratory studies of flow driven hydrodynamic instabilities relevant to astrophysics.

REFERENCES

1. J. Canto et al., AAP, **192**, 287 (1988).
2. D.D. Ryutov et al., APJ, **518**, 821 (1999).
3. B. Remington et al., Science, **284**, 1488 (1999).
4. D. R. Farley et al., Phys. Rev. Let., **83**, 1982 (1999).
5. K. Shigemori et al, Phys. Rev. E., **62**, 8838 (2000).
6. S.V. Lebedev et al, APJ, **564**, 113 (2002)
7. S.V. Lebedev et al., Phys. Plasmas **6**, 2016 (1999).
8. S.V. Lebedev et al., Phys. Plasmas, **8**, 3734 (2001).
9. J.P. Chittenden et al., Phys. Plasmas **8**, 2305 (2001).
10. D.E. Post et al., At. Data Nucl. Data Tables **20**, 398 (1977).

Deflection of Supersonic Plasma Jets by Ionised Hydrocarbon Targets

D.J. Ampleford, S.V. Lebedev, S.N. Bland, A. Ciardi,
M. Sherlock, J.P Chittenden, M.G. Haines

Blackett Laboratory, Imperial College, London, SW7 2BW, UK

Abstract. A new application of the wire array Z-pinch is the investigation of astrophysical phenomena. Here highly supersonic jets (mach number, M~20) formed using over-massed tungsten wire arrays in a conical arrangement [1] on the MAGPIE facility are to study the interaction with plasma targets. Self X-ray emission from the jet formation region is used to ablate various off-axis targets (CH foams, C and CHO foils). Interferometer, shadowgram and gated soft X-ray images show a deflection of the jet by the target plasma and shocks propagating within the jet. Bending angles up to 25° were observed, depending on the initial target position. Hydrodynamic simulations suggest that this deflection is unlikely to be caused by the static density gradient of 5×10^{17} cm^{-4} observed in the target plasma, but rather by momentum exchange from a sonic wind flowing from the target.

INTRODUCTION

Previous experiments [1] performed on the Magpie generator and simulations [2] show that precursor flow from a wire array in a conical arrangement produces a highly collimated supersonic jet. A typical jet produced by this method (from 16x18mm tungsten wires as used in experiments detailed below) will have a velocity 200 km s^{-1}, radius 0.5 mm, temperature 50 eV and ion charge, Z ~10 giving a Mach number (ratio of velocity to sound speed), M~20. In these experiments we use the plasma jet formed by this arrangement to study the interaction with various plasma targets formed by ablation of CH foams and C and CHO foils.

The primary purpose of these experiments is to model astrophysical jet interactions with ambient material. Given certain scaling relations based on invariants in the Euler Equations and MHD equations [3, 4, 5] it is possible to model astrophysical phenomena in the laboratory. The method of jet production used in the experiments described here has the scaling parameters described at the beginning of table along with those typical of Young Stellar Objects. Also shown in this table are predicted scaling parameters [2] for the Z-generator and those found for some typical laser produced jet experiments [6, 7]. As can be seen from the table, the different types of experiments cover contrasting regimes of scaling parameter space. The ability to benchmark simulations of plasma jets is also improved by increasing the variety in laboratory jet experiments.

TABLE 1. Scaling parameters of different jets.

Parameter	Astrophysical (YSO)	Magpie generator	Z-generator (predicted)	Farley (Nova)	Shigemori	
Material		W array	W array	Gold	CH	Gold
Mach Number	>10	20+	40	15	2-8	10-50
Density Contrast	≥ 1	>> 100	>> 100	>> 100	>> 100	>> 100
Cooling Parameter	<1	≤ 1	≤ 1	1	~ 40	0.7

EXPERIMENTAL SETUP

In these experiments on the Magpie generator (1MA peak current is reached in 240ns) a conical array of sixteen 18mm tungsten wires (thus an over-massed, non-imploding array) is used, with a 30° opening angle, 16mm lower radius and 1cm length. For this arrangement the JxB force acting on the coronal structure surrounding each wire will be perpendicular to the wire, and hence have an axial component. This results in the formation of a conical shock [2], as compared to the cylindrical shock (precursor plasma column) found in a standard cylindrical wire array. The result of this shock is to absorb the radial component of the velocity of the streams, and convert it into an axial flow (by dissipation of radial kinetic energy), which emerges from the top of the array. Radiative cooling collimates this flow into a highly supersonic jet.

Initial jet interaction experiments were performed with a CHO foil placed perpendicular to the jet, approx 15mm above the anode plate. The results of these experiments show that the radiation pulse from the wire array significantly ablates the foil and, given a sufficiently low target density, the jet can propagate through this ambient background with no major loss of collimation [1]. There is also a clear effect of jet foil interactions, with shocks seen in the ablated foil.

Based on these results, an experiment was developed with plastic targets placed parallel to the jet, approximately 2mm off axis and 5-10mm above the anode plate, as shown in Figure 1. This arrangement is used as an approximate scaling of jet interactions with asymmetric ambient clouds.

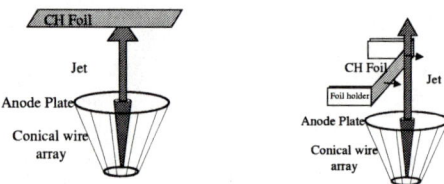

FIGURE 1. Experimental arrangements (a) for foil perpendicular to jet and (b) for foil parallel to jet.

RESULTS

Figure 2 shows typical interferometry, shadowgraphy and gated soft X-ray images for a jet passing an off-axis target[1]. It can be seen that the jet is bent away from the target, and maintains its collimation. Further experiments show that the jet deflection angle is reduced as the radial distance to the target is increased (and hence the cloud density, density gradient and probably flow speed is changed). Different target plasmas have varying effects on the jet, such as CHO foils causing a greater deflection than CH foams.

From interferometry it is determined that a CH cloud that deflects the jet by six degrees has, in the vicinity of the bend point, an ion density of 6.3×10^{17} cm^{-3}, density gradient of 4.2×10^{17} cm^{-4} and a wind velocity greater than 10 km s^{-1} (taken from the time for material to reach the jet).

A similar experiment, but with two foils on opposite sides of the jet, have also been performed. In this, the jet is seen to bend first to the right and then back to towards the axis as the two foils were at slightly different axial positions. It also appears in this arrangement that the ambient material compresses the jet radially. Currently experiments are also being attempted with two foams in this arrangement, but above a cylindrical rather than conical array (so no significant jet is present), with the intention to observe a shock produced by the collision of plasmas expanding from the targets.

In some experiments where the jet is bent, shocks are seen in two different locations. Firstly, within the jet a shock is seen level with the target plasma. Also in some experiments a shock-like boundary is seen between the jet and the ablated target material.

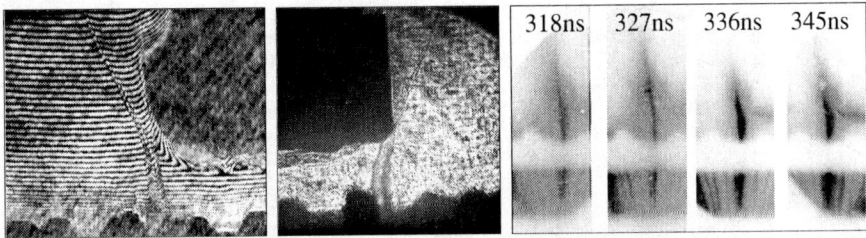

FIGURE 2. A deflected jet by CH foil at 308ns (a) interferogram, (b) Shadowgram (at skewed viewing angle) and (c) framed XUV pinhole images

Simulations have been performed in an attempt to understand the behavior of the jet in these situations, and determine the mechanism by which it is bent. Both 2D MHD [2] and 2D kinetic ion, fluid electron hybrid (developed by M. Sherlock) codes have been used. By using the experimentally determined values for jet parameters, and density, density gradient and likely wind velocity in these simulations it has been

---

[1] The gated X-ray is from a different shot to the other images due to experimental limitations (the inherent symmetry in normal Z-pinch experiments means that Magpie is arranged such that the laser probing and XUV framing camera are perpendicular to each other).

determined that the bending is not the result of the pressure gradient; the simulations do show that this bending can be produced by the wind expanding from the target.

SUMMARY & FUTURE WORK

In this experiment we have observed strong deflections of the supersonic plasma jet propagating through non-uniform plasma expansion from the targets. Despite the deflection by up to 25º, the jet retains a high degree of collimation.

Experiments have determined the physical parameters of the ambient cloud that deflects the jets. Simulations show that the mechanism for this deflection is momentum transfer from a sonic wind rather than pressure gradients within the cloud, although further experimental verification must be performed.

The current aim is to extend the lab-astrophysics application of the wire array Z-pinch by using photo-ionised gas targets and investigating the effect of an external linear B-field perpendicular to the direction of jet propagation. We may also continue our initial studies into the collision of counter propagating jets using twin conical arrays.

ACKNOWLEDGEMENTS

Work supported by Sandia National Laboratories and the U.S. Department of Energy (Contract No. DE-FG03-98DP00217).

REFERENCES

1. Lebedev, S.V., Chittenden, J.P., Beg, F.N., Ciardi, A., Ampleford, D.J., Hughes, S., Haines, M.G., Frank, A., Blackman, E.G., Gardiner, T., *ApJ* **564**, 113-119 (2002).
2. Ciardi, A., Lebedev, S.V., Chittenden, J.P., Bland, S.N., to be published in Laser and Particles beams (2002).
3. Blondin, J.M., Fryxell, B.A. & Königl, A., *ApJ* **360**, 370-386 (1990).
4. Ryutov, D., Drake, R.P., Kane, J., Liang, E., Remington, B.A. & Wood-Vasey, W.M., *ApJ* **518**, 821-832. (1999).
5. Ryutov, D., Remington, B.A., Robey, H.F. & Drake, R.P. *Phys. Plasmas*, **8**, 1804-1816 (2001).
6. Farley, D.R., Estabrook, K.G., Glendinning, S.G., Glenzer, S.H., Remington, B.A., Shigemori, K., Stone, J. M., Wallace, R.J., Zimmerman, G.B. & Harte, J.A..Phys. Rev. Lett., **83**, 1982-1985 (1999).
7. Shigemori, K., Kodama, R., Farley, D.R., Koase, T., Estabrook, K.G., Remington, B.A., Ryutov, D.D., Ochi, Y., Azechi, H., Stone, J. & Turner, N. Phys. Rev. E, **62**, 8838-8841 (2000).

Plastic Deformation and Perforation of Metal using Metallic Jet

Partha Sarkar, Shashank Chaturvedi, Anurag Shyam, Rajesh Kumar, Deepak Lathi, Vilas Chaudhari, Rishi Verma, Jaswant Sonara, Kunal Shah, Biswajit Adhikary

Institute for Plasma Research, Bhat, Gandhinagar – 382428, Gujarat, INDIA

Abstract. Pulsed underwater electrical discharges have been used in the past to generate pressures of the order of several tens of kilobars, for applications such as rock fragmentation and metallic jet production. Preliminary results for a metallic jet system have been reported earlier. A modified design for a metallic jet production system is reported here. With this arrangement, we are able to perforate 11 mm thick aluminium sheet. Such a system, at higher energy levels, could be used for oil and gas well perforation.

INTRODUCTION

Capacitor-bank driven pulsed underwater electrical discharges can create high-pressure shocks/pressure waves of up to a few GPa. This technique has been used in the past for various applications, such as rock fragmentation and metallic jet production. A low-inductance discharge circuit allows the production of short-duration discharges, whose effect is similar to that of detonation of chemical explosives, but at lower energy and pressure levels.

The geometry of the system, as shown in Figure 1, is similar to that of conventional shaped charges which uses chemical explosive. Here, the energy stored in the capacitor bank is discharged into a fluid-filled cavity. The discharge creates a high-pressure pulse in the cavity by vaporizing the fluid in a small region. This pressure pulse acts on a conical metallic liner, kept in contact with the fluid, producing radial collapse as well as axial acceleration similar to that in shape charge systems.

Early results from such a system have been reported in [1].

EXPERIMENTAL SETUP

The experiment is driven by a 100 kJ, 44 kV capacitor bank. The bank consists of two capacitors of 50 kJ each. The capacitor bank is charged by a 100 mA, 45 kV DC power supply. The discharge is triggered by a rail-gap switch. The rail-gap switch is triggered by a trigger generator having an output dV/dt of 6 kV/nano-sec. The trigger

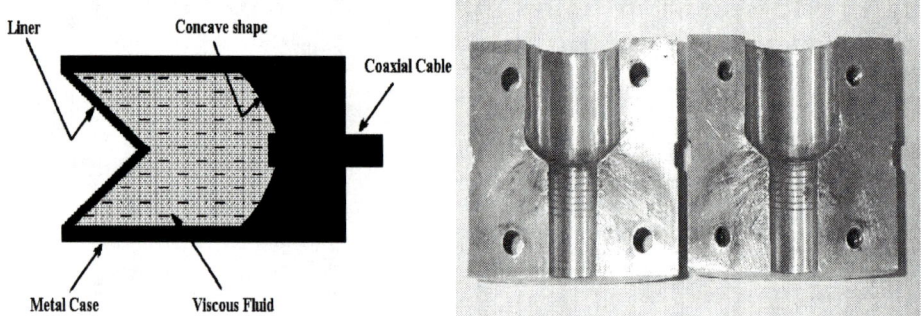

FIGURE 1. Geometry of the perforating system and sectional view of the casing.

generator forms the sub-master trigger generator. The initial trigger pulse comes from a master trigger generator consisting of a pulse transformer. Power is carried to the load by a heavy-duty RG-218 coaxial cable, through a collector plate. In order to reduce the inductance and to increase the current carrying capacity, eight coaxial cables have been used from the rail-gap switch to the collector plate. From the collector plate, a flat-plate transmission line connects to the load. Finally, a single RG-218 coaxial cable, 300 mm long, delivers power into the cavity. A schematic of the system is shown in Figure 2.

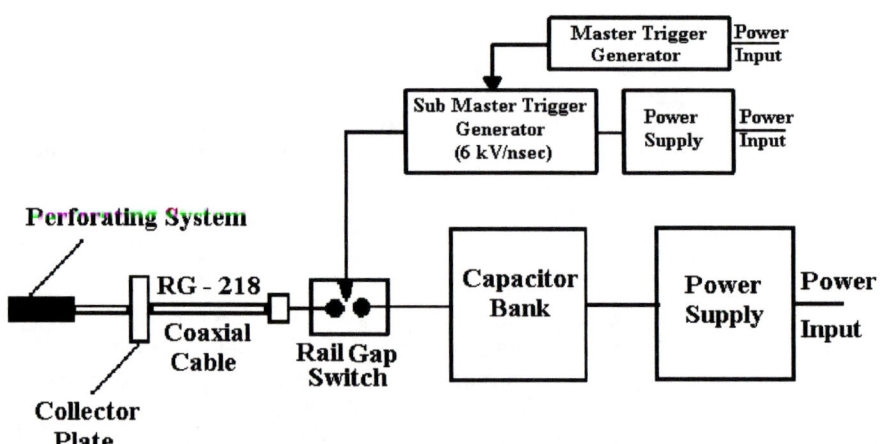

FIGURE 2. Schematic diagram of the system.

The casing of the perforating system is made up of SS-304, 20 mm thick, as shown in Figure 1. A cavity is formed inside the casing, 42 mm in both length and diameter. The cavity is open at one end and blocked at other end, by coaxial cable RG-218. The open end of the cavity is bounded by a conical liner. The cavity is filled with viscous fluid, which is a mixture of ordinary water and bentonite. The high viscosity facilitates accurate placement of the liner, and also allows casing orientations other than vertical.

The conical liner is made of copper. Typical liner dimensions are a base diameter of 38-42 mm, height of 38-40 mm and 0.19 mm thick. Aluminium sheets of varying thickness have been used as targets for perforation.

RESULTS AND DISCUSSION

The charging voltage has been varied from 13 kV to 22 kV. The peak current obtained at 22 kV was 288 kA. Given that the maximum allowed current from each capacitor is 150 kA, it is not desirable to go beyond 22 kV. We have used 160 µm and 190 µm thick copper liners. The 190 µm copper liner gave better results, which are shown in Table 1. Typical discharge current and voltage waveforms are shown in Figure 3. Figure 4 shows some perforated sample pieces.

TABLE 1. Test Results

Charging Voltage (kV)	Discharge Current (kA)	Target Thickness (mm)	Stand-off distance (mm)	Perforation	Energy (kJ)	Velocity (km/sec)
13.0	170.0		55.0	No	8.8	0.45
13.0	169.6	5.5	50.0	Yes	8.8	--
13.0	169.6		50.0	No	8.8	--
19.5	258.0	6.5	45.0	Yes	19.8	0.53
19.5	258.0		45.0	No	19.8	--
21.0	273.0	8.0	40.0	Yes	22.9	0.58
22.0	288.0	11.0	40.0	Yes	25.0	0.60

The stand-off distance is the distance between the target and the open end of the cavity. Table-1 shows that optimal results are obtained with a stand-off distance of 1 to 1.25 times the liner base diameter. As the distance increases, penetration thickness reduces. This may be due to break up of the jet, although this cannot yet be confirmed. The jet velocity has been measured by a foil breaking technique [1]. The velocity thus measured may be somewhat lower than the actual value. This is because, as the jet strikes the first foil, it loses some of its momentum, and thus takes a longer time to reach the second foil.

These experiments were earlier being performed using a 60 kJ, 15 kV capacitor bank, which had yielded a maximum penetration of 6.5 mm of aluminium. The use of the new bank (100 kJ, 44 kV) allows shorter duration discharges (from 200 µs to 46 µs), which have yielded higher penetration. The physical reason for higher penetration is not clear – it could be due either to higher jet velocities or to greater lengths. A fast framing camera is required, both to verify the velocity measurement, and to determine the shape and size of the jet. Another important observation is that the higher the energy, the more is the depth of penetration. This issue has to be studied in detail. This requires enhancement of the capacitor bank, and also the fabrication of larger loads.

We have also noticed that a certain amount of copper gets deposited around the periphery of the hole, which is formed due to penetration of the jet in the target. So it can be concluded that only a small portion of the liner mass goes into the jet formation.

In case of 11 mm thick target, the hole (perforation) is not so pronounced as in other cases. The energy has to be increased for achieving the desired result.

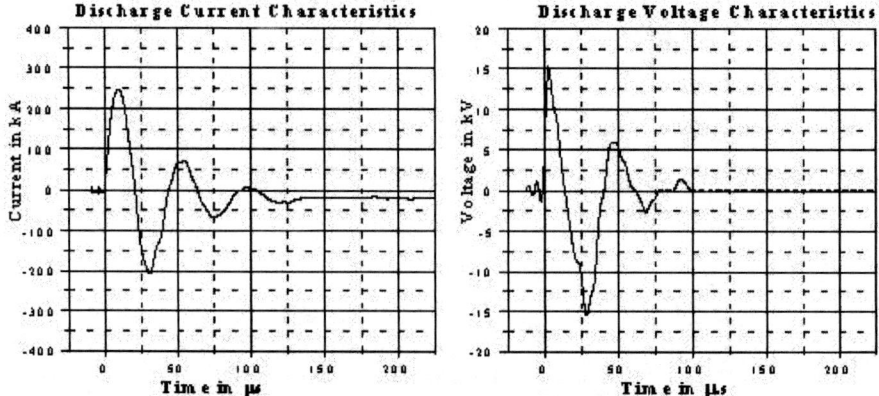

Figure 3. Discharge current and voltage characteristics.

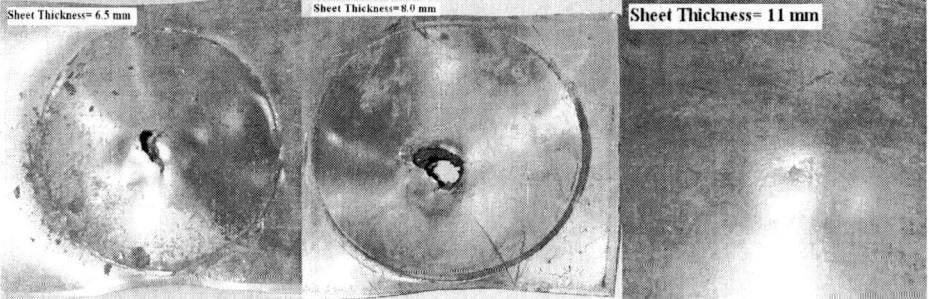

Figure 4. Perforated aluminium sheets of thickness 6.5 mm, 8.0 mm and 11.0 mm.

CONCLUSION

These sets of experiments are aimed at exploring the feasibility of using metal jets produced by underwater discharges for perforation. Up to 11 mm thick aluminium sheet has been perforated and with a jet velocity of 0.6 km/sec. We are now planning studies with a bigger capacitor bank and heavy-metal liners.

REFERENCES

1. P. Sarkar et al. "Metallic Jet Production using Pulsed Electrical Discharges in Water", in *Digest of Technical Papers, PPPS-2001, IEEE Pulsed Power Plasma Science Conference*, Las Vegas, U.S.A., June 17 – 22, 2001, pp.1378-1381.

Anomalous Resistivity Change in NiFe$_2$O$_4$ Nanosized Powders Synthesized by Pulsed Wire Discharge

Hisayuki Suematsu, Kazuhiro Ishizaka, Yoshiaki Kinemuchi, Tsuneo Suzuki, Weihua Jiang and Kiyoshi Yatsui

Extreme Energy-Density Research Institute, *Department of Electrical Engineering, Nagaoka University of Technology, Nagaoka 940-2188, Japan

Abstract. Nanosized powders of nickel ferrite (NiFe$_2$O$_4$) were synthesized by a powder synthesis method of pulsed wire discharge. Anomalous change in electrical resistivity in NiFe$_2$O$_4$ was observed above 200 °C in sintered NiFe$_2$O$_4$ nanosized powders. Here we show possible strong spin interactions in NiFe$_2$O$_4$.

INTRODUCTION

Nickel ferrite (NiFe$_2$O$_4$) is known as a ferrimagnetic material and has widely been used in magnetic recording media and magnetic fluids. Crystal structure of NiFe$_2$O$_4$ is inverse spinel type, which is formed by substituting half of octahedral Fe sites in Fe$_3$O$_4$ for Ni. It is well known that temperature dependence of resistivity in Fe$_3$O$_4$ exhibits an abrupt resistivity change at 120 K[1]. This phenomenon was firstly explained by Verwey et al. as an order-disorder transition from Fe^{2+} and Fe^{3+} ions to two Fe$^{2.5+}$ ions[2]. Since all Fe ions in stoichiometric NiFe$_2$O$_4$ are trivalent, similar transition was not observed. Recently, a new powder synthesis method of pulsed wire discharge has been developed[3]. By using this method, various nanosized powders[4,5], including NiFe$_2$O$_4$[6], were synthesized. Since this method produces the nanosized powders by quenching high temperature plasma, either anion deficient or cation-disordered compounds can be formed. Thus, NiFe$_2$O$_{4-x}$ having Fe^{2+} ions may be synthesized by pulsed wire discharge. In the present study, NiFe$_2$O$_4$ nanosized powders were synthesized by pulsed wire discharge and electric properties of the powders were measured.

EXPERIMENTAL PROCEDURE

Nickel and iron wires with diameters of 0.20 and 0.30 mm, respectively, and with length of 25 mm were wound and connected with electrodes in a chamber which was filled with Ar+33%O$_2$ mixed gas at a pressure of 600 Torr. The electrodes were also connected to a 20 μF capacitor bank through a gap switch. Then, the capacitor was charged at the voltage of 6 kV. By closing the gap switch, pulsed current was driven through the wound wires. The pulse width was 20 μs. The wires were evaporated by the Joule's heat. The metal vapor expanded and cooled in the gas to form nanosized powders. The powders floating in the chamber were collected by pumping through a membrane filter. Phases in powders synthesized with different conditions were identified by powder X-ray diffraction (XRD). Volume fraction of secondary phases was determined by relative intensities of XRD peaks using a working curve, which was obtained from XRD results for NiFe$_2$O$_4$ and NiO mixed powders with known compositions. Then, powders synthesized with different conditions were sintered at 600 °C in air for 1 h. Electric resistivity of the sintered bulk was measured by a four-terminal method from 150 to 350 °C in air.

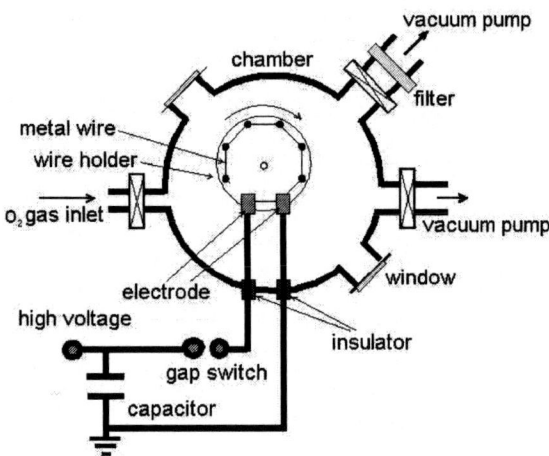

FIGURE 1. Experimental setup for NiFe$_2$O$_4$ nanosized powder by PWD.

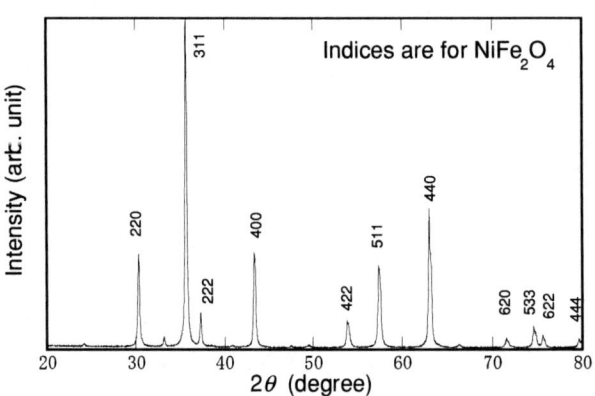

FIGURE 2. XRD patterns for NiFe$_2$O$_4$ nanosized powder synthesized by PWD

RESULTS AND DISCUSSION

Average particle size of the synthesized nanosized powder was estimated by specific surface area measurements to be approximately 50 nm. The XRD pattern for the collected powder is shown in Fig.2. All peaks correspond to those for a $NiFe_2O_4$ phase. Relative intensities of some peaks are slightly higher than those reported in a JCPDS card. Since the peaks exactly coincide to those for a NiO phase, NiO particles may be contained in the powders. Volume fraction of the NiO phase was determined to be 2 vol % in the powder.

The powder was sintered to produce a bulk sample for resistivity measurements. Temperature dependence of resistivity for the sample is shown in Fig.3. Resistivity gradually decreases with the increase in temperature up to 285 °C. Between 285 and 295 °C, however, resistivity drastically falls down by two orders of magnitude. Above 295 °C, resistivity gradually decreases. The temperature at the drastic decrease in resistivity was defined as transition temperature.

Temperature dependence of the resistivity for $NiFe_2O_4$ is compared with the reported results. Resistivity of $NiFe_2O_4$ single crystals[7] and polycrystals[8] had been reported by many researchers to be continuously decreased with increasing temperature. Although deflection of the resistivity curves around Néel temperatures was also observed in $NiFe_2O_4$ nanosized powders[9], drastic resistivity change, which is shown in Fig. 3, has not been reported.

Possible explanations of the drastic resistivity change are discussed here. Samples of Fe_3O_4 exhibit similar drastic resistivity change. As described before, this

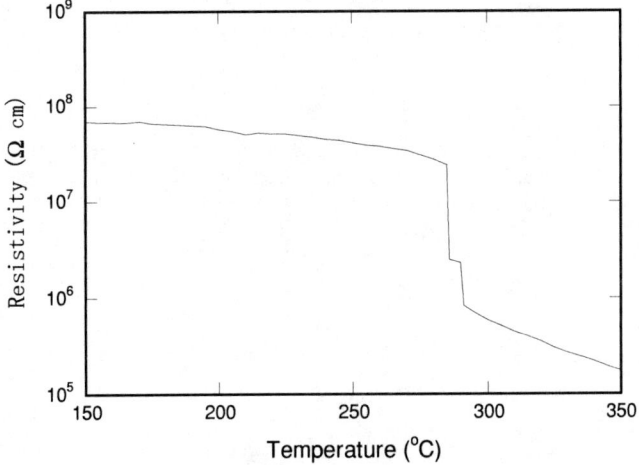

FIGURE 3. Temperature dependence of resistivity in $NiFe_2O_4$ nanosized powders sintered at 600 °C for 1h.

phenomenon used to be explained by a simple order-disorder transition model from Fe^{2+} and Fe^{3+} to two $Fe^{2.5+}$ ions. Since the present $NiFe_2O_4$ powders were synthesized by quenching of high temperature plasma, oxygen vacancies which induced formation of Fe^{2+} may be introduced in the crystal structure. However, the transition temperature for Fe_3O_4 is 120 K, which is much lower than that of the present $NiFe_2O_4$ nanosized powder. If the drastic resistivity change accompanies the similar order-disorder transition to that of Fe_3O_4, coupling between Fe^{2+} and Fe^{3+} in the present $NiFe_2O_4$ is much stronger than those of Fe_3O_4. The transition in the present $NiFe_2O_4$ may be controlled by other mechanism. For example, the smallness in particle size of the present $NiFe_2O_4$ powders may affect the electric properties since spin correlation in $NiFe_2O_4$ was changed with decreasing particle size or with decreasing isostatic pressure[10].

ACKNOWLEDGMENTS

The authors acknowledge Prof. Ito at Tokyo Institute of Technology and Prof. Inaguma in Gakushuin University for helpful discussion.

REFERENCES

1. Miles, P. A., Westphal, W. B. and Hippel, A. v., *Rev. Mod. Phys.* **29**, 279-307 (1957).
2. Verwey, E. J., Haayman, P. W. and Romeijn, F. C., *J. Chem. Phys.* **15**, 181-187 (1947).
3. Jiang, W. and Yatsui, K., *IEEE Transactions on Plasma Science* **26**, 1498-1501 (1998).
4. Kinemuchi, Y., Suzuki, T., Jiang, W. and Yatsui, K., *J. Am. Ceram. Soc.* **84**, 2144-46 (2001).
5. Sangurai, C., Kinemuchi, Y., Suzuki, T., Jiang, W. and Yatsui, K., *Jpn. J. Appl. Phys.* **40**, 1070-1072 (2001).
6. Kinemuchi, Y., Ishizaka, K., Suematsu, H., Jiang, W. and Yatsui, K., *Thin Solid Films* **407**, 109-113 (2002).
7. Whall, T. E., Salerno, N., Mirza, K. and Brabers, V. A. M., *Adv. Ceram.* **15**, 341-346 (1985).
8. Fayek, M. K., Mostafa, M. F., Sayedahmed, F., Ata-Allah, S. S. and Kaiser, M., *J. Mag. Mag. Mater.* **210**, 189-195 (2000).
9. Whall, T. E., Yeung, K. K., Proykova, T. G. and Brabers, V. A. M., *Phil. Mag.* **50**, 689-707 (1984).
10. Ma, Y. G., Jin, M. Z., Liu, M. L., Chen, C., Sui, Y., Tian, Y., Zhang, G. J, and Jia, Y Q., *Mater. Chem Phys* **65**, 79 84 (2000).

Reduction of Micrometer Size Al Particles in Nanosize AlN Powder Synthesized by Pulsed Wire Discharge

Chuhyun Cho, Yoshiaki Kinemuchi, Hisayuki Suematsu, Weihua Jiang, and Kiyoshi Yatsui

Extreme Energy-Density Research Institute, Nagaoka University of technology, Niigata, 940-2188, Japan

Abstract. Experimental studies were carried out on the synthesis of nanosize powders of aluminum nitride (AlN) by pulsed wire discharge (PWD). Efforts were made to produce high purity AlN powder by reducing the number of micrometer-size particles. The deposited energy in the wire before the explosion is significantly increased when the discharge energy was increased. The highest AlN content of the powder synthesized in N_2 gas mixed with NH_3 gas was 98 wt.%. The number of micrometer-size particles observed by scanning electron microscopy is decreased with increasing discharge energy.

INTRODUCTION

Various kinds of metallic and ceramic nanosize powders have been synthesized by the pulsed wire discharge [1-5]. The average particle size of these powders were several tens of nanometers. However, the particle size distribution was broad, and showed two distinct peaks in the size distribution. One of them was in the tens of nanometer range, while another one in the micrometer range [1-2, 4]. The origin of the micrometer-size particles may be attributed to the liquid droplets produced during the process of melting and vaporizing the wire.

AlN nanosize powder has been also synthesized by the pulsed wire discharge [5], where nitrogen gas mixed with ammonia was used as the ambient gas. In the previous research, a large number of micrometer-size particles were observed in the powders, which were identified as Al particles by energy dispersive spectroscopy (EDS) analysis. In this research, the energy deposited in the Al wire was increased by reducing circuit inductance and increasing charging energy to produce high purity AlN powder with narrow size distribution.

EXPERIMENTAL SETUP AND CONDITIONS

Figure 1 shows the schematic of the experimental setup. A wire feeder was installed inside a chamber. Aluminum wire was fed from a reel to the gap between the electrodes by rotating two rollers tightly contacted each other. The electrodes were made of aluminum to avoid contamination. The rollers can either be rotated by hand or by an electric motor. A membrane filter was installed between the chamber and a vacuum pump, and the chamber was evacuated by the vacuum pump. After evacuation, the chamber was filled with atmospheric gases of mixed N_2-NH_3. Electrical energy was stored in capacitors and released through a gap switch to the wire installed between two electrodes.

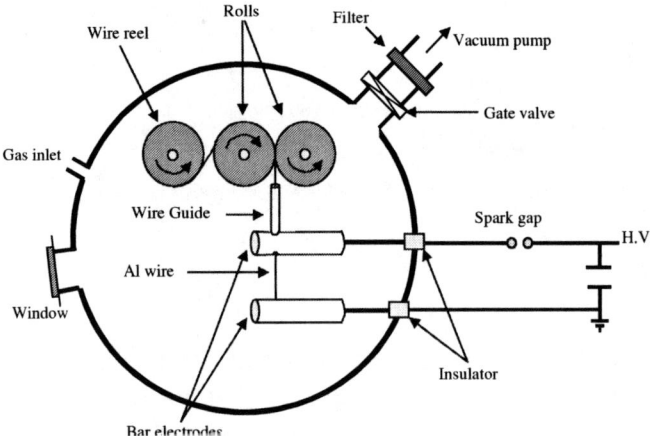

Figure 1. Schematic of experimental setup.

The electrical energy was concentrated in the wire because of the relatively high resistance of the wire. Thus, only the wire between the electrodes was heated, vaporized, and turned into plasma. Consecutive shots could be carried out in the same manner. The powders were collected on the surface of the filter by evacuating the chamber through the filter.

The diameter and length of the aluminum wire were 0.25mm and 25mm, respectively. Mixed N_2 gas with NH_3 was used as the atmospheric gas. The pressure of the N_2 and NH_3 were 600 and 150 Torr, respectively. The capacitance was $20\mu F$, and total circuit inductance was $0.7\mu H$.

EXPERIMENTAL RESULTS AND DISCUSSIONS

The AlN content in the synthesized powders was estimated by the ratio of integrated peak intensity of AlN 100 to Al 111 of XRD. Figure 2 shows the AlN content and crystallite size as a function of discharge energy. The AlN content was evaluated to be 98 wt.% for the powder at 360 J of discharge energy without any additional processes.

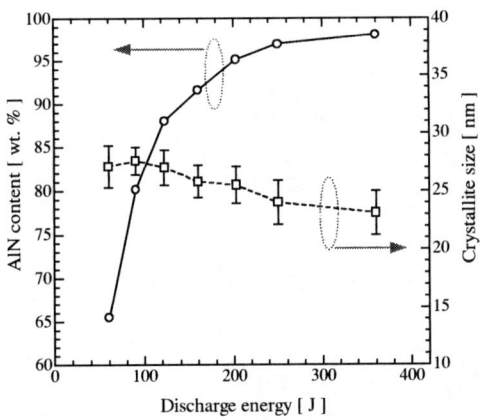

Figure 2. AlN content and crystallite size as a function of discharge energy

On the other hand, average crystallite size obtained by the broadening of the XRD peaks was estimated to be 23 nm for the powder synthesized at 360 J. The specific surface area was measured to be 65.5 m^2/g by Brunauer-Emmett-Teller (BET) method for the powder synthesized at 360 J. Assuming that the particles were spherical in shape, the average diameter was calculated to be 28 nm. Small difference between average crystallite size and average particle size may be attributed to the non-spherical shape of AlN powder and surface oxidation of particles by oxygen and humidity in air. Oxidized layers were probably amorphous in state, which gave smaller average crystallite size obtained by XRD method than average particle size measured by BET method.

By comparing the results of crystallite size by XRD and average particle size by BET, it can be estimated that there were few micrometer size particles in the powder synthesized at 360 J, which was confirmed by SEM observations.

The current waveforms of the exploding wires could be distinguished into ohmic heating and gas discharge [6-7]. The wires are mainly turned to the vapor by ohmic heating. The energy deposited in the wire during ohmic heating can be estimated by integration of the current waveform [7]. Figure 3 shows the deposited energy until the wire exploded, that is ohmic heating energy, (E_e), as a function of discharge energy.

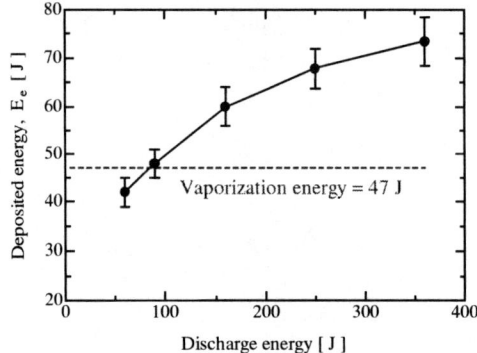

Figure 3. Energy deposited before the explosion as a function of charging energy, where the capacitance and inductance were 20 μ F and 0.7 μ H, respectively.

The energy required to vaporize the wire, E_v, is 47 J. It can be seen that the energy E_e exceeds the vaporization energy in Fig. 3. The ratio E_e/E_v is increased up to 1.6 at the charging energy of 360 J, when the capacitance and inductance are 20 μ F and 0.7 μ H, respectively. This is called superheating caused by inertia effect [7]. At the high superheating region, the wire is evenly vaporized. Thus AlN content increased, and the number of micrometer size particles is decreased.

REFERENCES

1. Umakoshi, M., Ito, H., and Kato, H., *Yogyo-Kyokai-Shi* **95**, 124-129(1987).
2. Umakoshi, M., and Yoshitomi, T., *J. Mater. Sci.* **30**, 1240-1244(1995).
3. Jiang, W. and Yatsui, K., *IEEE Trans. Plasma Science* **26**, 1498-1501 (1998).
4. Kotov, Yu. A. et al, *Key Engineering Materials* **132-136**, 173-176 (1998).
5. Sangurai, C. et al, *Jpn. J. Appl. Phys.* **40**, 1070-1072 (2001).
6. Nash, C.P., and Olsen, C.W. , *Phys. Fluids* **7**, 209-213(1964).
7. Chace, W.G., *Phys. Fluids* **2**, 230-235(1959).

THEORY AND MODELING

Pitfalls in Radiation Modeling of Z-Pinch Plasmas[+]

J. Davis,[a] J.L. Giuliani,[a] J.P. Apruzese,[a] R.W. Clark,[a] J.W. Thornhill,[a] K.G. Whitney,[a] A. Velikovich,[a] Y. K. Chong,[a] C.A. Coverdale,[b] C. Deeney,[b] and P.D. LePell[c]

[a] *Plasma Physics Division, Naval Research Laboratory, Washington, DC 20375 USA*
[b] *Sandia National Laboratories, Albuquerque, NM 87185 USA*
[c] *K-Tech Corporation, Albuquerque, NM 87106 USA*

Abstract. Over the last three decades there has been a quantum jump in the production of x-rays from pulsed power driven Z-pinch plasmas. Total radiative yields have gone from a few kilojoules to almost two megajoules. This increase occurred as a result of higher current drivers coupled with improvements in our understanding of the issues most relevant to good load design. Critical analyses of experimental data have led to a better understanding of the load dynamics, which includes all phases of load evolution extending from the cold start to the final collapsed phase and the emission of the x-ray pulse. A Z pinch is a deceptively simple device that has a very complex plasma dynamics. It can be a platform for demonstrating a variety of textbook plasma instabilities. However, its primary application in the present context is as an intense source of x-ray radiation. Therefore it is attractive both as a direct source of x-rays and for creating hohlraum conditions for plasma fusion experiments. After a few historical comments are offered on how radiation has been treated in modeling Z pinches, some of the methodologies and models that are employed in this endeavor are discussed. These include both nonLTE and LTE ionization dynamic models and escape probability radiation transport and LTE radiation diffusion models. To illustrate their use, comparisons are made between experimental data from a stainless steel wire array pinch implosion and 1-D MHD calculations that employ these models. The consequences that stem from the compromises and trade-offs that result from the different approximations used in these models are addressed. We will explore the role that radiation plays in the dynamic evolution of a Z-pinch and demonstrate the need for as near a self-consistent radiation-hydrodynamics treatment as possible.

INTRODUCTION

Pulsed power driven Z-pinch plasmas have been studied for over 50 years both experimentally and theoretically. The recent sustained interest in and fascination with Z-pinch plasmas is due in large measure to the success of the Z facility's performance in rapidly heating dense wire array and gas puff plasmas to temperatures high enough to produce an intense burst of K- and L- shell x-rays from moderate atomic number elements. It is also due to an improved understanding of load performance and load design. The linear Z-pinch is a deceptively simple device that exhibits a complicated plasma dynamics. It finds application in conducting indirect drive fusion, x-ray lasers, and hohlraum physics experiments. However, for our purposes here, we focus on its primary application as an intense source of soft x-ray radiation producing upwards of about 2 megajoules of total radiation for a number of material loads.

The production and transport of radiation in Z-pinch produced plasmas are often modeled using a radiative diffusion approximation and assuming Local Thermodynamic Equilibrium (LTE). Since these plasmas are neither in LTE nor entirely opaque or transparent to the radiation, LTE and radiation diffusion are erroneous assumptions and only provide poor descriptions of the energetics, emission spectra and yield from Z-pinch plasmas. The application of inappropriate models easily leads to the misinterpretation of experimental data and to misunderstanding of the plasma dynamics and load performance. Radiation modeling of Z-pinches (and laser produced plasmas) usually does not take the "high road" and rationalizes traversing the "low" road in the name of computational efficiency, which oftentimes is full of potholes, pitfalls and wasted computational resource.

The power radiated by a Z-pinch plasma consisting of multi-charged atomic ions can have important consequences on the plasma dynamics. The radiated power depends both on the detailed atomic structure of the ions and on the properties of the plasma such as T, N and opacity. Therefore, the calculation of the radiated power of a dynamically evolving Z-pinch represents a complex problem. In this presentation we discuss and assess some of the more common approximations employed to describe the radiative behavior of Z-pinch plasmas. Through the use of examples and illustrations, we show that in calculations with identical initial plasma conditions, atomic level structure and rate coefficient data, the change of radiation models has a strong affect on the dynamic evolution and hydrodynamic history of the plasma, often promoting an erroneous interpretation of experimental results. To illustrate the degree of error, we generate an emission spectrum using a 1-D radiation MHD model that self-consistently coupled to a circuit representing the Sandia Z-facility. Comparison between several radiation models for a multi-wire titanium array also illustrate the kinds of errors and pitfalls that are possible.

POPULATION KINETICS MODEL

The population kinetics model determines the occupation of atomic levels and ionization stages of the material load as a function of temperature, density, and opacity. The level dynamics is usually described by either Local Thermodynamic Equilibrium (LTE). Collision Radiative Equilibrium (CRE), or Time dependent (CRT) non-equilibrium, Corona Equilibrium (CE), or some form of Average Atom (AA) model with hydrogenic or averaged real atomic data. The level dynamics is important because it provides plasma opacity and emissivity data to the radiation transport models. This transport can be carried out using a variety of approximations from the simplistic Holstein escape factor to the solution of detailed moment equations for the emitted flux or intensity. We will focus on radiation diffusion and multi-cell, probability of escape, methods. For a fuller discussion of these see Davis, et. al.[1]

The average charge state, $Z_{Average}$, shown on Figure 1 for a gold plasma as a function of temperature for an ion density of $10^{20} cm^{-3}$ illustrating the large differences that are possible depending on the ionization dynamic model used (a) and temperature and density (b) for LTE, CRE and CE. For example, At a temperature of 1 keV the average charge state can differ by as much as 30 ionization stages at a density of 10^{20} ions. This result clearly demonstrates the need to determine the correct distribution of charge states because a number of parameters such as transport coefficients and the "color" of emitted photons depend on $Z_{Average}$. The only reason for adopting an LTE accounting of the level occupations is to simplify the calculations. Unfortunately, LTE is a bad approximation leading one has to ask why it is used and what does it provide. Figure 1 showing $Z_{Average}$ as a function of temperature and density illustrates the large errors involved in using an LTE model at low densities.

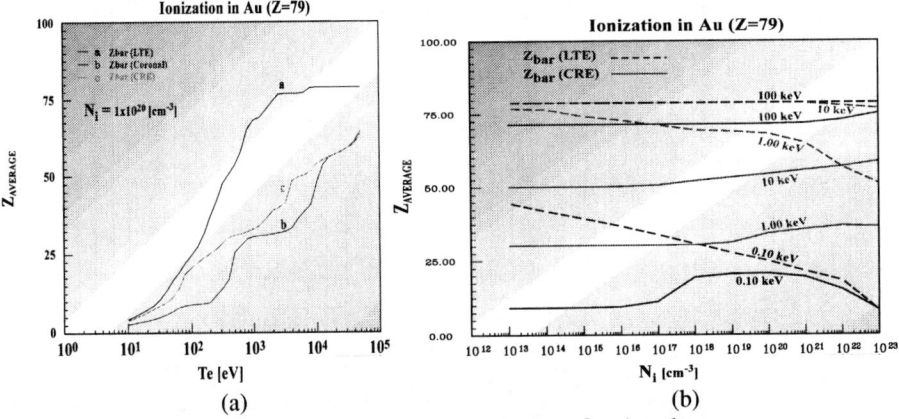

Figure 1. Average charge state for Au plasma

MODELING AND SIMULATION

The implosion dynamics of a titanium load on Sandia's Z facility was calculated with a one-dimensional (1D) Lagrangian MHD model using some commonly employed descriptions for the ionization dynamics and radiation transport. The simulations include LTE with single- and multi- group diffusion, CRE and TCRE[2] with on-the-spot transport. A summary of the total and K-shell calculated yields for six different cases are shown on Figure 2 along with experimental data for 3 loads. The numbers on the bar graph for the TCRE and NLCRE results represent the number of lines carried in the calculation. There are only minor differences between the calculated yields since the 126 line calculation was chosen judiciously. The experimental results with their shot number are also shown. The single- and multi-group diffusion results for the total yield are similar, but the single group cannot produce a K-shell yield. Figure 3 isolates specific results from Figure 2 and exhibits the variation of the yield as a function of photon energy for the non-local and multigroup LTE diffusion simulations. A quick view of these results suggests that LTE in conjunction with single- or multi-group diffusion agrees well with experiment for the K-shell and total radiative yields. This could lead us to the conclusion that LTE and radiation diffusion adequately represents the radiation yield and that the hydrodynamic profiles have evolved correctly to produce these results. However, as we shall see below that nature has conspired to mislead us to this conclusion because the plasma conditions are calculated to produce these results are very different. This is one of the unsuspected pitfalls! The transport has significantly changed the hydrodynamics to give comparable x-ray emissions.

Figure 3 isolates shot # 303 and shows a comparison between the multi-group based on a 1200 spectral lines from the full model and the non-local CRE for atomic models containing 126 and 483 radiating spectral lines, respectively. As the atomic model improves with increasing structure the yield results also improve. The experimental value is about 990 kJ's and appears midway between the diffusion and probability of escape. Once again, the pitfall is not looking beyond the yields and comparing with other experimental observations.

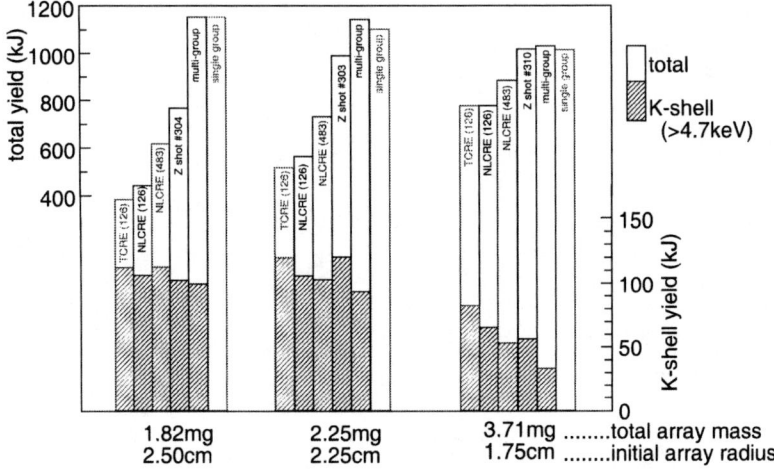

Figure 2. Comparison of total and K-shell yields for Ti array implosions.

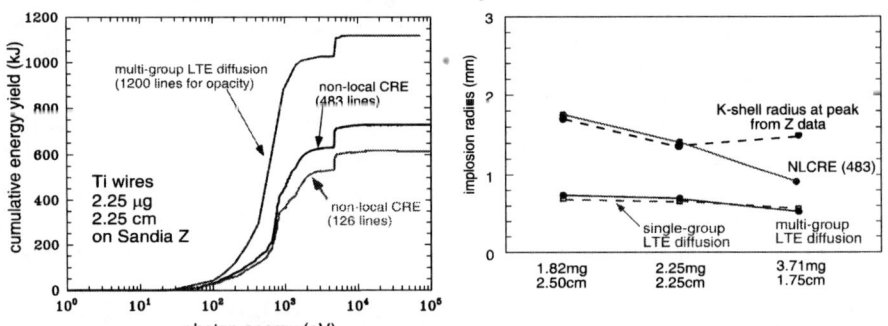

Figure 3. LTE diffusion transport   Figure 4. Implosion radii

X-ray pinhole pictures provide information on the size of the emitting plasma. The experimental observations of the radius of the emitting region at peak emission are exhibited in Figure 4 along with the results of the simulations. The LTE diffusion calculations pinch down to smaller radii for all 3 loads due to the large radiative losses making it possible to compress a cooler plasma. The radii obtained in the non-

LTE simulations are in better agreement with the observations in support of the other plasma properties such as temperature and density inferred from experiment. The LTE K-shell emitting radius deviates from the non-LTE estimates is difficult to understand by itself. However, one needs to compare the L-shell emitting region which occurs at larger radii and raises questions of plasma stability and a clear and concise observable radius. This behavior will be discussed in a forthcoming paper. The main point here is that the diffusion models disagree with the observations on the size of the minimum radius.

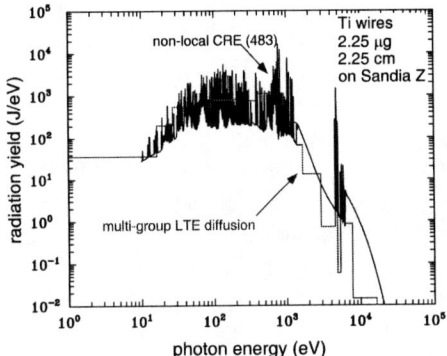

Figure 5. Ti spectra and multi-group LTE emission model

Figure 5 shows the calculated Ti spectra superposed on the LTE multi-group diffusion result. With a judicious choice of opacity groups it is possible to get a reasonable representation of the emission spectra from diffusion methods. Just how one determines the number of groups and their wavelength range potentially can be another pitfall!

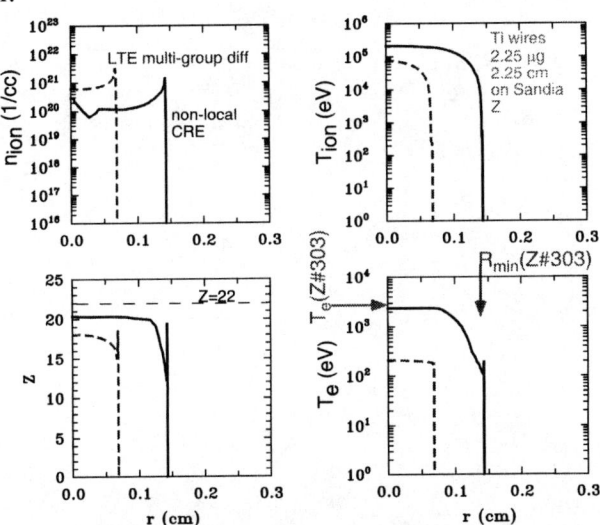

Figure 6. Hydrodynamic profiles for Ti shot #303

The hydrodynamic profiles of electron and ion temperature, average charge state and ion density are displayed in Figure 6 at about the time of peak emission. The results are self-explanatory in view of what has been stated above. The electron temperature, density and charge state can be inferred from the experimental data. The radius of the emitting region can be measured from the pinhole pictures. The comparison of these parameters with the non-LTE simulations are in good agreement. The LTE multi-diffusion hydrodynamic results are not! The LTE diffusion plasma cools at a greater rate lowering the temperature and charge state and reaching higher compressions leading to higher densities, behavior that is in contradiction to the observed measurements.

CONCLUSIONS

Modeling the radiation of Z-pinch plasmas, even for the experienced modeler focused more on the MHD phenomenology than on the radiative behavior of the plasma, oftentimes is pressured to make model choices that lead down a path of self-deception by making the code predictions agree with experiment. In the limited space provided in these proceedings we have provided some examples and consequences associated with substituting LTE multi-group diffusion methods for non-LTE and radiation transport methods in order to describe the plasmas' radiative evolution with high computational speed but low computational accuracy. The consequences result in a misinterpretation of the experimental observations and a model incapable of addressing issues relevant to the tailoring of x-ray production and of evaluating load performance. We are in the process of incorporating the non-LTE methods into multidimensional Z-pinch models to investigate how to best calculate the total and K-shell radiative yields and to evaluate the effects of plasma instabilities on x-ray production.

ACKNOWLEDGEMENTS

We would like to thank Dr. Paul Kepple for assembling the Au atomic database and organizing it into a useful rate table. Portions of this work were supported jointly by the Defense Threat Reduction Agency and the Department of Energy.

REFERENCES
1. Davis, J., et. al., Laser and Particle Beams, **19**, 557 (2001).
2. Thornhill, J. W., et. al., Physics of Plasmas, **8**, 3480 (2001.

Why do Wire-Array Z-Pinches give such a Sharp and Efficient X-Ray Pulse?

M.G. Haines, S.V. Lebedev, J.P. Chittenden, F.N. Beg, S.N. Bland and M. Sherlock

Blackett Laboratory, Imperial College, London SW7 2BW

Abstract. The Z-pinch is intrinsically unstable to MHD modes, yet the wire array Z-pinch yields sharp and reliable Z-pinch pulses. The physics of the various phases of wire array implosions, i.e. the early heating, melting and vaporization, plasma formation, uncorrelated m=0 instabilities on the wires, inward jetting to form a precursor column and next what appears to be a snowplough-like implosion are studied experimentally, computationally, and theoretically. The X-ray pulse occurs mainly at the stagnation on axis when up to 100keV kinetic energy of the ions is thermalised.

INTRODUCTION

Z-pinches are a text-book example of an unstable MHD plasma. Dynamic Z-pinches are in addition unstable to the Rayleigh-Taylor instability. Yet the wire-array Z-pinch appears to be so well behaved that it can yield a sharp 5 ns pulse of soft X-rays with high conversion efficiency [1]. Experiments, theory and simulations at many laboratories over the past 5 years have yielded much understanding but there are still many questions to be answered.

The following sections discuss briefly the time history of a typical wire array implosion.

EARLY PHASE

X-ray radiography has revealed that the wires expand into a mixed molten-vapour phase. This is followed by a breakdown of the vapour surrounding each core. X-radiography of single wires was first explored at Cornell [2]. Figure 1 shows how at Imperial College [3] an X-pinch driven in a return current post was used to backlight an array with 2-4 keV X-rays revealing an interior structure of a mixed liquid and vapour phase in the heated wire. A B-dot probe in a nested wire array configuration measures the current following in the inner wire array. The relative mutual inductance would indicate that 6% of the current should flow in the inner array, but after 15ns there is a marked increase which is interpreted as current shunting as the outer array heats up and becomes more resistive as it forms a liquid-vapour mixture. At 20ns there is a drop of current to zero in inner array; this marks the breakdown and plasma formation in the vapour surrounding the outer wires. The X-ray radiography shows an

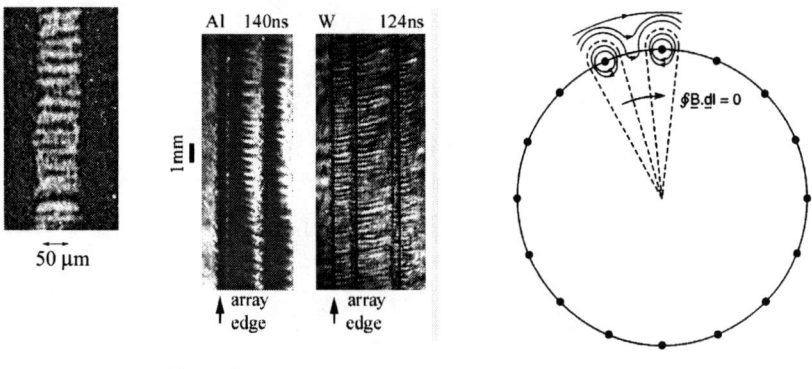

Figure 1 **Figure 2** **Figure 3**

expanding plasma of 250μm diameter in the outer array wires but even at 128ns the inner array wires have only a <30μm diameter.

INSTABILITIES ON THE WIRES, JETTING AND THE PRECURSOR PLASMA

The coronal plasma around each wire then carries most of the current which is rising in time. Axial structures, thought at first to be MHD m = 0 instabilities, appear, as shown in Fig.2. The wavelength of the instability is about 0.5mm for aluminium and 0.25mm for tungsten; ka is about 1.5 where k is the wave number and a is the core radius. The axial nonuniformities might well provide the seed for later instabilities. Inward jetting of the plasma due to the influence of the global magnetic field follows the magnetic field topology with separatrices as shown in Fig.3. This is illustrated by end-on laser probing and by the side on optical streak image shown in Fig.4. The inward accelerated jets of plasma stagnate on axis forming a precursor plasma which is usually a straight symmetrical and stable cylinder of radiating plasma and appears to carry little current (Fig.5). Pressure balance is achieved through the dynamic pressure of the inward flowing plasma balancing the plasma pressure. It can be seen in Fig.5

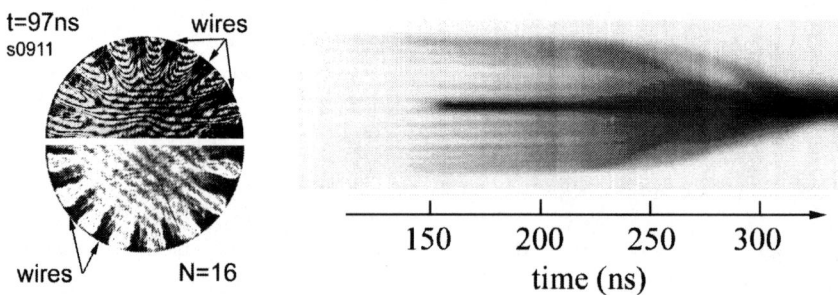

Figure 4

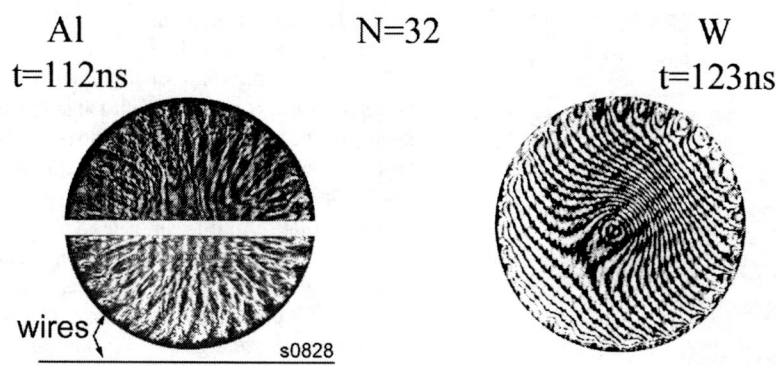

Figure 5

that the aluminium precursor plasma flow appears to have much structure associated with shocks and arising from collisions between the individual jets. In contrast tungsten appears to be collisionless with the separate jets interpenetrating. Kinetic modelling of the ions [4] using the Fokker-Planck equation confirms this, and indeed in the accumulation of stagnated plasma on the axis the mean-free-path for tungsten is greater than the radius of the precursor column (~0.3mm) until 150ns, while for aluminium the ~1 mm radius of the precursor column is collisional by 115ns for typical MAGPIE conditions.

MAGNETIC REYNOLDS' NUMBER

The plasma around the wires seems to be accelerated inwards by the global $\underline{J} \times \underline{B}$ force leaving the field and current behind. This is consistent with a low magnetic Reynolds' number which will occur if there is strong thermal coupling of the coronal plasma to the relatively cold wire cores [5]. This coupling by electron and possibly radiative heat flow is essential to provide the renewal of the plasma by ablation of the cores, and naturally leads to a low temperature corona and hence low magnetic Reynolds' number. But if the magnetic Reynolds' number (or Lundqvist number) is low it will not support an ideal m = 0 MHD instability. So perhaps the axial wave number mode is an electrothermal instability associated with the nonlinear heat flow [6]. This instability requires the mean-free-path to be less than the collisionless skin depth.

FINAL IMPLOSION

When the cores are completely ablated at various axial positions, the current and plasma become frozen in an inward moving shell, the details of which are still being explored. A 2-D simulation [7] shows in Fig.6 how the wire is essentially stationary as the inward plasma jetting takes up the $\underline{J} \times \underline{B}$ axially directed force, the final implosion only occurring when the wire core has disappeared. In 3-D this occurs at first in gaps

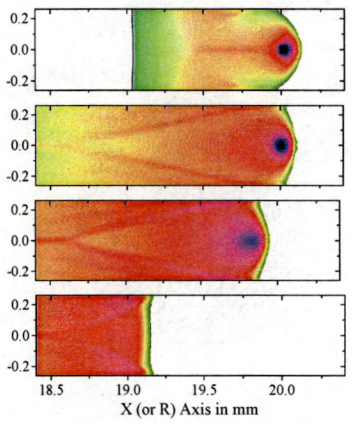

Figure 6

in each wire which at this time are correlated and emit X-rays [8]. This is rather like a snowplough implosion as it sweeps up the Rayleigh-Taylor instability, because the current piston moves with almost constant velocity as illustrated in Fig.7. Debris left behind by the implosion may also play a significant role in shunting current away from the imploding snowplough [9,10]. If the wire separation is small the coronal plasmas around each wire could merge significantly and at this point an interesting question is whether the precursor flow is stopped.

STAGNATION AND X-RAY PULSE

At stagnation the plasma is sufficiently dense for the energy to be transferred from the ions to the electrons and thence on to radiation in a few nanoseconds. The final current – carrying stagnated pinch is MHD unstable to both m = 0 and m = 1. The presence of this final instability may be beneficial in allowing a larger surface area for the radiation to escape. Disruption accompanied by electron and ion beams will occur later. The use of nested wire arrays has various beneficial effects upon each of these phenomena [1].

At least three modes of operation of nested arrays are possible [11,12,13]. The earliest model assumed that each array had merged to form a shell and the impact of the outer on the currentless inner would lead to a mitigation of Rayleigh-Taylor instabilities. However it seems questionable, especially for tungsten, as to whether merger occurs at all. Thus if no current is flowing in the inner wire as the outer array approaches, it will essentially pass through the inner array, but at this time the current will transfer to the inner array, this being the path of lowest inductance. The inner array will then implode onto the stagnated earlier outer array plasma.

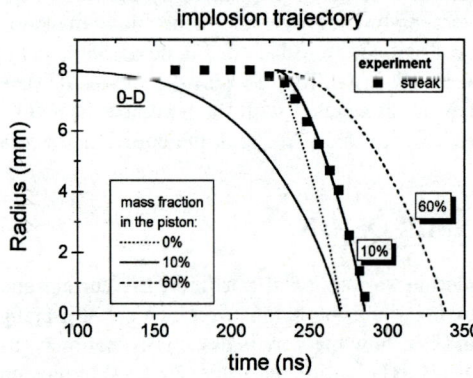

Figure 7

Interestingly this early outer stagnation in the absence of any $\underline{J} \times \underline{B}$ force or Joule heating yields little X-radiation, and the main sharply rising X-rays occur at the final implosion of the inner array. The third mode is when significant current flows in the inner wire (low mutual inductance case); then the outer array compresses the magnetic field associated with the inner array and drives it ahead to the axis.

There is much scope for pulse shaping as a result of these various modes and also by using different materials.

CONCLUSIONS

There is now a much greater understanding of the physics of wire array implosions. The earlier ideas of plasma wire expansion leading to merger [14] and the formation of a shell which is accelerated to the axis with Rayleigh-Taylor instabilities growing has been questioned by the scenario of inward jetting leading to a precursor column, followed by a relatively stable snowplough implosion. The lack of current in the precursor jets implies a low magnetic Reynolds' number and hence the coronal wire plasma instabilities might be electrothermal associated with the nonlinear heat flow to the cores rather than MHD in origin as originally thought. Whether instabilities occur on the inner of a nested array is an interesting question, but it would seem that the versatility of a nested array together with the much smaller contribution from Rayleigh-Taylor instabilities will permit both a sharp X-ray pulse plus the pulse shaping needed for inertial confinement fusion.

ACKNOWLEDGMENTS

We gratefully acknowledge Sandia National Laboratory, Cornell University, the U.S. Department of Energy and AWE for their finical support.

REFERENCES

[1] C. Deeney et al., Phys. Rev. Lett. **81**, 4883 (1998).
[2] D. Kalantar and D. Hammer, Phys. Rev. Lett. **71**, 3806 (1993).
[3] S.V. Lebedev et al., Phys. Rev. Lett. **85**, 98 (2000).
[4] M. Sherlock et al., BEAMS-DZP 2002, this Proceedings (2002).
[5] M.G. Haines, IEEE Trans. Plasma Science, to be published (2002).
[6] M.G. Haines, Phys. Rev. Lett. **47**, 917 (1981).
[7] J.P. Chittenden et al., Phys. Plasmas **8**, 2305 (2001).
[8] S.V. Lebedev et al., Phys. Plasmas **6**, 2016 (1999).
[9] S.V. Lebedev et al, Phys. Plasmas **9**, 2293 (2002).
[10] J.P. Chittenden et al, BEAMS-DZP 2002, this Proceedings (2002).
[11] J. Davis et al., Appl. Phys. Lett **70**, 170 (1997).
[12] S.V. Lebedev et al., Phys. Rev. Lett. **84**, 1708 (2000).
[13] J.P./ Chittenden et al, Phys. Plasmas **8**, 675 (2001)
[14] M.G. Haines, IEEE Trans. Plasma Science **26**, 1275 (1998).

Calculated Evolution of Side-on and End-on X-ray Images of Wire and Gas Puff Implosions on Z

J. P. Apruzese,[a] J. W. Thornhill,[a] C. Deeney,[b] C. A. Coverdale,[b] J. Davis,[a] A. L. Velikovich,[a] H. Sze,[c] P. L. Coleman,[d] B. H. Failor,[c] J. S. Levine,[c] and K. G. Whitney[e]

[a]*Radiation Hydrodynamics Branch, Plasma Physics Division, Naval Research Laboratory, Washington, DC 20375 USA*
[b]*Sandia National Laboratories, Albuquerque, NM 87185 USA*
[c]*Titan Systems Corporation, Pulse Sciences Division, San Leandro, CA 94577 USA*
[d]*Alameda Applied Sciences Corp., San Leandro, CA 94577 USA*
[e]*Berkeley Scholars, Inc., Springfield, VA 22150 USA*

Abstract. As $J \times B$ forces implode a Z-pinch load, a variety of phenomena that affect its radiative behavior may occur. These include snowplows and shocks, instabilities, amplification of initial load asymmetries, and high-density condensations (possibly caused by a local radiative collapse). On the axis, arrival of the main load mass occurs, but it may be preceded by a precursor plasma. An important aim of experimental diagnostics is to determine the degree and importance of such physical processes as a function of the properties of the load and the driving generator. X-ray spectroscopy and imaging are well-established components of most Z-pinch diagnostic suites. The purposes of the present work are to determine the x-ray signatures of some of these phenomena and to ascertain the x-ray energies that provide the clearest and least ambiguous indicators of these processes. To accomplish this, we have carried out radiation hydrodynamics calculations of argon gas puff and titanium wire array loads imploded on Sandia's Z generator. These calculations provide a large database of x-ray spectra and images. Some clear image signatures are seen in the calculations. These include intensity enhancement near the outer edge of the pinch, which is characteristic of the heating and compression that occurs in the snowplowed region of the pinch prior to its final on-axis assembly. Spatial image intensity profiles in the lower energy x-rays are generally indicative of the density profile of the pinch. The profile of the image at different photon energies within the K-shell spectral region is related to the presence of an electron temperature gradient.

INTRODUCTION

X-ray pinhole imaging of Z pinches has been used to assess and improve pinch "quality". This diagnostic has also provided quantitative measurements of the diameter of Z pinches to apply spectroscopy to diagnose their average temperature and density (see, e.g., Ref. 1). Undoubtedly, much more information is contained in such images, especially with regard to spatial variations of temperature and density. Routine, accurate diagnosis of these variations as a function of load and the current that drives it would be likely to provide valuable insight into the physical processes that are thought to create such variations. These include: formation of "snowplowed" regions, shocks, various plasma and hydrodynamic instabilities, on-axis assembly and compression, and radiative collapse. The fundamental purpose of our calculations is to connect the x-ray images of Z pinches to the conditions that produce the images, thereby optimizing the extractable information from pinhole images and their diagnostic cousin, spatially resolved spectroscopy.

MODEL AND CALCULATIONS

The implosion dynamics of argon and titanium loads on Sandia National Laboratories' Z generator [2] were calculated with a one-dimensional (1D) Lagrangian magnetohydrodynamic (MHD) model [3] in which detailed configuration atomic models and tabular collisional-radiative equilibrium radiation transport [4] are used to calculate the evolution of the ionic species, level populations, and spectra. As described in Ref. 3, enhanced viscosity is employed to model multidimensional effects such as turbulent flow. For 15-30 ns surrounding peak compression, the radiative transfer was computed on expanded numerical grids of 40 rays and ~ 2200 photon frequencies [5], to allow for calculation of detailed pinhole images at virtually any x-ray energy or specified band of energies. Both side-on and end-on images can be calculated.

The initial load conditions were chosen to approximately correspond to those of two very successful shots on Z. For Ar, a mass load of 0.8 mg/cm was distributed in two broad peaks centered near radii of 15 and 35 mm. This load is similar to that of Z shot 663, which produced 274 kJ of Ar K-shell radiation [6]. For the Ti calculation, a mass load of 1.1 mg/cm was assumed, 75% of which was concentrated at a radius of 22.5 mm, and the remainder distributed nearly uniformly in the interior. These conditions are similar to those of the single array Z shot 303, which produced 120 kJ of Ti K-shell radiation. The main purpose of analyzing the calculations was not detailed comparison with experiment, but to enhance the diagnostic value of pinhole imaging and spatially resolved spectroscopy. The calculations do reproduce the experimental K-shell yields within 35%, the diagnosed average temperatures within 20%, and the corresponding densities to within a factor of 2. However, the calculated K-shell pulse widths are too short by factors of 2-3. For the Ti load, a 3 ns pulse width was calculated vs. 7 ns measured, and for Ar the corresponding values are 5 ns calculated vs. 12 ns measured. These are compensated by greater calculated radiating load masses and powers, giving better agreement with measured K-shell yields.

RESULTS

The calculated density profiles for the Ar and Ti pinches at three times during the implosion are shown in Figs 1a and b. Note that a softer, more uniform density implosion results from the initially filled-in Ar puff load than from the mostly hollow single-array Ti load. For Ar, peak compression and K-shell power occur in the time frame 140-143 ns, and for the Ti load, these events occur at 121-123 ns. The electron temperature profiles at peak compression (not shown), are fairly similar. For both loads, a core temperature of ~ 2.5 keV remains relatively spatially flat out to about 75-90% of the total radius, at which point a sharp decline in temperature is followed by an increase due to ohmic heating in the further out, current-carrying plasma skin, then (for Ti) a decline at the very outer edge occurs. Figs. 2 a and b show the end-on L-shell images for the two loads, calculated by integrating the intensity from 0.4-1.0 keV for Ar, and 0.7-1.7 keV for Ti. Note that these intensity profiles are similar to the corresponding density profiles shown in Figs. 1 a and b. The corresponding K-shell image intensity profiles do not track the density profiles due to the temperature sensitivity of the higher energy x-ray emission.

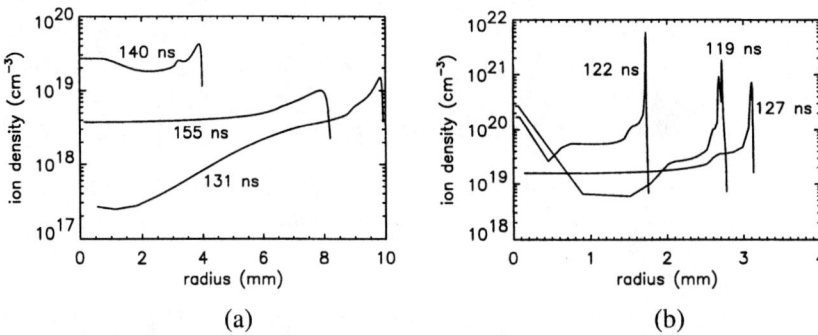

Fig. 1. Calculated ion density profiles at three times during the implosions of (a) Ar and (b) Ti.

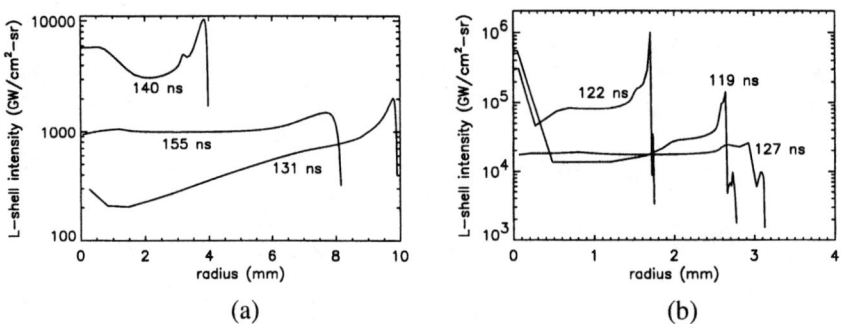

Fig. 2. Calculated L-shell intensity profiles at three times during the implosions of (a) Ar and (b) Ti.

Fig. 3 shows the side-on K-shell image of the Ti pinch near peak compression. There are two noteworthy features. One is the strong self-reversed core of the He-α line.

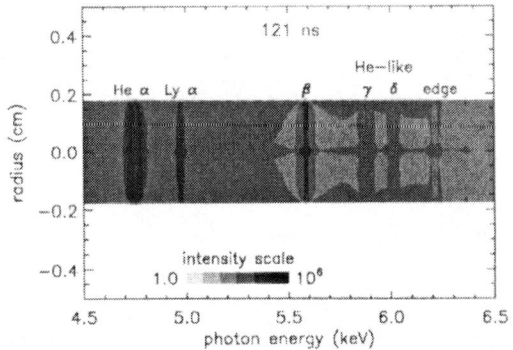

Fig. 3. The calculated K-shell side-on image of the Ti pinch near peak compression for hν=4.5-6.5 keV

This is due to the sharp decline in electron temperature at the very outer edge of the pinch. The second feature is an enhancement of intensity just inside the outer edge of the pinch (seen at photon energies of 5.5 to 6.2 keV). This reflects the increase in electron temperature that occurs just inward of the outer radius. The simultaneous collection of spectrally and spatially resolved data has the potential to be a powerful diagnostic of the pinch's temperature profile.

In summary, detailed calculations of the x-ray images of imploding Ar and Ti Z pinches have shown that x-ray "bright spots" are the result of corresponding enhancements of temperature, density, or both. Images of bright spots at lower photon energies are more likely to be the result of density, rather than temperature, enhancements. Self-reversals in the profiles of strong lines are sensitive to temperature gradients, as are spatial intensity profiles in the higher energy, K-shell regions of the x-ray spectrum. When the density falls with increasing distance from the pinch axis, it is found that the image is larger in the spectral lines than in nearby continuum regions.

ACKNOWLEDGMENT

This work was supported by the U.S. Defense Threat Reduction Agency and by Sandia National Laboratories, U.S. Department of Energy.

REFERENCES

1. Whitney, K. G., *et al.*, Phys. Plasmas **8**, 3708 (2001).
2. Spielman, R. B., *et al.*, Phys. Plasmas **5**, 2105 (1998).
3. Thornhill, J. W., *et al.*, Phys. Plasmas **1**, 321 (1994).
4. Thornhill, J. W., *et al.*, Phys. Plasmas **8**, 3480 (2001).
5. Apruzese, J. P., *et al.*, Phys. Plasmas **7**, 3399 (2000).
6. Sze, H., *et al.*, Phys. Plasmas **8**, 3135 (2001).

How 3D Effects Limit X-ray Power in Wire Array Z-pinches

J.P. Chittenden[a], S.V. Lebedev[a], M.E. Cuneo[b], C.A. Jennings[a], and A. Ciardi[a]

[a] *Imperial College, Blackett Laboratory, Prince Consort Road, London, SW7 2BZ, U.K.*
[b] *Sandia National Laboratory, Albuquerque, New Mexicio, NM 87185-1193 USA.*

Abstract. The effect of the complex 3D structure of imploding wire array Z-pinch plasmas on their X-ray power production is investigated using a 3D resistive MHD code. The break up of the wires at the start of the implosion is found to produce "debris" which shunts the current away from the implosion and limits the level of plasma compression at stagnation and hence the radiative power.

INTRODUCTION

The achievement of 5ns bursts of soft X-rays, with yields of 2MJ and power levels up to 250 TW, from imploding wire array Z-pinches at Sandia National Laboratory, [1] represents a major milestone towards achieving high yield thermo-nuclear fusion in the laboratory. If this energy could be extracted over even shorter time-scales, the increase in power would result in substantial reductions in the scale of Z-pinch facility required to achieve fusion ignition. In this respect it is instructive to consider how the power obtained compares to one-dimensional simulations of such imploding objects, which can be regarded as the fundamental limit of what could be achieved in the absence of any instability or in homogeneity. For a 2cm diameter 6mg/cm tungsten pinch on the 'Z' facility, the results described below indicate that the imploding plasma object approaches the axis with a velocity of ~6×10^5 m/s (i.e. 1MJ/cm of kinetic energy) and a radial width of less than 0.6mm. The kinetic energy is thus dissipated in less than one nano-second, with an X-ray power output of more than 1000 TW/cm. This compares with 100TW/cm obtained in experiments for this configuration. Understanding why higher power levels are not achieved represents a major challenge to Z-pinch theory. A number of potential limiting factors are being explored and are described throughout these proceedings. These include the effects of finite radiative and ionization rates, the effects of radiation trapping, the effects of finite ion mean free paths on shock widths in the imploding plasma and the effects of the magneto-Rayleigh-Taylor instability on the width of the imploding plasma. In this paper, we will examine the 3D inhomogeneous structure of the wire array plasma and in particular how this affects the X-ray power by limiting the ability of the current to continue accelerating the implosion right down to the axis. This work was supported by AWE Aldermaston, by Sandia National Laboratory and by the US Department of Energy.

HYDRODYNAMIC, MHD AND MAGNETO-RAYLEIGH-TAYLOR DESCRIPTIONS OF EXPERIMENTS

Recent experimental observations [2] of imploding tungsten wire array Z-pinchES on the 'Z' facility have confirmed that the structure and dynamics of the implosion is qualitatively similar (if quantitatively different) to that observed on smaller scale wire array experiments [3]. Optical streak photography measurements reveal that the implosion trajectory is delayed compared to that of a thin shell, implying that material is being injected into the array interior from early times. This is confirmed by end-on gated X-ray images showing the existence of an early precursor plasma on axis.

The apparent width of the X-ray emitting region observed end-on is 2-3mm with a final velocity of ~4×10^5 m/s just prior to impacting the precursor. By itself, this data appears almost sufficient to explain the parameters of the X-ray power pulse obtained (~4ns rise, ~6ns FWHM, ~100TW peak). For example, using the 2D(r,z) MHD code described in reference 4, (here run in a 1D radial fashion), we can approximately reproduce the X-ray pulse by initializing the code with a Gaussian density profile of FWHM 3.5mm and a constant velocity of 5×10^5 m/s and allowing this to impact the axis. This only works, however, if the 18MA of current present in the experiment is ignored. With the addition of the drive current to the simulation it becomes impossible to sustain a density profile of such large width. The **jxB** force both accelerates the plasma to higher velocity and compresses the density profile down to sub millimeter widths. The kinetic power density bombarding the axis, and hence the radiative output power, are then several times that of the purely hydrodynamic simulation. In addition the current continues to compress the plasma after stagnation driving a radiative collapse to far higher densities and smaller pinch sizes than observed in experiments. It is entirely possible that such narrow density profiles indeed exist in experiments and that the apparent width of the X-ray emitting region is causeD by the different optical depths observed when viewing the plasma end-on.

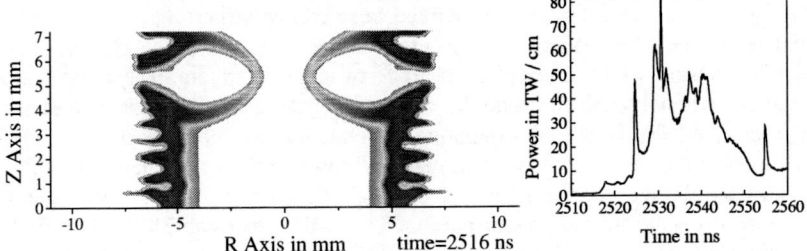

FIGURE 1. Density contours from a 2D simulation of an imploding shell and the X-ray power output.

For situations were the plasma is approximated to an imploding thin shell, the growth of Rayleigh-Taylor instabilities go some way to reducing the average X-ray by broadening the radial density profile as the implosion approaches the axis. Under these conditions, the X-ray power pulse is characterized by a series of high intensity bursts from the collapse of a series magnetically accelerated bubbles which impact the axis forming transient hot-spots of extremely high density and emissivity. Again these results are at odds with experimental results, which show a much smoother X-ray power pulse and a more homogenous source.

RADIAL AND AZIMUHTAL STRUCTURE

The radial distribution of mass in wire array Z-pinches is significantly different from that described above. X-ray radiographic studies at Cornell University [5] and Imperial College [6] reveal that despite the high electrical power, the phase transitions in the metallic wires are neither instantaneous nor homogeneous and as a consequence the wires ablate relatively slowly. The steadily produced plasma is electro-magnetically accelerated by the drive current producing supersonic plasma streams which are injected into the interior of the arrays and converge on the axis forming a precursor plasma.

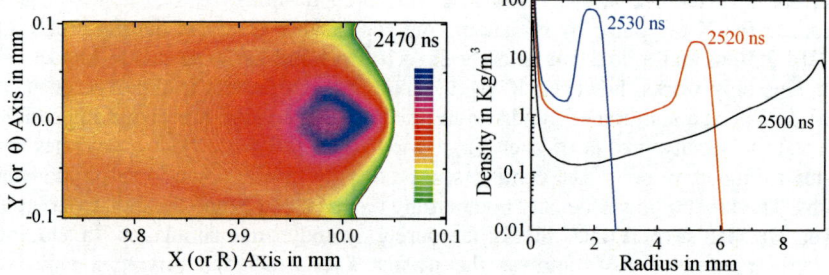

FIGURE 2. Density contours from a 2D(x,y) simulation of a single wire in a 300 wire array and a density profile from a 1D simulation of the ensuing implosion.

Figure 2 shows results from a simulation of one of the wires in an array of 300, using the 3D resistive MHD code described in reference 4 (here run in 2D with 6µm spatial resolution). The dense cold material of the core is ablating slowly, steadily producing plasma which is accelerated to form a radial stream. The computational domain extends only 300µm inside the wires. We make use of the plasma parameters on the left-hand side of the 2D simulation as boundary conditions for a 1D simulation which models the evolution of the plasma flow down to the axis. The final implosion initiates once the wire core runs out of material at 2490ns and is similar to that of a snowplow sweeping up the pre-fill plasma. The **jxB** force establishes a piston at the vacuum/plasma boundary with a shock propagating a finite distance ahead into the pre-fill. The piston-shock separation is then set by the γ (the ratio of specific heats) of the plasma. For $\gamma \sim 5/3$, a fairly wide separation is obtained, but for tungsten γ is reduced by ionization and radiative losses and the piston-shock separation is sub-millimeter by the time it reaches the axis. Thus the kinetic power density delivered to the axis is again large resulting in Peta-Watts of power. This implies that the modified azimuthal and radial structure observed in wire arrays is not solely responsible for the X-ray power obtained.

3D STRUCTURE OF THE FINAL IMPLOSION

Short wavelength variations in the plasma density along the length of each wire are visible in experiments from early times, indicating a modulation in the rate of wire ablation. As a consequence, at the start of the implosion phase, some fractions of the length of the wires are fully ablated and some are not. The fully ablated regions begin to implode, and thus the wire plasmas begin to break up [4]. Current bypasses the breaks and starts to drive the snowplow implosion through the pre-fill. The high density regions of wire plasma thus become current free and are left behind by the implosion. The effect of this behavior on the final implosion can be illustrated using a 3D resistive MHD simulation of a 32 wire array in which each wire is given a 4 period perturbation of random period and amplitude. The contour at 191ns shows a series of breaks in each wire and debris left behind as the current drives an interior snowplow surface towards the axis. Despite being broken and turbulent the debris field is large in cross section and so competes with the imploding snowplow for the generator current. Thus as the snowplow radius decreases the current is gradually shunted into the debris, so that when the implosion reaches the axis, it does so with insufficient current to drive a rapid radiative collapse to high density. The flow of current through the debris then accelerates further material towards the axis and at the same time the gaps between pieces of debris enlarge and therefore the current fraction in the debris decreases. The gradual increase in current transferred to the axis further compresses the stagnated plasma and starts to drive MHD instabilities. The result of this behavior is a slowly rising X-ray power pulse which starts with the stagnation of the snowplow, continues to rise with the current transferred to the axis and is limited by the onset of instabilities.

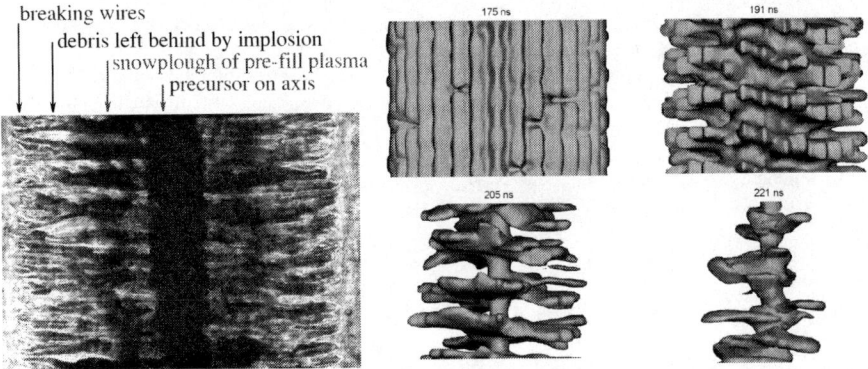

FIGURE 3. Laser shadowgram of a MAGPIE wire array and density surfaces from a 3D simulation.

REFERENCES

1. C. Deeney, et. al., Phys. Rev. Lett. **81**, 4883 (1998).
2 M.E. Cuneo, et al. Bull. Am. Phys. Soc. **46** p234 (2001).
3 S.V. Lebedev, *et al.* Phys. Plasmas **6** 2016 (1999)
4 J.P. Chittenden et al., Phys. Plasmas **8**, 2305 (2001).
5 T. A. Shelkovenko, et. al. Rev. Sci. Instrum. **70**, 667 (1999).
6 S.V. Lebedev et al., Phys. Rev. Lett. **85**, 98 (2000).

Modeling Enhanced Energy Coupling of Z-pinches to Pulsed-power Generators

K. G. Whitney,[a] J. W. Thornhill,[b] C. Deeney,[c] C. A. Coverdale,[c] J. P. Apruzese,[b] J. Davis,[b] A. L. Velikovich,[b] and L. I. Rudakov[a]

[a]*Berkeley Scholars Assoc., P. O. Box 852, Springfield, VA 22150 USA;* [b] *Radiation Hydrodynamics Branch, Plasma Physics Division, Naval Research Laboratory, Washington DC 20375 USA;* [c]*Sandia National Laboratories, P. O. Box 5800, Albuquerque, NM 87185 USA.*

Abstract. It has been observed over the years that the energy coupled to the load in many z-pinch experiments is larger than can be accounted for by the sum of the **jxB** work and classical Ohmic heating. Moreover, this energy enhancement appears to be a function of the generator design, increasing as the risetime of the current is increased. In recent experiments on the Saturn generator, for example, which was operated at current risetimes in excess of 160 ns, observed energy enhancements were factors of 2 to 4 times the energy input expected from JxB work alone. When Saturn operates with risetimes of less than 90 ns, much smaller energy enhancements over the JxB energy are seen. In the past, it was conjectured that some form of anomalous resistivity was needed to account for the extra energy input, while recently, a new idea was proposed based on the buildup of internally generated tubes of magnetic flux energy.[1,2] It was hypothesized that the growth of the Rayleigh-Taylor instability at the surface of the z-pinch plasma would generate bubbles of magnetic flux-tube energy that deposit their energy in the plasma at a current-to-the-third-power rate. While 0-D modeling of the Saturn experiments shows that an anomalously high load resistance can input the required energy needed to match the x-ray data, an alternate mechanism than magnetic flux-tubes exists for anomalous heating that is based on the production of micro-instabilities at the pinch surface. Both this and the flux-tube model are phenomenological and require guidance from experiments to be implemented. Several issues that arise from these enhanced energy coupling mechanisms are discussed in this paper.

SATURN EXPERIMENTS

Various aluminum array shots in which the load mass, the array length, and the array radius were varied were carried out on the Saturn generator at Sandia National Laboratories. In these experiments, Saturn was operated with current risetimes that varied between 140 to 230 ns depending on the load parameters. In all these experiments, large amounts of x rays were emitted. Generally, the total energy radiated was 2 to 4 times the energy imparted to the load by the **jxB** forces. Consequently, for a calculation to produce this much x-ray energy, it must impart in excess of this amount to the plasma by means that go beyond the conventional hydrodynamic modeling and that are generally considered to originate from some form of anomalous plasma heating.

The two total x-ray output pulses that are shown in Fig. 1 illustrate the variable nature of this added energy input. These pulses were generated in two shots that had identical load parameters: 180 wires of 2 cm length initially at a radius of 2 cm with a load mass of 616 µg/cm. The x-ray pulse from shot 2693 had a peak power approximately twice that of shot 2636. Moreover, shot 2693 had a late-time foot that followed the main pulse while shot 2636 did not. Fig. 2 contains two running time integrations of the power curves in Fig. 1. They show that the energy radiated in the main pulse was approximately the same in each shot -- roughly 500 kJ. However, a significant amount of late-time emission occurs in shot 2693 (~700 kJ), which is absent in shot 2636. For these energies to be radiated from the pinch, they must be supplied to it over the two, very different, time intervals of the emission. If the pinch is regarded as a dynamic circuit element having both a time dependent resistance, $R_{\ell oad}(t)$, and a time dependent inductance, $L_{\ell oad}(t)$, then this energy is supplied either through L-dot work or through resistive heating with the late-time emissions coming from late-time resistive heating.

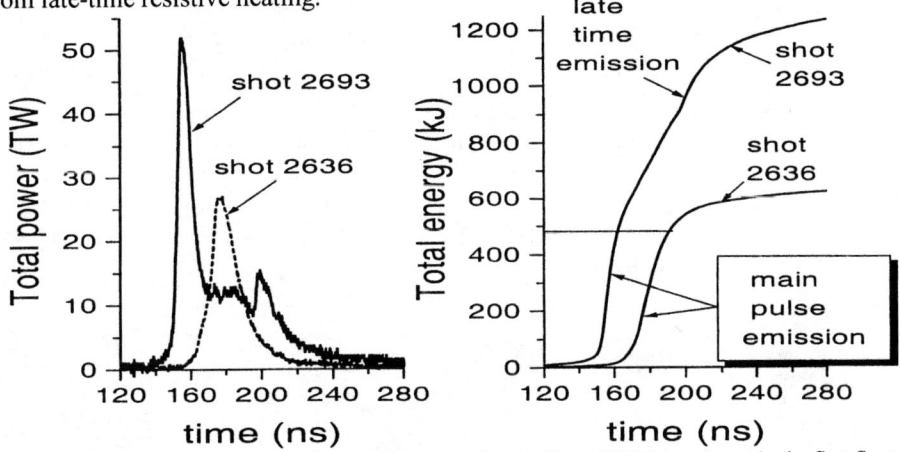

FIGURES 1 AND 2. The total x-ray power pulses for shots 2693 and 2636 are shown in the first figure and their running time integration in the second.

0-D COUPLING

The simplest way to model the L-dot and resistive heating energetics of shots 2636 and 2693 is to use a 0-D kinetics model of the load that is coupled to the circuit model for Saturn appropriate to its long current-risetime mode of operation. In this case, the load inductance, $L_{\ell oad}$, is given by $L_{\ell oad}(t) = (2\ell/c^2)\ln(r_{rc}/b(t))$, where ℓ is the length of the pinch, r_{rc} is the radius of the return current path, and $b(t)$ is the location of the outer boundary of the imploding pinch as a function of time. The energy equation of the circuit in a 0-D model describes the buildup of magnetic field energy, kinetic energy, and Ohmic heating by the discharge of the generator:

$$\frac{d}{dt}\left(\frac{1}{2}L_{gen}I^2 + \frac{1}{2}L_{\ell oad}I^2\right) + \left(R_{gen} + R_{\ell oad} + \frac{1}{2}\frac{dL_{\ell oad}}{dt}\right)I^2 = V(t)I(t),$$

where the dL_{load}/dt resistive term in this equation corresponds to the build-up of kinetic energy in the load:

$$\frac{d}{dt}\left(\frac{1}{2}m\ell\left(\frac{db}{dt}\right)^2\right) = \frac{1}{2}\frac{dL_{load}}{dt}I^2.$$

In these equations, m is the mass per unit length of the z-pinch load, $L_{gen} = 11.8$ nH is the generator inductance, $R_{gen} = 0.068$ Ω is the generator resistance, and $V(t)$ is the voltage drive of the Saturn generator. These equations are solved for b and the current, I. In calculations with these equations, dL_{load}/dt is obtained from b and db/dt, while $R_{load}(t)$ must be specified.

Solutions to 0-D equations can be used to determine energy inputs to loads only during the implosion phase of the dynamics. Hence, these energies can only be compared to total energies radiated during the main x-ray pulse. Some guidance for specifying R_{load} is provided in Ref. [3]. In sub-mega-ampere experiments, the load resistance was observed to grow rapidly to a maximum value late in the implosion. The calculations of the Ohmic and kinetic energy inputs to different massed loads, which are shown in Fig. 3, were guided by these experiments. The load resistance was turned on rapidly during the run-in to the maximum values that are shown in the figure prior to the termination of the implosion at a radius of 1.5 mm. The implosion calculations were begun at a radius of 2 cm.

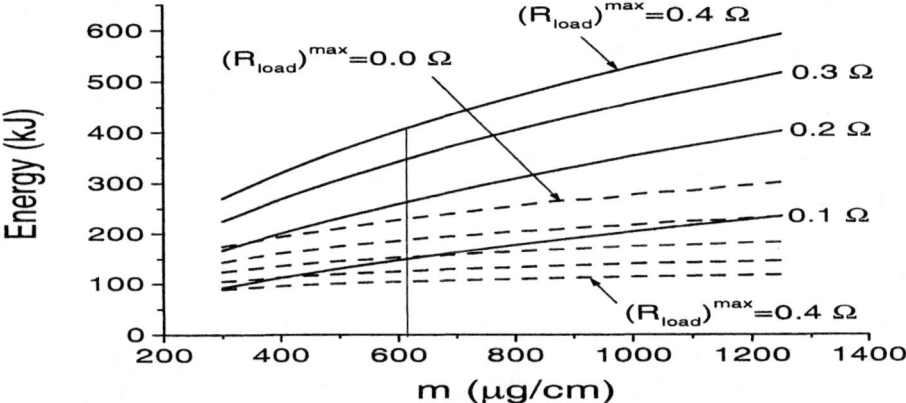

FIGURE 3. The kinetic and Ohmic energies imparted to different massed loads during the implosion phase of the z-pinch dynamics are shown for four values of the maximum load resistance. The dashed curves are kinetic energies and the solid curves, Ohmic energies. As the load resistance is increased, the kinetic energy generated in the implosion progressively decreases.

As seen in these 0-D calculations, the kinetic energy generated during implosion for the 616 µg/cm load, when there is no load resistance, accounts for only half of the energy radiated during the main x-ray pulses of shots 2636 and 2693; a load resistance peaking at values in excess of 0.4 Ω is needed to impart the necessary energy. Moreover, as the load resistance is increased, the amount of kinetic energy generated during implosion goes down. Both Ohmic and kinetic energy inputs to the load increase as the load mass, and consequently the load implosion time, is increased. This behavior as load mass is increased was also seen in the x-ray outputs of the Saturn shots shown in Table 1 in which only the load mass was varied.

Table 1 Variable Mass Experiments

Shot #	Array Mass (μg/cm)	Implosion Time (ns)	Total X-ray Output (kJ)	Total X-ray Output (main pulse) (kJ)
2640	328	134	777	334
2636	616	175	629	458
2637	887	183	736	382
2641	1050	184	845	546
2702	1576	208	1040	652

1-D AND 2-D COUPLING

Since Spitzer resistivity is too small to account for the Ohmic heating needed to produce the energy outputs (and thus the required energy inputs) seen in the Saturn experiments, two ways have been proposed for increasing the resistivity of a z-pinch load. One, proposed in Refs. [1] and [2], involves the generation of added amounts of magnetic field energy within the pinch plasma that is decoupled from the load inductance. The process by which this energy is generated is proposed to be related to the generation of Rayleigh-Taylor instabilities at the surface of the pinch. The subsequent dissipation of this magnetic field energy within the plasma then adds resistivity to the pinch. However, there is also a second mechanism for producing anomalous resistivity in a pinch. It occurs by way of micro-instability generation at the pinch surface.

An expression for R_{load} in terms of the surface micro-instability resistivity can be derived by calculating the power flow at the termination of the transmission line into the z-pinch load in a 1-D geometry. This power flow, P^{Load}, produces increases in the energy stored within the z-pinch plasma and radiated from it:

$$P^{Load} = \frac{d}{dt}\left(U_{Fluid} + \frac{1}{8\pi}\int_{Plasma} dV B_\theta^2\right) + P^{Rad},$$

where $U_{Fluid} \equiv \int_{Plasma} dV(u_{KE} + u_{Thermal} + u_{Ioniz})$, the total plasma energy, is the sum of kinetic, thermal, and ionization energy parts and P^{Rad} is the total x-ray power emitted from the plasma. From the total energy equation of the fluid equations in a 1-D geometry, one also finds that P^{Load} can be expressed in terms of the Poynting flux of the transmission line electromagnetic energy through the moving surface, $b(t)$, of the pinch:

$$P^{Load} = Lim_{\varepsilon \to 0} 2\pi\ell r \left\{ \frac{dr}{dt}\frac{B_\theta^2}{8\pi} + \frac{c}{4\pi}E_z B_\theta \right\}\Bigg|_{r=b-\varepsilon},$$

Finally, to be consistent with the circuit equation, one must also have that $P^{Load} = \left(\frac{1}{2}\frac{dL_{\ell oad}}{dt} + R_{\ell oad}\right) I^2$. On inserting the surface values for B_θ and E_z, which are obtained from Ampere's law: $B_\theta = 2I(t)/cb$ and from Ohm's law:

$$E_z = -\frac{1}{c}\frac{db}{dt}B_\theta + \frac{1}{n_e e}\left(\alpha_\perp \frac{j_z}{n_e e} + \beta_\wedge \partial_r (k_B T_e)\right),$$

into the Poynting vector expression for P^{Load} and by comparing the resulting expression for P^{Load} with the circuit equation expression, one finds that $L_{\ell oad}(t) = (2\ell/c^2)\ln(r_{rc}/b(t))$ in agreement with 0-D modeling, and one obtains the following expression for $R_{\ell oad}$:

$$R_{\ell oad} = \frac{\ell \eta_0}{A_{th}} \left(\frac{A_{th}}{A_j} \hat{\alpha}_\perp + \frac{1}{2} x_{tr} \hat{\beta}_\wedge \right).$$

The areas, A_{th} and A_j, appearing in this equation are defined by $A_{th} \equiv I/n_e e v_{th}$ and $A_j \equiv I/j_z(r=b)$, where the thermal velocity, v_{th}, is defined by $v_{th} \equiv \sqrt{2k_B T_e/m_e}$. The dimensionless transport quantities, $\hat{\alpha}_\perp$, $\hat{\beta}_\wedge$, and x_{tr}, are defined by $\hat{\alpha}_\perp \equiv 1 - (\alpha_1' x_b^2 + \alpha_0')/\Delta$, $\hat{\beta}_\wedge \equiv x_b(\beta_1'' x_b^2 + \beta_0'')/\Delta$, and $x_{tr} \equiv v_{th}\tau_e \partial_r \ln(T_e)$. The coefficients in these definitions are given in Ref. [4] and $x_b \equiv eB_\theta \tau_e/(m_e c)$. The resistivity, η_0, is defined by $\eta_0 \equiv m_e/(n_e e^2 \tau_e)$.

All of the quantities appearing in the expression for $R_{\ell oad}$ are evaluated at the surface of the pinch. At this surface, micro-instabilities can be generated because Ohmic heating drives up the plasma temperature and drives down the plasma density. As the density decreases, the current drift velocity increases and a point is reached at which micro-instabilities can onset. To account for this process, one needs to replace the classical resistivity, $\hat{\alpha}_\perp \eta_0$, by one with an additional micro-instability contribution: $\hat{\alpha}_\perp \eta_0 \to \hat{\alpha}_\perp \eta_0 + \eta_{Micro}$. In Ref. [5], it was postulated that lower hybrid drift waves, which have been detected in a theta pinch sheath, could also be important in adding anomalous resistance to z-pinches. In this case, a formula for η_{Micro} was offered that, unlike η_0, which is essentially density independent, has an inverse dependence on the surface density.

SUMMARY

In the absence of well-developed theories, one must rely on phenomenological models to compute the enhanced energy inputs to a z-pinch that are needed to account for the large x-ray outputs that are observed from pinches in multi-mega-ampere and/or long current-risetime generators. Moreover, the parameters in such models must be set with guidance from experimental data when theoretical guidance is missing. Since more than one explanation is often possible of observed z-pinch behavior, experimental data must also be used to determine the presence or absence of the underlying phenomena that each explanation assumes in accounting for the observed behavior. The data most relevant to this task to date has been the early-time rate of increase of the diode current, the onset times of the x-ray emissions, the times to peak emission, the pulseshapes of the total x-ray emission and the K-shell emissions, and the total and K-shell x-ray yields.

The problem of incorporating a phenomenological flux-tube model of enhanced energy coupling into a 1-D MHD z-pinch calculation provides an example of the kind of guidance that is needed to implement the model. Neither the dynamics by which a Rayleigh-Taylor instability at the surface of a pinch generates magnetic flux-tube energy nor the fluid dynamics by which this energy is subsequently thermalized has been demonstrated. Nevertheless, experimental data from the shots in Table 1 suggest two important features of this proposed enhanced coupling mechanism. One, a significant amount of flux-tube energy generation appeared to be needed during the run-in in order to account for the observed implosions times of the shots. However, the thermalization of this energy had to be delayed until just prior to pinch assembly in order to get agreement with the observed x-ray emission onset times.

Calculations carried out using a 0-D model of the Saturn generator support the need for a load resistance that grows rapidly to a maximum value prior to load assembly on axis. Maximum load resistances in excess of 0.4 Ω are needed for energy inputs to exceed 500 kJ for a 616 µg/cm load, in accord with the x-ray outputs in the main pulse that were observed in shots 2636 and 2693 on Saturn. However, the coupling of a z-pinch plasma to a transmission line occurs by virtue of a flow of electromagnetic energy from the vacuum diode through the plasma surface, where it is subsequently transported into the plasma by magnetic field diffusion, heat conduction, radiation transport, and compressive heating or shock wave propagation. How an enhanced amount of Ohmic heating is generated by these processes in a 1-D fluid calculation by micro-instabilities at the surface of a pinch has yet to be determined.

ACKNOWLEDGMENTS

This work was sponsored in part by Sandia National Laboratories and in part by the Defense Threat Reduction Agency.

REFERENCES

1. Rudakov, L. I., et. al., Phys. Rev. Lett., **84**, 3326 (2000).
2. Velikovich, A. L., et. al., Phys. of Plasmas, **7**, 3265 (2000).
3. Labetsky, A. Yu., et. al., IEEE Transact. on Plasma Sci., to be published.
4. Braginskii, S. I., in *Reviews of Plasma Physics*, edited by M. A. Leontovich (Consultants Bureau, New York, 1965), Vol. 1, pp. 205-311.
5. Robson, A. E., Phys. of Fluids B, **3**, 1461 (1991).

Numerical Studies of Neon Gas-puff Z-pinch Dynamic Process

Cheng Ning[*], Zhenhua Yang, Ning Ding

Institute of Applied Physics and Computational Mathematics, P. O. Box 8009, Beijing 100088

Abstract. The dynamic process of neon gas-puff Z-pinch, which may produce high temperature and density plasma, were numerically studied by one-dimensional lagrangian radiation magneto-hydrodynamic code developed by the authors. The spatio-temporal distributions of some plasma parameters in the process were obtained, and the dynamic process was reproduced. The results also show the zippering effect in gas-puff Z-pinch, and suggest that the main energy transformation process is Coulomb collision process between ion and electron, secondly is photoelectric process and bremsstrahlung radiation process. The energy transformation in Compton scattering process may not be taken into account. The x-ray energy radiated by Z-pinch plasma is produced mainly in photoelectric excitation and recombination process, secondly in bremsstrahlung radiation process. The energy consumed by Ohm heating is not larger than 10% of total energy delivered to load.

I. INTRODUCTION

Although heart-stirring high level (20MA, 290TW, 1.8MJ, 4ns) has been achieved in Z-pinch experiments on Saturn,[1] higher power and energy (about to 60MA, 1000TW, tens MJ) of x-ray radiated by Z-pinch plasma is desired for its further applications in astrophysics, radiation material studies, equation of state, and inertial confinement fusion etc. Furthermore, there are yet some problems not to have been well understood about physics of Z-pinch plasma, including coupling between electric and magnetic energy and load, mechanism of x-ray radiation, detailed physical process of wire-array plasma formation and its implosion, producing and developing mechanism of Magneto Rayleigh-Taylor (MRT) instability and measure to restrain its increasing. Especially, numerous experiments carried out in the past twenty years have showed the remarkable fact that the x-ray energy can be much greater than the kinetic or magnetic energy coupled to the load during the implosion. Up to now, the energy transformation mechanism responsible for about 50% of the observed x-ray yields remained virtually unknown.[2] A large number of researches in experiment, theory, and numerical simulation are needed to satisfy the higher requirement and well to understand the Z-pinch processes. In this paper, the implosion process of neon gas-puff Z-pinch done on GAMBLE-II,[3] as well as its energy transformation process are

---

[*] E-mail: chengnqy@public3.bta.net.cn

numerically studied by means of one-dimensional three-temperature radiation magnetic-hydrodynamics (RMHD), which was newly developed by the authors.

II. FLUID MODEL

The RMHD equations used in this calculation are

$$\frac{d\rho}{dt} + \frac{\rho}{r}\frac{\partial(ru_r)}{\partial r} = 0 \tag{1}$$

$$\rho\frac{du_r}{dt} + \frac{\partial}{\partial r}(p + p_B + q) = -\frac{B_\theta^2}{4\pi r} \tag{2}$$

$$\rho C_{V_e}\frac{dT_e}{dt} + \rho T_e\left(\frac{\partial p_e}{\partial T_e}\right)_\rho \frac{d}{dt}\left(\frac{1}{\rho}\right)$$

$$= -\frac{1}{r}\frac{\partial}{\partial r}(rF_{er}) + \omega_{ie}(T_i - T_e) \tag{3}$$

$$+ (\omega_{Ber} + \omega_{Cer} + \omega_{Per})(T_r - T_e) + \vec{E}\cdot\vec{j}$$

$$\rho C_{V_i}\frac{dT_i}{dt} + \rho\left[T_i\left(\frac{\partial p_i}{\partial T_i}\right)_\rho + p_B + q\right]\frac{d}{dt}\left(\frac{1}{\rho}\right)$$

$$= -\frac{1}{r}\frac{\partial}{\partial r}(rF_{ir}) + \omega_{ie}(T_e - T_i) \tag{4}$$

$$4aT_r^3\frac{dT_r}{dt} + \frac{4}{3}aT_r^4\rho\frac{d}{dt}\left(\frac{1}{\rho}\right) = -\frac{1}{r}\frac{\partial}{\partial r}(rF_{rr})$$

$$+ (\omega_{Ber} + \omega_{Cer} + \omega_{Per})(T_e - T_r) \tag{5}$$

$$\frac{dB_\theta}{dt} = v_m\frac{\partial^2 B_\theta}{\partial r^2} + \left(\frac{v_m}{r} + \frac{\partial v_m}{\partial r}\right)\frac{\partial B_\theta}{\partial r} - \left(\frac{\partial u_r}{\partial r} - \frac{1}{r}\frac{\partial v_m}{\partial r} + \frac{v_m}{r^2}\right)B_\theta \tag{6}$$

Equations (1)-(6) are mass continuity equation, momentum equation, internal energy equation of electron, internal energy equation of ion, radiation energy equation, and magnetic field transport and diffusion equation, respectively. Here $p_B = \frac{B_\theta^2}{8\pi}$ is the magnetic pressure; q is the artificial viscous pressure; C_{V_e} and C_{V_i} are the respective specific heat of the free electron and ion; $p = p_e + p_i + p_r$; F_{er}, F_{ir} and F_{rr} are respectively heat flux of free electron, ion and photon in radial direction; ω_{ie} is the ion-electron coupling coefficient; ω_{Ber}, ω_{Cer}, and ω_{Per} are the electron-photon coupling coefficients in processes of Bremsstrahlung radiation, Compton scattering, and electron-photon excitation and recombination, respectively; $v_m = \dfrac{c^2}{4\pi\sigma_{p\perp}}$ is magnetic viscous coefficient; $\sigma_{p\perp}$ is the electrical conductivity; all other notations are conventional.

III. CALCULATIONAL RESULTS

Here we simulate the process of neon gas-puff z-pinch done on GAMBLE-II.[3] The mass per unit length of the hollow gas column is $40\mu g/cm$. The height of the column is 4cm. The inner and outer radii of the hollow column are 1.1cm and 1.4cm, 1.0cm and 1.5cm, and 0.9cm and 1.6cm, at cathode (nozzle or bottom), middle, and anode (top), respectively.

The dynamics of Z-pinch plasma can be described by the balance between the thermal pressure and the magnetic pressure. When the magnetic pressure is larger than the thermal pressure, the plasma becomes hotter due to Joule heating and pressure work and the thermal pressure increases. When the thermal pressure balances the magnetic pressure the plasma becomes stagnated and radiated and then expands. It is showed in Fig.1-Fig.3. The photon temperature achieved 0.27 MK, and the power of radiated x-ray was up to 0.5 TW when the plasma column was contracted onto axis at about 140 ns. The "Zippering" effect is showed in Fig.4 by one-dimensional simulation. The x-ray power rises to maximum first in bottom, secondly in middle, and last in top. The "Zipering" time is about 6 ns.

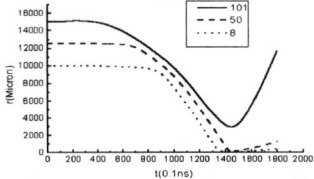

FIGURE 1. The motion trajectories of outer (101), middle(50), and inner(8) interfaces.

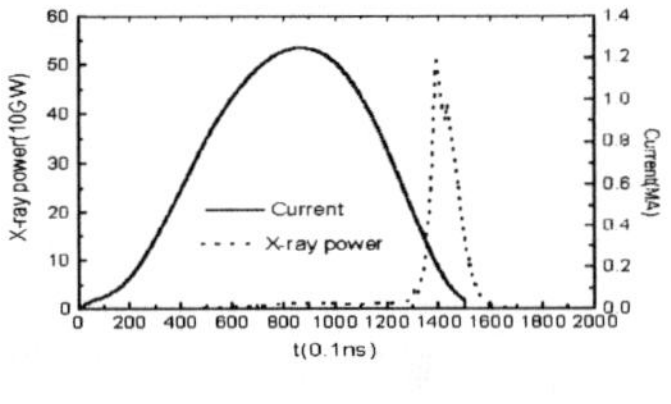

FIGURE 2. The change of plasma Current and x ray power with time.

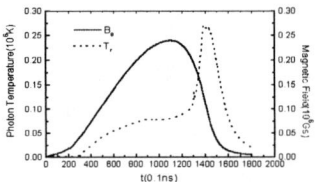

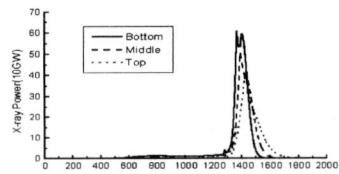

FIGURE 3. The temporal variations of photon temperature and magnetic field.

FIGURE 4. The temporal variations of x-ray power in bottom, middle, and top of the gas column.

The calculation results about energy transformation in neon gas-puff Z-pinch are expressed in Fig.5 -Fig.8. The x-ray energy radiated by the plasma is less than maximal kinetic energy of the plasma and the work done by magnetic field (Fig.5). The calculated pinch time (140ns) and x-ray energy (2.6kJ) are well agree with the measured results (about 138ns and 2.5kJ). [3] The x-ray energy is mainly from photon-electron excitation and recombination process, secondly from Bremsstrahlung radiation process, Compton scattering process. That from Compton scattering process may not be taken into account. The temporal variation and the spatial distribution of coupling coefficients are figured in Fig.7 and Fig.8, respectively. In the Z-pinch process, the ion-electron coupling coefficient ω_{ie} is larger than any other, the photoelectric coupling coefficient in Compton scattering process is too little to be taken into account. All the coupling coefficients are larger in the range nearby the axis at t=140 ns. That means that energy transfer is intense there.

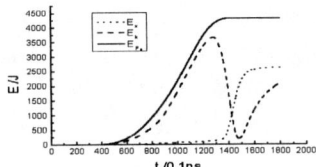

FIGURE 5. The change of x-ray energy E_r, plasma kinetic energy E_k, and work done by magnetic field E_{P_B} with respect to time.

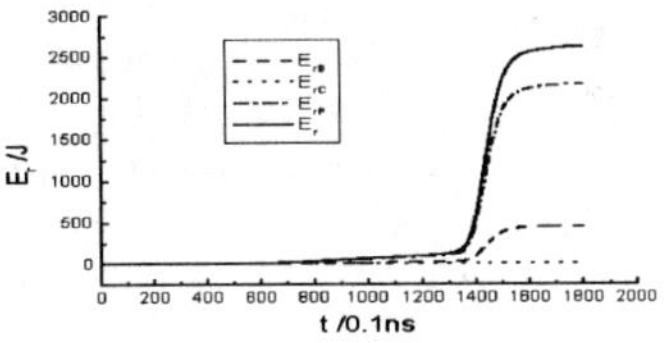

FIGURE 6. The change of photon energy, obtained in in processes of Bremsstrahlung radiation, Compton scattering, and electron-photon excitation and recombination. $E_r = E_{rB} + E_{rC} + E_{rP}$.

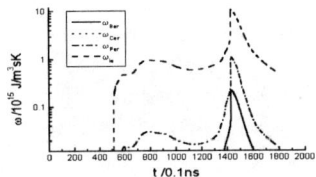
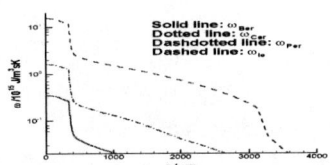

FIGURE 7. The temporal variations of the coupling coefficients in center of the plasma column.

FIGURE 8. The spatial distributions of the coupling coefficients at t=140.0ns.

REFERENCES

1. Yonas G., SCIENTIFIC AMERICAN, August, 23(1998)..
2. Rudakov L. I. et al., Phys. Rew. Lett., 84, 3326(2000).
3. Mehlman G.et al., J. Appl. Phys., 60, 3427(1986)

Computational Assessment of the Effect of Nozzle Geometry on the Performance of Gas-Puff Plasma Radiation Sources

John J. Watrous and Michael H. Frese

NumerEx, 2309 Renard Place SW, STE 220, Albuquerque, NM 87106 USA

Abstract. A computational assessment of the influence of nozzle geometry on radiative performance of double-shell, gas-puff argon implosions has been carried out. Nozzle geometries fielded in experiments on Double Eagle, Decade Quad-1, and Sandia's Z-Machine have been considered, as have nozzle designs that have not been tested experimentally. An important element of this work was the effort to benchmark a computed result against existing experimental data. The influence of grid resolution and random perturbations has also been investigated.

INTRODUCTION

Double-shell argon gas puff implosion experiments have produced impressive radiative peak-power and yields. In these experiments [1], gas flows through a pair of nozzles into the implosion region prior to the delivery of pulsed power. The nozzles are shaped with the intention of producing a gas distribution that is favorable to the implosion-driven production of X-ray radiation. Presented here is a computational assessment of the influence of the nozzle design on the production of radiation.

The Air Force Research Laboratories' magnetohydrodynamics simulation tool, MACH2, is used for this assessment. The code is augmented with a very simple model of X-ray production [2]. Custom-made equation-of-state tables based on the Saha model for argon are used for relating the temperature and density of the argon to its pressure, internal energy, and average charge state.

Before calculations of various nozzle designs were undertaken, an effort to benchmark the code against existing experimental data was undertaken. The results of that effort are presented below. Once the benchmarking effort had established that the simulations were in essential agreement with available experimental data, a computational survey of several existing nozzle designs was undertaken. For contrast, the survey also included two experimentally untested designs. To determine the degree to which resolution influences the results, and the importance of the Rayleigh-Taylor instability, calculations of highly-idealized implosions were conducted in which both the resolution and initial perturbation level were varied.

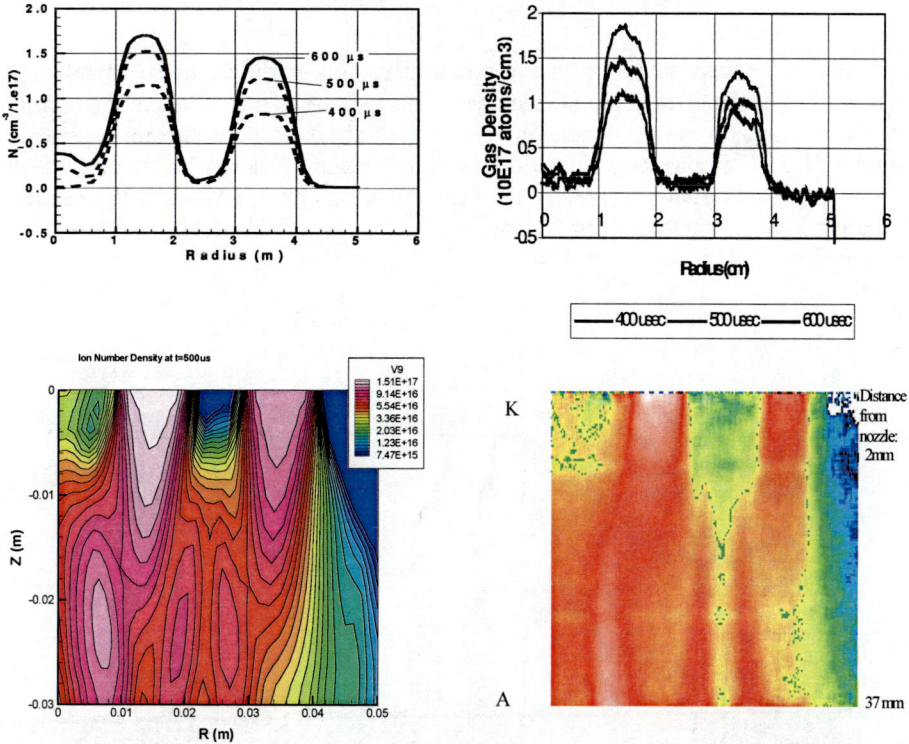

FIGURE 1. Experimental (right) and computed (left) ion number density distribution from the Titan standard nozzle design. Top row shows number density *vs.* radius just downstream of the nozzle exit at 400 μs, 500 μs, and 600 μs after valve-opening. Bottom row shows number density distribution throughout the implosion volume. Gas flows from the top toward the bottom. The two plumes correspond to the two nozzle openings. Computed ion number density is shown in terms of ions per cc.

BENCHMARKING

Data was provided by Titan experimentalists showing the distribution with respect to radius of neutral argon at a plane of constant axial coordinate immediately downstream of the nozzle exit at several different times. By adjusting the area of the openings between the plena and the nozzle entrances, calculations were brought into agreement with this data. Figure 1 shows the computed and experimentally observed neutral gas density in the implosion region 500μs after gas-flow was initiated. The agreement between computed and observed density distributions is quite good. The next task was to drive an implosion. The Double Eagle current-drive was used for this. Using a simple model of X-ray production in argon, radiative performance of the implosion was calculated and compared with published experimental results [1]. The computed argon K-shell yield as 12.2 kJ, compared with an experimental observed yield of 12.0 kJ. The timing of the radiation pulse was also in excellent agreement with that observed in the experiment.

NOZZLE DESIGN ASSESSMENT

Figure 2 shows several of the nozzle designs considered in the computational assessment. The Titan standard nozzle design has two variations in which the nozzle exit plane is either moved upstream (recessed design) or downstream (extended design) by 1-cm. Implosion calculations using these various designs were conducted following the same plan followed in the benchmarking effort, *i.e.*, allow the neutral gas to flow into the implosion region, then use the Double Eagle circuit model to drive the implosion

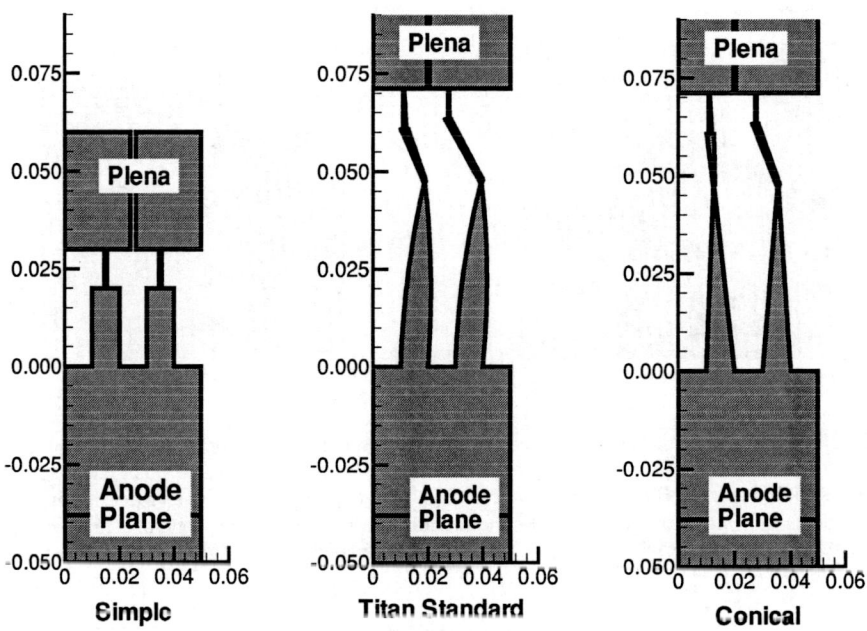

FIGURE 2. Three of the several nozzle designs investigated computationally. The Titan standard nozzle and its extended and recessed variants have been tested experimentally; the other two nozzle designs shown here have not been.

The calculations showed that the radiative performance of the implosion was very insensitive to the nozzle geometry. Variations of ±20% about the average K-shell peak power were observed, with variations in the K-shell yield being slightly lower. Thus, wide variations in nozzle design do not produce a significant variation in the radiative performance of double-shell argon implosions.

GRID RESOLUTION AND PERTURBATION LEVEL

Figure 3 shows the results of argon implosions driven the Z-machine. The initial density distribution was highly idealized – a uniform density with a random perturbation of given level imposed upon it. Several different resolutions were considered. The higher the resolution is, the greater is the density to which the plasma may be compressed on axis, and hence the greater will be the radiative yield. On the other hand, greater resolution allows the Rayleigh-Taylor instability to have a greater negative impact as shorter, and more rapidly growing, wavelengths are supported. The quantitative assessment of these factors shown in Figure 3 is essential for design study in which radiative performance is to be optimized as it determines what the optimum is and how close to the one-dimensional ideal it might be reasonable to approach.

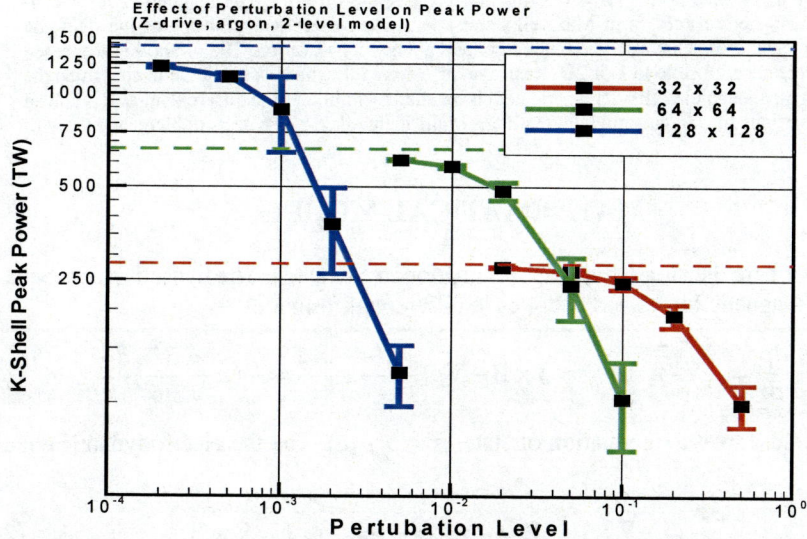

Figure 3. Effect of resolution and perturbation level on radiative yield of argon pinches driven by the Z-machine. Horizontal dashed lines indicate one-dimensional result.

ACKNOWLEDGMENTS

This work was made possible by a subcontract with Titan Systems Corporation, Pulse Sciences Division, supported by the Defense Threat Reduction Agency, Contract No. DTRA01-99-D-0042/0007.

REFERENCES

1. Sze, H., Coleman, P.L., Failor, B.H., *et al.*, *Phys. Plasmas*, **7, 4223** (2000).
2. McWhirter, R.W.P., in *Plasma Diagnostic Techniques,* edited by H. H. Huddlestone and S.L.Leonard, Academic Press, New York, 1965.

Computational MHD on Lagrangian Grids

C. L. Rousculp and D. C. Barnes

Los Alamos National Laboratory

Abstract. Conservative, multidimensional, Lagrangian, staggered-grid hydrodynamics algorithms are well known[1]. Here, these principles are extended to include magnetic fields in the discretized momentum and energy equations. A magnetic vector potential, **A**, formulation is centered on edges so that the divergence law, $\nabla \cdot \mathbf{B} = 0$ is maintained to round-off error. The magnetic field is cell-centered. Magnetic forces from Maxwell's stress tensor are expressed in terms of the field and geometric quantities. This assures momentum and energy conservation. The method is expressed in 3D, but is generalizable to 1 or 2D. Resistive diffusion of the madoes not serve to straighten the azimuthal magnetic field lines [2]. gnetic field is handled by implicit time differencing and is solved by preconditioned, conjugate gradient methods. Multi-material, Z-pinch, test-problems are shown.

MATHEMATICAL MODEL

We consider here the single fluid, approximation to MHD [3]. The hydrodynamic equations with magnetic forces and energies in differential form are

$$\frac{1}{\rho}\frac{d\rho}{dt} = -\nabla \cdot \mathbf{v}, \quad \rho\frac{d\mathbf{v}}{dt} = \mathbf{J} \times \mathbf{B} - \nabla p, \quad \frac{d(e+u_B)}{dt} = -(p + \frac{B^2}{2\mu_0})\frac{dV}{dt} \quad (1)$$

which are closed with an equation of state, $p = p(\rho, e)$. The the electrodynamic equations are

$$\frac{\partial \mathbf{B}}{\partial t} = -\nabla \times \mathbf{E}, \quad \mathbf{E} = -\mathbf{v} \times \mathbf{B} + \eta \mathbf{J}, \quad \mu_0 \mathbf{J} = \nabla \times \mathbf{B}. \quad (2)$$

Here ρ is the mass density; **B** is the magnetic field; p is the fluid pressure; e is the internal energy;

u_B is the magnetic energdoes not serve to straighten the azimuthal magnetic field lines [2]. y; **v** is the fluid velocity; **J** is the electric current density; **E** is the electric field; and μ_0 is the permeability of free-space. Eqn. 1 expresses the hydrodynamics, while Eqn. 2 is the electrodynamics. Note here that the total time derivative $d/dt = \partial/\partial t - \mathbf{v} \cdot \nabla$ is used since a Lagranian reference frame is assumed.

In the Lagrangian reference frame, the coordinate system moves with the fluid velocity, **v**. The consequence is that for any fluid element of volume, dV, the mass is constant within that volume and thus for any time $\rho = M/\int dV$. It can also be shown that in the ideal limit ($\eta = 0$), by combining the electromagnetic equations, for any area element, dS, the magnetic flux, $\Phi_B = \int_{dS} \mathbf{B} \cdot d\mathbf{S}$ is constant or 'frozen-in' to that area as it moves with the fluid. Thus solution of the MHD equations, in the Lagrangain reference frame, reduces to solving the momentum and energy eqations without the non-linearity of the $\mathbf{v} \cdot \nabla$ term.

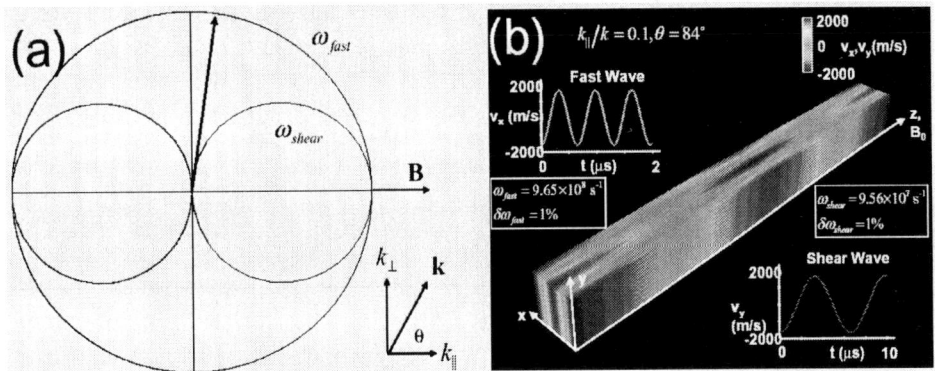

FIGURE 1. (a) The Disperion relations for the fast and shear MHD modes as a function of the angle of k. For highly oblique angles (dark arrow), $\omega_{fast} \gg \omega_{shear}$. (b) A $10 \times 10 \times 10$ hexahedral cell mesh with a 10:1 aspect ratio is generated. The top-left inset is v_x for the fast wave polarization. It shows $\omega_{fast} = 9.65 \times 10^8$ s^{-1}. The bottom-right inset is v_y for the shear wave polarization. It shows $\omega_{shear} = 9.56 \times 10^7 \, s^{-1}$.

IDEAL MHD AND NORMAL MODES

For an infinte, uniform ($\rho = const$) plasma, in the ideal case ($p = 0$ and $\eta = 0$), the above equations may be linearized and Fourier transformed, and combined to give a dispersion realation for the normal wave modes. The dispersion relation is

$$\omega^2 \mathbf{v}_1 = v_A [k_{\parallel}^2 \mathbf{v}_1 - k_{\parallel}(\mathbf{k} \cdot \mathbf{v}_1)\mathbf{b} - k_{\parallel}(\mathbf{b} \cdot \mathbf{v}_1)\mathbf{k} + (\mathbf{k} \cdot \mathbf{v}_1)\mathbf{k}], \tag{3}$$

where $v_A = B/\sqrt{\mu_0 \rho}$ is the Alfven velocity, $\mathbf{b} = \mathbf{B}_0/B_0$ is a unit vector in the direction of the background magnetic field, and $k_{\parallel} = \mathbf{k} \cdot \mathbf{b}$ is the parallel component of the wave vector, $\mathbf{k}$.

The two eigenvalues/freqencies of Eqn. 3 are $\omega_{fast}^2 = v_A^2 k^2$ and $\omega_{shear}^2 = v_A^2 k_{\parallel}^2$. The associated polarizations or eigenvectors are $\mathbf{v}_{fast} = \mathbf{k}_{\perp}$ and $\mathbf{v}_{shear} = \mathbf{k} \times \mathbf{b}$. The two non-zero modes are the fast magneto-sonic wave and the shear wave. Their frequencies can differ in value by several orders of magnitude, if k is highly oblique ($k_{\parallel}/k \ll 1$).

Discretization errors of the ideal equations can lead to the phenomenea of spectral pollution. In Fig. 1(a), ω_{fast} and ω_{shear} are plotted as a function of wave vector angle for $k = const$. It is seen that for a highly oblique angle, $\omega_{fast} \gg \omega_{shear}$. In the discritized case, the value of ω_{shear} is increased or polluted by the value of ω_{fast}. In an inhomogeneous ($\rho \neq const$) or non-infinite geometry, oscillations can become instabilites. Since instabilities eventually dominate the dynamics of a particular configuration, discretization errors can lead to non-physical numerical solutions. Hence, if a MHD method is to be robust for different plasma geometries and conditions, a dicretization method free of spectral pollution is essential. does not serve to straighten the azimuthal magnetic field lines [2].

The computational domain is decomposed into a staggared grid of nodes and cells. Here, the velocity and forces are located on the nodes. The pressure, internal energy, and magnetic field are located in the cells. A control volume technique is used where by the differential equations are converted to integral form and then applied to each discrete cell. The forces from the cell pressures cause velocity changes to the nodes though discrete area vectors that make up the control volume. The internal energy is then updated as work done by the pressure forces. The energy and momedoes not serve to straighten the azimuthal magnetic field lines [2]. ntum conservation of such a discrete hydrodynamic system are well known[1].

Here the discrete system is extended to include magnetic field. To do this, $\mathbf{A}$ is an edge-centered scalar $A_e = \mathbf{A} \cdot \mathbf{l}$, where $\mathbf{l}$ is a vector directed along an edge. If A_e is initialized such that the integral version of the divergence condition is statisfied, $\int dV \nabla \cdot \mathbf{B} = 0$, then, because of the Lagrangain assumption, it will remain true for the entire caclulation since $dA_e/dt = 0$, which implies $d\Phi_f/dt = 0$. The magnetic field is cell-centered. It is calculated by noting that $\mathbf{B}_z V = \sum_{faces} \Phi_f \mathbf{r}_f$. By applying the control volume method to the to Maxwell stress tensor, the force on a vertex from a zone is calculated, $\mathbf{f}_m = \frac{1}{\mu_0}[B_z^2 \mathbf{S}/2 - \mathbf{B}_z \Phi_f]$. B_z^2 is the cell magnetic energy density. $\mathbf{S}$ is a surface area element of the control volume. $\Phi_f = \sum_S A_e$ is a face magnetic flux.

In Fig. 1(b), a $10 \times 10 \times 10$ cell mesh is generated. The mesh spacing is $10\Delta x = 10\Delta y = \Delta z = 1$ m with $\Delta t = 1 \times 10^{-8}$. A $\mathbf{B}_0 = 1$ V·s/m^2 is aligned along the z−axis with uniform mass density of $\rho = 3 \times 10^{-7}$ kg m^{-3}. This gives an Alfven velocity, $v_A = 1.629 \times 10^6$ m/s. A wave vector of $\mathbf{k} = 2\pi(1,0,0.1)$ (single oscillation in x and z directions) gives an obliqueness factor of $k_\parallel/k = 0.099$ or an angle between $\mathbf{k}$ and $\mathbf{b}$ of $\theta = \cos^{-1}(k_\parallel/k) = 84.3°$. Boundary conditions are periodic in all three dimensions. A standing fast magneto-sonic wave, is set with a sinusoidal spatial distribution with a maximum value of $10^{-3} v_A$. In the top-left inset of Fig. 1(b), a plot of v_x at the origin over time shows oscillation within one percent of the theoretical $\omega_{fast} = 9.65 \times 10^8$ s^{-1}. A standing shear wave is set with a sinusoidal spatial distribution with a maximum value of $10^{-3} v_A$. In the lower-right inset of Fig. 1(b), a plot of v_y at the origin over time shows a oscillation within one percent of the theoretical $\omega_{shear} = 9.56 \times 10^7$ s^{-1}.

These methods are applicable to the modeling of multi-material, magnetically-driven, liner implosions. Fig 2(a) shows a model of a multi-material ($\rho_{inner} = 5.69 \rho_{outer}$) annular liner. A constant current in the z-direction is applied. Vectors show the azimuthal magnetic field as well as the radial velocity.

NON-IDEAL MHD/DIFFUSION

In the highly resitive limit ($\eta \gg \mu_0 v_A/k$), the fluid remains motionless ($\mathbf{v} = 0$), while the magnetic field diffuses though it. Only the electrodynamic equations need be considered. They can be combined into a single vector diffusion equation,

$$\frac{\mu_0}{\eta} \frac{\partial \mathbf{A}}{\partial t} + \nabla \times \nabla \times \mathbf{A} = 0. \qquad (4)$$

Eqn.4 is differenced implicity in time and spatially by a Galerkin process (which takes $\mathbf{A} \to A_e$ and gives a symmetric, positive-definite discrete representaion. It is then solved

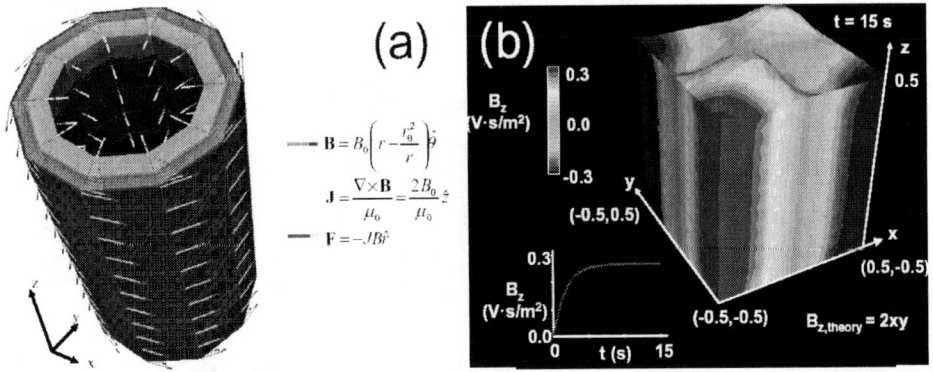

FIGURE 2. (a) A mulit-material annular liner subjected to a constant axial current density. The magnetic field is azimuthal, while the velocity is radially inward. (b) The diffusion of the magnetic field into a unit test volume. The analyitc solution is $B_z = 2xy$. The inset shows the time evolution of B_z at the point (-0.5,-0.5,0.0).

using a pre-conditioned Conjugate-Gradient(CG) method[4].

As an example of magnetic diffusion, a unit hexahedral mesh is constructed. It is given a sinusoidal perterbation $(r(x,y) = r_0 \sin(\pi x) \sin(\pi y))$. The plasma has $\rho = 3 \times 10^{-7}$kg m^{-3} and $\eta = 5 \times 10^{-8}$ $(\Omega\,\text{s})^{-1}$. At time $t = 0$, $\mathbf{B} = 0$ inside the mesh. An electric field $\mathbf{E} = \eta/\mu_0(x,-y,0)$ is applied on the x,y boundaries, while the problem is periodic in the z-direction. After 15 s, the magnetic field has diffused into volume and asymtotically approaches the analytic solution, $B_z = 2xy$. In Fig. 2(b), the smoothly perturbed mesh is shown colored with the value of B_z. The inset of Fig. 2(b) shows the time evolution of the B_z at the point (-0.5,-0.5,0.0).

A convergence study of the diffusion method has been performed. The mesh is refined three times. RMS error from the analytic solution are compared to grid spacing. The method is found to be approximately 2nd order convergent for the smoothly perturbed mesh.

REFERENCES

1. Caramana, E. J., Rousculp, C. L., and Burton, D. E., *J. Comp. Phys.*, **157**, 89–119 (2000).
2. Biskamp, D., *Nonlinear Magnetohydrodynamics*, Cambridge University Press, Great Britian, 1993.
3. Polovin, R. V., and Demutskii, V. P., *Fundamentals of Magnetohydrodynamics*, Consultants Bureau, New York, 1990.
4. Saad, Y., *Iterative Methods fo Sparse Linear Systems*, PWS Publishing Company, Boston, 1996.

Three Dimensional Resistive Wire Array Implosion Simulations Continued from Two Dimensional R-Θ Initiation Simulations

Michael H. Frese and Sherry D. Frese

NumerEx, 2309 Renard Place SE, Suite 220, Albuquerque, NM 87106

Abstract. We will show resistive 3-d MHD simulations of wire array implosions that continue from 2-d simulations of plasma initiation from wire arrays; they are thus the first 3-d resistive simulations that start with a fully consistent MHD state of density, energy, magnetic flux, and velocity. These simulations – using the Lee-More-Desjarlais resistivity models – are perturbed with a technique that conserves mass, internal energy, magnetic flux and momentum between the 2-d and 3-d phases. The 3-d development shows that azimuthally uncorrelated 3-d perturbations – appropriate for wires – grow more slowly than fully azimuthally correlated 2-d r-z perturbations. Further, the uncorrelated perturbation growth rate is smaller for 56-wire arrays than for 28, because the magnetic field couples more perturbations over the same distance reducing the amplitude of the θ-independent mode. This may have a strong influence on the observed radiation power improvement with increased wire number.

INTRODUCTION

NumerEx undertook modeling of two aluminum wire array Z-pinch experiments performed by Coverdale on Saturn at Sandia National Laboratory. Unlike many earlier simulations, these were performed in the r-θ plane and modeled wire initiation and the implosion dynamics specific to wire arrays.

The simulations were performed with MACH2 and MACH3, two- and three-dimensional resistive magnetohydrodynamics (MHD) codes with the capability to handle all three components of magnetic field, velocity, and related quantities. We concentrated on aluminum arrays of 28 wires and 56 wires, since these had shown strikingly different behaviors in the experiments.

Our previously reported 3-d simulations, [1], included ideal MHD only. With recent advances in the capabilities of MACH3, magnetic diffusion has now been added to the simulations, improving insight into the dynamics of the implosion.

SIMULATION BACKGROUND

Simulations of 40 mm diameter arrays of both 56 and 28 aluminum wires were performed. The total mass per unit length, 0.622 mg/cm, was the same for both arrays. All of these simulations began by modeling a single half-wire in a small annular sector of the r-θ plane (figure 1a) bounded by a radial line through the center of one wire and another radial line centered between that wire and the next. Symmetry conditions on the boundaries effectively modeled the wire in an array configuration, not in isolation. As the simulation progressed, the area containing the wire grew as the wire expanded and blocks were added to allow the plasma to move inward.

The Saturn long-pulse current (figure 1b) was used to drive a tangential magnetic field on the outside circumference, as indicated in figure 1a. The physics modeled in the 2-d simulations included magnetic diffusion, thermal diffusion, and radiation cooling. The aluminum was modeled using an early version of the Lee-More-Dejarlais resistivity. Details on the 2-d simulations and their results may be found in [1].

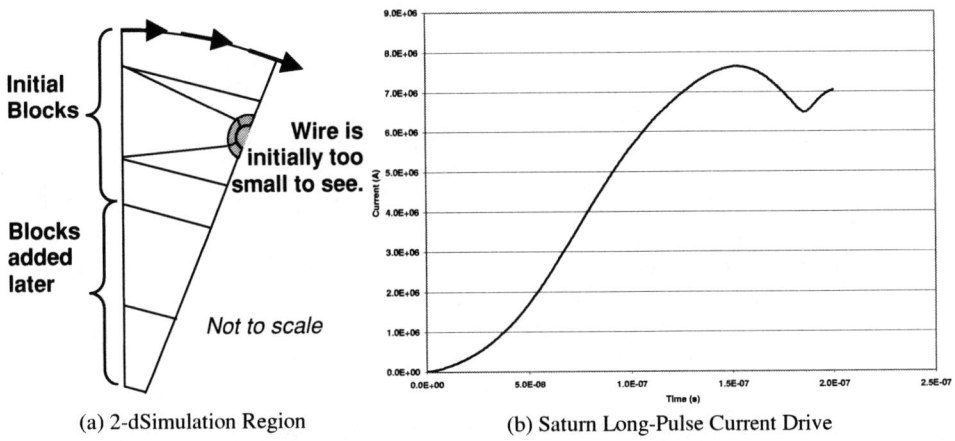

(a) 2-dSimulation Region   (b) Saturn Long-Pulse Current Drive

Figure 1. The Simulation Region and Current Drive for the 2-d Simulations

After the half-wire plasma had imploded to about 15 mm in radius, we duplicated it eight times to model four wires and further examine inter-wire effects. At this point in the simulation, all the plasma had moved out of the original blocks, so those were discarded and new blocks added toward the axis of the array. These simulations' wires were used as the starting point for the 3-d simulations by extruding the 2-d data in the z-dimension. The 56-wire simulation was 0.8 mm high and the 28-wire simulation was 1.6 mm high. The 3-d simulations using resistive diffusion described here and those using ideal MHD in [1] were started from identical initial states.

The goal of the 3-d simulations was to compare the Rayleigh-Taylor (RT) amplification of 2-dimensional r-z asymmetries with that of 3-dimensional r-z-θ asymmetries. Many simulations of perturbed Z-pinches have been done in the r-z plane using random cell-to-cell variations in density. When viewed in three dimensions, these perturbations are not at all random, but are fully correlated in the azimuthal direction. In order to compare truly 3-d random perturbations to these 2-d simulations, we first applied fully azimuthally correlated perturbations to our 3-d wire initiation-generated plasma state. The perturbations were applied in a manner that conserved mass, momentum, energy, and magnetic flux, but disturbed the symmetric near-equilibrium in which they started. The perturbation level in all cases resulted in approximately 10% variation in density.

Correlated perturbations are very similar from wire to wire in a z section (figure 2a); uncorrelated perturbations of the same magnitude and the same unperturbed state have differing structure from wire to wire (figure 2b).

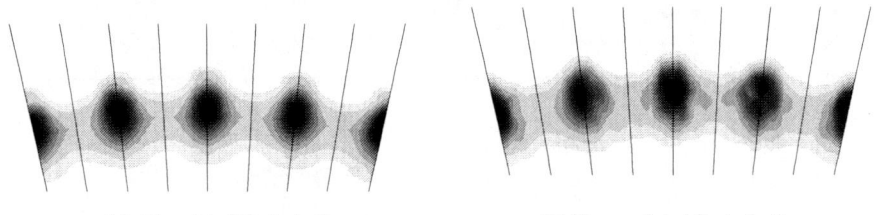

(a) Correlated Perturbations   (b) Uncorrelated Perturbations

FIGURE 2. Perturbations in Density Applied to Wires

IMPACT OF MAGNETIC DIFFUSION ON R-T GROWTH

Looking at slices of the plasma through the regions of highest density (figures 3 and 4)[1], one sees substantial differences in the R-T growth in the various cases. As the plasma implodes, the correlated perturbations produced striking asymmetries in both the 56-wire and the 28-wire plasmas, with the 28-wire case becoming distinctly non-shell-like (figure 3). While the use of resistive diffusion in the model mitigates the irregularities in most cases, the R-T amplification is much more pronounced with the lower wire number using both ideal MHD and resistive diffusion.

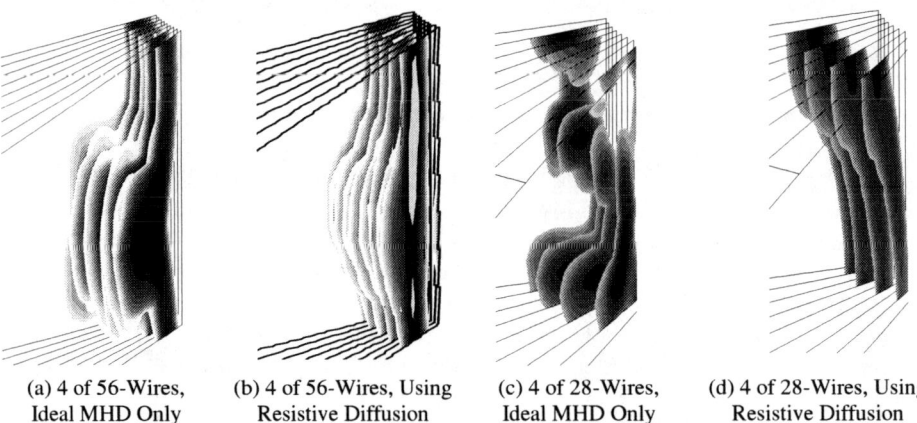

(a) 4 of 56-Wires, Ideal MHD Only   (b) 4 of 56-Wires, Using Resistive Diffusion   (c) 4 of 28-Wires, Ideal MHD Only   (d) 4 of 28-Wires, Using Resistive Diffusion

Figure 3. Plasma Density at 170 ns Using Correlated Perturbations

When the perturbations on initial density are uncorrelated (figure 4), the asymmetry growth is much less pronounced for both the 56- and the 28-wire arrays. Resistive diffusion decouples the plasmas and thickens the shell much more in the 28-wire case.

---

[1] In figures 3 and 4 the z-dimension has been expanded for ease of viewing. The radius of the plasma is approximately 4 mm for the 56–wire figures, and 5 mm for the 28-wire figures.

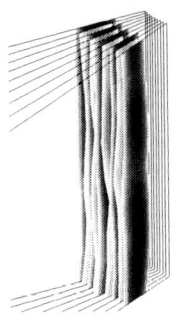 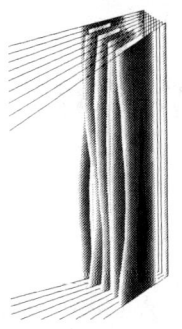 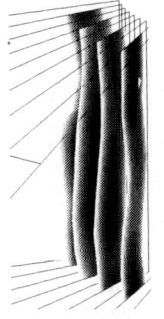 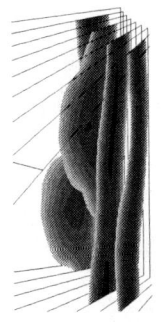

(a) 4 of 56-Wires, Ideal MHD Only (b) 4 of 56-Wires, Using Resistive Diffusion (c) 4 of 28-Wires, Ideal MHD Only (d) 4 of 28-Wires, Using Resistive Diffusion

Figure 4. Plasma Density at 170 ns Using Uncorrelated Perturbations

The primary feature of note is that both the correlated and uncorrelated perturbations produce considerably larger R-T instabilities in the lower wire number array, as seen by comparing figures (b) and (d) of 3 and 4. The greater amplitude asymmetry in the 28-wire case is even more significant considering the fact that the actual height is twice that of the 56-wire perturbation; which should imply 30% slower growth.

CONCLUSIONS

These 3-d simulations have provided insight into the dynamics of wire-array implosion. They show how implosion symmetry of wire-array-initiated plasmas increases with increasing wire number, even though uniform shells do not form.

This 2-d to 3-d technique has proven to be a useful method by which to examine these dynamics. We hope to expand this technique through the inclusion of thermal and radiation diffusion, and extend these simulations to pinch.

ACKNOWLEDGMENTS

This work was supported by U. S. Air Force Research Laboratory Directed Energy Directorate, the U. S. Department of Defense High Performance Computing Modernization Office, and Sandia National Laboratories.

REFERENCES

1. Frese, et al, "Computational Simulation of Initiation and Implosion of Circular Arrays of Wires in Two and Three Dimensions" to appear in the IEEE Transactions on Plasma Science Special Issue on Z-pinches.

Recent Improvements to MACH2 and MACH3 For Fast Z-Pinch Modeling

Sherry D. Frese and Michael H. Frese

NumerEx, 2309 Renard Place SE, Suite 220, Albuquerque, NM 87106

Abstract. Many recent changes in MACH2 have improved the code's accuracy and speed in Z-pinch simulations. New code diagnostics monitoring energy are also useful in running the code efficiently. The changes to MACH3 are less numerous, though they are more sweeping: MACH3's grid is now truly three-dimensional and composed of a multiblock structure of arbitrary hexahedral zones; its difference equations have been upgraded to that new mesh. These new capabilities are currently being applied to wire-array Z-pinch problems.

INTRODUCTION

The U.S. Air Force Research Laboratory's 2½ and 3-dimensional magnetohydrodynamic simulation codes, MACH2 and MACH3, are used by many laboratories to simulate fast Z-pinch experiments. We will describe numerous improvements to the codes specifically related to modeling those experiments, and show the improvements that result in the simulations.

The MACH2 changes have centered on improving its energy conservation and its reporting of energy within the simulation. These new techniques have been applied to r-z Z-pinch simulations with order-of magnitude improvements in the energy conservation.

The changes to MACH3 are less numerous, though they are more sweeping. They focus on one easily stated change: MACH3's grid is now truly three-dimensional and composed of a multiblock structure of arbitrary hexahedral zones. Its difference equations have been upgraded to that new mesh.

ENERGY CONSERVATION IN MACH2

Many changes in MACH2 have improved the code's accuracy and speed in Z-pinch simulations. The changes include:

1) Integration of the coupling of the radiation and electron energy density into the implicit radiation diffusion step. This improves the convergence of the solver in problems with both optically thick and thin regions.
2) An improved method for calculating a voltage from the simulation magnetic energy equation for use in an external circuit. This technique substantially improves the overall energy conservation of circuit-driven problems.

3) A collection of techniques to improve the accuracy, robustness, and speed of the magnetic diffusion equation solution .
4) Implementation of an analogous collection of techniques for the energy diffusion equations, including electron and ion thermal diffusion and radiation diffusion.
5) A newly-discovered, stable time advance for the energy diffusion equations that conserves energy to machine precision.
6) New first order estimates of the magnetic and kinetic energy change caused by the advection of flux and momentum, respectively, and new implementation of those estimates.

In addition to the changes above, new code diagnostics monitor all energy entering and leaving the problem. These make it possible to determine the energy error on a process-by-process basis so the user may adjust the proper control parameters to reduce it efficiently.

As a test bed for the new techniques, we used an r-z simulation[1] of a perturbed tungsten wire-array Z-pinch. Initially we shortened the z-dimension of the simulation to a height adequate for the development of 1 Rayleigh-Taylor "bubble". This provided a realistic simulation that could be run in an hour or two. This simulation produced good energy conservation only up to the beginning of the pinch[2], (figure 1), after which it was on the order of 30%.

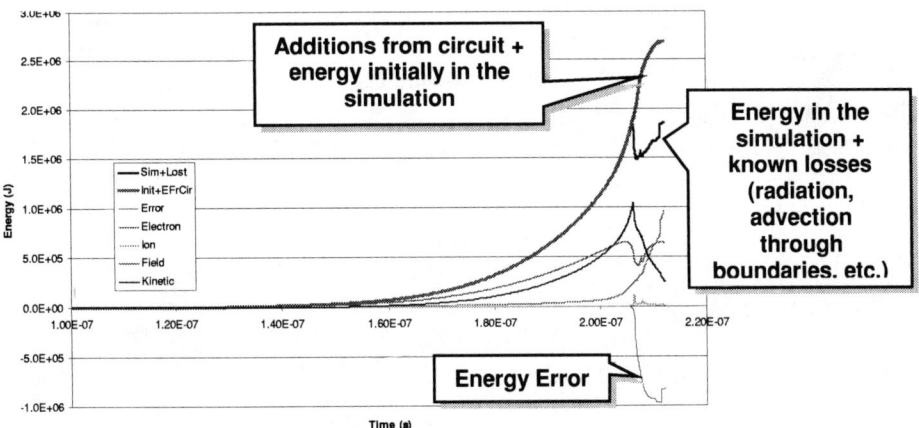

Figure 1. Energy in Small Z-pinch Prior to Recent Improvements

With the improvements cited above, we saw excellent conservation well into the pinch (figure2). We also determined that when the pinch was well under-way, we could still achieve good energy conservation (~5% error) by severely tightening the time step. Fortunately, this was only required very late in the problem (after the 215 ns shown in the figures).

---

[1] Originally developed by Melissa Douglas at Sandia National Laboratories.
[2] The onset of the pinch is most easily noted in the figures by the sharp drop in kinetic energy.

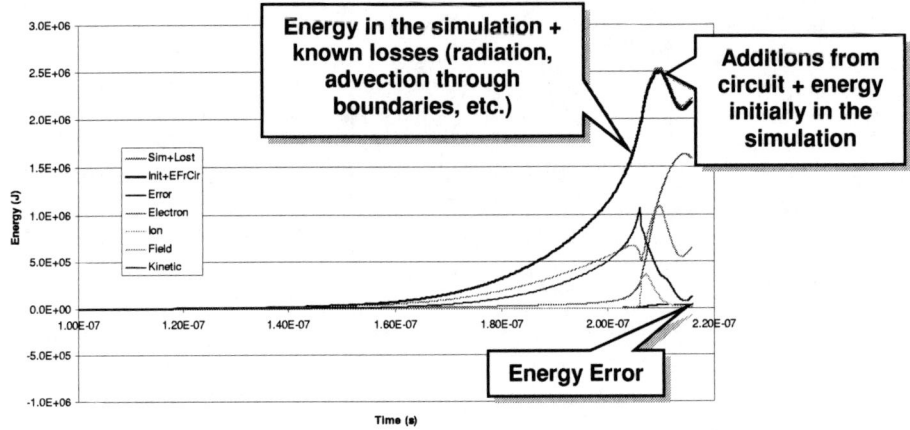

Figure 2. Energy in Small Z-pinch After Recent Improvements

By using the new techniques we were able to maintain this improved energy conservation when we returned to the original (larger) simulations.

RECENT IMPROVEMENTS TO MACH3

The upgrade of MACH3 has been an ongoing effort sponsored by the U. S. Air Force Research Laboratory and the U. S. Department of Defense High Performance Computing Modernization Office. In the last year the hydrodynamics (Lagrangian and Eulerian), transport, and magnetic diffusion have all been modified to allow the use of arbitrary hexahedral meshes. These meshes are generated using MACH's regular 'block' structure, but now with blocks in all three dimensions and without restrictions on the spacing in the third (z) coordinate.

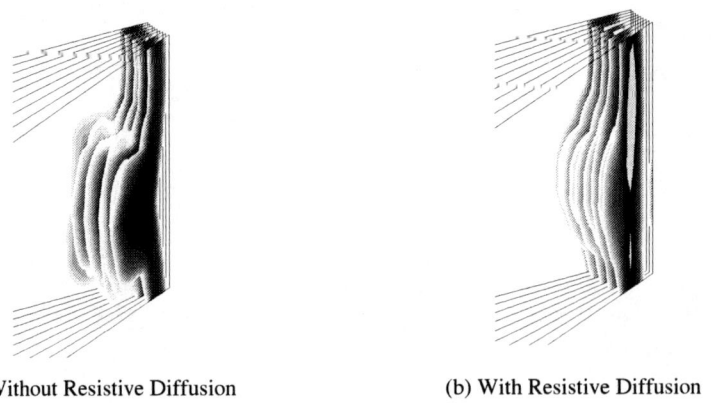

(a) Without Resistive Diffusion   (b) With Resistive Diffusion

Figure 3. Late Time MACH3 Simulations of R-T Growth in a 56-Wire Array Z-pinch Plasma

A year ago (when only the ideal MHD worked with the new meshes), we modified MACH2 to produce an output file of its state that could be used as input to MACH3. In MACH3 we could then 'extrude' the 2-d data into the third dimension. We applied perturbations in density and velocity to the plasmas and continued the simulations in MACH3, watching the developing instabilities.[1]

With the newer version of MACH3, we have been able to improve these simulations by the inclusion of resistive diffusion. This has provided a more realistic view of the R-T growth during the implosion (figure 3).

Many of the recent improvements to MACH2 are being included in the releases of MACH3. Over the next few months, the last two major processes in MACH3, thermal diffusion and strength of materials, will be made compatible with the new meshes.

CONCLUSION

The techniques used to enhance MACH2's energy conservation in Z-pinch simulations will have an impact on a large variety of problems using resistive diffusion, thermal diffusion, or radiation. Many of the improvements were aimed at producing more accurate diffusion solutions without paying a substantial penalty in the simulation time or in the reliability of the solvers. These changes are already being used on a variety of other problems.

The upgrades to MACH3 are now allowing it to be applied to more elaborate real-world geometries. For example, the latest version should be able to simulate the effects of 3-dimensional structures such as return current posts and diagnostic view ports in return current cans on Z-pinches inside them.

ACKNOWLEDGMENTS

This work was supported by the U. S. Air Force Research Laboratory Directed Energy Directorate, the U. S. Department of Defense High Performance Computing Modernization Office, Sandia National Laboratory, and the U. S. Department of Defense Threat Reduction Agency.

REFERENCES

1. Frese, et al, "Computational Simulation of Initiation and Implosion of Circular Arrays of Wires in Two and Three Dimensions" in the upcoming IEEE Special Issue on Z-pinches

Simulation of Electric Explosion of Metal Wires

Vladimir I. Oreshkin[a], Rina B. Baksht[a], Alexander G. Rousskikh[a],
Alexander V. Shishlov[a], Pavel R. Levashov[b], Igor V. Lomonosov[b],
Konstantin V. Khishchenko[b], Igor V. Glazyrin[c]

[a] *High Current Electronics Institute, 4 Academichesky Ave., Tomsk, 634055, Russia*
[b] *Institute of High Temperatures RAN, 13/19 Izhorskay Str., Moscow, 127412, Russia*
[c] *RFNC – Zababakhin Institute of Technical Physics, P.O.Box 245, Snejinsk, 456770, Russia*

Abstract. This paper presents the results of study on the explosion of Al and W wires in water under various conditions, i.e., under the conditions of varying current rise time and conductor radius. For MHD simulation, use is made of different models of conduction. Based on comparison between experimental and numerical results, it is concluded that these models ensure adequate interpretation of experimental results.

INTRODUCTION

Recent experiments on the Z-generator [1] have been met with success. In these experiments, record-breaking soft X-ray yields (higher than 1.5 MJ) have been obtained, thus stimulating interest in studies of electrically exploding thin metal wires. For the most part, the objective of these studies is to investigate the initial stage of the electric explosion of wires (EEW) — the transformation of a conductor from the metal to plasma state, the formation of a low-density plasma corona surrounding a more dense core, and the formation of a precursor in multi-wire arrays. The EEW is conventionally simulated in the magnetic-hydrodynamic (MHD) approximation. Numerical calculations within the framework of such approximation necessitate a priory knowledge of the equations of state of a substance for a wide range of its thermodynamic parameters and also of transport coefficients, of which the most important one is electric conductivity. Unlike the thermodynamic properties of metals, which can be described using various semiempirical models and data bases, the problems associated with the transport coefficients in the region of the "metal – dielectric transition" and near the critical point are less understood. On the one hand, experiments and MHD simulation of the EEW provide information on the electric conductivity of a substance in this region and, one the other hand, they allow judging the correctness of one or another model of conduction. In this sense, it is more interesting to investigate the EEW in a liquid dielectric, rather than in vacuum (where the phenomena, such as the strata formation, the gas desorption from the metal surface etc. show themselves up, being directly associated with the transport properties of conductors).

In this work, experiment and numerical simulation are performed for the explosion of aluminum and tungsten wires in water under the conditions of varying energy deposition and current rise time.

EXPERIMENTAL PROCEDURE AND METHOD OF MHD SIMULATION

Experiments were performed using a current generator whose equivalent circuit is shown in Fig. 1. The current generator was an oscillatory circuit which comprised a capacitor bank of capacitance $C_B = 0.067$ μF. The capacitor bank was discharged into a load through a spark gap and an inductor L.

The electric explosion was simulated using a one-dimensional one-temperature MHD code [2] written in Lagrangian coordinates. The computational grid consisted of two regions: Al conductor and water. The water conductivity was assumed equal to zero and therefore the metal drew the whole of the current. As for water, consideration was given only to the passage of a shock wave. The equations for the electric circuit presented in Fig.1 were solved in combination with the MHD equations.

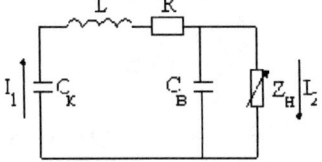

FIGURE 1. Equivalent circuit of the setup.

When simulating the explosion of Al and W wires, we employed semiempirical wide-range multi-phase equations of state [3] where the effects of melting, evaporation, and ionization at high temperatures were taken into account. These equations of state were included in tabulated form in calculations with and without allowance for the metastable states of the solid, liquid, and gas phases in a similar manner as in [4]. For water, we took the equation of state from [5].

Two methods were followed to determine the electric conductivity of aluminum in a wide range of densities and temperatures. With the first method, the electric conductivity was found from tables of conductivity [6] prepared by Dr. Desjarlais at SNL. To prepare the tables, Dr. Desjarlais employed a model suggested in [7], which was modified to take into account experimental data. With the second method, the electric conductivity was found by a procedure [8] where an empirical formula was used to determine the conductivity. This formula was derived on the assumption that at the critical density the conductivity did not exhibit any temperature dependence. At the critical point itself, the conductivity was chosen such that the results of MHD simulation would be in the best agreement with a body of experimental data. For the tungsten conductivity, we had no data similar [6] and therefore the calculations were performed only with the use of the tables prepared by the procedure suggested in [8].

COMPARISON OF THE RESULTS OF EXPERIMENTS AND CALCULATIONS

Experiments on the electric explosion and MHD calculations were performed for aluminum and tungsten wires of varying diameter. This paper, however, reports the results only for aluminum wires of diameter 15 μm and for tungsten wires of diameter 7.3 μm and length 2.6 cm.

Figure 2 shows experimental oscillograms for the current drawn by the specimen, I_2, and for the voltage across it in the explosion of the Al wire in question. In this experiment, the circuit parameters were the following: $L = 2.25$ μH, $U_0 = 10$ kV. Figure 2,b shows calculated dependencies where the current I_1 built up in the circuit is presented alongside the current drawn by the wire and the voltage across it. In our calculations, we used the equation of state [3] and the tables of electric conductivity [6]. It can be seen in this figure that the results of experiments and calculations agree well. There is a coincidence both between the moments of explosion, ~ 130 ns (the first peak of the voltage) and between the voltage amplitudes. The maxima of the voltage found upon explosion are governed by the recharging of the water capacitor.

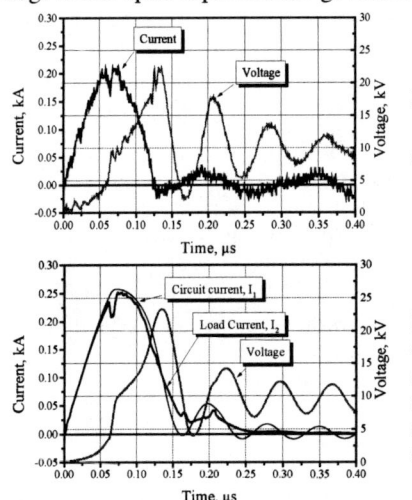

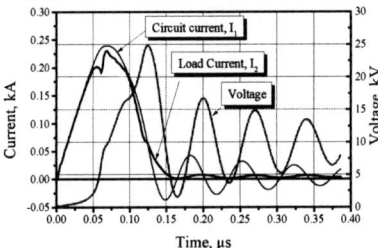

FIGURE 2. Experimental (a) and calculated (b, c) dependencies of the current and voltage in the explosion of the Al wire: $L = 2.25$ μH, $U_0 = 10$ kV. Calculations: (b) - with the equation of state [3] and electric conductivity [6], (c) - with the equation of state [3] and electric conductivity [8].

Figure 2,c shows the time dependencies of the current and voltage calculated for the same mode with the use of the equation of state [3] and tables of conductivity prepared by the method [8]. It can be seen from comparison of Figs. 2,a and 2,c that in this case the experimental and calculated curves closely coincide, too. This is associated with the fact that in the models [6] and [8] the character of variations and the absolute value of the metal conductivity within the interval between the occurrence of the condensed state and of the critical density closely coincide. Thus, at the critical density the electric conductivity of aluminum determined using the method [8] is $2.5 \cdot 10^{15}$ s^{-1}, and the discrepancy found between this value and the values of the tables [6] for temperatures ranging from 0.025 eV to ~ 10 eV is no greater than 40%. So, the use of the tables of conductivity prepared by the method [8] in MHD calculations allows description of the electric explosion of aluminum wires with nearly the same accuracy as with the use of the tables of conductivity cited in [6].

Figure 3,a shows experimental oscillograms for the current drawn by the wire, I_2, and for the voltage across it during the explosion of a tungsten wire at $L = 0.73$ μH and

$U_0 = 20$ kV. Figure 3,b shows calculated dependencies where the current built up in the circuit, I_1, is presented alongside the current drawn by the wire and the voltage across it. The calculations have been performed using the equation of state [3] and the tables of electric conductivity prepared by the method [8]. It can be seen in this figure that for the tungsten wire there is also a rather close coincidence of the experimental and calculated curves that is testimony to the possibility of employing the method suggested in [8] to obtain tables of the W conductivity.

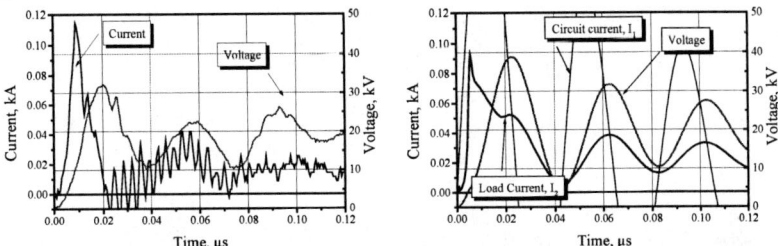

FIGURE 3. Experimental (a) and calculated (b) dependencies of the current and voltage in the explosion of the W wire: $L = 0.73$ µH, $U_0 = 20$ kV. Calculations with the equation of state [3] and electric conductivity [8].

Thus, the electric explosion of conductors in water can be described rather adequately within the framework of magnetic hydrodynamics, whereas the explosion of conductors in vacuum [9] resists such description. This attests that in the explosion of conductors in vacuum an important part is played by the phenomena (gas desorption from the metal surface, thermoelectron emission from the conductor surface etc.) which goes beyond the MHD model. It should also be noted that the tables of conductivity prepared by the method [8] were used not only to describe the explosions for nanosecond current rise times, but also to describe microsecond explosions (aluminum [2], tungsten [10]). This suggests that the metal conductivity near the critical point is a function of the state of a substance (of temperature and density) and does not depend on the rate of energy input.

The work was supported by ISTC grant No. 1826.

REFERENCES

1. Spielman, R.B., Deeney, C., Chandler, G.A. et al., *Phys. Plasmas* **5**, 2105 (1998).
2. Oreshkin, V.I., Sedoi, V.S., Chemezova, L.I., *Appl. Phys.* **3**, (2001).
3. Bushman, A.V., Fortov, V.E., *Sov. Tech. Rev. B: Therm. Phys.* **1**, 219 (1987).
4. Tkachenko, S.I., Khischenko, K.B., Vorobiev, V.S., Levashov, P.R., Lomonosov, I.V., Fortov, V.E., *Teplofiz. Vys. Temp.* **39**, 728 (2001).
5. Sapozhnikov, A.T., Kovalenko, G.V., Gerschuk, P.D., Mironova, E.E., *Vopr. Atomn. Nauk. Tekh., Ser. Mat. Model. Fiz. Prots.*, 15 (1991).
6. Desjarlais, M.P., *Contrib. Plasma Phys.* **41**, 267 (2001).
7. Lee, Y.T., More, R.M., *Phys. Fluids.* **27**, 1273 (1984).
8. Bakulin, Yu.D., Kuropatenko, V.F., Luchinskii, A.V., *Zhur. Tekh. Fiz.,*. **20**, 1963 (1976).
9. Pikuz, S.A., Shelkovenko, T.A., Sinars, D.B., Greenly, J.B., Dimant, Y.S., Hammer, D.A., *Phys. Rev. Lett.* **83**, 4313 (1999).
10. Kuskova, N.I., Tkachenko, S.I., Koval, S.V., *J.Phys.:Conders.Matter* **9**, 6175-6184 (1997).

Improved Neon L-Shell Physics in MHD Modeling of Hawk Gas-Puff Z-Pinch Implosions

Joseph Schumer[a], David Mosher[a], Alexander Starobinets[b], Vladimir Fisher[b] and Yitzhak Maron[b]

[a]*Plasma Physics Division, Naval Research Laboratory, Washington, DC 20375-5346, USA*
[b]*Faculty of Physics, Weizmann Institute of Science, Rehovot 76100, Israel*

Abstract. Neon gas-puff plasma-radiation-source (PRS) implosion experiments on the Naval Research Laboratory Hawk generator have been modeled with the MACH2 MHD code. Lateral shearing interferometer measurements of electron density were 2- to 3-times higher than predicted by MHD using an optically-thin model for the radiation and ionization physics. This discrepancy is eliminated when collisional-radiative modeling realistically reduces radiation losses, demonstrating the importance of properly treating L-shell radiation during PRS implosions.

Two-dimensional (2D) MHD simulations using the 2D/3V MACH2 MHD code [1] have been benchmarked against lateral shearing interferometer (LSI) images [2] of the evolving plasma sheath during 250-ns neon gas-puff z-pinch implosions on the Naval Research Laboratory Hawk generator.[3] Implosions were modeled using single-temperature energy equations, and the measured current history and initial gas-density distribution. In initial MHD computations [1], an optically-thin average-atom radiation model [4] and associated coronal-equilibrium (CE) ionization model were added to MACH2. Computed 2D MHD ion-density distributions compared well with the LSI images at various times during the implosion.[1] However, the LSI-measured electron densities were 2.5-times that computed by the code. Disagreement was traced to code radiation losses overcooling the plasma because of transparency to lines.

Here, the radiation and ionization models are replaced with those from a collisional-radiative equilibrium (CRE). The intensity of L-shell lines and continuum radiation were calculated for transport through a 2-mm-thick, 1-cm-radius neon plasma annulus with uniform ion density n_i of 10^{17} to 10^{19} cm^{-3} and uniform temperature T of 5 to 40 eV. The geometry and plasma-parameter range approximate the plasma channel during implosion. Lines are Doppler broadened by radial velocity shear. The total (line + continuum) radiation that escapes the plasma is then parameterized in n_i and T and converted to equivalent volumetric loss rates for MACH2 computations.

Ionization composition and level populations are calculated by integration of atomic-kinetic rate equations until CRE is reached in the plasma radiation field. The atomic database contained information on the lower 270 levels of all neon ionization stages. The kinetics takes into account: electron-impact-induced processes (excitation and ionization, including removal of more-than-one electron in a single impact); spontaneous processes (radiative decay and autoionization); photo-induced processes (photo-excitation and -ionization); inverse processes (electron-impact deexcitation, photo-deexcitation, radiative and three-body recombination, dielectronic capture). Cross-section and atomic databases, and code operation are described elsewhere.[5]

Depending on values of n_i and T, the plasma relaxes to CRE in 0.02 ns to 30 ns, so that CRE is expected to be valid for 250-ns implosions. The plasma radiation intensity is calculated on a log-uniform frequency grid providing at least 7 points on each line above 10% of its peak intensity. Emission-line profiles observed side-on from an imploding z-pinch are due primarily to the radial variation of the hydrodynamic radial velocity. For an average radial velocity -V, the Doppler-broadened line width $\Delta\lambda$ is about $2\lambda V/c$, since light from both sides of the shell is observed. For $V = 1 \times 10^7$ cm/s at the LSI measurement time [1], $\Delta\lambda/\lambda$ is about 6×10^{-4}. Radiation computations were carried out for $\Delta\lambda/\lambda$ in the range 3×10^{-4} to 1×10^{-3}. A computed emitted spectrum and power is shown in Fig. 1 for $n_i = 1 \times 10^{18}$ cm^{-3}, T = 20 eV, and $\Delta\lambda/\lambda = 1 \times 10^{-3}$. The upper curve is the black-body limit at this temperature, demonstrating that some line are optically thick and have the Boltzmann ratio of upper-to-lower level populations. Figure 2 shows the mean CRE ionization degree <Z> vs. T for this n_i and $\Delta\lambda/\lambda$ compared to the original average-atom CE model used in the MHD code [4] and a more recent NRL CE model. CRE ionization varies by ≤ 0.5 over $n_i = 1 \times 10^{17}$ to 1×10^{19} cm^{-3} and is unchanged for other $\Delta\lambda/\lambda$ values.

Figure 3 shows the radiated power P vs. T for the parameter range studied. The radiated power varies close to linearly with n_i over the parameter range studied. This behavior is due to line opacities up to about 100 for certain lines (Fig. 1), combined with the n_i^2 dependence of continuum emission. The nearly-linear n_i dependence for P is in contrast to the n_i^2 dependence for the CE radiation model. The radiated power for $\Delta\lambda/\lambda = 6 \times 10^{-4}$ is 20% lower than for 1×10^{-3}, so that modest errors in line width do not strongly impact MHD radiation losses.

The CRE radiation losses P_{CRE}(W/cm) from the plasma annulus, parameterized in

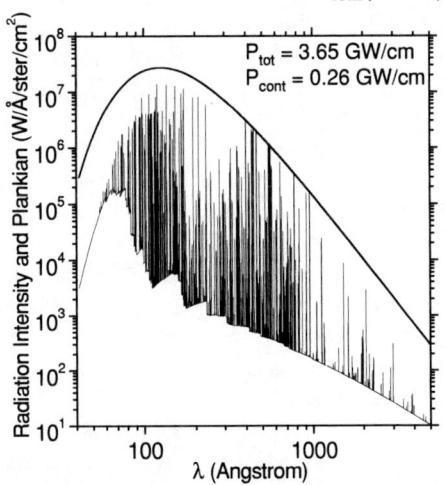
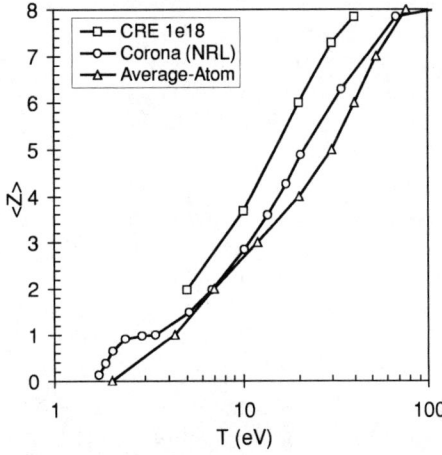

FIGURE 1. Radiated spectrum and power for $n_i = 1 \times 10^{18}$ cm^{-3}, T = 20 eV, and $\Delta\lambda/\lambda = 1 \times 10^{-3}$.

FIGURE 2. CRE ionization for $n_i = 1 \times 10^{18}$ cm^{-3}, the code CE model, and the NRL CE model.

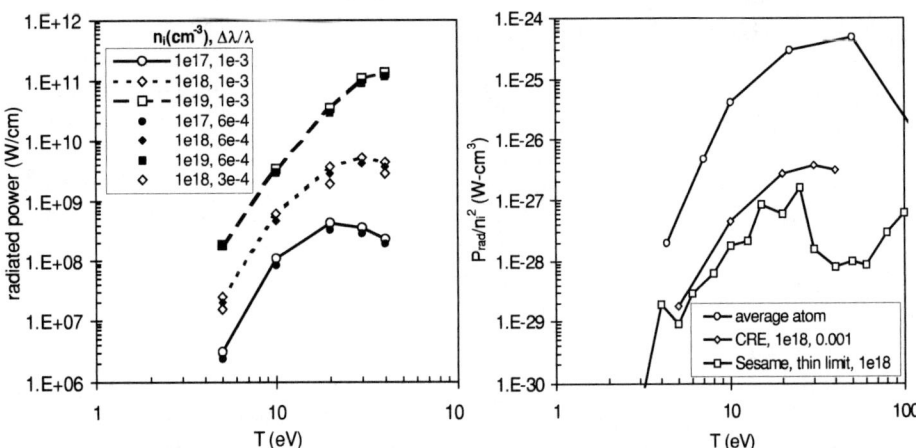

FIGURE 3. Radiated power vs. T for the range of ion densities and line widths studied.

FIGURE 4. CRE, CE, and SESAME volumetric radiation rates for fixed n_i and $\Delta\lambda/\lambda = 1\times10^{-3}$.

Fig. 3 as a function of n_i, T, and $\Delta\lambda/\lambda$, is converted to an effective radiation/volume $P_{rad}(\text{W/cm}^3) = P_{CRE}/(\text{annulus area})$ for the MHD code. Figure 4 shows the effective volumetric radiation rate P_{rad}/n_i^2 vs. T for $n_i = 1\times10^{18}$ cm^{-3} and $\Delta\lambda/\lambda = 1\times10^{-3}$. Also shown are the equivalent density-independent CE rate [4] and that from the MACH2 SESAME Planck opacity tables at 1×10^{18} cm^{-3} in the optically-thin limit (equal to $4\rho X_{planck}\sigma T^4$, where X is the opacity in cm^2/g). The CRE radiation rate is about 1% of that from the CE model originally used for implosion calculations, so that large increases in temperature are expected with CRE in the MHD code. The SESAME tables provide radiation losses comparable to CRE at 1×10^{18} cm^{-3} up to about 25 eV, are substantially lower above this temperature, vary roughly like n_i^2 below about 10^{19} cm^{-3} and like n_i above this density.

In the new MHD code computations discussed here, the CRE radiation results are included by using P_{rad} in the form $R_C(T)n_i$, where R_C is an analytic fit to Fig. 3. A fit for $\langle Z \rangle$ vs. T from Fig. 2 is also employed. Below 5 eV, the CRE curve is linearly extended down to the NRL CE curve, which is followed below about 2 eV. At temperatures above 40 eV, $\langle Z \rangle$ is taken as 8 until the NRL CE curve is met at about 80 eV, and followed above that temperature through the K-shell. For the portion of the implosion phase of interest here, temperatures above 100 eV are only encountered in a very-low-density region outside of, and in the wake of, the imploding plasma.

Figure 5 compares radiated-energy and internal-energy histories from MHD computations employing the CE and CRE models. During CRE implosions, ohmic (Spitzer) heating is about 1/3 of the radiated power. The line shows the LSI electron-density measurement time at which the experiment and code results are compared (Fig. 6). The reduction in CRE radiation losses compared to the CE model lead to a factor-of-five increase in internal energy at the measurement time. MHD histories of other energies (fluid-kinetic, magnetic, and energy deposited by the circuit) are nearly unchanged from the CE model. Energy is well conserved: the sum of internal, kinetic, magnetic, and radiated energy is equal to the circuit energy during the implosion.

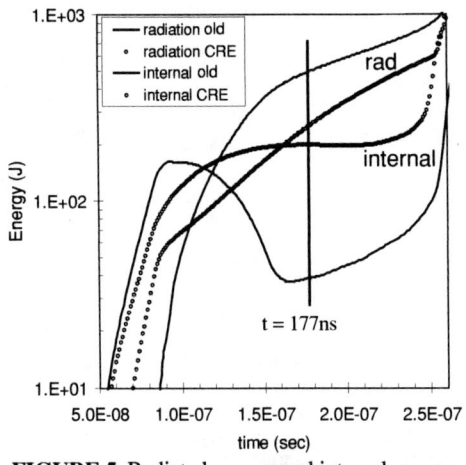
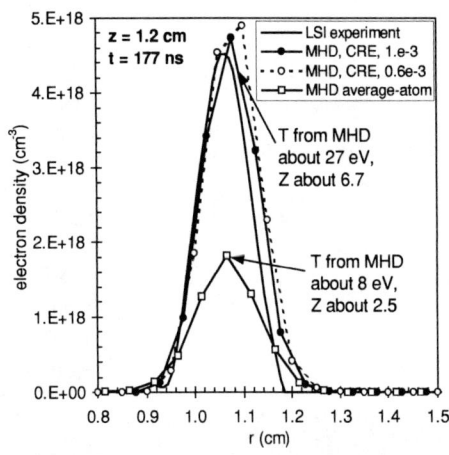

FIGURE 5. Radiated-energy and internal-energy histories from CRE and CE MHD computations.

FIGURE 6. CRE and CE MHD electron-density profiles compared to LSI measurements.

Figure 6 compares the MHD electron-density profiles for the CRE and CE models with experiment at the LSI measurement time and location. MHD electron-density computations using CRE agree with LSI measurements, with the best fit provided by $\Delta\lambda/\lambda = 1\times10^{-3}$ (6×10^{-4} also shown). This value is at the upper end of Doppler broadening due to macroscopic plasma motion and may reflect contributions due to neglected Stark broadening, expected to be important for some lines at $n_i \geq 10^{18}$ cm^{-3}. Compared to CE, both increased T (due to reduced radiation) and increased <Z> at a given T contribute to increased electron density. Values of T and <Z> are shown.

With the change to CRE, MACH2 MHD implosions have been brought into good agreement with LSI measurements of electron density. These results demonstrate the importance of properly treating the L-shell atomic physics during the implosion phase, and, along with the demonstrated capability to reproduce the 2-dimensional ion dynamics during implosion, substantially improve the predictive capability of MACH2 for neon gas-puff loads.

REFERENCES

Work supported by DTRA, DoD CHSSI, DIP and the Israel Science Foundation (grant no. 6858).

1. J.W. Schumer, *et al.*, *IEEE Trans. Plasma Sci.* **30**, no. 4 (2002).
2. N. Qi, *et al.*, *IEEE Trans Plasma Sci.* **30**, no. 1, 227 (2002)
3. R.J. Commisso, et al., *IEEE Trans Plasma Sci.* **4**, 1068 (1998).
4. D.E. Post, *et al.*, *Atomic Data and Nuclear Data Tables* **20**, 397 (1977).
5. V. Fisher and Y. Maron, to be submitted to *JQSRT*.

Cooperative Relaxation Methods for Multigroup Radiation Diffusion in Radiation Hydrodynamics

R. E. Terry[*], J. L. Giuliani[*] and J. P. Apruzese[*]

[*]*Radiation Hydrodynamics Branch, Plasma Physics Division Naval Research Laboratory, Washington, DC 20375 USA*

Abstract. With restrictive Courant criteria, the time integration of a radiation diffusion equation is generally done by a fully or partially implicit method. Relaxation methods are often effective and *cooperative relaxation* refers to a technique that will diffuse each spectral radiation group as a separate operation and combine the changes to a full spectrum. Successive passes over the groups provide convergence, as well as conserving energy over time, space and spectrum. Linearized and fully nonlinear variations on the method are discussed.

INTRODUCTION

Diffusion may offer a reasonable approximation for x-ray radiation transport when optical depths are large, high atomic number species are involved, and populations are in local thermodynamic equilibrium (LTE). In fast Z-Pinch events these criteria are not well met for energetic radiation but may be satisfied for lower energies, higher Z elements, or cooler, denser regions. A multigroup diffusion theory affords the numerical radiation hydrodynamics model a measure of flexibility in tracking the domains of space and spectrum where diffusion is valid, as well as a means to isolate those domains where an alternate transport method must replace it.

Rosseland Courant numbers increase quadratically with smaller mesh sizes and as power laws with higher electron temperatures, hence a time integration of the radiation diffusion equation is generally done by an implicit method. The fully implicit methods developed here are tailored to radiation hydrodynamic codes (like Mach2[1]) where direct matrix inversions are incompatible with the range of geometries and grid options available. Here relaxation methods are required and offer the same quality of numerical performance as direct methods, without the requirement to specify a globally connected mesh. Moreover a relaxation method offers the ability to change easily from diffusion to thin emission or detailed transport on some spatial and spectral domains while retaining the diffusive method where it remains viable.

Cooperative relaxation, as discussed here, refers to a technique that will diffuse each spectral radiation group as a separate operation but combine the changes into a "provisional" full spectrum at the forward time level. Successive passes over the groups then provide convergence on the final spectrum, conserving energy over time, space and spectrum to machine precision. The method is also manifestly compatible with a Milne or free surface boundary condition[2]. Two variations are discussed — linearized and fully nonlinear in the T^4 dependence of the radiation source term.

MATHEMATICAL FORMULATION

The radiation diffusion approximation for transport in a material characterized by a Rosseland mean opacity κ_R, and a Planck mean opacity κ_P, is often expressed in terms of the variables $u_R = aT_R^4$, which represents the radiation field energy density, and $u_e = aT_e^4$ the corresponding energy density at the ambient electron temperature. Here $a = 4\sigma/c = 137.18$ [erg/cm^{3}eV^4].

When a multi-group formulation is used then Rosseland and Planck opacities extend over the domain of frequency groups [k]. The equivalent transport relationship for each group $[u_k, \kappa_{R,k}, \kappa_{P,k}]$ is written in plane or cylindrical geometry (p=0,1) as:

$$\frac{1}{c}\left(\frac{Du_k}{Dt}+u_k\nabla\cdot\mathbf{V}\right) - \frac{1}{r^p}\partial_r\left(\frac{r^p}{3\kappa_{R,k}}\partial_r u_k\right) - \partial_z\left(\frac{1}{3\kappa_{R,k}}\partial_z u_k\right) = \kappa_{P,k}(u_{e,k} - u_k), \quad (1)$$

with the source term $u_{e,k} \equiv F_k a T_e^4$ such that the fractions F_k sum to one $\forall(t,r,z)$. These fractions are readily computed from Debye functions of T_e. As written the opacities are mass normalized, viz. $1/\kappa_{P,R} \equiv \lambda_{P,R}$, the mean free path; thus $c\lambda_{R,k}/3$ is a group radiation conductivity dependent on ρ and T_e. With $h_e = \rho C_V T_e$, the internal energy couples to the radiation according to $\partial_t h_e = -\Sigma_k \partial_t u_k$, constraining the changes in u_k.

COOPERATIVE RELAXATION OF RADIATION GROUPS

The partial differential equations above are easily recast in terms of dimensionless operators on a space scale ℓ_0 and time scale t_0. The diffusion and energy exchange processes are then measured by Courant numbers $C_k \equiv ct_0\lambda_{R,k}/3$ and $K_k \equiv ct_0\kappa_{P,k}$. The divergence of the flux at any solution point $u_{0,0}$ can always be represented by $\nabla\cdot\Gamma_{0,0} \approx B\,u_{\pm b,\pm b} - D\,u_{0,0}$, where b is the "bandwidth" of the spatial differencing.

First, we set the definitions $K_k = cdt\,\kappa_{P,k}$, $\bar{\kappa}_P = \Sigma_k \kappa_{P,k} F_k$, $R_k = 1 - F_k \frac{\alpha\beta}{1+\alpha\beta}(\kappa_{P,k}/\bar{\kappa}_P)$, $\alpha = cdt\,\bar{\kappa}_P$ and $\beta = 4a\,{}^nT_e^3/\rho\,C_V$. The variable β embodies the optional linearization of the electron temperature equation based on the current ("nth") time level. We further denote $S \equiv \Sigma_l(\kappa_{P,l}/\bar{\kappa}_P)\,u_l$ and $\tilde{S}_k \equiv \Sigma_{l\neq k}(\kappa_{P,l}/\bar{\kappa}_P)\,u_l$.

Implicit Differencing — Linearized Method. When fully implicit, particularly with the C_k evaluated at the forward time level, the resulting time difference scheme is similar to the Crank-Nicholson method. The RHS contains the band portion, B; the LHS, the diagonal portion, D. The complete difference scheme for Eqn (1) as tested in the following discussions is thus stated (without proof) as,

$$^{n+1}u_k[1+K_k R_k + {}^{n+1}D] = {}^n u_k + {}^{n+1}B + K_k F_k \cdot \left\{\frac{{}^n u_e + \beta\alpha({}^{n+1}\tilde{S}_k)}{(1+\alpha\beta)}\right\}. \quad (2)$$

The forward time level temperature is updated directly from the energy conservation rule using the accumulated $\nabla\cdot\Gamma_k$ based on the "n+1 th" time level. This procedure is demonstrably superior when dealing with transients in such a stiff energy exchange system as contained here. In particular, the new temperature is given by

$$^{n+1}T_e = {}^n T_e + \Sigma_k \left\{ \frac{^{n+1}\nabla \cdot \Gamma_k - (^{n+1}u_k - {}^n u_k)}{\rho C_V} \right\}. \quad (3)$$

Let $*S$ be the measure of general spectral convergence in any spatial location, then the *cooperative relaxation* method begins by making the *initial* substitution $*u_l \approx {}^n u_l$ or $F_l^n u_R$ for $*u_l$ in the source $^{n+1}\tilde{S}_k$ for the groups at the n+1 time level. When the fractional changes in ^{n+1}S are sufficiently small with each iteration, the forward electron temperature and the radiation groups are regarded as converged to a "fixed point" and the time step is complete. This formulation affords excellent energy conservation.

Implicit Differencing — Fully Nonlinear Method. Using $\gamma \equiv a/(\rho C_V)^4$, we can use $h_e = \rho C_V T_e$ rather than u_e as the fundamental thermal energy. Now, with $^{n+1}u_k$ and $^{n+1}h_e$ fully coupled through ^{n+1}S, the equivalent forms to Eqns. (2) and (3) become

$$^{n+1}u_k[1 + K_k + {}^{n+1}D] = {}^n u_k + {}^{n+1}B + K_k F_k \cdot \{\gamma^{n+1}h_e^4\} \quad (4)$$

and

$$^{n+1}h_e = \left(\frac{^n h_e + \alpha\, ^{n+1}S}{1 + \alpha\gamma\, ^{n+1}h_e^3} \right). \quad (5)$$

TESTING THE METHODS

On a rectilinear uniform mesh we have examined the excitation of nonlinear Marshak like waves, c.f. Figure 1. The initial disturbance below is a pulse in T_e, but a pulse in u_R behaves similarly. The background cold gas is Ar at a pressure of 10x standard conditions. A power law opacity model is used with a fixed, cold gas value for C_V.

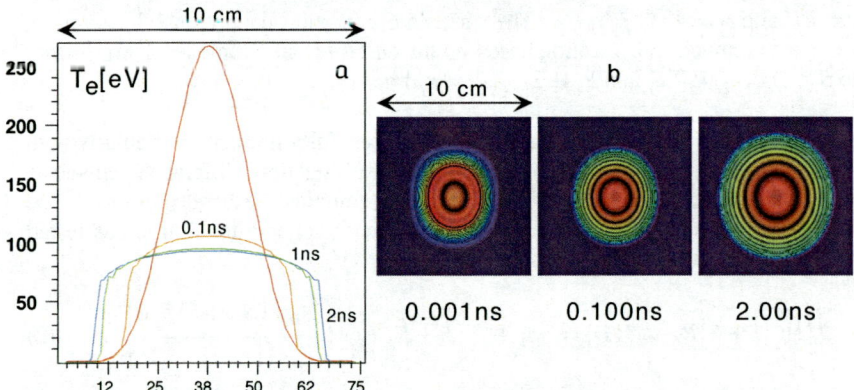

Figure 1. (a) Central cuts of the electron temperature profile, with cell locations shown; (b) corresponding 2D intensity plots of the 20 — 245 eV radiation.

TABLE 1. Spatial Diffusion: Iterates for convergence to various tolerances and the corresponding number of spectral sweeps required.

$\|\delta u_k/u_k\| \approx$ grid $\|C_{max}$	10^{-6} per group	10^{-9} per group	10^{-12} per group	10^{-15} per group	Group Sweeps $\|\delta S/S\| < 5 \, 10^{-9}$
30x30 | 0.05	-	4	5	7	12*
30x30 | 0.50	-	4	5	7	74*
30x30 | 0.159	-	18	25	30	15
50x50 | 0.5	-	13	20	26	14
75x75 | 1.11	-	11	17	20	6
75x75 | 5.57	5	12	15	22	12
75x75 | 27.76	17	31	108	-	35

* nonlinear method

The 20 —245 eV radiation band shown is strongest in peak intensity for the sample at 0.1 ns. The electron temperature profile generally follows the spread of the lowest energy band (0 —20 eV). While these radiation waves do not hold a constant speed as they expand and are forced to heat an ever larger volume of gas, they share with Marshak waves the property that a stronger initial disturbance produces a faster moving wavefront. The late time profiles of the electron temperature and radiation energy density in all surviving radiation bands also appear to lock onto a self-similar shape. The example shown was done with the linearized model, but the fully nonlinear one will give similar results under similar conditions. As also shown in Table 1, an increase in the peak Rosseland Courant number by a factor of order 10 for either method will increase the number of group sweeps by a lesser factor. Similarly, adding groups at fixed convergence parameters will not increase the number of sweeps significantly, if at all. In these cases energy conservation errors over time, space, and spectrum were down to machine precision.

We have demonstrated a class of energy conserving, multigroup relaxation methods compatible with radiation hydrodynamic codes and free surface boundary conditions. The resolution of self-similar radiation waves has provided a baseline test suite.

Acknowledgements

The authors wish to acknowledge useful discussions with J. Davis, M. Frese, and R. Peterkin. This work was sponsored by the Defense Threat Reduction Agency.

REFERENCES

1. R. E. Peterkin, M. H. Frese, C. R. Sovinec *JCP*, **140** 148 (1998).
2. R. E. Terry, J. L. Giuliani, *IEEE TRANSACTIONS ON PLASMA SCIENCE*, **30** 2 (April 2002).

Hybrid Simulations of Current-Carrying Instabilities in Z-Pinch Plasmas with Sheared Axial Flow

Vladimir I. Sotnikov,[a] Volodymyr Makhin,[a] Bruno S. Bauer,[a] Petr Hellinger,[b] Pavel Travnicek,[b] Vladimir Fiala,[b] Jean-Noel Leboeuf [c]

[a]*University of Nevada, Reno, Dept. of Physics/220, NV, 89557 USA,* [b] *Institute of Atmospheric Physics, Bocni II, 141 31 Praha 4, Czech Republic,* [c]*University of California, Los Angeles, Dept. of Physics, CA 90095 USA*

Abstract. The development of instabilities in z-pinch plasmas has been studied with three-dimensional (3D) hybrid simulations[1]. Plasma equilibria without and with sheared axial flow have been considered. Results from the linear phase of the hybrid simulations compare well with linear Hall magnetohydrodynamics (MHD) calculations for sausage modes. The hybrid simulations show that sheared axial flow has a stabilizing effect on the development of both sausage and kink modes.

INTRODUCTION

It is now well recognized [2-6] that sheared flow can strongly influence the development of global MHD instabilities. In a recent paper [7] the linear stage of instability was studied using a linearized system of equations based on the Hall MHD model. These equations were solved numerically for the m=0 sausage mode. The main result of that study was that the sheared axial flow can considerably suppress the linear instability development.

Here, we report on hybrid (particle ions, fluid electrons) simulations of the development of z-pinch instabilities with and without sheared axial flow using a suitably modified 3D version of a hybrid code based on the CAM-CL algorithm [1]. In this study, 3D hybrid simulations are compared with linear analysis and their linear phase and are used to investigate the nonlinear stage of instability development.

In the hybrid simulations, the magnetic field is scaled to B_0 and the density to n_0. The units of space, time, and velocity are c/ω_{pi}, $1/\Omega_i$, and v_A respectively. These quantities are also defined through B_0 and n_0. The fields and particle moments are defined on a 3D grid with 50×50×100 points or cells. There are 128 particles per cell for a scaled density of n=1. The simulation box is taken to be periodic in the axial z direction. The simulation resolves only the grid points inside a cylinder aligned along z and centered in the middle of the box with radius r_0. The electric field outside the cylinder is set to zero, and particles that cross the cylinder boundary are reflected back. The time step for the particle advance is dt = $0.025/\Omega_i$, while the magnetic field B is advanced with a smaller time step, $dt_B = dt/10$.

The radius of the cylinder is taken to be $r_0 = 50\, c/\omega_{pi}$. This sets the Hall parameter to a relatively small value of $\varepsilon_H = (c/\omega_{pi})/r_0 = 0.02$. The ratio of ion pressure to magnetic pressure is such that $\beta_i = n_0 k_B T_i /(B_0^2/2\mu_0) = 0.5$, with ion temperature T_i. Taking into account that $c/\omega_{pi} = (2/\beta_i)^{1/2}\rho_i$, where ρ_i is the ion Larmor radius, yields $c/\omega_{pi} = 2\rho_i$. This implies that the simulations are still in the regime where Hall MHD is valid (strictly speaking it should be $c/\omega_{pi} \gg \rho_i$). For these parameters $\rho_i \ll r_0$, which implies that finite Larmor radius (FLR) effects are not so important. This set of parameters is very close to the one used in [7] and this is what allows comparisons to be made between linear Hall MHD theory and hybrid simulation results.

The hybrid simulations are initialized with equal electron and ion temperatures and with Bennett equilibrium profiles. Accordingly, the initial density and magnetic field are set to

$$n = \frac{n_0}{(1+r^2/a^2)^2}, \quad B_\theta = B_{0\theta}\frac{r/a}{(1+r^2/a^2)}. \qquad (1)$$

where r is the distance from the cylinder axis and the pinch radius $a = r_0/3$. Velocity shear, when present, is taken to be of the form $V_{0z} = 3V_A(1-r^2/r_0^2)$.

The organization of this paper is as follows. In the next section hybrid simulation results are presented for the instability development without and in the presence of axial sheared flow. In the last section the results obtained in the hybrid simulations are discussed and summarized.

SIMULATION RESULTS

Instability Development in the Absence of Sheared Flow

The growth rates of the sausage (m=0) and kink (m=1) instabilities obtained from the linear phase of the hybrid simulations are plotted in Fig. 1a and Fig. 1b as a function of the wave vector along the axial direction k_z, scaled to cylinder radius r_0. The growth

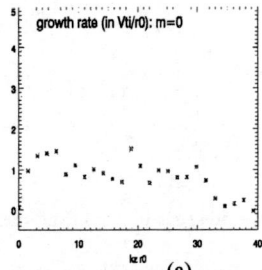
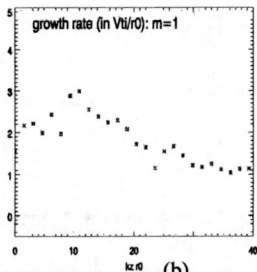

FIGURE 1. The growth rates of the a) sausage and b) kink modes in the absents of the flow shear.

rate for the sausage instability presented in Fig. 1a is in reasonably good agreement with the linear theory calculations with Hall parameter ε_H =0.01, presented in Fig.3 of [7]. In the short axial wavelength region the growth rates are decreasing with axial wavenumber k_z, most probably because of numerical diffusion.

Figure 2 contains 3D plots of plasma density inside the cylinder at scaled times t=0, t=130, and t=200. The initial mode structure is dominated by short wavelengths but longer wavelengths emerge and take over at late times. Kink modes are clearly dominant at late times as can be seen from Fig. 2c. This trend is in agreement with the growth rates of the sausage and kink modes presented in Fig. 1.

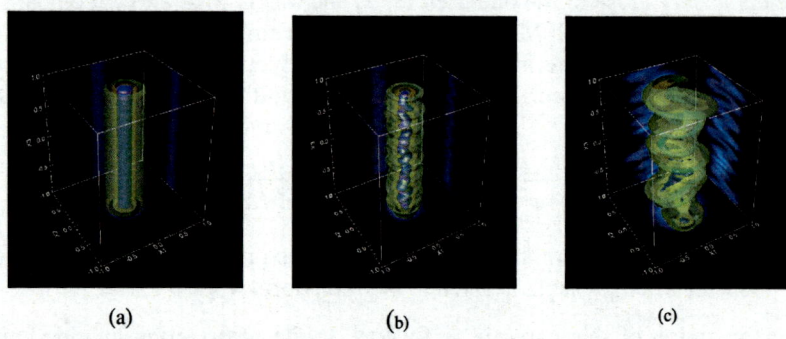

(a) (b) (c)

FIGURE 2. 3D plots of plasma density at moments a) t=0, b) t=130 and c) t=200.

Instability development in the presence of sheared flow

The growth rates of the sausage (m=0) and kink (m=1) instabilities from hybrid simulations with axial sheared flow are plotted in Fig. 3a and Fig. 3b. A comparison of Fig. 1 to Fig. 3 shows that the growth rates of both the m=0 and m=1 modes are substantially reduced in the presence of sheared flow. For short wavelengths, the growth rate of the m=0 mode is in good agreement with the linear theory results presented in Fig. 3 of [7], while it is slightly larger than linear theory predictions for long wavelengths.

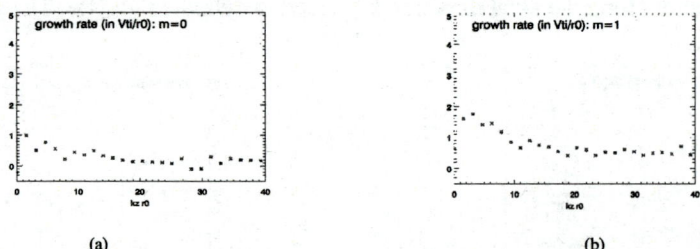

(a) (b)

FIGURE 3. The growth rates of the a) sausage and b) kink modes in the presence of axial flow shear.

Figure 4 contains 3D plots of plasma density inside the cylinder at scaled times t=0, t=130, and t=200. A comparison of Fig.2 to Fig.4 clearly shows that instability development for both sausage and kink modes is suppressed in the presence of axial sheared flow.

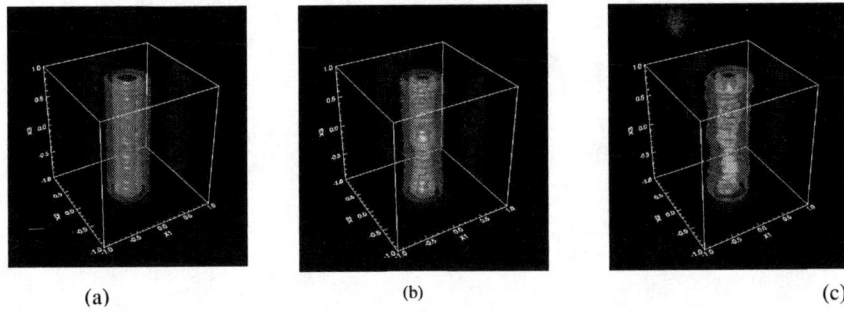

(a) (b) (c)

FIGURE 4. 3D plots of plasma density at the moments a) t=0, b) t=130, and c) t=200.

SUMMARY, DISCUSSION, AND CONCLUSIONS

In this paper we have examined the development of the instabilities of current-carrying z-pinch plasmas using 3D hybrid simulations with and without axial sheared flow included in the model. The simulations show that axial sheared flow can substantially suppress both sausage and kink modes. For the m=0 sausage modes, the linear growth rates and their trends obtained from the hybrid simulations are in reasonable agreement with the linear Hall MHD calculations of [7].

ACKNOWLEDGMENTS

The authors wish to acknowledge valuable discussions with I. Lindemuth, L. Rudakov, P. Sheehey and R. Siemon. We also express our gratitude to R. A. Fonseca and F. Tsung from IST and UCLA for permission to use the OSIRIS_Analysis scientific visualization package developed in the context of laser and beam-plasma interactions. This work was supported by the United States Department of Energy under Grant No. DE-FG03-01ER54617 at the University of California at Los Angeles and Grant No. DE-FC08-01NV14050 at the University of Nevada at Reno.

REFERENCES

1. Matthews, A. P., *Journ. Comp. Physics* **112**, 102 (1994).
2. Winterberg, F., *Beitr. Plasmaphys.* **25**, 117 (1985).
3. Shumlak, U., Hartman, C.W., *Phys. Rev. Letters* **75**, 3285 (1995).
4. Arber, T.D., and Howell, D.F., *Physics of Plasmas* **3**, 554 (1996).
5. Appl, S., and Camenzind, M., *Astronomy and Astrophysics* **354**, 256 (1992).
6. DeSouza-Machado, S., Hassam, A.B., and Ramin, Sina, *Physics of Plasmas* **7**, 4632 (2000).
7. Sotnikov, V.I., Paraschiv, I., Makhin, V., Bauer, B.S., Leboeuf, J.N., Dawson, J.M., *Physics of Plasmas* **9**, 913 (2002).

A Comparison of Radiation Transport Models for a Ti Z-Pinch

John L. Giuliani, Robert W. Clark, J. Ward Thornhill, and Jack Davis

Radiation Hydrodynamics Branch, Plasma Physics Division
Naval Research Laboratory, 4555 Overlook, Ave. SW, Washington, DC 20375

Abstract. One dimensional simulations of titanium (Ti) wire array Z-pinches on the Z generator are compared with experimental data using four models of radiation transport: single and multi-group diffusive transport, tabulated and non-local collisional radiative equilibrium (CRE) transport. While the multi-group diffusion model can reasonably predict the total and K-shell radiative yields, there are significant discrepancies in the plasma properties at implosion between the diffusion approach and data. The present CRE for Ti models match the K-shell yield and the plasma properties, but the total yield increases with the number of emission lines transported.

INTRODUCTION

A Z-pinch implosion driven by a high current generator efficiently converts mechanical energy into radiative energy. Hence the model employed to treat the radiation physics in a Z-pinch simulation can affect the calculated dynamics. The most accurate model incorporates a self-consistent coupling of the ionic level populations with the radiation field through kinetic rate equations and non-local photo-processes. Such a general approach, termed the non-local collisional radiative equilibrium (NLCRE) model, is practical for a 1-D radial implosion of low atomic number plasmas (Z<40) [1]. In multi-dimensional simulations the predominant model in use is the three temperature approximation: ion and electron temperatures together with a radiation temperature [2,3]. The latter is calculated with a single group radiation diffusion (SGRD) equation wherein the input tabular opacities are determined from local thermodynamic equilibrium (LTE) conditions and a Planckian radiative source function is assumed. A natural extension of this approach is the multi-group radiation diffusion (MGRD) treatment which has the advantage over a single group of providing spectral content of the pinch's radiative yield. An alternative approximation is the tabulated collisional radiative equilibrium (TCRE) model recently proposed by Thornhill, et al. [4].

The objective of the present paper is to compare and contrast the above four models of the radiation physics (SGRD, MGRD, TCRE, and NLCRE) within the context of 1-D simulations for titanium (Ti) Z-pinches on the Z-generator at Sandia National Laboratory. This study will briefly indicate successes and some pitfalls of the diffusion approach.

RADIATION MODELS

Both the diffusion models and the CRE models are prepared from the same Ti atomic data base composed within the Radiation Hydrodynamics Branch at NRL. For the single and multi-group diffusion approaches, the atomic level populations were determined by LTE and the absorption coefficient included 1200 lines in all ionization stages of Ti as well as bound-free and free-free continuum processes. The multi-frequency absorption data was then averaged according to the standard formulas for single and multi-group Planck and Rosseland opacities [5]. Twenty groups were judiciously chosen to capture important spectral features, with eight resolving the K-shell emission (>4.7 keV). The standard Lagrangian 1-D hydromagnetic were solved along with a radiation diffusion equation for each group. The latter include an extension of a single group flux limiter [6] to each group, and a Milne relation is used to specify the group radiation boundary condition at the moving outer edge of the plasma.

Details of the TCRE and NLCRE models have been described previously. We only note here that the tabular data used in the TCRE and NLCRE models have either 126 lines or 483 lines. The transport coefficients were chosen to optimize agreement with the K-shell yield from three shots on Z, vis., #303, 304, and #310. The same coefficients and initial grid were used in all the calculations.

TABLE 1. Results from different radiation transport models for Ti Z-pinches on the Z generator.

Z-shot #304 2.50 cm, 1.82 µg	Single group	Multi-group	TCRE (126)	NLCRE (126)	TCRE (483)	NLCRE (483)	Exp data
Total yield (kJ)	1150	1150	384	431	476	612	764
K-shell yield (kJ)	-	100	110	105	124	113	102
Z-shot #303 2.25 cm, 2.25 µg							
Total yield (kJ)	1100	1140	512	558	602	720	990
K-shell yield (kJ)	-	93	116	106	122	103	120
Z-shot #310 1.75 cm, 3.71 µg							
Total yield (kJ)	1010	1030	738	780	778	884	1008
K-shell yield (kJ)	-	32	80	66	80	53	55

RESULTS AND DISCUSSION

The total yields, K-shell yields, and implosion radii are given in Table 1 for various radiation models along with the corresponding experimental data [7]. The model results for the total yield are taken after the first bounce while the same yield from the experiments is determined from the product of the peak power and the FWHM. The SGRD and MGRD models give very similar total yields, which exceeds the observed value at the low mass shot, but equals it for the high mass one. Both the TCRE and NLCRE models give too little total yield, but one can see an increase in the total yield for both models as the number of transported emission lines (number in parentheses) increases. This indicates that the L-shell radiation in the CRE models depends upon the number of emission lines transported in the calculation.

The K-shell yields for all the models, except SGRD, are fairly accurate, within a factor of two of the experimental results. The largest variation occurs for the high mass case where MGRD is too low and the TCRE results are too large. An estimate of the K-shell yield for the SGRD models was made by considering the emission at each time to be the Planck function at the radiation temperature of the outer zone. The result was less than a joule for each mass. The origin of the K-shell emission in the MGRD models is deep inside the plasma. Such high energy radiation groups are optically thin and nearly free stream outward to the plasma surface.

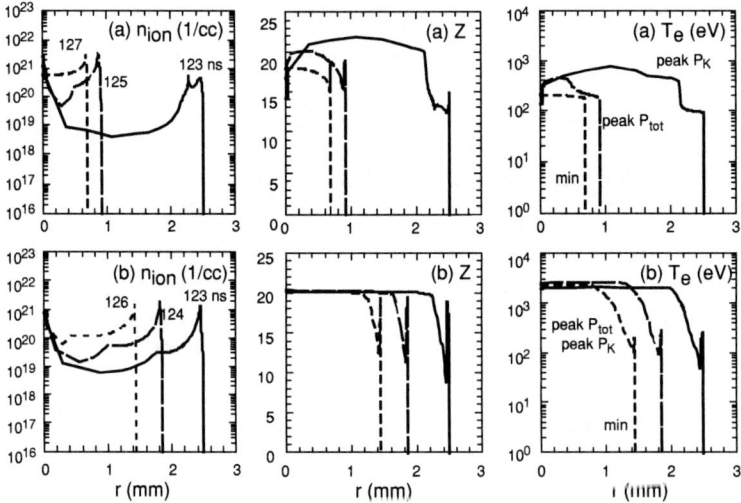

FIGURE 1. (a) Profiles of the ion density, mean charge state, and electron temperature at the time of peak K-shell power (123 ns), the time of peak total power (125 ns), and the minimum radius (127 ns) for the MGRD model. The initial radius of the Ti wire array was 2.25 cm and the total mass is 2.25 mg. (b) The same quantities for the NLCRE(483) model at similar times. The peak total and K-shell power occur at the minimum radius.

The agreement in the gross radiation features between the diffusion results and the experiments does not follow through to the dynamic features. We note four problems with the LTE radiation diffusion models. Simulation results of the pinch dynamics at three times near the implosion are displayed in Fig.1 for the intermediate mass case. First one should note that the minimum radius of the MGRD model is about half that of the NLCRE model, which is close to the observed value of 1.4 mm [7]. This high compression ratio (~30) is typical for the diffusion treatments of radiation, and it explains the large total yields obtained with such models: a deep implosion translates to a high kinetic energy in the pinch which is efficiently radiated by an LTE plasma because the source function is Planckian. Second, the electron temperature, T_e, is smaller in the MGRD model than for the NLCRE one, especially at the minimum radius. Analysis of experimental data of wire arrays on the Z generator [8] indicate T_e ~3000 eV, again closer to the NLCRE result. Third, the rapid decrease in T_e in the MGRD implosion predicts that the peak K-shell emission occurs prior that of the total power. This is opposite to the observed relation [8]. Finally, the same change in T_e causes the mean charge state, Z, to rapidly decrease over ~1 ns due to the LTE condition. This is unphysical, as the dominant recombination process under the pinch conditions, dielectronic recombination, gives a time scale of >10 ns to recombine from hydrogen to helium-like Ti.

In conclusion, the total yields from the 1-D diffusion models follow from the large compression ratio and the subsequent conversion of a large coupled kinetic energy to the plasma. We found that the TCRE model fairly well matched the NLCRE results, including the density, Z, and T_e profiles, though it is 3 to 4 orders of magnitude faster in run time. It is remarkable that the MGRD model reproduced the K-shell yields to within a factor of two given that the plasma conditions of T_e and Z are so disparate from the CRE results. The K-shell yields for the MGRD model result from compensating effects. For a given T_e, the ionization level for LTE is higher than for CRE, and the amount of K-shell emission would be excessive except that the assumption of a Planckian source function leads to strong radiative cooling and a low T_e.

ACKNOWLEDGMENTS

The authors gratefully acknowledge the support of the U.S. Defense Threat Reduction Agency.

REFERENCES

1. Davis, J., Clark, R., Blaha, M., and Giuliani, J.L., *Laser and Particle Beams*, **19**, 557-577 (2001).
2. Peterson, D.L., Bowers, R.L., Brownell, J.H., Greene, A.E., McLenithan, K.D., Oliphant, T.A., Roderick, N.F., and Scannapieco, *Phys. Plasmas*, **3**, 368-381 (1996).
3. Douglas, M.R., Deeney, C., Spielman, R.B., Coverdale, C.A., Roderick, N.F., Haines, M.G., *Phys. Plasmas*, **7**, 2945-2958 (2000).
4. Thornhill, J.W., Apruzese, J.P., Davis, J., Clark, R.W., Velikovich, A.L., Giuliani, J.L., Chong, Y.K., Whitney, K.G., Deeney, C., Coverdale, C.A., Cochran, F.L., *Phys. Plasmas*, **8**, 3480-3489 (2001).
5. Mihilas, D., and Mihalas, B.W., *Foundations of Radiation Hydrodynamics*, Oxford University Press, New York, 1984, pp.363-365.
6. Levermore, C.D., *J. Quant. Spectrosc. Radiat. Transfer*, **31**, 149-160 (1984).
7. Denney, C., private communication.
8. Deeney, C., C.A. Coverdale, Douglas, M.R., Nash, T.J., Spielman, R.B., Struve, K.W., Whitney, K.G., Thornhill, J.W., Apruzese, J.P., Clark, R.W., Davis, J., Beg, F.N., Ruiz-Camacho, J., *Phys. Plasmas*, **6**, 2081-2088 (1999).

Z-scaled K-shell Dielectronic Recombination Rate Coefficients

Arati Dasgupta, Paul Kepple, and Jack Davis

*Naval Research Laboratory, Plasma Physics Division,
Radiation Hydrodynamics Branch, Washington, DC 20375, USA*

Abstract. We present total as well as state-specific dielectronic recombination (DR) data such as singly and doubly-excited state energies, autoionization and radiative rates and DR branching ratios for recombination from H-like to He-like and from He-like to Li-like ions. The data obtained are calculated using a detailed Hartree-Fock calculation with relativistic corrections. Explicit calculations for Al, Ti, and Kr are performed to cover a range of ions such that scaling relations obtained using these data can be used to interpolate the DR data for any ion of interest within this range and may be somewhat beyond. Since accurate and detailed state-specific knowledge of DR rate coefficients are crucial for atomic model development of any species, these present data will be extremely relevant for analysis of z-pinch physics.

INTRODUCTION

Dielectronic recombination (DR) is an important process for ionization balance in a non-LTE plasma of moderate to high-Z ions. The dielectronic satellite lines appearing on the long–wavelength side of the resonance lines as a result of this recombination can be used as plasma temperature and density diagnostics and analysis of x-ray spectra. DR is a two-step ion-electron resonant collision process in which an electron undergoes radiationless capture into an autoionizing doubly excited state followed by radiative decay into a singly excited state of the recombined ion. These doubly excited states can also autoionize to the ground or other possible excited states of the initial ion.

In the low-density corona approximation, the DR rate coefficients α^{DR} from initial state i to final state k through intermediate autoionizing state j at temperature T is given by

$$\alpha^{DR}(i,k) = \left[\frac{4\pi\Re}{kT}\right]^{3/2} \frac{a_0^3}{2g_i} \sum_j F_{ijk} \exp(-\varepsilon_j/kT), \tag{1}$$

where

$$F_{ijk} = \frac{g_j A^a_{ji} A^r_{ik}}{\sum_{i'} A^a_{ji'} + \sum_{k'} A^r_{jk'}}. \tag{2}$$

Here g_i and g_j are the statistical weights of the initial and final states, kT is the electron temperature, ε_j is the energy of the free recombining electron, A^a_{ji} and A^r_{jk} are the autoionization and radiative rates from state j to states i and k respectively.

These DR branching ratios also provide a direct measure of the intensity of satellite lines produced due to this recombination and they can be used for plasma diagnostics of electron temperature:

$$I(j,k) = N_e E_{jk} \left[\frac{4\pi \Re}{kT} \right]^{3/2} a_0^3 \frac{e^{-\varepsilon_j/kT}}{2} \sum_i \frac{N_i}{g_i} F_{ijk}. \qquad (3)$$

In this equation, N_e is the electron density, N_i is the density of initial state and E_{jk} is the energy of the satellite line.

Most of the published works present only total DR rates as a function of temperature. In this work we present state-specific DR rates and the total rate can easily be obtained by summing these rates. More importantly, the Z dependence of various DR data are determined using detailed calculations of Al, Ti, and Kr ions. These scaling relations can then be used to obtain DR rates and satellite line data for any ion between Al and Kr and maybe somewhat beyond with little effort.

ATOMIC MODEL AND Z SCALING

Evaluation of detailed DR data requires considerations of many doubly and singly excited levels. Although in some instances it is necessary to work at the fine-structure levels, this leads to the generation of a superabundance of atomic data. In order to calculate the ionization balance of non-coronal plasma it is sometimes more practical to work at the configuration level. If necessary, one can determine multiplet emissions from the configuration populations as a next step.

Our atomic model for recombination from H- to He-like ions consist of the initial $1s$ ground state, 33 doubly-excited He-like states of which there are 31 unlumped $2lnl'$ and two lumped $2sn"l"$ and $2pn"l"$ states and 11 final singly-excited states of $1snl'$ configurations. Here l takes on values 0 and 1, $2 < n < 7$, $0 < l' < 7$, $7 < n" < 20$, and $l" < 7$. We consider an equivalent atomic model for our calculations of DR data for recombination from the He- to Li-like recombination. The calculations of all the quantities relevant for obtaining F_{ijk} in Eq. (2) were obtained using the HFR (Hartree-Fock with relativistic corrections) code of R. D. Cowan [1]. The theoretical methods used in these calculations are given in detail in Ref. 2. At configuration level, i, j, and k in Eq. (2) go over to the configuration averaged labels a, b, and c respectively and the configuration averaged DR branching ratios F_{abc} and intensity $I(b,c)$ can be expressed in terms of the configuration-averaged energies, rates and statistical weights by replacing i, j, and k in Eq. (2) and (3) by a, b, and c respectively.

The scaling of F_{abc}, $\alpha^{DR}_{(a,c)}$ and $I(b,c)$ is accomplished by scaling A_{bc}^a, A_{ba}^r, and ε_b. These rates A_{bc}^a, and A_{ba}^r depend on Z through the doubly- and singly-excited states energies ε_b and ε_a. We found that these quantities scale as

$$\varepsilon_b = Z^2(b_0^{**} + b_1^{**} + b_2^{**} + b_3^{**})_b \qquad (4)$$

$$\varepsilon_c = Z^2(b_0^* + b_1^* + b_2^* + b_3^*)_c \qquad (5)$$

$$A_{bc}^r = Z^4(b_0^r + b_1^r/Z + b_2^r/Z^2 + b_3^r/Z^3)_{bc} \qquad (6)$$

$$A_{ba}{}^a = (b_0{}^a + b_1{}^a/Z + b_2{}^a/Z^2 + b_3{}^a/Z^3)_{ba} \tag{7}$$

Accommodating the scaling behaviors of both A^r and A^a, the DR branching ratio is expressed as a four-coefficients polynomial :

$$F_{abc} = Z^2(b_0{}^F + b_1{}^F/Z + b_2{}^F/Z^2 + b_3{}^F/Z^3) \tag{8}$$

A partial list of scaling coefficients for the doubly excited states energies and the branching ratios F_{abc} for recombination from H- to He-like ions are given in Tables I and II respectively. We do not present any scaling data for He- to Li-like recombination.

TABLE I. Scaling Coefficients for energies of doubly-excited He-like states

b	state	b_0^{**}	b_1^{**}	b_2^{**}	b_3^{**}
1	2s2p	5.458E-01	-1.879E+00	3.686E+01	-2.112E+02
2	2p^2	5.604E-01	-2.525E+00	4.881E+01	-2.796E+02
3	2s3p	6.913E-01	-2.313E+00	4.224E+01	-2.417E+02
4	2p3p	7.063E-01	-3.025E+00	5.452E+01	-3.121E+02
5	2p3d	7.094E-01	-3.149E+00	5.693E+01	-3.256E+02
6	2s4p	7.429E-01	-2.533E+00	4.494E+01	-2.573E+02
7	2p4s	7.554E-01	-3.118E+00	5.467E+01	-3.119E+02
8	2p4p	7.575E-01	-3.221E+00	5.664E+01	-3.239E+02

TABLE I. Scaling Coefficients of DR branching ratios F_{abc} for He-like states

b	c	b_0^F	b_1^F	b_2^F	b_3^F
1 (2s2p)	1 (1s 2s)	-1.032E-01	8.302E+00	-1.255E+02	6.060E+02
2 (2p^2)	2 (1s2p)	-4.363E-01	3.251E+00	-5.099E+02	2.339E+03
3 (2s3p)	1 (1s2s)	-3.255E-02	2.524E+00	-3.753E+01	1.738E+02
4 (2p3p)	2 (1s2p)	-2.371E-02	1.626E+00	-1.102E+01	-4.836E+01
4 (2p3p)	4 (2s2p)	-8.160E-02	5.427E+00	-1.711E+01	-3.752E+02
5 (2p3d)	5 (1s3d)	3.123E-03	-5.255E+00	4.835E+01	-3.699E+02
7 (2p4s)	6 (1s4s)	2.375E-03	-2.376E+00	1.260E+01	-8.643E+01
8 (2p4p)	2 (1s2p)	1.251E-03	-1.445E+00	8.567E+00	-7.771E+01

The total DR rates for each ion are obtained by summing the state-specific rates $\alpha^{DR}(a,c)$ over all the final singly excited recombined state c. In calculating the DR rates to each of these singly excited state, we explicitly calculated all the data for $n<20$ and employed the usual $1/n^3$ extrapolation procedure for all bound Rydberg levels of the captured electron with $n>20$. These total DR rates are shown in Fig. 1 as a function of temperature.

In order to investigate the reliability of this scaling, we performed detailed calculations for selenium ions and compared the DR rates obtained with those evaluated using the scaling coefficients and these are also presented in Fig. 1. We see that the scaled DR rates for H- and He-like Se obtained using the scaling coefficients given in Tables I and II (only partial data are included here) and the scaling relations given above are very close to those calculated explicitly.

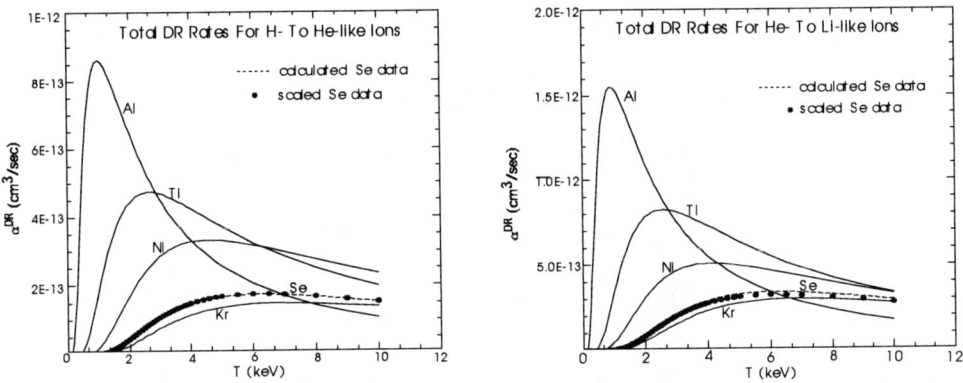

FIGURE 1. Total dielectronic recombination rates for H-to He-like and He- to Li-like ions. Shown are also the calculated as well as scaled DR rates for selenium.

SUMMARY

The reliability and scalability of atomic data used in the ionization calculations of K-shell moderate-Z elements are of great concern. For ionization balance calculations, data obtained at the configuration-averaged level are quite reliable although the accuracy of these data should be checked against other published configuration-averaged data. The number of dominant channels included in our calculations combine to give 95% or more of the total DR rates for recombination from H- and He-like ions. All the atomic quantities scale very smoothly From the data presented for selenium in Fig. 1, we conclude that the DR rates obtained using the scaling coefficients, one can very accurately predict the DR dates for any ion between argon and krypton without any detailed and difficult calculations. It will be instructive to investigate the scalability of these DR rates beyond the range of Z considered here by extrapolating these calculated data [3].

ACKNOWLEDGMENTS

This work was supported by the Defense Threat Reduction agency.

REFERENCES

1. Dasgupta, A., and Whitney, K. G., *Phys. Rev. A* **42**, 2640-2652 (1990).
2. Cowan,R. D., in *Book The Theory of Atomic Structure and Spectra*, University of California Press, Berkeley, CA., 1981.
3. Dasgupta, A., and Whitney, K. G., *J. Phys. B* **28**, 515-529 (1995).

A Kinetic Description of Ions in Aluminium Wire-Array Precursor Plasma

M. Sherlock, J.P. Chittenden, S.V. Lebedev and M.G. Haines

Plasma Physics Group, Blackett Laboratory, Imperial College of Science, Technology and Medicine, Prince Consort Rd, London SW7 2BZ, U.K.

Abstract. During the early stages of a wire-array Z-pinch implosion, low density plasma streams toward the axis by virtue of the Lorentz force. This streaming precursor plasma is initially highly collisionless and therefore cannot be modeled using standard fluid theory. A hybrid method (fluid electrons and particle ions) capable of modeling collisionless behaviour is presented. We show that the axial stagnation of the plasma flow occurs once the density becomes sufficiently high to initiate a non-linear rise in collisionality. Radiation and electron-ion energy exchange effects then result in the attainment of a dense, long-lived precursor column on axis, as observed experimentally. The column is held in place by the kinetic pressure of the streaming precursor plasma, which is balanced by the thermal pressure of the plasma in the column at the column's edge.

INTRODUCTION

Experiments designed to study the implosion phase of wire-array z-pinches have revealed the formation and implosion of a thin shell does not occur [1,2]. Instead, rapid Ohmic heating of the wires leads to the formation of expanding coronal plasma around each wire core. The radially directed Lorentx force accelerates this plasma into the inside of the array, giving rise to a series of high-speed plasma streams that interact on and around the axis. In the experiments performed by Lebedev *et al.* [2,3] on the MAGPIE facility at Imperial College, the precursor plasma forms a narrow, uniform and relatively stable plasma column on axis. End-on interferometry gave an estimate of $1.5 \times 10^5 \text{ms}^{-1}$ for the radial plasma speed and the shape of the contours suggested the azimuthal speed was approximately ¼ of the radial speed. Initially, the precursor plasma is low density ($\sim 10^{22} \text{m}^{-3}$) and the typical ion streaming mean-free-path is comparable to or larger than the system. Hence the initial development of the precursor involves collisionless behaviour, which cannot be resolved by fluid theory. A recent 3D model of wire ablation [4] suggests the magnetic Reynold's number is less than unity so that the streams do not drag the magnetic field into the array. In this paper we will assume B=0. The wires begin to implode only after ~80% of the total implosion time and the presence of precursor plasma could affect x-ray production.

THE MODEL

A 2D hybrid r-θ code has been developed which allows for a kinetic treatment of the ions in three velocity dimensions via a Monte-Carlo collision algorithm [5]. Here we concentrate only on the one-dimensional (r) predictions, which clearly show the basic physical processes at work in precursor development.

The ions are advanced in configuration and velocity space in the usual Particle-In-Cell manner, with collisions being augmented by the random pairing and scattering of ions within each computational cell. For each ion-pair, a scattering angle Θ is chosen at random from a Gaussian distribution of angles in such a way so that over a large number of such choices, the variance in the quantity $\delta \equiv \tan(\Theta/2)$ is given by

$$\langle \delta^2 \rangle = \frac{Z^4 e^4 n_i \ln \Lambda}{8 \pi \varepsilon_0^2 m_{ii}^2 u^3} \Delta t \qquad (1)$$

where Z is the ion charge state, e is the electron charge, n_i is the background ion density, $\ln \Lambda$ is the Coulomb logarithm, ε_0 is the permittivity of vacuum, m_{ii} is the reduced ion mass, u is the relative velocity of the pair and Δt is the time-step size. In the limit $\Delta t \to 0$, the method is equivalent to a Fokker-Planck description. We ensure Δt is much smaller than the characteristic ion-ion collision time. Hence ion-ion collisions are accurately modeled and the correct velocity-space diffusion coefficients can be reproduced. The ions also experience a force $-\nabla P_e/n_i$ due to the electron pressure gradient.

The equations of charge neutrality, $n_e = Z n_i$ and $Z e n_i \mathbf{u}_i - e n_e \mathbf{u}_e = 0$, imply $\mathbf{u}_i \approx \mathbf{u}_e$, enabling the electron energy equation,

$$\frac{3}{2} n_e \frac{\partial T_e}{\partial t} = -\frac{3}{2} n_e (\mathbf{u}_e . \nabla) T_e - n_e T_e \nabla . \mathbf{u}_e - \nabla . \mathbf{q}_e + Q_{ei} - Q_c \qquad (2)$$

to be solved for the electron temperature. The thermal diffusion term involving the heat flow $\mathbf{q}_e$ is solved implicitly. Electron-ion energy exchange is represented by $Q_{ei} = n_e v_{ei}(T_i - T_e)$ and is mirrored in the ions by the application of the grid-based force found in reference [6], modified to give the correct ion velocity (v_i) dependence in the high ion temperature limit:

$$\mathbf{F}_{ie} = v_{ie}(T_e, \mathbf{v}_i) \frac{[T_i - T_e]}{[\langle \mathbf{v}_i^2 \rangle - \langle \mathbf{v}_i \rangle^2]} [\langle \mathbf{v}_i \rangle - \mathbf{v}_i] \qquad (3)$$

where $v_{ie}(T_e, v_i) = 16\pi^{1/2} Z^2 e^4 n_e \ln \Lambda_{ie}/M^2 ([2kT_e/m_e]^2 + [v_i - \langle v_i \rangle]^2)^3$. A simple blackbody limited radiation-ionisation model is used to obtain the cooling rate Q_c [7].

The 0-D model of wire ablation given in reference [3] is used to obtain the time-dependence of the ion density at the boundary (r=R):

$$n_i(r = R, t) = \left[\frac{\mu_0 I_{max}^2}{8 \pi m_i u_r^2 R^2} \right] \sin^4 \left(\frac{\pi t}{2 t_{max}} \right) \qquad (4)$$

where tmax is the time at which the peak current Imax flows in the wires and the radial flow velocity ur is maintained at $-1.5\times10^5 ms^{-1}$. The plasma is injected with a typical temperature of 20eV and an azimuthal velocity spread u_θ/u_r of 1/4.

ALUMINIUM PRECURSOR PLASMA

The work presented here focuses on attempts to model 16mm diameter Al wire-array implosions performed on the MAGPIE facility (T_{max}=250ns, I_{max}=1MA), for which there exists detailed experimental data of the implosion phase and precursor development [2,3].

Following a dwell time of ~40ns, low density plasma fills the inside of the array and first reaches the axis at ~85ns. At 125ns, a broad, axially-peaked density profile exists (Fig.1a) which is a consequence of the convergent flow rather than collisional stagnation. Examination of the ion phase-space distribution at this early time reveals a significant ion component travelling with positive radial velocity in the central region of the array, indicating collisionless flow. During the ensuing 50ns, the density on axis increases by two orders of magnitude and the width of the peak contracts, giving rise to a tight precursor column with radius ~300-400μm. The rapid formation of the column is due to the increase in density of the streams, which leads to a decrease in mean-free-path below the density-peak scale length of ~1mm. The streaming plasma is then collisional and begins to compress the plasma on axis to high densities, initiating a non-linear increase in collisionality as is evident from the dramatic drop in mean-free-path through almost four orders of magnitude between 150ns and 160ns. This phase continues until the thermal pressure in the column ($n_eT_e+n_iT_i$) is sufficient to withstand the kinetic pressure of the streams (ρV^2), so that the two are balanced at the column's edge (Fig.1b). The confinement of the plasma is, then, purely hydrodynamic and there is no need to invoke a magnetic field to explain the phenomenon.

A simple model of energy fluxes into, within and out of the precursor column seems to describe its development. The streaming ions thermalise rapidly upon impact with the dense column, maintaining an energy deposition region with high ion temperature (~500eV) and an initial width of 1mm, but decreasing to ~300μm once the column becomes stable. The electrons, being much lighter, gain fractionally little energy upon stagnation and therefore act as an energy sink for the ions. Once the column is formed, its high density ensures that electron-ion thermal equilibration occurs fast enough to maintain a relatively low ion temperature (~50eV), preventing the column from expanding hydrodynamically. The electrons maintain a low temperature by radiatively cooling.

The final stable stage is reached after ~180ns, when the temperature in the column remains constant so that the ratio $\rho V^2/(n_eT_e+n_iT_i)$ remains constant and the column density can no longer rise. The column therefore slowly increases in diameter as it accrues the incoming plasma, at a rate of ~6.5m/ns, in good agreement with an analytic model of accretion. Experiments indicate a column radius of ~1mm, which remains roughly constant in time. This final phase of slow growth continues for ~100ns, after which the main array implodes onto the column.

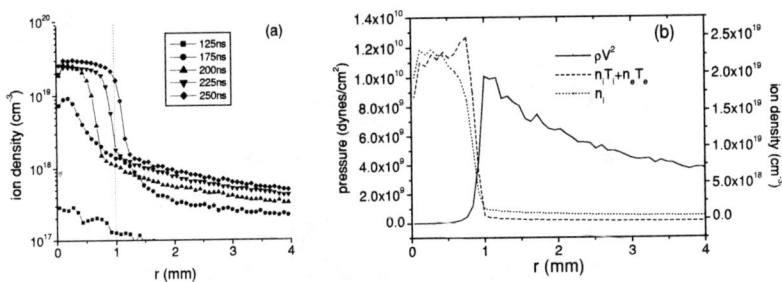

FIGURE 1. (a) The ion density profile at various times. The vertical dotted line represents the approximate experimentally observed column radius. (b) The kinetic (ρV^2) and thermal ($n_e T_e + n_i T_i$) pressures at 200ns. The dotted curve represents the ion density at this time.

CONCLUSIONS

We have shown that the precursor plasma is initially collisionless and becomes collisional once the density becomes sufficiently high. A tight, dense precursor column is formed by compression of the plasma on axis and is held in place by the kinetic pressure of the streams. The incoming ion energy is transferred to the electrons, cooling the ions and preventing the stagnated plasma from expanding hydrodynamically. The column retains its shape over ~100ns with an average radius of ~1mm, in good agreement with experiment.

The code has also been used to model W and Al in 2D as well as 1D interactions between a simple foam target and precursor plasma. These results will be submitted for publication in the near future.

ACKNOWLEDGMENTS

This work was funded by the EPSRC (U.K.) and Sandia National Laboratories (U.S.).

REFERENCES

1. Aivazov, I. K. et al., *Sov. J. Plasma Phys.* **14**, 110 (1988).
2. Lebedev, S. V. et al., *Phys. Rev. Lett.* **81**, 4152 (1998).
3. Lebedev, S. V. et al., *Laser and Particle Beams* **19**, 355 (2001).
4. Haines, M. G., private communication.
5. Takizuka, T. G. and Abe, H., *J. Comp. Phys.* **25**, 205 (1977).
6. Jones, M. E. et al., *J. Comp. Phys.* **123**, 169 (1996).
7. Tarter, C. B., *J. Quant. Spectrosc. Radiat. Transfer* **17**, 531 (1977).

Theoretical Development of M-shell Spectroscopy for Z-Pinch Plasma Diagnostics

Alla S. Shlyaptseva,[a] Safeia M. Hamasha,[a] Stephanie B. Hansen,[a]
Nicholas D. Ouart,[a] Ulyana I. Safronova [b]

[a]*Physics Department/220, University of Nevada, Reno, NV 89557 USA*
[b]*Department of Physics, University of Notre Dame, Notre Dame, IN 46566 USA*

Abstract. Tungsten wire explosions are being intensively studied at Sandia National Laboratories. Any available x-ray spectral data accumulated in Z experiments with appropriate theoretical modeling can lead to better understanding of plasma evolution during a wire explosion. The present work focuses on the theoretical development of M-shell spectroscopy of W ions in the spectral range from 4 up to 8 Å. The majority of line emissions in this spectral region is composed of $3l \rightarrow 4l'$, $5l''$ transitions. Atomic data were calculated using Cowan and MBPT codes for all isoelectronic sequences contributing into this spectral range. The non-LTE kinetic model was developed based on these atomic data. The sensitivity of this model to the number of included ions, configurations and levels, the electron density, ionization balance, and electron distribution function is discussed. The complete modeling of this spectrum allows a detailed diagnostic of a hotter plasma core in z-pinch experiments involving heavy ions.

INTRODUCTION

Recently, there has been renewed interest in studying fast, dynamic Z pinches as x-ray sources for fusion applications. The SNL-Z facility, the largest Z pinch facility in the world, produces very high x-ray power and plays a very important role in these studies. In particular, tungsten wire explosions are being intensively studied at Sandia National Laboratories. Results of tungsten wire-array Z-pinch experiments and modeling have been published elsewhere [1,2]. High-resolution x-ray spectral data have been accumulated in tungsten experiments on Z [3], which require a development of appropriate theoretical modeling. Recently, M-shell line radiation of Au plasmas has been actively studied. For example, the density, temperature and charge state distribution have been determined in highly ionized non-LTE Au plasmas [4]. Laser heated Au microdots buried in Be foil have reached temperatures of 2 keV and ionized into the M-shell. It was indicated that only emission measurements and calculations of 5f-3d transitions were presented. The present paper reports on progress of the development of M-shell diagnostics for W ions.

SPECTROSCOPIC MODEL

X-ray spectra of W produced on the SNL-Z accelerator indicate the spectral region of interest between 4 and 8 Å [3]. The spectra are immensely rich and the majority of line emissions in this spectral region is composed of $3l \rightarrow 4l'$ and $3l \rightarrow 5l'$ transitions. Identification of this spectral region is a complex problem even for Tokamak plasmas [5]. Recently, we initiated M-shell studies of W ions at the LLNL EBIT. These studies will allow us to break down this very complicated spectrum into spectra produced by separate W ions to benchmark advanced atomic structure and ionization balance calculations.

The most intense transitions radiate in the spectral range from 5 and 6 Å. These lines are due to 3d-4f transitions and most likely are optically thick. The less intense transitions below 5 Å are $3l \rightarrow 5l'$ transitions and most likely are optically thin. To calculate atomic data of all needed isoelectronic sequences matching this spectrum, two codes have been used: Cowan and MBPT codes. Cowan code is a well-known MCHF code [6], and the MBPT code is a fully relativistic code based on a many-body perturbation theory which includes the Breit interaction [7]. Eleven ionization stages contribute into the spectral region from 5 and 6 Å. The sensitivity of this spectrum to the number of included configurations and levels has been studied. For example, for Cu-like ions, fourteen even and twelve odd configurations of excited states are included in calculations. Some of the excited configurations have autoionization levels. Fig. 1 illustrates the importance of contributions of autoionization states into detailed spectra calculations. Synthetic spectra of all ionization stages from Cr- to Se-like W ions are presented in Fig. 2 for two different resolutions of 0.002 Å (left) and 0.02 Å (right). The most intense 542 lines of Cr-like, 308 lines of Mn-like, 699 lines of Fe-like, 6 lines of Ni-like, 550 lines of Cu-like, 89 lines of Zn-like, 202 lines of Ga-like, 220 lines of Ge-like, 127 lines of As-like, and 50 lines of Se-like W ions have been included. The detailed contribution of these eleven stages is shown in Fig. 3, where the abundances of the ions were taken to match an experimental spectrum [3].

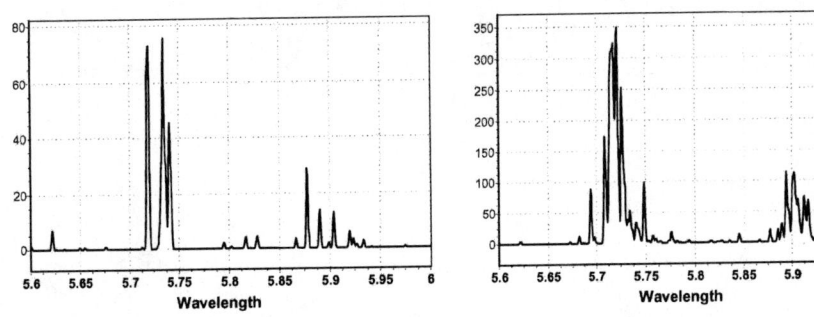

FIGURE 1. Synthetic spectra of $3d^{10}4l$-$3d^{9}4l'4l''$ transitions in Cu-like W ions. Upper configurations include $3d^9$ 4p4d, $4d^2$ (left) and $3d^9$ 4p4d, 4s4f, 4p4f, $4d^2$, 4d4f, and $4f^2$ (right).

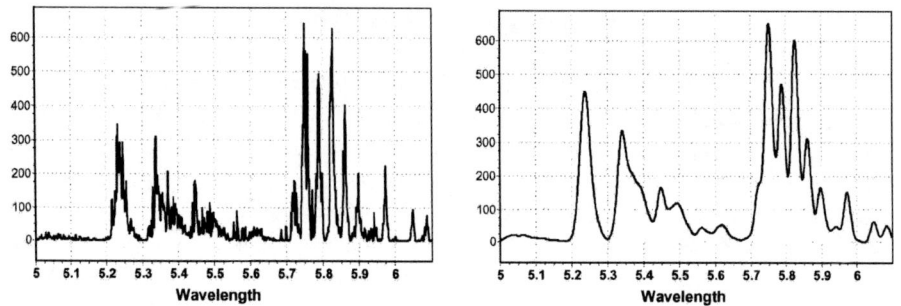

FIGURE 2. Synthetic spectra of 3d-4l transitions in Cr- to Se-like W ions. Resolution is 0.002 Å (left) and 0.02 Å (right).

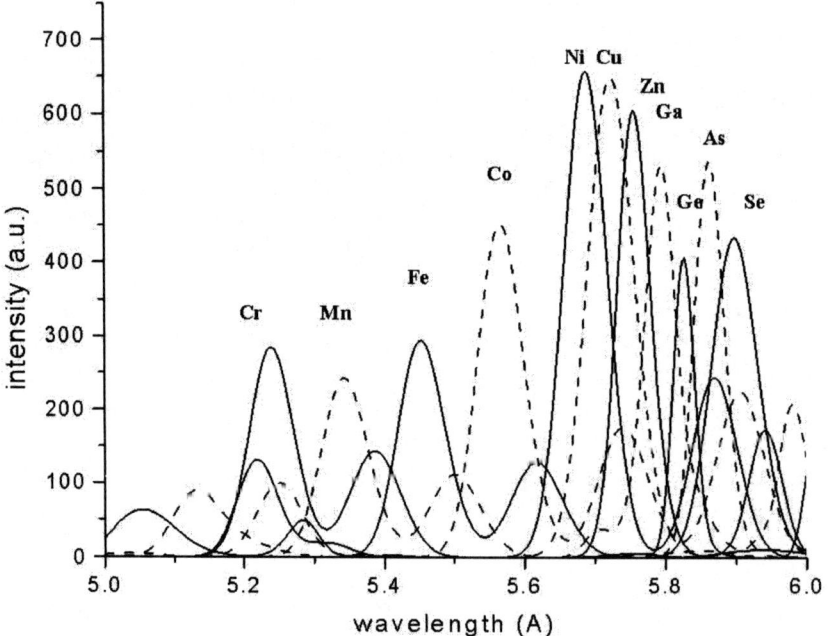

FIGURE 3. Contributions of Cr- to Se-like W ions into a spectral region from 5Å to 6Å.

The collisional-radiative atomic kinetic model has been developed which includes the ground states of every ionization stage of W from the bare ion with no electrons to neutral W with 74 electrons. Ionization potentials were taken from published tables [8,9]. Detailed atomic structure is included for ionization stages from Cr-like to Se-like W. Each fine structure state is linked to other states within its ionization stage via collisional excitation, collisional de-excitation, and radiative decay. All states of ions are linked via collisional ionization, three-body recombination, and radiative recombination. The Van Regemorter formula is used to calculate the excitation cross sections of optically allowed transitions. A modified Lotz formula [10] is used to calculate collisional ionization cross sections. Radiative recombination cross sections are calculated with Kramer's approximation. Preliminary ionization calculations of a coronal model with a maxwellian electron distribution function have been performed. They indicate that at a temperature of 2 keV the maximum abundance belongs to Ge-like ions, and as a temperature increases Ni-like ions dominate above 2.8 keV. For a monoenergetic beam, Ni-like ions start to dominate at much larger temperature close to the value of the ionization potential. The work is in progress to apply the present model to the SNL W spectra.

ACKNOWLEDGEMENTS

The present work was supported by the Sandia National Laboratories, the DOE-NNSA/NV Cooperative Agreement DE-FC08-01NV14050, and UNR.

REFERENCES

1. Spielman R.B. et. al., *Phys. Plas.* **5**, 2105-2111 (1998).
2. Deeney, C. et. al., *Phys. Plas.* **6**, 3576-3586 (1999).
3. Bailey, J. (private communication).
4. Foord, M.E. et. al., *JQSRT* **65**, 231-241 (2000).
5. Neu, R. et. al., *J. Phys. B* **30**, 5057-5067 (1997).
6. Cowan, R.D., *The Theory of Atomic Structure and Spectra* (Berkeley, 1981).
7. Safronova, U.I. et al., *Phys. Rev. A* **62**, 052505 (2000).
8. Carlson, T.A. et. al., *Oak Ridge National Laboratory Report* **4562** *(1970)*.
9. Fournier, K.B., *ADNDT* **68**, 1-48 (1998).
10. Bernshtam, V.A. et. al., *J. Phys. B* **33**, 5025-5032 (2000).

Modeling of Capillary Discharge Plasma for X-ray lasers, XUV Lithography and other Applications

V.N.Shlyaptsev[1], J.Dunn[2], S.J. Moon[2], K.B.Fournier, A.L.Osterheld[2], J.J.Rocca[3], J.Filevich[3], M.Marconi[3], E. Jankowska[3], E. C. Hammarsten[3], S. Sakadzic[3], A. Rahman[3], M. Frati[3], F.G. Tomasel[3], N.Fornaciari[4], D.Buchenauer[4], H.A.Bender[4], S.Karim[4], M.Kanouff[4], J.Dimkoff[4], G.Kubiak[4], G.Shimkaveg[5], W.T.Silfvast[5]

[1]*UC Davis-Livermore, LLNL, Livermore, CA, 94551,* [2]*LLNL, Livermore, CA, 94550,* [3]*Colorado State University, Ft.Collins, CO, 80523,* [4]*Sandia National Labs, Livermore, CA, 94551,* [5]*School of Optics/CREOL, University of Central Florida, 4000 Central Florida Boulevard, Orlando FL 32816-2700*

Abstract. It is long ago recognized that Z-pinches represent very natural medium for x-ray lasers (XRL) due to its favorable geometry and achievable high densities and temperatures. They also are very efficient x-ray sources. One of their variants, the capillary discharges, attracted attention of plasma physics researchers for almost two decades. It has been used for hot dense plasma formation and x-ray lasers[1,2], for transportation of laser beams and XUV radiation generation in x-ray lithography[3,4], for basic Z-pinch research and some others. The combination of efficiency, simplicity and low cost of capillary electrical discharges allowed to scale capillary x-ray lasers to table-top dimensions. In this paper we show the modeling results for next, 3-4 times shorter wavelength x-ray lasers.

As an efficient x-ray source of line and continuum radiation it can be used for many practically important application in science and technology. In particular, the capillary discharge can appear as powerful potential candidate for emerging XUV microlithography. We present here the results of numerical modeling of spectra and density of Xe EUV source which involved plasma heating and dynamics, detailed atomic kinetics and radiation transport and material ablation physics.

CAPILLARY DISCHARGE X-RAY LASERS AND APPLICATIONS

During last decade has been made substantial progress in parameters of materials and drivers for capillary discharges which are able to reach electric currents of 200KA in risetime of about 10 ns. It allowed to reach electron temperatures up to 500 eV, densities $(1-3) \, 10^{20}$ cm^{-3}, 300 times density compression ratios, 1 mJ of XRL energy per pulse, high-rep rate of operation, table-top dimensions. Last its quality represented the major goal of XRL development during these years because it makes them affordable for many different applications. Utilized right now mostly for scientific applications, they are already expanding many traditional scientific methods allowing to extract more detailed information with better accuracy. Among such applications are

- Shadowgraphy. Capillary x-ray laser can be used for shadowgraphy due to its very large peak brightness of the order of 10^{25} ph. mm^{-2} mrad^{-2} s^{-1} (0.1% BW)$^{-1}$. The modeling and the shadowgraphy experiments with argon 46.9nm capillary x-ray laser radiation passing through elongated plasma formed with driving current via another capillary was done in this work [5]. Same kind of setup can in principle be applied for diagnostics of NIF hohlraum and other fusion experiments. It is also important that this method has exponential sensitivity because of exponential dependence of transparency on absorption coefficients.

- Soft x-ray imaging. The images of plasma in different regions of spectra for wide range of currents and initial plasma diameters sizes, for gas-, vapor-filled or evacuated capillaries were obtained in numerous experiments. Specifically interesting was near- and far-field imaging of output characteristics of capillary x-ray laser. In this case experiments and its numerical modeling allow to reveal detailed data on plasma evolution and amplification dynamics because, as with shadowgraphy, this method of diagnostics also has exponential sensitivity (this time on gain instead of absorption coefficient) [6].

- Spectroscopy. It is primary and widely used method, but combined with x-ray lasers the spectroscopy can demonstrate exponential sensitivity via influence of different factors on linewidth and hence gain. It can be used not only for temperature and density diagnostics of plasma column, evaluate amplification of x-ray laser etc but also to find extremely weak plasma processes like Zeeman effect [7] or isotopic hyperfine splitting in spectral lines [8].

- Interferometry. The interferometry promise to be the key most important application of x-ray lasers because it utilizes their unique property of coherency. Recently the plasma interferometry with capillary XRL was demonstrated which then has been used for plasma diagnostics applications [9]. Already first such experiments allowed to observe unexpected 2D features not seen before of laser produced plasma used as an object for this probing[10]. Combined with high brightness of capillary lasers, the interferometry will bring incredible precision to applications in material science, x-ray holography, metrology and fusion applications. So, with capillary XRL we got in our hands extremely powerful and sensitive diagnostic tool with great potential in many future applications.

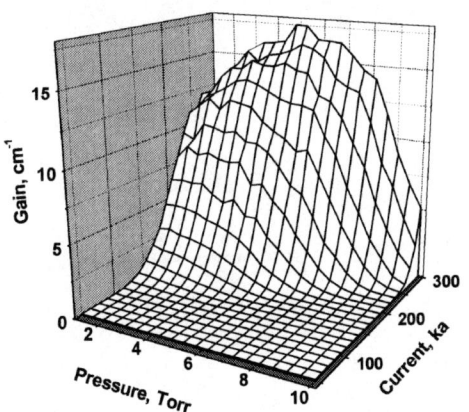

Fig. 1 Small signal gain on 4d-4p transitions in Ni-like AgXX as a function of current and pressure. This RADEX optimization data was obtained with over 400 hydro/atomic kinetics runs.

Substantial experimental and theoretical efforts are now devoted to extend this kind of XRL to shorter wavelengths. The RADEX simulations and spectroscopic diagnostics experiments of plasma indicate that the electron density $N_e > 10^{20}$ cm^{-3}

and temperature $T_e \sim 200\text{-}400$ eV at higher currents ~150-200 kA can be achieved in capillary discharge of high-Z plasma[2,11]. At such temperatures and densities the collisional XRL scheme on Ni-like ions can work particularly well. The atomic elements suitable for lasing are in the range A=42 - 50 which are lasing at wavelengths down to ~100 Å. For Ni-like CdXXI, the gain is calculated ~1-2 cm^{-1}, which is in qualitative agreement with current experimental data [11]. Fig.1 shows the results of numerical calculations of small signal gain for Ni-like AgXX ions which is substantially larger than for Cd. We plan the experiments with Ni-like AgXX and possibly PdXIX in near future.

CAPILLARY DISCHARGE EUV SOURCE FOR MICROLITHOGRAPHY

The sources for EUV lithography must satisfy the large amount of specific practical, technological and environmental requirements of complex processes of microchip production. Among the requirement are the achievement of efficient x-ray conversion in specific spectral range, high average power, pulse-to-pulse stability, large source lifetime etc. Extremely high spatial and temporal stability of capillary plasma (the jitter in radial position of dense plasma column can be 20 microns or less, comparable to laser plasma), the relatively large temperatures achievable with small currents, simplicity and efficiency attracted attention of researchers to capillary discharge as radiation source.

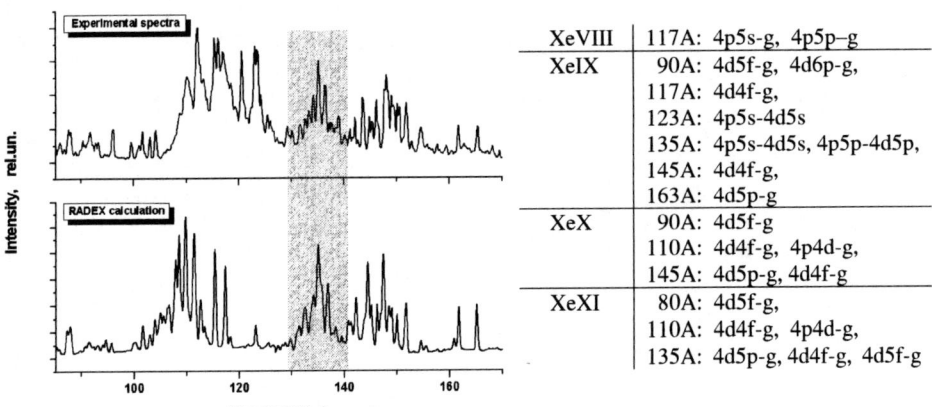

Fig. 2 The experimental (top) and code RADEX modeling spectra (bottom) of capillary discharge. The Table summarizes transitions which contribute to these spectra, "g" here denoted ground state(s)

The experiments with different capillary materials, currents and gas fills were performed in CREOL and Sandia National Labs [4,12,13]. For calculations of gas filled capillary discharge in Xe we use the same rad-MHD and atomic kinetics code RADEX developed for the x-ray laser modeling [1] but this time discharge was driven by much smaller 3-6kA and longer 1-2 µs currents pulses. The parameters of this discharge are somewhat intermediate between capillary XRL case and one of 300 micron microcapillary investigated previously in [5]. Both optical interferometry of capillary discharge plasma experiment performed at CREOL and RADEX simulations show that when current rises the electron density is reaching the maximum on axis due to Lorentz force and agrees well with value of density, the time it has been reached and its sharp increase near the wall. Still some disagreements also exist, specifically for cases of Xe+He mixtures where inter-diffusion of separate components was neglected though code well predicts the temperatures and spectrum and hence ion composition in Xe+He mixtures.

Most computationally difficult in spectra calculations with high-Z gases like Xe is atomic kinetics and radiation transport. Just atomic data for several most abundant in our conditions ion stages XeVIII-XeXI is of the order of 1GB with number of atomic levels of the order of 10^4 and lines approaching 10^6. Main source of radiation in such systems is due to numerous often many fold overlapped atomic lines. The spectra obtained from 5kA 1.4µs diamond capillary discharge in Xe at 2 Torr in comparison with modeling spectra are shown in Fig.2. The table marks the major transitions contributing into different spectral bands. Some missing transitions in RADEX spectra around 120A are due to omission in the calculations of lower Z stages Xe V-VII radiating at much lower temperatures. The calculations done with wide range parameters clearly support the observations that Ru-like ions XeXI are major contributors into 135A region of interest for microlithography. Also falling into the same region are several 4p5s-4d5s transitions of Pd-like ions XeIX and many weak $4d4f-4d^9$ transitions of Rh-like ions XeX total contribution of which is relatively small. Neighboring strong clusters of transitions at 110A are of all XeVII - XeXII while at 145A spectra mostly belong to Rh-like ions XeX. Still further modeling work is needed for better understanding the dynamics of this source, to optimize its spectral power, heat and radiation load on the capillary walls etc

References

1. J.J.Rocca, V.Shlyaptsev, F.G.Tomasel, O.D.Cortazar, D.Hartshorn, J.L.A.Chilla, Phys. Rev. Lett., **73**, 2192 (1994).
2. J.J. Gonzalez, M. Frati, J.J.Rocca, V.N.Shlyaptsev, A.L.Osterheld, Phys.Rev.E **65**(2), 026404 (2002)
3. Y.Ehrlich, C.Cohen, and A.Zigler, J.Krall, P.Sprangle, and E.Esarey, *Phys. Rev. Lett.* **77**, 4186 (1996)
4. M.A. Klosner, H. Bender, W.T. Silfvast and J.J. Rocca, Optics Letters, **22**, 34, (1997).
5. M.C. Marconi, C.H. Moreno, J.J.Rocca, V.N.Shlyaptsev and A.L.Osterheld. Physical Review E **62**, 7209 (2000)
6. C.H.Moreno, M.C.Marconi, V.N.Shlyaptsev, B.R.Benware,C.D.Macchietto, J.L.A.Chilla, J.J.Rocca, A.L.Osterheld, Phys.Rev.A, **58**(2), 1509 (1998)
7. F.G. Tomasel, V.N. Shlyaptsev and J.J. Rocca, Phys.Rev.A, **54** ,2474, (1996).
8. J.N.Nilsen, J.Koch, J.H.Scofield, B.J.MacGowan, J.C.Moreno, L.B.DaSilva, *Phys. Rev. Lett.* **70**, 3713 (1993).
9. J. Filevich, K. Kanizay, M.C. Marconi, J.L.A. Chilla, and J.J. Rocca. Optics Lett. 25, 356, (2000).
10. J. Filevich, J.J. Rocca, E. Jankowskaa, E.C. Hammarsten, M.C. Marconi, S.J. Moon, V.N. Shlyaptsev, submitted to Phys.Rev.Lett
11. S. Sakadzic, M. Frati , F.G.Tomasel, A.Rahman, J.J.Rocca, V.N. Shlyaptsev, Proc.SPIE, Vol.**4505,** "X-ray lasers and applications" , 134 (2001)
12. N.R.Fornaciari, H.Bender, D.Buchenauer, M.P.Kanouff, S.Karim, C.D.Moen, K.D.Stewart, W.T.Silfvast, G.M.Shimkaveg, Proc. SPIE Vol.**4688**, "Microlithography-2002" (in press).
13. J.Dimkoff, N.Fornaciari, D.Buchenauer, S.Karim and H.Bender, Proc.SPIE, Vol.**4688**, "Microlithography-2002", Santa Clara (2002).

The Z-pinch Structure Generation by the Evolution of the Nonquasineutral Electron Vortex

Alexander V. Gordeev,[a] Tatiana V. Losseva [b]

[a]*Russian Research Center "Kurchatov Institute", 1 Kurchatov Square 123182, Moscow, Russia*

[b]*Institute of Geospheres Dynamics RAN, 38 Leninsky prospect, bldg. 6, 119334, Moscow, Russia*

Abstract. A new scenario of the z-pinch generation as a result of the vortex dynamics with the azimuthal magnetic field from the diapason $4\pi n_e m_e c^2 \ll B_\theta^2 \ll 4\pi n_i m_i c^2$ is considered. The presence of the electron density peak on the axis of the filament vortex structure leads to the ion acceleration towards the axis and results in the forming of a dense and hot core of the z-pinch. The final ion energy in the core is essentially restricted by collisions of the energetic ions with the magnetized electrons. It is shown that the dense and hot plasma core could be disrupted on the picosecond times by the inclusion of the ion-ion collisions, which is in a qualitative agreement with the modern high-resolution measurements.

INTRODUCTION

Until now, the best information on the z-pinch plasma condition corresponded only to the spatial resolution about 1 mm and temporal resolution of the order 10 ns [1,2]. However, the contemporary diagnostics of the z-pinch revealed that the processes in the plasma core occur on the 10 ps time scales and close to 1 μm in size at peak plasma density [3]. In theoretical papers the investigation of the z-pinch evolution are usually limited to the complicated numerical calculations, where along with both the electrodynamic processes and the dissipative plasma processes also the radiation effects are taken into account [2]. Meanwhile for a better understanding of the z-pinch phenomenum for the small space and time scales it is very desirable to investigate the electrodynamic processes in the simplest approximation. In the initial stage of the z-pinch evolution the role of the applied electric field E_z is essential. However, if the magnetic field B_θ is increased, so that $h = \sigma B_\theta/(en_e c) > 1$, the electron vortex structures can be generated by the time on the order of ω_{pe}^{-1} quite analogous to the Weibel instability. In this vortex, the radial Hall electric field E_r arises at the Debye magnetic radius $r_B \sim B_\theta/(4\pi e n_e)$, so that the electric potential difference in the vortex is usually much more as compared with the applied voltage difference [1]. As the electric field E_r is negative, the ions are accelerated towards the axis with the characteristic time on the order of ω_{pi}^{-1}. The presence of the initial ion pressure prevents from the unlimited compression and manifests itself in the unloading shock wave for the radial ion velocity, which results in the forming of the dense and hot ion core. It is shown that the final dense ion core should be destroyed by the inclusion of the ion-ion collisions.

PHYSICAL MODEL

By the describing of the nonquasineutral vortex structure one can use the relativistic equation for electrons in a modified form

$$\frac{\partial \vec{p}_e}{\partial t} + \nabla \gamma m_e c^2 = -e\vec{E} - \frac{e}{c}[\vec{v}_e \times \vec{\Omega}_e], \qquad \vec{\Omega}_e = \vec{B} - \frac{c}{e} rot \vec{p}_e, \qquad (1)$$

and the Maxwell equations

$$rot\vec{B} = \frac{4\pi e}{c}(Zn_i \vec{v}_i - n_e \vec{v}_e) + \frac{1}{c}\frac{\partial \vec{E}}{\partial t}, \qquad div\vec{E} = 4\pi e(Zn_i - n_e). \qquad (2)$$

Here m_e and $\vec{v}_e$ are the electron mass and velocity, $\vec{p}_e = \gamma m_e \vec{v}_e$ is the electron momentum, $\gamma = 1/\sqrt{1-\vec{v}_e^2/c^2}$, n_e is the electron density, $\vec{E}$ is the electric field, $\vec{B}$ is the magnetic field. Taking rot of Eq (1) and making use the electron continuity equation one can obtain the equation for the Lagrangian invariant I

$$\frac{\partial I}{\partial t} + \vec{v}_e \nabla I = 0, \qquad I = \frac{\Omega_{e\theta}}{rn_e}. \qquad (3)$$

The ion dynamics can be described by the following hydrodynamic equations

$$\frac{\partial n_i}{\partial t} + div(n_i \vec{v}_i) = 0, \qquad m_i \frac{d\vec{v}_i}{dt} = Ze\vec{E} + \frac{Ze}{c}[\vec{v}_i \times \vec{B}] - \frac{1}{n_i}\nabla p_i, \qquad (4)$$

where n_i and $\vec{v}_i$ are the ion density and velocity, p_i is the ion pressure.

In the limit of the introduced equations the plasma dynamics can be considered in the following diapason of the magnetic field

$$4\pi n_e m_e c^2 \ll B_\theta^2 \ll 4\pi n_i m_i c^2. \qquad (5)$$

The l.h.s. of this inequality corresponds to the nonquasineutrality of the initial vortex at the Debye magnetic radius $r_B \sim B_\theta/4\pi e n_e$. As the characteristic temporal scale is the inverse ion plasma frequency $t_0 \sim \omega_{pi}^{-1}$, so the r.h.s. of the inequality (5) results in the appearance of the small parameter $\varepsilon = r_B/ct_0 \ll 1$, what means the quasistatic approximation for electrons. This allows to neglect nonstationary term in Eq. (1), and also the ion motion along the z-axis.

MAIN EQUATIONS

Taking into account the smallness of the parameter ε and introducing the dimensionless quantities

$$r = \rho\sqrt{\frac{m_e c^2}{4\pi e^2 n_\infty}}, \; B_\theta = b\sqrt{4\pi m_e c^2 n_\infty}, \; v_{ez} = vc, \; n_e = v n_\infty, \; n_i = n\frac{n_\infty}{Z}, \; I = 4\pi e i$$

one can obtain the system of the equations that describes the quasistatic electron vortex structure (see [4])

$$\frac{1}{\rho}\frac{\partial}{\partial \rho} = -vv, \quad \gamma^3 \frac{\partial v}{\partial \rho} = \rho vi - b, \quad v = \frac{n\gamma^3 + b^2}{\gamma + \rho ib}. \qquad (6)$$

In such a setting of the problem the vortex dynamics is determined only by a slow ion motion in the electric field of the vortex. In this case, block of the nonstationary dimensionless equations, that describes the dynamics of the current structure by the account of the ion motion, takes the form

$$\frac{\partial n}{\partial t} + \frac{1}{\rho}\frac{\partial}{\partial \rho}(\rho n u) = 0, \qquad \frac{\partial u}{\partial \tau} + u\frac{\partial u}{\partial \rho} = -\frac{\partial \gamma}{\partial \rho} - i\frac{\partial}{\partial \rho}(\rho b) - \frac{1}{n}\frac{\partial p}{\partial \rho}, \qquad (7)$$

$$\frac{\partial i}{\partial \tau} + \left[\frac{n}{v}u + \frac{1}{v}\frac{\partial}{\partial \tau}\left(\frac{\partial u}{\partial \tau} + u\frac{\partial u}{\partial \rho} + \frac{1}{n}\frac{\partial p}{\partial \rho}\right)\right]\frac{\partial i}{\partial \rho} = 0. \qquad (8)$$

Here the dimensionless time τ, the ion radial velocity u, and the pressure p with the isentropic exponent equal 2 are introduced

$$t = \tau\sqrt{\frac{m_e}{4\pi e^2 Z n_\infty}}, \quad v_{ir} = uc\sqrt{\frac{Zm_e}{m_i}}, \quad p_i = pn_\infty m_e c^2, \quad p = \lambda\frac{n^2}{2}.$$

Below, the initial conditions n(τ=0) =1 and u(τ=0)=0, and the boundary conditions n(ρ=∞)=1 and u(ρ=0)=u(ρ=∞)=0 will be used. Here the calculations for several values of λ are performed.

RESULTS OF NUMERICAL CALCULATIONS

Figure 1 demonstrates the initial equilibrium of the electron vortex structure, which corresponds to the undisturbed ions. One can see that near the axis there exists the peak of the electron density that originates the negative electric field. The relativistic electron motion along z-axis generates the azimuthal magnetic field, and besides the electrons drift in the crossed electric and magnetic fields. The ions are not magnetized at this size scale, therefore the electric field results in the radial ion acceleration. Figure 2 shows the evolution of the ion density and radial velocity. Beginning with the time τ = 1 the unloading shock wave for the radial ion velocity is formed. In this shock wave the velocity of the accelerated ions is slowed-down near the shock-wave front to the zero value. The immediate cause of the ion deceleration consists in the accumulation of the ions beyond the front of the shock wave. Thus, the final dense and hot equilibrium core near the axis is formed by the account of the magnetic and electric fields

$$r^2(B_\theta^2 - E_r^2) + 8\pi\int_0^r dr\, r^2\frac{\partial p_i}{\partial r} = 0. \qquad (9)$$

CONCLUSIONS

The final ion energy in the dense core is about of the several MeV and at first glance does not meet the experimental results [3]. However, as it follows from [2], by the correct estimation the Doppler-broadened lines are indicative of ion kinetic energies of about 1 MeV. The characteristic time of the energy transfer from the ions with the energy $\varepsilon_i > (m_i/m_e)T_e$ to the electrons is equal to $t^{i/e} = \sqrt{m_i}/(\pi\sqrt{2}e^4 Z^2)(\varepsilon_i^{3/2}/\lambda n_e)(m_e/m_i)$ and for the electron density $n_e \sim 10^{22}$ cm^{-3} less than 1 ps [5]. This stops the large increase of the ion energy in the core and protects from the strong nonquasineutrality in the final equilibrium according to $Zen_i E_r = \partial p_i/\partial r$.

By the inclusion of the ion-ion dissipation the resulting equilibrium can be destroyed. In order to show that, we consider the magnetic field evolution under the influence ion-ion dissipation [6]

$$\frac{\partial B_\theta}{\partial t} + \frac{\partial}{\partial r}(v_{ir} B_\theta) = \frac{c^2}{4\pi} \frac{\partial}{\partial r}\left[\frac{(\eta_1 \xi_2 - \eta_2 \xi_1)^2}{\xi_1 \xi_2} \frac{h^2}{1+h^2} \frac{1}{r\sigma} \frac{\partial}{\partial r}(rB_\theta)\right], \qquad (10)$$

where η_k and ξ_k the mass and charge fraction of the ion components (k=1,2), $\sigma(T_i)$ – the ion-ion conductivity. By the satisfying the inequalities ($\vec{v}_i = \eta_1 \vec{v}_1 + \eta_2 \vec{v}_2$ -- the mass velocity) $|\vec{v}_i| \gg |\vec{v}_1 - \vec{v}_2| \gg (m_e/Z^2 m_i)^{1/4} |\vec{j}|/(en_e)$, when one can consider the values η_k and ξ_k to be constant, and by the fast equalization the ion temperature T_i the Eq. (10) can be presented in the self-similar form relative to the variable $\zeta = r^2/t$. Then for h=h(ζ) by the account of the hydrodynamic equations analogous to (4) in the quasineutral plasma is correct

$$n_e = t^{-1} N(\zeta), \qquad B_\theta = t^{-1} B(\zeta), \qquad v_{ir} = t^{-1/2} V(\zeta), \qquad (11)$$

The inclusion of the ion-ion collisions for the electron density with $n_e \sim 10^{23}$ cm^{-3} will occur by the ion temperature $T_i \sim 10$ keV, where the disruption time of the ion core with the size of about 1 μm is equal to 1ps, what is considerably less than that for the electron-ion collisions. This corresponds to the calculations performed in paper [3].
This paper is supported in part by RFBR grants N 00-02-16305 and N 00-15-96599. Authors are grateful to Dr. S.A. Pikuz for exposing to new experimental results.

References
1. Yu.L. Bakshaev, P.I.Blinov, A.S.Chernenko et al., International Conference ReserchApplicationof plasma, Warsaw, Poland, 19-21 September, PS-22.
2. T.A.Shelkovenko, D.B.Sinars, S.A.Pikuz, and D.A.Hammer, Phys. of Plasmas, **8**, 1305-1318 (2001).
3. S.A.Pikuz, D.B.Sinars, T.A. Shelkovemko et al., Phys.Rev.Lett. (in press).
4. A.V.Gordeev, T.V.Losseva, JETP Lett., **70**, 684-690 (1999).
5. B.A. Trubnikov, in: Review of Plasma Physics, v.1, ed. M.A.Leontovich (Consultant Bureau, New York, 1965).
6. A.V.Gordeev, Plasma Physics Reports, **27**, 659-668 (2001).

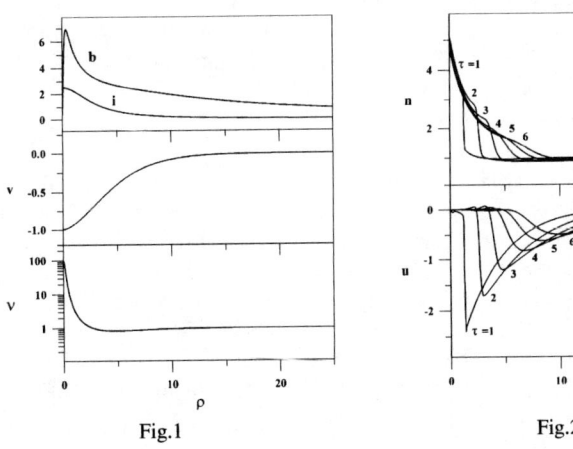

Fig.1　　　　　　　　　　　　　　Fig.2

On Stabilization of the Rayleigh-Taylor Instability for the Imploding Liner on Account of Ion-Ion Collisions

Alexander V. Gordeev

Russian Research Center "Kurchatov Institute", 1 Kurchatov Square 123182, Moscow, Russia

Abstract. The stabilization of the Rayleigh-Taylor instability for the imploding cylindrical liner in the limit of a low plasma density $\Pi = \omega_{pi}^2 \delta^2/c^2 \ll 1$ (δ -- the characteristic size of the current layer) is investigated, when the electron currents are much greater than the ion currents. The stabilization of the Rayleigh-Taylor instability for the parameter diapason $v_{ii}/\omega_{Bi} < (Z^2 M/m)^{1/2}$ is considered, when the plasma dissipation connected with the ion-ion collisions considerably superior the usual dissipation due to the electron-ion collisions. For the electric conductivity, caused by the ion-ion collisions and resulted in the minimum value $\sigma \sim enc/B$, the effect of the partial stabilization of the Rayleigh-Taylor instability is demonstrated.

INTRODUCTION

The Rayleigh-Taylor (RT) instability is responsible for the disruption of the effective imploding of the current layers by the self-magnetic field, thus lowering the parameters of the source of neutrons and electromagnetic radiation [1,2]. The RT instability exists both in the limit of the one-fluid magnetic hydrodynamics, when the Hall effect may be neglected for the small parameter $\Pi^{-1} \ll 1$, where $\Pi = 4\pi e^2 n \delta^2/Mc^2$ (δ- the characteristic dimension of the plasma layer), and for the case of the two-fluid magnetic hydrodynamics, when $\Pi \ll 1$ and the Hall effect is very essential. In recent time, it was shown in [3] that in the limit $\Pi \ll 1$ the linear equation for this instability may be exactly integrated for an arbitrary density and pressure profiles in the accelerated slab. In this approximation no influence of the density and pressure profiles on the RT increment was discovered. Further on, the two-fluid magnetic hydrodynamics of a special type will be considered – the Hall plasma model, where the ion component is not magnetized [4]. In addition, in the following consideration in contrast to [3] the influence of the kinetic pressure will be omitted. This means that below the acceleration of the plasma slab will be considered away from the axis, where the final liner implosion occur. At this stage of the liner acceleration the plasma temperature is less than $T \sim 10^2$ eV, what is connected with a very strong radiation from a nonequilibrium plasma. Therefore, the ignoring of the kinetic pressure as compared with the magnetic field pressure will be the main assumption in the equations considered below.

HYDRODYNAMIC EQUATIONS

As the starting point, according to the papers [5,6] will be chosen the continuity equation

$$\frac{\partial \rho}{\partial t} + div(\rho \vec{V}) = 0, \quad \rho = M_1 N_1 + M_2 N_2, \quad \vec{V} = \eta_1 \vec{V}_1 + \eta_2 \vec{V}_2, \tag{1}$$

the equation for the ion mass fraction of the species number 1

$$\frac{d\eta_1}{dt} + div(\rho \eta_1 \eta_2 \vec{w}) = 0, \quad \eta_1 = \frac{M_1 N_1}{\rho}, \quad \eta_1 + \eta_2 = 1, \tag{2}$$

and the hydrodynamic equation for the mass velocity

$$\frac{dV}{dt} + \frac{1}{\rho} div_i (\rho \eta_1 \eta_2 w_i \vec{w}) = \frac{1}{\rho c}[\vec{j} \times B], \quad \frac{d}{dt} = \frac{\partial}{\partial t} + (\vec{V}\nabla) \tag{3}$$

Thus, in hydrodynamic equation (3) the term arises that in structure is analogous to that of the kinetic pressure, but with the relative ion velocity instead of the thermal velocity.

In the following, the ion–ion electrical conductivity σ will be introduced, so that the ion-ion friction coefficient will be equal to

$$\alpha_{12} = \frac{\xi_1 \xi_2 e^2 n^2}{\sigma}, \quad \sigma^{-1} = \frac{4\sqrt{2\pi}}{3} \frac{\lambda Z_1 Z_2 e^2}{T^{3/2}} \sqrt{\frac{M_1 M_2}{M_1 + M_2}}, \quad \xi_1 = \frac{Z_1 N_1}{n}, \quad \xi_1 + \xi_2 = 1. \tag{4}$$

Here, the electron density $n = Z_1 N_1 + Z_2 N_2$ is introduced according to the quasineutrality condition.

In the limits $\omega_{Bi} \gg \tau^{-1}, \alpha_{12} \gg \rho\tau^{-1}$ the introduced relative ion velocity $\vec{w} = \vec{V}_1 - \vec{V}_2$ with regard for the only ion–ion collisions can be presented in the following form

$$\vec{w} = \beta \frac{\mu}{c} B_\theta \frac{\vec{j}}{\alpha_{12}^2 + \beta^2} - \alpha_{12} \frac{\mu}{c} \frac{[\vec{j} \times B]}{\alpha_{12}^2 + \beta^2}, \tag{5}$$

where $\mu = \eta_1 \xi_2 - \eta_2 \xi_1, \beta = \xi_1 \xi_2 enB_\theta /c$.

MAGNETIC FIELD ION-ION DIFFUSION

As it was mentioned earlier, further on the case will be investigated, where the ion-ion collisions are only taken into account. This is because the ion-ion dissipation is much more essential as compared with the electron-ion dissipation by virtue of the fulfilment of the inequality

$$|\vec{w}| \gg \left(\frac{m}{Z^2 M}\right)^{1/4} \frac{|\vec{j}|}{en}. \tag{6}$$

By the account of the only ion-ion collisions the equation for the magnetic field diffusion can be presented in the following form

$$\frac{\partial B_y}{\partial t} + \nabla(\vec{V} B_y) = \frac{c}{4\pi} \frac{\partial}{\partial z} \left\{ \frac{(\mu^2/\xi_1\xi_2)(c\beta^2/\sigma)\partial B_y}{\alpha_{12}^2 + \beta^2} \frac{\partial B_y}{\partial z} + \frac{B_y}{en}\left[1 + \frac{(\mu^2/\xi_1\xi_2)\beta^2}{\alpha_{12}^2 + \beta^2}\right] \frac{\partial B_y}{\partial x} \right\} +$$

$$+ \frac{c}{4\pi} \frac{\partial}{\partial x} \left\{ \frac{(\mu^2/\xi_1\xi_2)(c\beta^2/\sigma)\partial B_y}{\alpha_{12}^2 + \beta^2} \frac{\partial B_y}{\partial x} - \frac{B_y}{en}\left[1 + \frac{(\mu^2/\xi_1\xi_2)\beta^2}{\alpha_{12}^2 + \beta^2}\right] \frac{\partial B_y}{\partial z} \right\}. \qquad (7)$$

Here we introduced the plain approximation, where the coordinate r is transformed into x and the coordinate θ -- into y. One should bear in mind that the magnetic field is frozen in electrons and the magnetic field diffusion is only connected with the ion-ion collisions. The mechanism by which the ion motion affects the magnetic field can be understood in terms of the quasineutrality condition, which implies that the relative ion motion changes the ion density and, accordingly, the magnetic field, which is frozen in the electron component.

In this case the dimensionless Hall parameter, that corresponds to the ion-ion dissipation, should meet the following condition

$$h = \frac{\beta}{\alpha_{12}} = \frac{\sigma B_\theta}{enc} \gg \left(\frac{m}{Z^2 M}\right)^{1/4}. \qquad (8)$$

In the Eq. (7) one can neglect the terms in l.h.s. by the fulfilment of the following inequalities

$$1 \ll h^2 \ll \frac{\kappa}{\Pi}, \quad \kappa = k\Delta, \qquad (9)$$

where k the wave vector by the Fourier transformation along the z-axis, Δ is the size of the current layer, accelerated by the magnetic field, and the parameter Π meets the condition $\Pi \ll 1$.

In the further investigation we consider the case, when the Hall parameter meets the inequality h > 1. Then Eq.(7) can be transformed to the following simplified form

$$\frac{\partial}{\partial z}\left(\frac{\partial B_y}{\partial z} + \gamma \frac{\sigma B_y}{enc} \frac{\partial B_y}{\partial x}\right) + \frac{\partial}{\partial x}\left(\frac{\partial B_y}{\partial x} - \gamma \frac{\sigma B_y}{enc} \frac{\partial B_y}{\partial z}\right) = 0, \qquad (10)$$

where $\gamma = 1 + \xi_1\xi_2/\mu^2$.

It follows from Eq. (5) that by the fulfillment of the condition

$$|\vec{V}| \gg |\vec{w}| \qquad (11)$$

one can consider that $d\eta_1/dt=0$ and $\eta_1=$ const. All other coefficients ξ_k and μ can be expressed through η_1, therefore one can also consider $\gamma =$ const.

Combining Eqs. (6) and (11), one can obtain the final condition for the parameter Π

$$1 \gg \Pi \gg \left(\frac{m}{Z^2 M}\right)^{1/2}. \qquad (12)$$

By the account of these assumptions the final equation for the dimensional density ν reads

$$\nu - bb_0 \frac{\nu_0'}{\nu_0} = -i \frac{\nu_0}{\overline{\eta}_0 \kappa} \hat{\Delta} b, \qquad \hat{\Delta} = \frac{d^2}{d\xi^2} - \kappa^2. \tag{13}$$

In the absence of the dissipation, when the r.h.s. term in Eq. (13) can be neglected, and by the use of the additional equation from [3]

$$\frac{d\nu}{d\xi} - \Omega^2 \nu = \hat{\Delta}(bb_0), \tag{14}$$

the usual RT instability can be obtained. By taking account of the dissipative term in (13) and inserting the expression for ν from Eq. (13) in Eq. (14), the final equation for the RT instability by the account of the ion-ion collisions can be represented in the following form relative to the variable $\varphi = bb_0/\nu_0$

$$\nu_0 \hat{\Delta} \varphi + \nu_0'(\varphi' + \Omega^2) = -i \left(\frac{d}{d\xi} - \Omega^2 \right) \frac{\nu_0}{\overline{\eta}_0 \kappa} \hat{\Delta} \left(\varphi \frac{\nu_0}{b_0} \right). \tag{15}$$

Here, the r.h.s. of Eq.(15) may be considered as the disturbance that originates from the ion-ion collisions. In addition, the corresponding solutions grow exponentially with ξ.

Now we must take into account that according to [3] in the absence of the dissipation the solutions of this equation meet the condition $\hat{\Delta} \varphi = 0$ with the exponential accuracy.

Therefore, the resulting dispersion equation takes the form

$$\Omega^2 + |\kappa| = 2i \frac{\nu_0}{\overline{\eta}_0 \nu_0'} \left(\frac{\nu_0}{b_0} \right) (\Omega^2 - |\kappa|). \tag{16}$$

One can see from this equation that there exists the partial stabilization of the RT instability.

CONCLUSIONS

The considered effect can be essential for the imploding liners. It was discovered in [7] that for the nested wire liner, when the metal wires were covered by an another metal, the compactness of the imploding liner increases. This effect can be connected with the ion-ion collisions.

This work is supported in part by RFBR grants N 00-02-16305 and N 00-15-96599.

References
1. Sce H., Coleman P., Failor B. et al., 13th Intern. Conf, On High-Power Particle Beams, Nagaoka, Japan, June 25-30, 2000, v.I, p.36.
2. Blinov P., Chernenko A.. Chesnokov A., et al., ibid, p.76.
3. Gordeev A.V., Plasma Physics Reports, **25**, 202 (1999).
4. Gordeev A.V., Kingsep A.S., and Rudakov L.I., Physics Reports, **243**, 215 (1994).
5. Gordeev A.V., Soviet Journal of Plasma Physics, **13**, 713 (1987).
6. Gordeev A.V. Plasma Physics Reports, **27**, 659 (2001).
7 Deeney C., Coverdale C.C., Douglas M.R., et al., In: 13th Intern. Conf. On High- Power Particle Beams, Nagaoka, Japan, June 25-30, 2000. Program and Abstract, p.336.

The Tentative Opinion of Modeling Plasma Formation in Metallic Wire Z Pinch

N. Ding

Institute of Applied Physics and Computational Mathematics, P. O. Box 8009 Beijing, 100088, People's Republic of China

Abstract. Numerous experiments in both single wire and in wire arrays have attracted much attention. For the wire array Z-pinch implosions the plasma formation is a key question and very important. By means of analyzing a number of single-wire and multi-wire experiments, we put forward a tentative opinion of modeling plasma formation in the metallic wire. We suggest two models to describe the behavior of a wire array Z-pinch in initial phase. In this phase each wire carries a rising current and behaves independently, in a way similar to that found in single wire Z-pinch experiments in which a comparable current in one wire is employed. Based on one- or/and two-dimensional magneto-hydrodynamics (MHD) theory, one model is used to simulate the electrical explosion stage of the metallic wire; another is used to simulate the wire-plasma formation stage.

I. INTRODUCTION

Present level of x-ray power and yield (280 TW, 4 ns, 1.8 MJ), achieved in wire array Z-pinch plasma radiation sources,[1,2] makes our great interest. Further increases of both x-ray and yield are needed for a number of applications, for example, inertial confinement fusion, which require a more detailed understanding of the physics involved. Any attempt to completely model all aspects of wire array Z-pinch evolution using a single calculation model would be completely impractical. As Chittenden *et al.*[3] pointed out, a more feasible approach is to model different phases of the evolution using different models and attempt to link them together to form a composite model of the whole wire array Z-pinch experiment. They have presented the results of a series of one-, two- and three-dimensional magneto-hydrodynamic (MHD) simulations of the different phases of wire array evolution leading up to the final stagnation. Numerous experiments and theoretical analyses have shown that for the wire array Z-pinch implosions the plasma formation is a key question and very important. Chastened *et al.* first used 2-D "cold-start" calculation[4] of single wires to illustrate the plasma formation processes present in wire array. However, there "cold-start" simulation begin at the temperature higher than the melting point of metallic wire.

According to the heuristic model[5] of the wire array Z-pinch suggested by Haines, the dynamics and behavior of the wire-array pinch can be divided into four distinct phases: (1) Each wire carries a rising current and behaves independently and in a way similar to that found in single wire Z-pinch experiments. The current through the cold wire, the wire is heated. The resistivity of the wire is appropriate to the temperature, the temperature rise, once pass solid and liquid two phases, the vaporization begins and the volume of the wire is expanded. So we define first phase in the phase with heating, fusion, evaporation and expansion. (2) After first phase, each wire can be postulated to form a plasma cloud that expands with an approximately constant radial velocity about its own axis. Merger of the wire plasmas will then occur after a time when the wire array as a whole has accelerated inwards from its initial radius to a small radius. At the merger, a plasma shell may be formatted and its width will be determined by the diameter of the wire plasmas that, in the simplest model, are cylindrical about each wire axis. However, under some conditions, the plasma shell does not formatted. Numerous experiments have observed a two-component structure with a low-density plasma corona surrounding a much higher density core that persists until late into the discharge. The implosion dynamics has shown to be driven by a competition between the implosion pressure, making the array converge to the axis as a set of individual plasma columns, and the tidal pressure that makes the wire merge, forming an annular conducting shell. Their relative roles are determined by the gap-to-diameter ratio. (3) If the shell is formatted, it is accelerated inwards under the influence of the $j \times B$ pinch force. But, if the two-component structure with dense and cold cores surrounded by a low density, hot coronal plasma is formed, the dynamics of the slow ablation of wire cores and the motion of the coronal plasma towards the array axis dominates the first 3/4 of the implosion. In these three phases, the magnetic energy supplied by the pulsed power driver is converted into plasma kinetic energy and plasma thermal energy. (4) Finally, at stagnation on the axis, the plasma energy is converted into x-ray radiation.

By analyzing a number of single-wire and multi-wire experiments, [6,7] we think that it is very important that the initial energy deposition, expansion behaviors and morphology of exploding wires within a very short time (although may be a few nanoseconds) prior to vaporization and ionization of the wire material.

II. MODELS

In the present many two dimensional (2-D) models are capable of explaining most of the effects seen in experiments. Such as the effects of instability growth on the radiation out put from cylindrical implosions is discussed in Ref.8. However, in many computational models, the plasma was initially taken to be a thin shell. It is not reality. A wire arrays load is not an annular plasma shell, at least initially, there are some important phenomena affecting the radiative performance, which cannot be described by assuming the plasma shell. It is very important to describe the transition from discrete solid wire to an imploding plasma ensemble. Chittenden et al. [4] firstly successfully described the plasma formation in metallic wire Z pinches using a two-dimensional resistive magneto-hydrodynamics code. Although Chittenden et al.'s "cold start" simulations of single wire experiments have illustrated some of the important processes in the plasma formation phase of wire arrays. However, their simulation began at the temperature higher than the melting point of metallic wire. To study the process of wire electrical explosion during initial current pulse, we must make to model from room temperature. If we consider the physical processes of the wire from the low temperature (room temperature) to the higher temperature of the plasma, the solid phase converted into the liquid phase is very important.

In order to investigate magneto-hydrodynamics of the electrical explosion of the wire, we adopt a MHD model. Assuming, after evaporation around the solid wire occurs, have no strong current discharge. So we can neglect the heat conduction and radiation translation. The model to describe the electrical explosion is as follows:

Mass continuity equation is given by

$$\frac{\partial \rho}{\partial t} + \nabla \cdot (\rho u) = 0. \tag{1}$$

Momentum equation is

$$\frac{\partial}{\partial t}(\rho u) + \nabla \cdot (\rho u u) = -\nabla P + \frac{1}{c} j \times B. \tag{2}$$

Where $P = p+q$, p is the pressure and q is artificial viscous stress.

Internal energy equation is

$$\frac{\partial}{\partial t}(\rho \varepsilon) + \nabla \cdot (\rho \varepsilon u) = -P \nabla \cdot u + j \cdot E. \tag{3}$$

We note that in the initial phase during wire exploding, main physical processes are the fusion and evaporation. The system's ionization and heating conduction do not concern us in this paper.

From Maxwell equations, a magnetic diffusion equation used to calculate the spatial and temporal evolution of the magnetic field can be derived. In general, the only component of the magnetic field used to describe Z pinch experiments is the theta component. In the some approximations, the evolution equation for B_θ is

$$\frac{\partial B_\theta}{\partial t} = \frac{\partial}{\partial z}\left(\frac{c^2}{4\pi}\eta \frac{\partial B_\theta}{\partial z}\right) + \frac{\partial}{\partial r}\left(\frac{c^2}{4\pi}\eta \frac{\partial r B_\theta}{\partial r}\right). \tag{4}$$

Where η is the electrical resistivity. The right two terms above Eq. (4) represent resistive diffusion of the magnetic field.

The resistivity includes data describing the solid and liquid regimes, but no plasma regime. Why have no the advection terms in Eq. (4)? If the transport of magnetic flux with the fluid occurs, in the hydrodynamics the magnetic field equation must have the part of the advection. In our consider system, the metal wire through the strong current fuses and evaporates from the wire's surface. The wire has a density core and its volume becomes large, that is, Ohmic heating expands the metal wire and increases the volume it. However, because the wire core has heavy density, we can think the wire's velocity very small, so in the Ohmic law,

$$j = \sigma \left(E + \frac{1}{c} u \times B\right), \tag{5}$$

we don't consider the second term right hand of Eq.(5).

Last, in order to model the dynamics of explosion wire on a pulsed power machine, the driver and wire configuration are approximate using an equivalent circuit. The pulsed power driver is represented by the open circuit voltage, $V(t)$, the line impedance, R_0, and the inductance, L_0. The dynamic wire is represented by the time varying inductance and resistance, $L(t)$ and $R(t)$. Adding the voltage drops across each element of circuit gives the external circuit equation

$$V(t) = IR_0 + IR(t) + \frac{d}{dt}[(L_0 + L(t))I]. \tag{6}$$

Here I is the current in the series circuit. The initial condition of the equivalent circuit equation is $I(0)=I_0, V(0)=V_0$.

Another important issue is the equation of state describing this system. During the solid and liquid phases, the wire material remains virtually stationary and holds together. In the condensed phases, exchange effect act as a binding force so that little no pressure are exerted by the electrons. Even in the vapor state, at high enough densities and low enough temperatures, the pressure and internal energy of the electron fluid are substantially reduced with respect to the perfect gas values. The behavior of the condensed states of matter can be emulated in a fluid simulation by modifying the equations of state to include degeneracy, nuclear potential and exchange effects.

In summary, given the suitable initial and boundary conditions, chosen the precise equation of state and the electrical resistivity, using a set of equations (1)-(4) and (6), the initial phase during the metal wire exploding can be described. Before the plasma formation, the pulse current effects on the wire density varying, its volume expansion, and the energy (or temperature) increase.

After investigating the initial stage of the wire exploding, we can study the plasma formation stage of the wire exploding. How do the plasma form during the large current through it, and how does the "core-corona" structure form and maintain. In order to model the plasma formation stage of single wire experiments including the dynamics of the core and the development of $m = 0$ instabilities in the corona, we can use the 2-D resistive MHD code provided by Chittenden.[4]

III. CONCLUSIONS

Numerous experiments and theoretical analyses have shown that for the wire array Z-pinch implosions the plasma formation is a key question and very important. We think that it is very important that the initial energy deposition, expansion behaviors and morphology of exploding wires within a very short time prior to vaporization and ionization of the wire material. We put forward a tentative opinion of modeling plasma formation in the metallic wire and suggest two models to describe the behavior of a wire array Z-pinch in initial phase. Based on MHD theory, one model is used to simulate the electrical explosion stage of the metallic wire; another is used to simulate the wire-plasma formation stage. The former may be the resistive MHD code neglecting heat conductivity, the latter can be the 2-D MHD model given in Ref.4.

REFERENCES

1. Sandford, T. W. L., Allshouse, G. O., et al., *Phys. Rev. Letters* **77**, 5063 (1996).
2. Spielman, R. B., Deeney ,C., et al., *Phys. Plasmas* **5**, 2105 (1998).
3. Chittenden, J. P., Lebedev, S. V., et al., *Phys. Plasmas* **8**, 2305 (2001).
4. Chittenden, J. P., Lebedev, S. V., et al., *Phy. Rev. E* **61**, 4370 (2000).
5. Haines, M. G., *IEEE Trans. P. S.*, **26**, 1275 (1998).
6. Pikuz, S. A., Shelkovenko, T. A., et al., *Phys. Rev. Letters* **83**, 4313 (1999).
7. Sinars, D. B., Min Hu, et al., *Phys. Plasmas* **8**, 216 (2001).
8. Peterson, D. L., Bowers, R. L. et al., *Phys. Plasmas*, **3**, 368 (1996).

APPENDICES

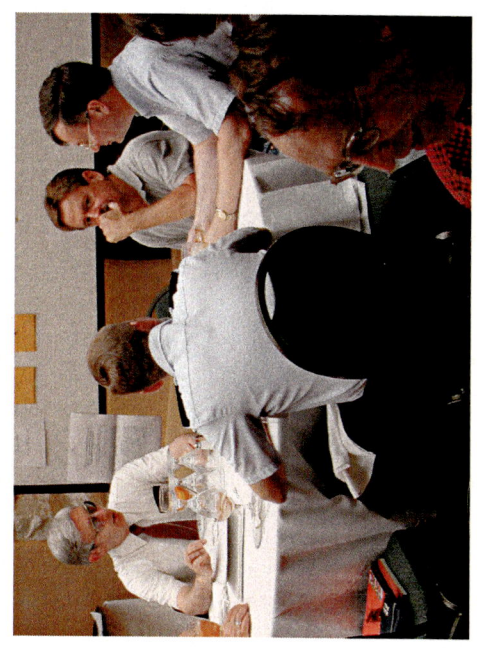

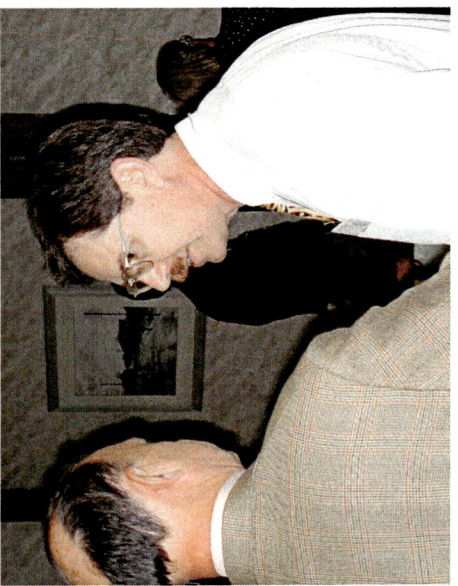

List of Participants

Aase, Nicholas
neaase@sandia.gov

Abubakirov, Edward
edward@appl.sci-nnov.ru

Abyar Monfared, Kashani Mehrdad
kashani@phys.cst.nihon-u.ac.jp

Allen, Raymond
allen@suzie.nrl.navy.mil

Alliot, Jean-Claude
alliot@onera.fr

Ampleford, David
davidampleford@ic.ac.uk

Ando, Ritoku
ando@plasma.s.kanazawa-u.ac.jp

Andrey/Petrovich, Orlov
ivanovsky@ntc.vniief.ru

Andrey/Vladimirovich, Ivanovsky
ivanovsky@ntc.vniief.ru

Apruzese, John
apruzese@ppd.nrl.navy.mil

Armijo, Joseph
jcarmijo@lanl.gov

Arzhannikov, Andrei
arzhannikov@inp.nsk.su

Bailey, Jim
jebaile@sandia.gov

Bailey, Vernon
vbailey@titan.com

Baksht, Rina
baksht@ovpe2.hcei.tsc.ru

Barker, Robert
robert.barker@afosr.af.mil

Barnea, Gideon
barnea@actcom.co.il

Baronova, Elena
baronova@nfi.kiae.ru

Bauer, Bruno
kari@physics.unr.edu

Beezhold, Wendland
beezholt@physics.isu.edu

Bell, David
david.bell@ieee.org

Belyaev, Vadim
VadimBelyaev@mtu-net.ru

Bezdek, Milan
mbezdek@vtupv.cz

Bland, Simon
sn.bland@ic.ac.uk

Blaugrund, Abraham
fnbla@weizmann.ac.il

Bliss, David
debliss@sandia.gov

Bloomquist, Douglas
ddbloom@sandia.gov

Bluhm, Hansjoachim
bluhm@ihm.fzk.de

Brinsmead, William
kari@physics.unr.edu

Brzosko, Jan
brzoskoj@diana0hitech.com

Burtsev, Vladimir
burtsev@niiefa.spb.su

Buttram, Malcolm
mtbuttr@sandia.gov

Bystritskii, Vitaly
vbystrit@uci.edu

Bystritsky, Viacheslav
bystvm@nusun.jinr.ru

Cagliostro, Dominic

Caporaso, George
caporaso1@llnl.gov

Carey, William
carey@apelc.com

Cartwright, Keith
Keith.Cartwright@kirtland.af.mil

Cassany, Bruno
cassany@bordeaux.cea.fr

Castagno, Scott
sgcasta@sandia.gov

Cevallos, Michael
mic-dana@swbell.net

Chaikovsky, Stanislav
stas@ovpe.hcei.tsc.ru

Chalise, Priya
chalise@hotta.es.titech.ac.jp

Chandler, Gordon
gachand@aol.com

Chandler, Katherinie
kmc25@cornell.edu

Chantrenne, Sophie
schantrenne@titan.com

Chapman, Randall
Randy.Chapman@arnold.af.mil

Chen, Yanqiao
yqchen@tsinghua.edu.cn

Chernenko, Andrey
kingsep@dap.kiae.ru

Chittenden, Jeremy
j.chittenden@ic.ac.uk

Choi, Junho
choi@kingtrabu.nrl.navy.mil

Choi, Song Hun
samchoi@unm.edu

Chong, Young
chong@ppdmail.nrl.navy.mil

Chornyi, Valentin
chorny@pht.univer.kharkov.ua

Christenson, Peggy
pjc@lanl.gov

Chuaqui, Hernan
hchuaqui@fis.puc.cl

Clark, Robert
clark@ppdmail.nrl.navy.mil

Clough, Stephen
stephen.clough@awe.co.uk

Cochran, Frederick
flc@lanl.gov

Coleman, Philip
plcoleman@mailaps.org

Commisso, Robert
commisso@suzie.nrl.navy.mil

Cook, Donald
colson@sandia.gov

Cooper, Graham
graham.cooper@awe.co.uk

Cooperstein, Gerald
cooperstein@nrl.navy.mil

Corcoran, Patrick
pcorcoran@titan.com

Cotter, Timothy
tim.cotter@arnold.af.mil

Coverdale, Christine
cacover@sandia.gov

Crotch, Ian
ian.crotch@awe.co.uk

Cuneo, Michael
mecuneo@sandia.gov

Curry, Randy

Dale, Gregory
daleg@missouri.edu

Dannenberg, Korbie
korbster@umich.edu

Dasgupta, Arati
dasgupta@ppdmail.nrl.navy.mil

Davanloo, Farzin
fdavan@utdallas.edu

Davis, Jack
davisj@ppdmail.nrl.navy.mil

Davis, Harold
davis@lanl.gov

Davis, Randolph
randy.davis@dtra.mil

De Groot, John
jsdegroot@ucdavis.edu

Deeney, Chris
cdeene@sandia.gov

Deng, Jianjun
bcys@caep.ac.cn

Desjarlais, Michael
mpdesja@sandia.gov

Dickens, James
jdickens@coe.ttu.edu

Dobbins, Aaron
ahdobbi@sandia.gov

Dolgachev, Georgy
kingsep@dap.kiae.ru

Dovbnya, Anatoliy
zakutin@kipt.kharkov,ua

Droemer, Darryl
Dwdroem @sandia.gov

Duselis, Peter
pud2@cornell.edu

Dutkowski, Eugene
dutkowski_joe@crane.navy.mil

Egorov, Oleg
egorov@triniti.ru

Ekdahl, Carl
cekdahl@lanl.gov

Elegure, Folorunso
statemin@yahoo.com

Engelko, Vladimir
engelko@niiefa.spb.su

Faehl, Rickey
rjf@lanl.gov

Faretto, Harold
kari@physics.unr.edu

Favre, Mario
mfavre@fis.puc.cl

Feduschak, Vladimir
ratakhin@ovpe.hcei.tsc.ru

Filatov, Alexander
Filatov@iep.uran.ru

Fisher, Amnon
amfisher@physics.technon.ac.il

Freeman, Bruce
bruce.freeman@ne.tamu.edu

Frese, Michael
Michael.Frese@numerex.com

Frese, Sherry
Sherry.Frese@numerex.com

Frey, Wolfgang
wolfgang.frey@ihm.fzk.de

Fridman, Boris
fridman@mail.infostar.ru

Fuelling, Stephan
fuelling@physics.unr.edu

Fuks, Mikhail
fuchs@eece.unm.edu

Gahl, John
gahlj@missouri.edu

Garanin, Sergey
sfgar@vniief.ru

Garasi, Chris
cjgarasi@sandia.gov

Genoni, Thomas
genoni@mrcabq.com

Gericke, Dirk
gericke@lanl.gov

Gilgenbach, Ronald
rongilg@umich.edu

Ginzburg, Naum
ginzburg@appl.sci-nnov.ru

Giuliani, John
giul@ppdmail.nrl.navy.mil

Glazyrin, Igor
I.V.Glazyrin@vniitf.ru

Goldsack, Timothy
tim.goldsack@awe.co.uk

Goldstein, Steven
goldstsa@nv.doe.gov

Golub, Tatiana
gng_nt@sprynet.com

Juan, Carlos
jhernand@ttacs.ttu.edu

Hinshelwood, David
ddh@suzie.nrl.navy.mil

Holt, Thomas
tinyball@yahoo.com

Hoppe, Peter
hoppe@ihm.fzk.de

Horioka, Kazuhiko
khorioka@es.titech.ac.jp

Hu, Min
mh99@cornell.edu

Hughes, Thomas
tph@mrcabq.com

Ivanovsky, Andrey
ivanovsky@ntc.vniief.ru

Jennings, Chris
christopher.jennings@ic.ac.uk

Jiang, Weihua
jiang@nagaokaut.ac.jp

Jobe, Daniel
djobe@sandia.gov

Johnson, David
djohnson@titan.com

Johnston, Mark
markdj@engin.umich.edu

Jungwirth, Karel
jungwirth@fzu.cz

Kablambaev, Beissengazy
ratakhin@ovpe.hcei.tsc.ru

Kaganovich, Igor
ikaganov@pppl.gov

Kalinin, Yuri
kalinin@dap.kiae.ru

Kamada, Keiich
kkamada@plasma.s.kanazawa-u.ac.jp

Kamada, Masaki
kokamada@plasma.s.kanazawa-u.ac.jp

Kantsyrev, Victor
victor@physics.unr.edu

Karpinski, Leslaw
lekarp@ifpilm.waw.pl

Kasuya, Koichi
kkasuya@es.titech.ac.jp

Kawata, Shigeo
kawata@cc.utsunomiya-u.ac.jp

Keely, Sean
kpkeely@sandia.gov

Kemp, Mark
makemp@sandia.gov

Khakberdiev, Ibragim
khakb_i@yahoo.com

Khishchenko, Konstantin
konst@ihed.ras.ru

Kiefer, Mark
mlkiefe@sandia.gov

Kim, Alexandre
kim@oit.hcei.tsc.ru

Kim, Hyoung Suk
hyoungskim@hotmail.com

Kindel, Joseph
jkindel@lanl.gov

Kingsep, Alexander
kingsep@dap.kiae.ru

Knudson, Marcus
mdknuds@sandia.gov

Koidan, Vasili
koidan@inp.nsk.su

Kolb, Alan
alan.kolb@sbr.global.net

Konkashbaev, Isak
isak@anl.gov

Kononenko, Vladimir
bratchikov@five.ch70.chel.su

Korenev, Sergey
sergey@korenev.com

Kormilitsyn, Alexei
bratchikov@five.ch70.chel.su

Korolev, Yuri
korolev@lnp.hcei.tsc.ru

Korovin, Sergei
korovin@hcei.tsc.ru

Koshelev
Vladimir
koshelev@lhfe.hcei.tsc.ru

Koustantin, Khishchenko
konst@ihed.ras.ru

Kovalchuk, Boris
kim@oit.hcei.tsc.ru

Kovalev, Nikolay
kovalev@appl.sci-nnov.ru

Krastelev, Evgueni
e_krastelev@mail.ru

Krauz, Vyacheslav
krauz@nfi.kiae.ru

Kravarik, Jozef
kravarik@fel.cvut.cz

Krompholz, Hermann
hermann.krompholz@coe.ttu.edu

Kruecken, Thomas
thomas.kruecken@philips.com

Kubes, Pavel
kubes@fel.cvut.cz

Kubyshkin, Alexander
kub@sch2.ru

Kunze, Hans-Joachim
Hans-Joachim.Kunze@ruhr-uni-bochum.de

Kussse, Bruce
brk2@cornell.edu

Kuznetsov, Sergey
sfgar@md08.vniief.ru

Kwan, Thomas
tjtk@lanl.gov

Laroussi, Mounir
laroussi@jlab.org

Lassalle, Francis
lassallf@cegramat.fr

Lauer, Eugene
lauer1@llnl.gov

Le Galloudec, Nathalie
kari@physics.unr.edu

LeBeau, Hank
kari@physics.unr.edu

Lebedev, Sergey
s.lebedev@ic.ac.uk

Leeper, Ramon
rjleepe@sandia.gov

Lemaire, Jean-Louis
jlemaire@cea.fr

Lemke, Raymond
rwlemke@sandia.gov

LePell, Paul
pdlepel@sandia.gov

Levine, Jerrold
jlevine@titan.com

Ligachev, Aleksander
ligachev@msk.sitek.net

Lindemuth, Irvin
irl@lanl.gov

Lockner, Thomas
trlockn@sandia.gov

Loza, Oleg
loza@fpl.gpi.ru

Lu, George
lu@ao.dtra.mil

Lucero, Robert
rjlucer@sandia.gov

MacFarlane, Joseph
jjm@prism-cs.com

Maenchen, John
jemaenc@sandia.gov

Makhin, Volodymyr
kari@physics.unr.edu

Mardahl, Peter

Markovits, Meir
meirm@rafael.co.il

Maron, Yitzhak
fnmaron@wisemail.weizmann.ac.il

Masugata, Katsumi
masugata@eng.toyama-u.ac.jp

Mathuthu, Manny
mathuthu@science.uz.ac.zw

Matzen, M. Keith
mkmatze@sandia.gov

Mayes, Jon
mayes@apelc.com

Mazarakis, Michael
mgmazar@sandia.gov

McCarrick, James
mccarrick1@llnl.gov

McDaniel, Dillon

McGurn, John
jsmcgur@sandia.gov

Mehlhorn, Tom
tamehlh@sandia.gov

Mendel, Clifford
mendel@swcp.com

Merle, Eric
eric.merle@cea.fr

Mesyats, Gennady
mesyats@pran.ru

Meyer, Ryan
rmmeyer@sandia.gov

Miklaszewski, Ryszard
rysiek@ifpilm.waw.pl

Mitchell, Marc
mdm46@cornell.edu

Mix, Louis
lpmix@sandia.gov

Miyamoto, Tetsu
miyamoto@asrl.org

Molina, Isidro
imolina@sandia.gov

Mond, Michael
mond@menix.bgu.ac.il

Mosher, David
mosher@suzie.nrl.navy.mil

Mueller, Georg
georg.mueller@ihm.fzk.de

Musk, Jeffrey
jhm29@cornell.edu

Myers, Matthew
myers2@ccf.nrl.navy.mil

Navon, Itamar
navoni@barak-online.net.il

Neuber, Andreas
andreas.neuber@ttu.edu

Nielson, Dan
nielson5@covad.net

Ning, Cheng
chengnqy@public3.bta.net.cn

Ning, Ding
ding_ning@mail.iapcm.ac.cn

Novoselov, Yuri
nov@iep.uran.ru

O' Malley, John
jomalley@awe.co.uk

Oleinik, Gueorgui
oleinik@triniti.ru

Oliver, Bryan
boliver@mrcabq.com

Olson, Craig
clolson@sandia.gov

Onishchenko, Ivan
onish@kipt.kharkov.ua

Oreshkin, Vladimir
oreshkin@ovpe.hcei.tsc.ru

Orlov, Andrey
mailbox@ntc.vniief.ru

Ossakow, Sidney
ossakow@ccf.nrl.navy.mil

Ottinger, Paul
ottinger@suzie.nrl.navy.mil

Ouart, Nicholas
kari@physics.unr.edu

Oxner, Andrew
kari@physics.unr.edu

Palmer, James
james.palmer@awe.co.uk

Paraschiv, Ioana
ioana@physics.unr.edu

Pereira, Nino
pereira@speakeasy.org

Peskov, Nikolai
peskov@appl.sci-nnov.ru

Peterson, Darrell
dlp@lanl.gov

Peterson, Kyle
kyle.peterson@arnold.af.mil

Petrov, Peter
pvpetrov@snezhinsk.ru

Pikunov, Viktor
vmp@vmp.phys.msu.su

Pikuz, Sergei
clj1@cornell.edu

Plokhoim, Vladimir
V.V.Plokhoi@vniitf.ru

Pointon, Timothy
tdpoint@sandia.gov

Polevin, Sergey
polevin@lfe.hcei.tsc.ru

Powell, Charles
powellc@diana-hitech.com

Prestwich, Kenneth
krprestwich@cs.com

Presura, Radu
kari@physics.unr.edu

Price, David
price@titan.com

Putnam, Sidney
sputnam@titan.com

Quintenz, Jeffrey
jpquint@sandia.gov

Ramirez, Juan
mccasau@sandia.gov

Ratakhin, Nikolai
ratakhin@ovpe.hcei.tsc.ru

Rej, Don
drej@lanl.gov

Renk, Timothy
tjrenk@sandia.gov

Riordan, John
jriordan@titan.com

Rizakhanov, Rajudin
kercgor@dol.ru

Robledo-Martinez, Arturo
a.robledo@ieee.org

Roderick, Norman
roderick@unm.edu

Rodgers, Matthew
msrodge@sandia.gov

Rogowski, Sonrisa
strogow@sandia.gov

Rose, Evan
ear@lanl.gov

Rose, David
drose@mrcabq.com

Rosenthal, Stephen
serosen@sandia.gov

Rostomyan, Eduard
evrostom@irphe.am

Rostov, Vladislav
rostov@lfe.hcei.tsc.ru

Roth, Markus
m.roth@gsi.de

Rousculp, Christopher
rousculp@lanl.gov

Rousskikh, Alexander
russ@ovpe2.hcei.tsc.ru

Rovang, Dean
dcrovan@sandia.gov

Rudakov, Leonid
rudakovl@usam.umd.edu

Rukin, Sergei
rukin@iep.uran.ru

Ryutov, Dmitri
ryutov1@llnl.gov

Ryzhov, Victor
ryzhov@to/hcei.tsc.ru

Sachs, Bob
bob.sachs@teamsp.com

Sakamoto, Nobuhiro
nobusakamoto@mac.com

Sandoval, Joe
js_sandoval@lanl.gov

Sanford, Thomas
twsanfo@sandia.gov

Sarkar, Partha
psarkar@ipr.res.in
psarkar_70@yahoo.com

Sarkisov, Gennady
gssarki@sandia.gov

Sasorov, Pavel
sasorov@itep.ru

Savage, Mark
mesavag@sandia.gove

Schamiloglu, Edl
edl@eece.unm.edu

Schmidt, Jiri
schmidt@ipp.cas.cz

Schneider, Larry
lxschne@sandia.gov

Schoenbach, Karl
schoenbach@ece.odu.edu

Schultheiss, Christoph
christoph.schultheiss@ihm.fzk.de

Schumer, Joseph
schumer@calvin.nrl.navy.mil

Schwoebel, Paul
paul.schwoebel@sri.com

Seamen, Johann
jfseame@sandia.gov

Sethian, John
sethian@this.nrl.navy.mil

Shao, Hao
lwei@nint.ac.cn

Sheehey, Peter
pete@lanl.gov

Shelkovenko, Tatiana
clj1@cornell.edu

Sherlock, mark
mark.sherlock@ic.ac.uk

Shi, Jinshui
bcys@caep.ac.cn

Shimomura, Naoyuki
simomura@ee.tokushima-u.ac.jp

Shishlov, Alexander
ash@ovpe2.hcei.tsc.ru

Shkuratov, Sergey
sshkuratov@ppl.ee.tu.edu

Shlapakovski, Anatoli
shl@npi.tpu.ru

Shlyaptsev, Vyacheslav
slava@llnl.gov

Shlyaptseva, Alla
alla@physics.unr.edu

Shoup, Roy
roy.shoup@itt.com

Shpak, Valery
radan@ief.uran.ru

Shu, Ting
mrtingshu@yahoo.com.cn

Sinars, Daniel
dbsinar@sandia.gov

Sincerny, Peter
psincerny@titan.com

Sinclair, Mark
mar.sinclair@awe.co.uk

Sinitsky, Stanislav
sinitsky@inp.nsk.su

Smirnov, Valentin
VSmirnov@nfi.kiae.ru

Smith, John
smith@lanl.gov

Smith, Ian
ismith@titan.com

Sokovnin, Sergei
sokovnin@iep.uran.ru

Sotnikov, Vladimir
kari@physics.unr.edu

Soto, Leopoldo
lsoto@cchen.cl

Spencer, Thomas
Thomas.Spencer@kirtland.af.mil

Stallings, Charles
cstallings11@attbi.com

Steele, Jimmy
jim.steele@arnold.af.mil

Steen, Paul
psteen@titan.com

Stoltzfus, Brian
bsstolt@sandia.gov

Strasburg, Sean
stras@suzie.nrl.navy.mil

Strickler, Trevor
strickler@umich.edu

Struts, Vasily
struts@ephc.npi.tpu.ru

Struve, Kenneth
kwstruv@sandia.gov

Sturtz, Jason
jestur@sandia.gov

Sudan, Ravindra
clj1@cornell.edu

Sunka, Pavel
sunka@ipp.cas.cz

Swanekamp, Stephen
swane@calvin.nrl.navy.mil

Sweeney, Mary Ann
masween@sandia.gov

Sze, Henry
hsze@titan.com

Takasugi, Keiichi
takasugi@phys.cst.nihon-u.ac.jp

Talantsev, Yevgeniy
etalantsev@ppl.ee.ttu.edu

Tauschwitz, Andreas
A.Tauschwitz@gsi.de

Teramoto, Yusuke
tera@st.eecs.kumamoto-u.ac.jp

Terry, Robert
terry@ccs.nrl.navy.mil

Thoma, Carsten
cthoma@mrcabq.com

Thomas, Kenneth
Ken.Thomas@awe.co.uk

Thornhill, Joseph
thornhil@ppdu.nrl.navy.mil

Threadgold, Jim
jim.threadgold@awe.co.uk

Truesdale, Robert
bob.truesdale@arnold.af.mil

Turchi, Peter
peter.turchi@kirtland.af.mil

Tyo, J. Scott
tyo@eece.unm.edu

Velikovich, Alexander
velikov@ppdmail.nrl.navy.mil

Vezinet, Rene
vezinetr@cegramat.fr

Vidmar, Robert
rvidmar@unr.edu

Vikhrev, Victor
vikhrev@nfi.kiae.ru

Vladimir/Fedorovich, Ermolovich

Volkov, Nikolay
nbv@ami.uran.ru

Waisman, Eduardo
waisman@aasc.net

Walter, John
j.walter@ieee.org

Wang, Xinxin
wangxx@mail.tsinghua.edu.cn

Wang, Huacen
bcys@caep.ac.cn

Watrous, John
jwatrous@numerex.com

Weber, Bruce
weber@suzie.nrl.navy.mil

Welch, Dale
drwelch@mrcabq.com

Wemlinger, Erik
ecwemli@sandia.gov

Westenskow, Glen
westenskow1@llnl.gov

White, Roger
rogerw@titan.com

Whitney, Kenneth
whitney@ppdmail.nrl.navy.mil

Wong, Sik-Lam
swong@titan.com

Woodworth, Joseph
jrwoodw@sandia.gov

Xie, Weiping
bcys@caep.ac.cn

Yalandin, Michael
yalandin@ief.uran.ru

Yang, Libing
bcys@caep.ac.cn

Yatsui, Kiyoshi
yatsui@nagaokaut.ac.jp

Yu, Edmund
epyu@sandia.gov

Yusupov, Odil
khodjiev@online.ru

Zhong, Huihuang
hhzhong@nudt.edu.cn

Author Index

A

Adhikary, B., 325
Akiyama, H., 95, 131
Alexandrov, V. V., 29, 87, 91
Aliaga-Rossel, R., 55, 145, 233, 241
Alikhanov, S. G., 29
Altes, B., 43
Ampleford, D. J., 65, 71, 79, 317, 321
Apruzese, J. P., 101, 105, 339, 350, 358, 392
Aranchuk, L. E., 47
Asay, J. R., 299
Aubel, K., 161
Avrillaud, G., 51
Azizov, E. A., 29

B

Babineau, M. A., 43, 101, 109
Bachtin, V. H., 29
Bakshaev, Y. L., 33, 197
Baksht, R. B., 117, 123, 217, 384
Banaszak, A., 193, 255
Banister, J., 101, 109
Barnes, D. C., 372
Bartov, A., 33
Bauer, B. S., 279, 396
Bavay, M., 51
Bayol, F., 51
Beg, F. N., 83, 345
Bell, D., 101, 105, 109
Bender, H. A., 416
Bienkowska, B., 193, 255
Birstein, L., 229
Blackman, E., 317
Bland, S. N., 65, 71, 75, 79, 83, 291, 317, 321, 345
Blinov, P. I., 33, 197
Bliss, D. E., 23
Bohacek, V., 165
Boydston, J. C., 261
Buchenauer, D., 416

C

Castillo, F., 237
Chaikovsky, S. A., 59, 117, 123, 127, 225
Chamberlain, D., 185
Chandler, G. A., 269
Chandler, K. M., 141, 173, 221
Chaturvedi, S., 325
Chaudhari, V., 325
Chengwei, S., 309
Chernenko, A. S., 33, 197
Childers, K., 43
Chittenden, J. P., 65, 71, 75, 79, 83, 269, 291, 317, 321, 345, 354, 408
Cho, C., 333
Choi, P., 237, 241
Chong, Y. K., 339
Chrien, R. E., 269
Chuaqui, H., 15, 55, 145, 233, 237, 241
Chuvatin, A. S., 47
Ciardi, A., 291, 317, 321, 354
Clark, R. W., 101, 275, 339, 400
Clausse, A., 265
Coleman, P. L., 101, 113, 350
Corcoran, P., 43
Corley, J. P., 23
Cotter, T., 43
Coverdale, C. A., 101, 105, 339, 350, 358
Cuneo, M. E., 354

D

Dan'ko, S. A., 33, 197
Dasgupta, A., 404
Datsko, I. M., 153, 157
Davis, J., 101, 105, 275, 339, 350, 358, 400, 404
Deeney, C., 101, 105, 339, 350, 358
Demidov, V. A., 287
Deng, J., 135
Dimkoff, J., 416
Ding, B., 135
Ding, N., 135, 364, 428
Dolgachev, G. I., 29, 33

Dolgoleva, G. V., 287
Douglas, J., 43
Dunn, J., 416
Dunne, A. M., 291
Duselis, P. U., 205

E

Elizondo, J. M., 23
Enis, C., 101, 109

F

Failor, B. H., 101, 105, 109, 113, 350
Favre, M., 55, 145, 233, 237, 241, 245
Fedin, D. A., 177, 181, 185, 189
Fedulov, M. V., 87, 91
Fedunin, A. V., 117, 123
Ferguson, J. M., 261
Fiala, V., 396
Filevich, J., 416
Fisher, A., 101
Fisher, V., 388
Fornaciari, N., 416
Fortov, V. E., 37, 313
Fournier, K. B., 177, 416
Frank, A., 317
Frants, O. B., 149, 153, 157
Frati, M., 416
Freeman, B. L., 261
Frese, M. H., 368, 376, 380
Frese, S. D., 376, 380
Frolov, I. N., 87, 91
Frolov, O., 165
Fuelling, S., 189, 279
Fursov, F. I., 47

G

Gardiner, T., 317
Gasilov, V. A., 47
Geyman, V. G., 149, 153, 157
Gilliland, T. L., 101
Giuliani, Jr., J. L., 275, 339, 392, 400
Glazyrin, I. V., 384
Glukhikh, V. A., 29
Gomez, J. A., 145

Gordeev, A. V., 420, 424
Grabovskii, E. V., 3, 29, 87, 91
Gribov, A. N., 29
Gu, Y., 135

H

Hagen, E. C., 261
Haidong, L., 309
Haill, T. A., 299
Haines, M. G., 65, 71, 75, 83, 317, 321, 345, 408
Hallimullin, Y. A., 29
Hamann, F., 51
Hamasha, S. M., 412
Hammarsten, E. C., 416
Hammer, D. A., 141, 173, 221
Hansen, S. B., 177, 181, 185, 189, 412
Harjes, H. C., 23, 305
Hayashi, Y., 169
Hellinger, P., 396
Horioka, K., 169
Hotta, E., 169
Hu, M., 201, 221
Huang, X., 135
Huet, D., 47, 51

I

Idzorek, G. C., 269
Ishizaka, K., 329
Ivanova-Stanik, I., 193, 255
Ivanovsky, A. V., 287
Ivashov, R. V., 149, 153, 157
Ives, III, H. C., 23

J

Jakubowski, L., 193, 255
Jancarek, A., 165
Jankowska, E., 416
Jennings, C. A., 354
Jiang, W., 329, 333
Johnson, D. L., 23
Juha, L., 193
Jun, L., 309

K

Kai, O., 309
Kalinin, J. G., 29
Kalinin, Y., 33
Kanouff, M., 416
Kantsyrev, V. L., 177, 181, 185, 189
Karakin, M. A., 37
Karelin, V. I., 287
Karim, S., 416
Karpinski, L., 193, 197, 255
Kashani, M. A. M., 249
Katsuki, S., 95
Kenyon, V., 43, 101, 109
Kepple, P., 404
Khautiev, E. Y., 37
Khishchenko, K. V., 313, 384
Kies, W., 229, 265
Kinemuchi, Y., 329, 333
Kingsep, A. S., 29, 33
Kirkpatrick, R., 279
Kitterman, D. L., 23
Klír, D., 193, 197
Knudson, M. D., 299
Kohno, S., 95
Kokshenev, V. A., 47
Kolacek, K., 165
Kormilitcin, A. I., 29
Korolev, V. D., 197
Korolev, Y. D., 149, 153, 157
Kouchinsky, V. G., 29
Kovalenko, I., 33
Krása, J., 193
Krauz, V. I., 37
Kravárik, J., 193, 197, 255
Krishnan, M., 113
Krukovskii, A. Y., 47
Kubeš, P., 193, 197, 255
Kubiak, G., 416
Kumar, R., 325
Kurmaev, N. E., 47
Kurucz, P., 43
Kusse, B. R., 201, 205, 221
Kwek, H., 83

L

Labetsky, A. Y., 117, 123, 225
Lalle, B., 51

Landl, N. V., 149, 153, 157
Lassalle, F., 51
Lathi, D., 325
Lebedev, S. V., 65, 71, 75, 79, 83, 291, 317, 321, 345, 354, 408
Leboeuf, J.-N., 396
Leeper, R. J., 269
Lemke, R. W., 269, 299
Leñero, A. M., 237
LePell, P. D., 339
L'Eplattenier, P., 51
Levashov, P. R., 313, 384
Levashov, V. A., 29
Levine, J. S., 101, 105, 109, 113, 350
Li, Z., 135
Libing, Y., 309
Lindeburg, B., 261
Lindemuth, I., 279
Lobanov, A., 33
Lomonosov, I. V., 313, 384
Losseva, T. V., 420
Lotocky, A. P., 29
Luginbill, A. D., 261

M

Maenchen, J. E., 23
Makhin, V., 279, 396
Mangeant, C., 51
Marconi, M., 416
Maron, Y., 388
Maslennikov, D., 33
Masnavi, M., 169
Mazarakis, M. G., 23
McDaniel, D. H., 23, 209, 213
McGurn, J., 101
Medovschikov, S. F., 37
Mehlhorn, T. A., 299
Miklaszewski, R., 255
Mitchell, I. H., 55, 145, 233, 237, 241
Mitchell, M. D., 141, 173
Mitrofanov, K. N., 87, 91
Miyamoto, T., 249, 295
Mizhiritsky, V., 33
Mock, R. C., 269
Mokeev, A. N., 37
Monjaux, P., 51
Moon, S. J., 416
Morell, A., 51

Moreno, J., 161, 229, 265
Mosher, D., 101, 388
Muñoz, R. M., 229
Myalton, V. V., 37

N

Nakajima, M., 169
Narisawa, S., 131
Naz, N., 317
Nazarenko, A., 161
Nedoseev, S. L., 29, 37, 87, 91
Niimi, G., 169
Ning, C., 135, 364

O

Oleinik, G. M., 87, 91
Oreshkin, V. I., 117, 123, 217, 384
Orlov, A. P., 287
Osterheld, A. L., 416
Ouart, N. D., 177, 181, 185, 412

P

Paduch, M., 193, 197, 255
Palmer, J. B. A., 79
Pavez, C., 161, 233
Pechersky, O. P., 29
Peng, X., 135
Peterson, D. L., 269
Pikuz, S. A., 141, 173
Pismenniy, V. D., 29
Pointon, T. D., 23, 305
Porofeev, I. Y., 87, 91
Prasad, R., 113
Presura, R., 279

Q

Qi, N., 113

R

Rahman, A., 416
Ratachin, N. A., 153
Repin, P. B., 287
Rikovanov, G. P., 29
Ripa, M., 165
Robinson, A. C., 299
Rocca, J. J., 416
Rock, J. C., 261
Romanova, V. M., 193, 255
Rosenthal, S. E., 23, 209, 213
Rousculp, C. L., 372
Rousskikh, A. G., 117, 123, 217, 384
Rudakov, L. I., 47, 275, 358
Ruiz, C. L., 269
Ryć, L., 193

S

Saavedra, R., 229
Sadowski, M. J., 193, 255
Safronova, U. I., 177, 412
Sakadzic, S., 416
Sakamoto, N., 169
Samokhin, A. A., 87, 91
Sandford, T. W. L., 269
Sarkar, P., 325
Sarkisov, G. S., 209, 213
Sasorov, P. V., 87, 91, 209, 213
Savage, M. E., 305
Schmidt, H., 193, 255
Schmidt, J., 165
Scholz, M., 193, 197, 255
Schumer, J., 388
Selemir, V. D., 287
Shah, K., 325
Shashkov, A. Y., 33, 197
Sheehey, P., 279
Shelkovenko, T. A., 141, 173
Shemyakin, I. A., 149, 153, 157
Sherlock, M., 321, 345, 408
Shimkaveg, G., 416
Shimomura, N., 95
Shishlov, A. V., 117, 123, 217, 384
Shlyaptsev, V. N., 416
Shlyaptseva, A. S., 177, 181, 185, 189, 412

Shyam, A., 325
Siemon, R., V. V., 279
Silfvast, W. T., 416
Silva, P., 161, 229, 245, 265
Sinars, D. B., 141, 173
Sincerny, P., 43
Skobelev, I. Y., 173
Smirnov, V. P., 29, 33, 37, 87, 91
Smith, D. L., 23
Sonara, J., 325
Song, Y., 101
Sorokin, S. A., 59, 127
Sotnikov, V. I., 396
Soto, L., 161, 229, 265
Spence, P., 43
Spielman, R. B., 101
Starobinets, A., 388
Struve, K. W., 23, 87, 101, 209, 213, 299
Stygar, W. A., 23, 101
Suematsu, H., 329, 333
Susuki, F., 237
Suzuki, T., 329
Sweeney, M. A., 9
Sylvester, G., 229
Sze, H. M., 101, 105, 109, 113, 350
Szydlowski, A., 193, 255

T

Takasugi, K., 131
Teramoto, Y., 95
Terry, R. E., 392
Thornhill, J. W., 101, 105, 339, 350, 358, 400
Tkachenko, S. I., 313
Tomasel, F. G., 416
Tomaszewski, K., 193, 197, 255
Travnicek, P., 396
Tucker, T., 43
Tumanov, V. I., 197
Tutt, T. E., 261

U

Urakami, H., 95

V

Velikhov, E. P., 29
Velikovich, A. L., 101, 105, 275, 339, 350, 358
Verma, R., 325
Vinogradov, V. P., 37
Vitulli, S., 255
Volkov, G. S., 87, 91
Vorob'ev, V. S., 313
Vrba, P., 165
Vrbova, M., 165
Vucina, T., 161

W

Waisman, E. M., 113, 209, 213
Watrous, J. J., 368
Watt, R. G., 269
Weber, B. V., 101
Weinbrecht, E. A., 23
Whitehead, L., 43
Whitney, K. G., 105, 339, 350, 358
Worley, T., 101, 109
Wyndham, E. S., 55, 145, 233, 237, 241

X

Xianbin, H., 309

Y

Yang, Z., 135, 364
Yatsui, K., 329, 333
Yermolovich, V. F., 287

Z

Zambra, M., 229
Zepf, M., 291
Zhitlukhin, A. M., 29
Ziegler, L., 261
Zielinska, E., 255
Ziethen, G., 229
Zukakischvili, G. G., 87, 91
Zurin, M. V., 87, 91